Complex Analysis

Conformal Inequalities
and the Bieberbach Conjecture

MONOGRAPHS AND RESEARCH NOTES IN MATHEMATICS

Series Editors

John A. Burns
Thomas J. Tucker
Miklos Bona
Michael Ruzhansky

Published Titles

Application of Fuzzy Logic to Social Choice Theory, John N. Mordeson, Davender S. Malik and Terry D. Clark

Blow-up Patterns for Higher-Order: Nonlinear Parabolic, Hyperbolic Dispersion and Schrödinger Equations, Victor A. Galaktionov, Enzo L. Mitidieri, and Stanislav Pohozaev

Complex Analysis: Conformal Inequalities and the Bieberbach Conjecture, Prem K. Kythe

Computational Aspects of Polynomial Identities: Volume I, Kemer's Theorems, 2nd Edition Alexei Kanel-Belov, Yakov Karasik, and Louis Halle Rowen

Cremona Groups and Icosahedron, Ivan Cheltsov and Constantin Shramov

Difference Equations: Theory, Applications and Advanced Topics, Third Edition, Ronald E. Mickens

Dictionary of Inequalities, Second Edition, Peter Bullen

Iterative Optimization in Inverse Problems, Charles L. Byrne

Line Integral Methods for Conservative Problems, Luigi Brugnano and Felice Iavernaro

Lineability: The Search for Linearity in Mathematics, Richard M. Aron, Luis Bernal González, Daniel M. Pellegrino, and Juan B. Seoane Sepúlveda

Modeling and Inverse Problems in the Presence of Uncertainty, H. T. Banks, Shuhua Hu, and W. Clayton Thompson

Monomial Algebras, Second Edition, Rafael H. Villarreal

Nonlinear Functional Analysis in Banach Spaces and Banach Algebras: Fixed Point Theory Under Weak Topology for Nonlinear Operators and Block Operator Matrices with Applications, Aref Jeribi and Bilel Krichen

Partial Differential Equations with Variable Exponents: Variational Methods and Qualitative Analysis, Vicenţiu D. Rădulescu and Dušan D. Repovš

A Practical Guide to Geometric Regulation for Distributed Parameter Systems, Eugenio Aulisa and David Gilliam

Signal Processing: A Mathematical Approach, Second Edition, Charles L. Byrne

Sinusoids: Theory and Technological Applications, Prem K. Kythe

Special Integrals of Gradshteyn and Ryzhik: the Proofs – Volume I, Victor H. Moll

Special Integrals of Gradshteyn and Ryzhik: the Proofs – Volume II, Victor H. Moll

Forthcoming Titles

Actions and Invariants of Algebraic Groups, Second Edition, Walter Ferrer Santos and Alvaro Rittatore

Analytical Methods for Kolmogorov Equations, Second Edition, Luca Lorenzi

Geometric Modeling and Mesh Generation from Scanned Images, Yongjie Zhang

Groups, Designs, and Linear Algebra, Donald L. Kreher

Handbook of the Tutte Polynomial, Joanna Anthony Ellis-Monaghan and Iain Moffat

Microlocal Analysis on R^n and on NonCompact Manifolds, Sandro Coriasco

Practical Guide to Geometric Regulation for Distributed Parameter Systems, Eugenio Aulisa and David S. Gilliam

Reconstructions from the Data of Integrals, Victor Palamodov

Stochastic Cauchy Problems in Infinite Dimensions: Generalized and Regularized Solutions, Irina V. Melnikova and Alexei Filinkov

Symmetry and Quantum Mechanics, Scott Corry

MONOGRAPHS AND RESEARCH NOTES IN MATHEMATICS

Complex Analysis

Conformal Inequalities
and the Bieberbach Conjecture

Prem K. Kythe

University of New Orleans, Louisiana, USA

CRC Press
Taylor & Francis Group
Boca Raton   London   New York

CRC Press is an imprint of the
Taylor & Francis Group, an **informa** business

A CHAPMAN & HALL BOOK

First published 2016 by Chapman and Hall

Published 2019 by CRC Press
Taylor & Francis Group
6000 Broken Sound Parkway NW, Suite 300
Boca Raton, FL 33487-2742

© 2016 by Taylor & Francis Group, LLC
CRC Press is an imprint of Taylor & Francis Group, an Informa business

First issued in paperback 2019

No claim to original U.S. Government works

ISBN 13: 978-0-367-44578-2 (pbk)
ISBN 13: 978-1-4987-1897-4 (hbk)

Library of Congress Cataloging-in-Publication Data

Names: Kythe, Prem K.
Title: Complex analysis : conformal inequalities and the Bieberbach conjecture / Prem K. Kythe.
Description: Boca Raton : Taylor & Francis, 2016. | Series: Monographs and research notes in mathematics ; 17 | "A CRC title." | Includes bibliographical references and index.
Identifiers: LCCN 2015032007 | ISBN 9781498718974 (alk. paper)
Subjects: LCSH: Functions of complex variables. | Inequalities (Mathematics) | Mathematical analysis.
Classification: LCC QA331.7 .K95 2016 | DDC 515/.98--dc23
LC record available at http://lccn.loc.gov/2015032007

Visit the Taylor & Francis Web site at
http://www.taylorandfrancis.com

and the CRC Press Web site at
http://www.crcpress.com

To the memory of my Grandparents and Parents,
so wise, so learned, and so caring.
It was a challenge to emulate them.

Contents

Preface

The historical development of the subject shows that there was a sustained and broad interest in the Bieberbach conjecture and its impact on creative thinking of mathematicians of the last century. This single phenomenon has been responsible for development of certain beautiful aspects of complex analysis, especially in the geometric-function theory of univalent functions.

It was 99 years ago that Ludwig Bieberbach in his research paper [1916] mentioned the conjecture in a footnote. This conjecture deals with the bound on the coefficients of the class S of univalent (schlicht) functions $f : E \mapsto f(E)$, where E is the unit disk, with the series expansion $f(z) = z + \sum_{n=2}^{\infty} a_n z^n$, and normalized by $f(0) = 0$ and $f'(0) = 1$. This normalization is a scaling that also implies that the boundary of $f(E)$ cannot get close to the origin. The Bieberbach conjecture says that $|a_n| \leq n$ for all $n = 2, 3, \ldots$.

Although the Bieberbach conjecture still appears in the title of the book, it should, in fact, have been the 'de Branges Theorem' instead, as the conjecture was finally proved in 1984 after 68 years of intensive research. The approach that de Branges pursued was to prove the Milin conjecture which established the Robertson conjecture, which in turn established the Bieberbach conjecture. This book is devoted to the beautiful mathematical analysis created around a single conjecture. It occupied a large number of researchers, big and small, of the twentieth century, all competing in a race to 'get the prize.'

Overview

The geometric function theory started when Riemann [1851] published the theorem, now known as the *Riemann mapping theorem*, but he did not complete the proof, which was completed by Koebe [1912] and Carathéodory [1912]. The historical timeline around the Bieberbach conjecture is as follows:

- 1851. Riemann mapping theorem.
- 1909. Koebe proves compactness of the class S of schlicht functions.
- 1912. Carathéodory and Koebe prove the Riemann mapping theorem, independently.
- 1914. Gronwall proves the area theorem for functions in the class Σ.

- 1916. Bieberbach proves $|a_2| \leq 2$, and announces his conjecture $|a_n| \leq n$ for functions in the class S.

- 1917. Löwner (also spelled Loewner) proves $|a_n| \leq 1$ for convex functions.

- 1920. Nevanlinna proves $|a_n| \leq n$ for starlike functions.

- 1923. Löwner proves $|a_3| \leq 3$ for functions of the class S, and establishes what is now known as the Löwner theory.

- 1925. Littlewood proves that $|a_n| < e \cdot n$.

- 1931. Dieudonné and Rogosinski prove that $|a_n| \leq n$ for functions in the class S with real coefficients.

- 1932. Littlewood-Paley conjecture for odd functions is announced.

- 1933. Fekete and Szegö prove that $\max |a_5| = \frac{1}{2} + e^{-2/3} \approx 1.0134\ldots > 1$ for $f \in S_{\text{odd}}$.

- 1936 Robertson conjecture for odd functions in the class S is announced.

- 1939. Grunsky inequalities appear; these inequalities generalize Gronwall's area theorem.

- 1955. Hayman proves the regularity theorem.

- 1955. Reade proves $|a_n| \leq n$ for close-to-convex functions.

- 1955. Garabedian and Schiffer use the variation method to prove $|a_4| \leq 4$.

- 1960. Charzyński and Schiffer use Grunsky inequalities and derive $|a_4| \leq 4$.

- 1965. Milin shows numerically that $|a_n| < 1.243 \cdot n$.

- 1967. Lebedev-Milin inequalities are published.

- 1968. Pederson and Ozawa use the Grunsky inequalities and show independently that $|a_6| \leq 6$.

- 1970. Friedland proves the Robertson conjecture for $n = 4$.

- 1970. Brickman determines the extreme points for functions of the class S.

- 1971. Milin conjecture is announced.

- 1972. Fitzgerald proves that $|a_n| < \sqrt{7/6} \cdot n \approx 1.0801\ldots \cdot n$.

- 1972. Pederson and Schiffer prove that $|a_5| \leq 5$.

- 1974. Kulshrestha obtains the coefficient bounds for α-complex functions by solving the related Löwner equation in the complex plane.

- 1978. Horowitz refines Fitzgerald's bound and shows that $|a_n| < 1.0657 \cdot n$.

- 1984-1985. De Brages proves the Milin conjecture, and thus, the Bieberbach conjecture.

- 1989. Weinstein provides an alternate proof of the de Branges theorem.

Although de Branges theorem is deep and meritorious, a quote from Carl
H. Fitzgerald [1985] also deserves due credit: "It is important to note that the
Milin conjecture was achieved because of the earlier work on Grunsky inequalities and the Robertson conjecture, and Milin's earlier work. ... Although the
work of Fitzgerald and Pommerenke and all previous researchers provided significant progress, the insight of Louis de Branges claimed the ultimate prize
by solving Löwner's equation ... ."

Outline

The first two chapters provide a review of the theory of analytic functions in
the complex plane and of univalent functions and conformal mapping. Chapter 3 deals with Gronwall's and Lebedev-Milin's area theorems, distortion
theorems, Gronwall's inequality, Grunsky's inequalities and bounds, polynomial area theorem, univalency criterion, and the Schwarzian derivative. The
Löwner theory is discussed in Chapter 4, starting with the Carathéodory's kernel theorem, and derivation of Löwners differential equation, Löwner-Kufarev
equation, proof of the bound for the third coefficient a_3, and slit mappings.

Chapter 5 discusses the Schiffer's variaion method and the bounds for the
fourth and higher-order coefficients. Since the class S is a 'very nonlinear set'
(Hamilton [1986: 86]), one cannot simply add small perturbations to obtain
a valid variation unless one ensures that the boundary of a simply connected
domain is perturbed into a boundary of another simply connected domain.
The Garabedian-Schiffer's method on the bound of the fourth coefficient a_4
is described in detail since it provides a unique technique for solving a nonlinear first-order boundary value problem in the complex plane. The unitary
Grunsky matrix, used to solve the fourth coefficient by Pederson, is presented
in detail.

Chapter 6 deals with various subclasses of univalent functions, such as
the subclass S_0, subclass of typically real functions, of starlike functions, of
functions with real coefficients, functions of the classes $\mathcal{P}$, $\mathcal{T}$, and $\mathcal{S}_a$, and
distortion and rotation theorems are proved. Generalized convexity and the
class of α-convex functions are discussed in Chapter 7. The subject of numer-
ical estimates of the coefficients $|a_n|$ is covered in Chapter 8, starting with
the concept of mean modulus and the Hayman index of univalent functions
in the unit disk, exponentiation of inequalities, and the Fitzgerald inequality.

Chapter 9 is a summary of certain orthogonal polynomials, such as the
Jacobi, Gegenbauer, and Legendre polynomials, hypergeometric functions,
and Faber polynomials. Chapter 10 is all about de Branges theorem, which
starting with a detailed account of different conjectures, proves his famous
theorem; it also provides Weinstein's alternative proof as well as Ming-Qin's
generalization; and the de Branges' and Weinstein's special system of functions are compared. Chapter 11 is an epilogue: it discusses what types of

development have been occurring after de Branges' theorem and points to the direction these developments are progressing; theory of several complex variables and the complex space $\mathbb{C}^n$, quasi-conformal mapping, Loewner's theory, slit mappings, multivatiate holomorphic mappings, and Beurling and Cauchy transforms are some of the topics of modern research. This chapter also establishes that there *does not exist* a Bieberbach conjecture in the case of several complex variables by means of a counter-example in $\mathbb{C}^2$.

There are five appendices, dealing with more details about mappings, parametrized curves, Green's theorems, two-dimensional potential flows, and subordination principle. There are about 90 exercises with complete solutions, provided at the end of each chapter.

Intended Readers

This book assumes a basic but thorough knowledge of complex analysis and differential equations. It is mainly written at the graduate level for effective study of different topics related to the geometric-function theory. The majority of intended readers fall into one of the following three categories: firstly, they may be students ready for a graduate course in any of the topics discussed in the book; secondly, they may be graduate students engaged in analytical research in one or more topics in the book; and thirdly, they may be scientists and researchers in related areas of complex analysis in one or more complex variables.

Acknowledgments

I take this opportunity to thank Mr. Robert B. Stern, Former Executive Editor, Taylor & Francis/CRC Press, for his encouragement to complete this book. I thank Mr. Sarfraz Khan, Executive Editor, Taylor & Francis, for his support, and the Project Editor Michele A. Dimont for doing a great job of editing the text. Thanks are due to the referees and the reviewer who made some valuable suggestions. I thank Dr. Emre Celebi, Professor of Computer Science, LSUS, for providing timely logistic support. I also thank Mr. Karl M. Cambre, my former graduate student in advanced complex analysis at the University of New Orleans in the Summer of 1975, for his notes on earlier chapters. Finally, I thank my friend Michael R. Schäferkotter for help and advice freely given whenever needed.

Prem K. Kythe

Notations, Definitions, and Acronyms

A list of the notations, definitions, abbreviations, and acronyms used in this book is given below.

$(a)_n$, shifted factorial or Pochhammer symbol, $(a)_n = a(a+1)\cdots(a+n-1)$, $n \geq 1$, $(a)_0 = 1$; also $(a)_n = \Gamma(a+n)/\Gamma(a)$

a.e., almost every

$\mathbf{a} = (a_1, a_2, \ldots, a_n)$, an arbitrary point in $\mathbb{C}^n$

$\bar{A}$, closure of a set A

$A \backslash B$, complement of a set B with respect to a set A

$A \cup B$, union of sets A and B

$A \cap B$, intersection of sets A and B

$A \times B$, (Cartesian) product of sets A and B

$A \subset B$, set A is contained in the set B, or set B contains A

$A \subseteq B$, set A is contained in the set B or sets $A = B$

Area (G), area of a region G

$B^n(0,r) = \{z \in \mathbb{C}^n : \|z\| < r\}$, an open ball in $\mathbb{C}^n$

$B^n(0,1)$, unit ball in $\mathbb{C}^n$

$B^1(0,1)$, unit disc, same as $C(0,1)$, denoted by E

const, short for 'constant'

$C(a,r)$, open disk in $\mathbb{C}$ of radius r and center at a: $\{z : |z-a| < r\}$

$\bar{C}(a,r)$, closed disk of radius r and center at a: $\{|z-a| \leq r\}$

$\partial C(a,r)$, circle of radius r and center at a, also denoted by C_r

$C(0,1)$, open unit disk $\{z : |z| < 1\}$ (also denoted by E)

$C^0(D)$, class of functions continuous on a region D

$C^k(D)$, class of continuous functions with kth continuous derivative on a region D, $0 \leq k < \infty$

$C^\infty(D)$, class of continuous functions infinitely differentiable on a region D

C-function, same as a C^0-function; continuous function

$\mathcal{C}[a,b]$, set of all real-valued continuous functions defined on $[a,b]$

C_r, the circle $\partial C(0,r)$, i.e., $|z| = r$

$C_n^\mu(x)$, Gegenbauer (ultraspherical) polynomials

$\mathcal{C}$, class of close-to-convex functions in E

$\mathbb{C}^n$, n-dimensional complex space

$\mathbb{C}$, complex plane

$\mathbb{C}_\infty$, extended complex plane

$\mathrm{diam}(D)$, transfinite diameter of a set D

D, domain, usually in the z-plane

$\bar{D}$, closure of D; $\bar{D} = D \cup \partial D$

D^*, complement of a set D; , $D^* = \mathrm{Ext}\,(\Gamma)$ region exterior to Γ

E, open unit disk $B^1(0,1) \equiv C(0,1)$

E^*, region exterior of the unit circle $|z| = 1$ $(= \mathbb{C}\backslash\{E \cup \partial E\})$

$\bar{E}^*$, domain $E^* \cup \partial E^*$

$\mathrm{Ext}(\Gamma)$, region exterior to a closed contour Γ

Eq(s)., Equation(s) (when followed by an equation number)

$f \circ g$, composite function of f and g: $(f \circ g)(\cdot) = f\,(g(\cdot))$

$f(D,t)$, image of the domain D under the map $f(z,t)$ at time t

$\langle f, g \rangle$, inner product of f and g

$\|f\|_\infty$, norm of $f \in L^\infty$

$f'(z,t)$, partial derivative of f with respect to z, same as $\partial_z f(z,t)$

$\dot{f}(z,t)$, partial derivative of f with respect to t, same as $\partial_t f(z,t)$

f_μ, f_ν, short for $f\,(z_\mu,t)$, $f\,(z_\nu,t)$, respectively

$\mathbf{f} = (f_1, f_2, \ldots, f_n)$, a function in $\mathbb{C}^n$

$_2F_1(a,b;c;x) \equiv {}_2F_1\left(\begin{matrix} a,b \\ c \end{matrix}\,\middle|\, x\right) = \sum_{n=0}^{\infty} \frac{(a)_n (b)_n}{(c)_n} \frac{x^n}{n!}$, hypergeometric function

$_3F_2(a,b,c;d.e;x) \equiv {}_3F_2\left(\begin{matrix} a,\,b,\,c \\ d,\,e \end{matrix}\,\middle|\, x\right) = \sum_{n=0}^{\infty} \frac{(a)_n (b)_n (c)_n}{(d)_n (e)_n} \frac{x^n}{n!}$, hyper-

geometric function

$\bar{g}$, complex conjugate of an analytic function g

G, a domain in the z- or w-plane; often $G = f(E)$ in the w-plane

$\mathcal{G}(\cdot,\cdot)$, Green's function

H^1, Lipschitz condition

H^α, Hölder condition of order α

$\mathbb{H}^+$, right half-plane $\Re\{z\} > 0$, or $\Re\{w\} > 0$

$\mathbb{H}^-$, left half-plane $\Re\{z\} < 0$, or $\Re\{w\} > 0$

$\mathrm{Int}\,(\Gamma)$, region interior to a closed contour Γ

$I(\Gamma, z_0)$, index of a contour Γ (winding number)

$\mathbf{I}$, identity matrix

$\Im\{\cdot\}$, imaginary part of a complex quantity

J, Jacobian of the transformation $w = f(z)$

$K(z)$, Koebe function

$\mathcal{K}$ class of convex function in E

$l(\Gamma)$, length of a curve Γ

L^2, Hilbert space of square-integrable functions

L^∞, Hilbert space of 2π-periodic and bounded functions

$L^2(D)$, class of square-integrable functions defined on a region D

ℓ^p, the space ℓ^p, $1 \le p < \infty$

$\mathcal{M}(\alpha)$ class of alpha-convex functions in E

$P\{a = t_0 < t_1 < \cdots < t_n = b\}$, partition of an interval $[a, b]$

$P_n(x)$, Legendre polynomials

$P_n^{\alpha,\beta}(x)(z)$, Jacobi polynomials

$\mathcal{P}$ class of functions $p(z)$ with positive real part, $|z| < 1$

$\mathbb{Q}$, the set of rational numbers

$\mathcal{R}(f)$, range of f

$\Re\{\cdot\}$, real part of a complex quantity

$\mathbb{R}^n$, Euclidean n-space

$\mathbb{R}^+$, set of nonnegative real numbers

$\mathcal{S}$, class of univalent (schlicht) functions $f(z) = z + a_2 z^2 + \cdots$,
$$f(0) = 0, f'(0) = 1$$

$\mathcal{S}^\star$ class of starlike functions in E

t, a parameter $t \in \mathbb{R}$, usually representing 'time'

$T(z)$, bilinear transformation, defined by (6.1.5); analytic automorphism
     of E; sometimes denoted by $\mathrm{Aut}(f)$

$\mathcal{T}$ class of typically-real functions in E

$u_x = \dfrac{\partial u}{\partial x}; u_y = \dfrac{\partial u}{\partial y}; v_x = \dfrac{\partial v}{\partial x}; v_y = \dfrac{\partial v}{\partial y}$; partial derivatives of u and v
     with respect to x and y[1]

U, unitary matrix (complex square matrix)

$U^\star$, conjugate transform of unitary matrix U

$\mathbb{U}^+$, upper half-plane $\Im\{z\} > 0$

$w = f(z) = u + iv = \rho e^{i\phi}$, image of z under f

$\{w, z\}$, Schwarzian derivative

$(W, \|\cdot\|)$, Banach space

$x = \dfrac{z + \bar{z}}{2} = \Re\{z\}$, real part of z

$X \times Y$, Cartesian product of sets X and Y

$y = \dfrac{z - \bar{z}}{2i} = \Im\{z\}$, imaginary part of z

$z, = x + iy = r e^{i\theta}$, complex variable or number

$\bar{z}, = x - iy = r e^{-i\theta}$, conjugate of a complex variable z

$z^\star$, point symmetric to a point z

(z_1, z_2, z_3, z_4), cross-ratio

$\{z_k\}_1^\infty$, a set of distinct points in D

---

[1] In certain publications on Loewner chains, the notation f_t is used for the function $f(z, t)$. If such usage is not declared, be careful not to read f_t as $\partial f / \partial t$.

$\mathbf{z} = (z_1, z_2, \ldots, z_n)$, a point in $\mathbb{C}^n$

$\mathbf{z}^T$, transpose of $\mathbf{z}$ in $\mathbb{C}^n$

$\mathbb{Z}^+$, set of positive integers

γ, path, contour

$\gamma(t)$, continuous parametrized curve

$\tilde{\gamma}$, reparametrization of a path γ

$\dot{\gamma}(t)$, t-derivative (or time-derivative) of γ: $\dot{\gamma}(t) = \dfrac{d\gamma}{dt}$

$\Gamma = \partial D$, boundary of a region D (a simple closed curve or a Jordan contour)

$\Gamma(n)$, gamma function

$\Gamma(z) = \int_0^z t^{\alpha-1}(1-t)^\alpha\, dt$, $\Re\{z\} > 0$, gamma function

δ_{mn}, Kronecker delta, equal to 1 if $m = n$, 0 if $m \neq n$

$\mu(t)$, function of bounded variation such that $\int_{-\pi}^{\pi} d\mu(t) = 1$

$\pi_n(z)$, Bieberbach polynomial

$\Phi_n(w)$, Faber polynomials

σ_k, arcs for $k = 0, 1, \ldots, n$

Σ, class of functions $f(z) = z + b_0 + \dfrac{b_1}{z} + \dfrac{b_2}{z^2} + \cdots$, $|z| > 1$, $f(\infty) = \infty$ and
$\quad\quad f'(\infty) = 0$.

Σ_0, class of functions $f(z) \in \Sigma$, and $f(z) \neq 0$ for $z \in E^*$

$\varkappa$, Beltrami coefficient

Ω, domain in $\mathbb{C}^n$

$\partial_z, \bar{\partial}_z$, partial differential operators: $\partial_z f = \dfrac{\partial f}{\partial z}$, $\bar{\partial}_z f = \dfrac{\partial f}{\partial \bar{z}}$

$\partial_t f$, partial derivative of f with respect to t

$\dfrac{\partial f}{\partial t}$, or $\dfrac{\partial f}{\partial z}$, first partial derivatives of f with respect to t, or with respect to z

$\partial B(a, r)$, boundary of the ball $B(a, r)$ in $\mathbb{R}^n$

$\partial C(a, r)$, circle of radius r and center at a: $|z - a| = r$

$\partial_\infty D$, boundary of a set D in $\mathbb{C}_\infty$

$\partial_z f$, partial derivative of $f \in \mathbb{C}^n$ with respect to $z = \{z_1, \ldots, z_n\}$

∇, gradient vector $= \mathbf{i}\dfrac{\partial}{\partial x} + \mathbf{j}\dfrac{\partial}{\partial y}$

∇^2, Laplacian $\equiv \dfrac{\partial^2}{\partial x^2} + \dfrac{\partial^2}{\partial y^2} = 4\bar{\partial}_z\partial_z = 4\partial_z\bar{\partial}_z$

$\langle \nabla u, \nabla v \rangle$, inner (scalar) product

$g \prec f$, g is subordinate to f

$:=$, definition; defined by

1

Analytic Functions

Some basic concepts and results from complex analysis are presented. They include analytic functions, Cauchy's theorems, Cauchy kernel, Riemann mapping theorem, analytic continuation, Schwarz reflection principle, and univalent functions. Proofs for most of the results can be found in many textbooks.

1.1 Definitions

Let $\mathbb{R}^n$ denote the Euclidean n-space, and $\mathbb{R}^+$ the set of nonnegative real numbers. The complement of a set B with respect to a set A is denoted by $A \backslash B$ (or $\text{compl}\,(B)$ if the reference to set A is obvious), the product of the sets A and B by $A \times B$, and the closure of a set A by $\bar{A}$.

A complex-valued function f is said to belong to the class $C^k(D)$ if it is continuous together with its kth derivatives, in a domain D, $0 \leq k < \infty$. In this case we often say that f is a $C^k(D)$-function, or that f is a C^k-function on D, while a C^0-function is simply a continuous function. The function f in the class $C^k(D)$, for which all kth derivatives admit continuous continuations in the closure $\bar{D}$, form the class of functions $C^k(\bar{D})$. The class $C^\infty(D)$ consists of functions f which are infinitely differentiable on D, i.e., continuous partial derivative of all orders exist. These classes are linear sets. Thus, every linear combination $\lambda f + \mu g$, where λ and μ are arbitrary complex numbers, also belongs to the respective class.

Let $\mathbb{C}$ denote the complex plane. If $a \in \mathbb{C}$ and $r > 0$, then

$$C(a,r) = \{z \in \mathbb{C} : |z - a| < r\},$$
$$\bar{C}(a,r) = \{z \in \mathbb{C} : |z - a| \leq r\}, \tag{1.1.1}$$
$$\partial C(a,r) = \{z \in \mathbb{C} : |z - a| = r\},$$

denote, respectively, an open disk, a closed disk, and a circle, each of radius r and centered at a. We will often denote the circle $\partial C(0,r)$ by C_r and the open unit disk $C(0,1)$ by E. A connected open set $A \subseteq \mathbb{C}$ is called a domain.

The extended complex plane is denoted by $\mathbb{C}_\infty$. Then $\partial_\infty D$ is the boundary of a set D in $\mathbb{C}_\infty$, that is,

$$\partial_\infty D = \begin{cases} \partial D & \text{if } D \text{ is bounded,} \\ \partial D \cup \{\infty\} & \text{if } D \text{ is unbounded.} \end{cases}$$

For a domain $D \subset \mathbb{C}_\infty$, the following statements are equivalent: (a) D is simply connected, (b) $\mathbb{C}_\infty \backslash D$ is connected, and (c) $\partial_\infty D$ is connected. Domains that have more than one layer over the complex plane are called Riemann surfaces.

1.2 Jordan Contour

A simple closed curve Γ in $\mathbb{C}$ is a path $\gamma : [a, b] \mapsto \mathbb{C}$ such that $\gamma(t) = \gamma(s)$ if and only if $t = s$ or $|t - s| = b - a$. In what follows, a simple closed curve shall be called a *Jordan contour*. The *Jordan curve theorem* states that if Γ is a simple contour, then $\mathbb{C} \backslash \Gamma$ has two components, one called the *interior* of Γ, denoted by $\text{Int}\,(\Gamma)$, and the other called the *exterior* of Γ, denoted by $\text{Ext}\,(\Gamma)$, each of which has Γ as its boundary. Thus, if Γ is a Jordan contour, then $\text{Int}\,(\Gamma)$ and $\text{Ext}\,(\Gamma) \cup \{\infty\}$ are simply connected domains.

Let a continuous curve Γ, defined by $\gamma(t) = \alpha(t) + i\beta(t)$, be divided into n arcs $\sigma_k = z_{k-1}\overset{\frown}{} z_k$, $k = 1, \ldots, n$, where $z_k = \gamma(t_k)$ for $k = 0, 1, \ldots, n$, such that the end point of each arc, except the last one, overlaps the initial point of the next arc. If we join each segment $[z_{k-1}, z_k]$, $k = 1, \ldots, n$, by straight line segments, we obtain a polygonal line L inscribed in Γ. The segments of L are the chords joining the end points of the arcs σ_k, and

$$\text{length of } L = \sum_{k=1}^{n} |z_k - z_{k-1}|. \tag{1.2.1}$$

The curve Γ is said to be *rectifiable* if

$$\sup_{P} \sum_{k=1}^{n} |z_k - z_{k-1}| = l < +\infty, \tag{1.2.2}$$

where the least upper bound is taken over all partitions $P = \{a = t_0, t_1, \cdots, t_n = b\}$ of the interval $[a, b]$, $a \le t \le b$. The nonnegative number l is called the length of the curve Γ. The curve is said to be *nonrectifiable* if the sums (1.2.1) become arbitrarily large for suitably chosen partitions. More information on parametrized curves is given in Appendix B.

1.3 Cauchy-Riemann Equations

Let $z = x + iy$ be a complex number. Then $\bar{z} = x - iy$, $x = \dfrac{z + \bar{z}}{2}$, and $y = \dfrac{z - \bar{z}}{2i}$. We say that an arbitrary function $w = f(z)$ is *analytic* at a point z_0 in a domain D if the Taylor series of f at the point z_0, assuming single-valuedness, continuity and differentiability of f, is represented by

$$f(z) = f(z_0) + (z - z_0)f'(z_0) + \frac{(z - z_0)^2}{2!}f''(z_0) + \cdots, \quad z, z_0 \in D, \quad (1.3.1)$$

where the differentiation of a complex-valued function $f(z)$ is defined as follows: Since the derivative $f'(z)$ does not depend on the path, we take $\Delta z = \Delta x + i\,\Delta y$, so that $\Delta f \equiv \Delta w = \Delta u + i\,\Delta v$, and the let Δz approach zero by first letting $\Delta y \to 0$ and then letting $\Delta x \to 0$. This yields

$$f'(z) = \lim_{\Delta y \to 0} \frac{\Delta u + i\,\Delta v}{\Delta x + i\,\Delta y} = \lim_{\Delta x \to 0}\left(\frac{\Delta u}{\Delta x} + i\frac{\Delta v}{\Delta x}\right) = \frac{\partial u}{\partial x} + i\frac{\partial v}{\partial x}, \quad (1.3.2a)$$

whereas reversing the order, i.e., letting $\Delta x \to 0$ first and then $\Delta y \to 0$, we get

$$f'(z) = \lim_{\Delta y \to 0}\left(\frac{\Delta u}{i\,\Delta y} + i\frac{\Delta v}{\Delta y}\right) = \frac{\partial v}{\partial y} - i\frac{\partial u}{\partial y}. \quad (1.3.2b)$$

From (1.3.2a) or (1.3.2b) we get

$$|f'(z)|^2 = \left(\frac{\partial u}{\partial x}\right)^2 + \left(\frac{\partial v}{\partial x}\right)^2 = \left(\frac{\partial v}{\partial y}\right)^2 + \left(\frac{\partial u}{\partial y}\right)^2. \quad (1.3.3)$$

Equating real and imaginary parts of Eqs (1.3.2a) and (1.3.2b) we obtain the *Cauchy-Riemann equations*

$$u_x = v_y, \quad u_y = -v_x, \quad (1.3.4)$$

which in polar form ($z = re^{i\theta}$) become

$$u_r = \frac{1}{r}\,v_\theta, \quad v_r = -\frac{1}{r}\,u_\theta. \quad (1.3.5)$$

Thus,

$$f'(z) = u_x + i\,v_x = v_y - i\,u_y. \quad (1.3.6)$$

This means that a function $f : D \mapsto \mathbb{C}$ is analytic on D if and only if $\bar{\partial}_z f = 0$, which is equivalent to the *Cauchy-Riemann equations* for the function $w = f(z) = u(x, y) + iv(x, y)$. The Cauchy-Riemann equations are necessary conditions for $f(z)$ to be analytic on D. However, merely satisfying the Cauchy-Riemann equations alone is not sufficient to ensure the differentiability of $f(z)$ at a point in D.

If the four partial derivatives u_x, u_y, v_x, v_y of $w = f(z) = u(x, y) + iv(x, y)$ exist and are continuous throughout a domain D, then the Cauchy-Riemann equations are necessary conditions for the derivative $f'(z)$ to exist throughout

D. A function that is differentiable in this sense is an *analytic function* (also called a *holomorphic function*). The points at which f is not analytic are called the *singularities* of f; they are said to be *isolated* singularities if the points occur at isolated, single locations where they can be circumvented. The mapping $w = f(z)$ cannot be conformal at the points where $f'(z) = 0$ or ∞; these points are called *critical points*. Analytic functions have the following properties: If f and g are analytic in a domain D, so are the functions $f(z) \pm g(z)$; $f(z) \cdot g(z)$; $f(z)/g(z)$ where $g(z) \neq 0$; and $(f \circ g)(z) = f(g(z))$. Moreover, we have

$$\partial_z f \equiv \frac{\partial f}{\partial z} = \frac{1}{2}(f_x - i\, f_y), \quad \bar{\partial}_z f \equiv \frac{\partial f}{\partial \bar{z}} = \frac{1}{2}(f_x + i\, f_y). \tag{1.3.7}$$

The inverse of an analytic function $w = f(z)$, which is defined as $z = f^{-1}(w)$ is also analytic if $f(z)$ is analytic, and its derivative is the reciprocal of $f'(z)$, that is,

$$\frac{dz}{dw} = \frac{1}{dw/dz}. \tag{1.3.8}$$

The following results are obvious: $\partial_z (\log|z|) = \dfrac{1}{2z}$, $\bar{\partial}_z (\log|z|) = \dfrac{1}{2\bar{z}}$, $\overline{\partial_z f} = \bar{\partial}_z \bar{f}$, and the Laplacian is defined as

$$\nabla^2 \equiv \frac{\partial^2}{\partial x^2} + \frac{\partial^2}{\partial y^2} = 4\bar{\partial}_z \partial_z = 4\partial_z \bar{\partial}_z. \tag{1.3.9}$$

The Cauchy-Riemann equations (1.3.4) for the function $f(z) = u(x,y) + iv(x,y)$ satisfy the partial differential equations

$$u_x\, v_x + u_y\, v_y = 0,$$
$$\nabla^2 u = 0, \quad \nabla^2 v = 0. \tag{1.3.10}$$

Using the gradient vector $\nabla \equiv \mathbf{i}\dfrac{\partial}{\partial x} + \mathbf{j}\dfrac{\partial}{\partial y}$, the first equation in (1.3.10) can be written as the inner (scalar) product:

$$\langle \nabla u, \nabla v \rangle = 0. \tag{1.3.11}$$

Then the Cauchy-Riemann equations yield $|\nabla u| = |\nabla v| = |f'(z)|$. Eq (1.3.11) also signifies the orthogonality condition for the families of level curves defined by $u(x,y) = \text{const}$ and $v(x,y) = \text{const}$.

If $w = u + iv$, then

$$\Re\{z\}\cdot\Re\{w\} = \Re\left\{\frac{z^2 + |z|^2}{2z}\, w\right\}, \quad \Im\{z\}\cdot\Im\{w\} = \Im\left\{\frac{z^2 - |z|^2}{2iz}\, w\right\}. \tag{1.3.12}$$

Theorem 1.3.1. (Inverse function theorem) *If f is a continuously differentiable function of a single variable with nonzero derivative at a point a, then f is invertible in a neighborhood of a, the inverse is continuously differentiable, and*

$$\left(f^{-1}\right)'(b) = \frac{1}{f'(a)}, \quad b = f(a).$$

This theorem gives sufficient conditions for a function to be invertible in the neighborhood of a point in its domain, and also provides a formula for the derivative of the inverse function.

1.3.1 Holomorphic or Analytic Functions. A function that is defined and differentiable on an open set D in the complex plane $\mathbb{C}$ is said to be *holomorphic* on D.

It is more difficult for a function of a complex variable to be differentiable than it is for a function of a real variable. In $\mathbb{C}$ there are infinitely many ways to approach a point a, and the difference quotients $\Delta f / \Delta z$ must converge to a single (unique) limit no matter how we make the approach. On the other hand, in $\mathbb{R}$ there are only two different directions to consider, and a function has *less* to do if it is to be differentiable.

Theorem 1.3.2. *A function of a complex variable is analytic on an open set D if and only if it is holomorphic on D.*

This result is special to functions of a complex variable. In the case of a real variable, a function $f(x)$ can be infinitely differentiable without being analytic, or in other words, $f(x)$ can be unrelated to its Taylor series even though it has continuous derivatives of all orders throughout $\mathbb{R}$. For example, consider the function $f(x)$ defined for all real x by

$$f(x) = \begin{cases} 0 & \text{if } -\infty < x < 0, \\ e^{-(1/x^2)} & \text{if } 0 < x < +\infty. \end{cases} \tag{1.3.13}$$

This function is obviously continuous at $x = 0$, since $\lim\limits_{x \to 0} f(x) = \lim\limits_{x \to 0} e^{-(1/x^2)} = 0$. However, using L'Hospital's rule, we find that f is infinitely differentiable at $x = 0$, having $f^{(n)}(0) = 0$ for $n = 0, 1, 2, \ldots$; this function has derivatives of all orders at any point $x \neq 0$. Now, if f has a local power series expansion about the point $a = 0$ (i.e., near the point $a = 0$), it must be given by the Taylor series of f about a. But this Taylor series is degenerate: it converges to 0, i.e., $f(0) = f'(0)x + \dfrac{f''(0)}{2!}x^2 + \cdots = 0 + 0 \cdot x + 0 \cdot x^2 + \cdots = 0$ for every $x \in \mathbb{R}$. Thus, the sum $F(x)$ of the Taylor series of f about a is the function $F(x) = 0$, and the Taylor series does not give back the original function for all x near $x = 0$ even though f is infinitely differentiable throughout $\mathbb{R}$.

Three corollaries of Theorem 1.3.2 are as follows.

Corollary 1.3.1. *If f is analytic, so is its reciprocal $g = 1/f$, except at points z where $f(z) = 0$, since the reciprocal will not be defined there.*

Corollary 1.3.2. *If $w = g(z)$ is analytic at $z = a$ and if $\zeta = f(w)$ is analytic at the image point $b = g(a)$, then the composite function $h(z) = (f \circ g)(z) = f(g(z))$ is analytic at $z = a$.*

Corollary 1.3.3. *If $w = f(z)$ is an analytic, invertible mapping between open sets $f : D \mapsto G$, and if the derivative df/dz is non-zero on D, then the inverse mapping $f^{-1} : G \mapsto D$ is analytic.*

If we replace 'analytic' by 'holomorphic', the validity of these statements is clear from the definition of holomorphic functions and the formulas for differentiation of $(f \pm g)'(z)$, $(\alpha f)'(z)$ (α a fixed complex number), $(f \cdot g)(z)$; $(f/g)'(z)$ provided $g(z) \neq 0$, and $(1/f)'(z)$ provided $g(z) \neq 0$ remain valid. The following example shows that differentiability implies analyticity for functions of a complex variable.

Example 1.3.1. Let $f(z)$ be differentiable throughout an open set D and let a be a point in D. We will show that (i) the function f is analytic at a; and (ii) if d denotes the distance from a to the boundary ∂D, so that $0 < d < +\infty$, then show that the Taylor series for f about the point a converges at least on the open disk $C(a, d)$. In fact, by definition, the disk $C(a, r) \subset C(a, d)$, $r < d$, lies entirely within D. In fact, $C(a, d)$ is the largest open disk about a lying entirely within D. Also the boundary circles $\partial C(a, r)$ lies within D since $r < d$ (which may not be true if $r = d$), and we may parametrize them to get closed contours γ_r which move once counterclockwise around Γ_r.

Now consider the disk Γ_r. If ζ is a point on this isk, then $|\zeta - a| < r$ and

$$f(\zeta) = \frac{1}{2\pi i} \int_{\gamma_r} \frac{f(\zeta)}{z - \zeta} \, dz. \tag{1.3.14}$$

But $1/(z - \zeta)$ can be written as

$$\frac{1}{z - \zeta} = \frac{1}{z - a} \left[\frac{1}{1 - \left(\dfrac{\zeta - a}{z - a} \right)} \right] = \frac{1}{z - a} \sum_{n=0}^{\infty} \left(\frac{\zeta - a}{z - a} \right)^n = \sum_{n=0}^{\infty} \frac{(\zeta - a)^n}{(z - a)^{n+1}}, \tag{1.3.15}$$

which converges because $\left| \dfrac{\zeta - a}{z - a} \right| = \dfrac{|\zeta - a|}{r} < 1$ for z on Γ_r. Multiplying each term in the series (1.3.15) term-by-term by $f(z)$ we get

$$\frac{f(z)}{z - \zeta} = \sum_{n=0}^{\infty} f(z) \frac{(\zeta - a)^n}{(z - a)^{n+1}} \quad \text{for all } z \text{ on } \Gamma_r. \tag{1.3.16}$$

If we keep ζ fixed, the series (1.3.16), taken as a function of z, converges uniformly on the circle Γ_r (see Exercise 1.10.1). Thus, we may substitute this series into the integral (1.3.14) and interchange the operations of integration and summation, to get

$$
\begin{aligned}
f(\zeta) &= \frac{1}{2\pi i} \int_{\gamma_r} \left[\sum_{n=0}^{\infty} \frac{f(z)}{(z-a)^{n+1}} (\zeta - a)^n \right] dz \\
&= \sum_{n=0}^{\infty} \left(\frac{1}{2\pi i} \int_{\gamma_r} (\zeta - a)^n \frac{f(z)}{(z-a)^{n+1}} \, dz \right) \\
&= \sum_{n=0}^{\infty} \left(\frac{1}{2\pi i} \int_{\gamma_r} \frac{f(z)}{(z-a)^{n+1}} \, dz \right) (\zeta - a)^n = \sum_{n=0}^{\infty} a_n (\zeta - a)^n,
\end{aligned}
\tag{1.3.17}
$$

where the constants a_n do not depend on ζ, and therefore, (1.3.17) is valid for each ζ in D. The formula (1.3.17) establishes a power series expansion for $f(\zeta)$ about the point a, and this series converges at least on $C(a,d)$, which proves part (i). The coefficients of the terms $(\zeta - a)^n$ in (1.3.17) must agree with the Taylor series coefficients, by uniqueness of the power series coefficients, so that $a_n = \dfrac{f^{(n)}(a)}{n!}$ for $n = 0, 1, 2, \ldots$, which are valid for each of the disks D_r such that $r < d$. The Taylor series for $f(\zeta)$ about the point a converges at every point in each disk D_r, and so this series must converge to f at each point in the disk D_r, which is the union of the smaller disk D_r as r increases with $r < d$. This proves part (ii). ∎

1.4 Line Integrals

Let a piecewise smooth curve $\widetilde{\Gamma}$ be defined by $\widetilde{\gamma} = [\tilde{a}, \tilde{b}] \mapsto \mathbb{C}$. Then $\widetilde{\Gamma}$ is called a *reparametrization* of a curve Γ defined by $\gamma = [a, b] \mapsto \mathbb{C}$ if there exists a function $\alpha \in C^1$, $\alpha : [a, b] \mapsto [\tilde{a}, \tilde{b}]$ with $\alpha'(t) > 0$, $\alpha(a) = \tilde{a}$ and $\alpha(b) = \tilde{b}$, such that $\gamma(t) = \widetilde{\gamma}(\alpha(t))$, $t \in [a, b]$. Then $\int_{\Gamma} f = \int_{\widetilde{\Gamma}} f$ for any C-function f defined on an open set containing the image of γ (which is equal to the image of $\widetilde{\gamma}$).

Let f be a continuous function on an open set $D \subset \mathbb{C}$, and let Γ be a piecewise C^1-continuous curve in D. If $|f(z)| \le M$ for all points $z \in \Gamma$, i.e., for all $z = \gamma(t)$ for $t \in [a, b]$, where $M > 0$ is a constant, then

$$
\left| \int_{\Gamma} f \, dz \right| \le M \, l(\Gamma),
\tag{1.4.1}
$$

where

$$
l(\Gamma) = \int_a^b |\gamma'(t)| \, dt = \int_a^b \sqrt{x'(t)^2 + y'(t)^2} \, dt
\tag{1.4.2}
$$

is the arc length of the path Γ. In general,

$$
\left| \int_{\Gamma} f \, dz \right| \le \int_{\Gamma} |f| \, |dz| = \int_a^b |f(\gamma(t))| \, |\gamma'(t)| \, dt.
\tag{1.4.3}
$$

If $\gamma : [a, b] \mapsto \mathbb{C}$ defines a piecewise smooth contour Γ and F is a function defined and analytic on a domain containing Γ, then

$$\int_\Gamma F'(z)\,dz = F(\gamma(b)) - F(\gamma(a)). \qquad (1.4.4)$$

If $\gamma(a) = \gamma(b)$, then

$$\int_\Gamma F'(z)\,dz = 0. \qquad (1.4.5)$$

This results is known as the *fundamental theorem for line integrals (contour integration)* in the complex plane. Thus, if a function f is defined and analytic on a domain $D \subset \mathbb{C}$ and if $f'(z) = 0$ for all points $z \in D$, then f is a constant on D. If f is a C-function on a domain D, then the following three statements are equivalent:

(i) If Γ_1 and Γ_2 are two paths in D from a point $z_1 \in D$ to a point $z_2 \in D$, then $\int_{\Gamma_1} f(z)\,dz = \int_{\Gamma_2} f(z)\,dz$, i.e., the integrals are path-independent.

(ii) If Γ is a Jordan contour lying in D, then $\int_\Gamma f(z)\,dz = 0$, i.e., the integrals on a closed contour are zero.

(iii) There exists a function F defined and analytic on D such that $F'(z) = f(z)$ for all $z \in D$, i.e., there exists a global antiderivative of f on D.

A function $f(z)$, analytic inside a domain with boundary $\Gamma = \cup_{k=1}^n \Gamma_k$, is said to be *sectionally analytic* on Γ if $f(z)$ is continuous on each Γ_k from both left and right except at the end points where it satisfies the condition $|f(z)| \le \dfrac{C}{|z - c|^\alpha}$, where c is the corresponding end point of Γ_k, and C and α are real constant with $\alpha < 1$.

Let $f(t)$ be defined on a Jordan curve Γ (open or closed). If

$$|f(t_1) - f(t_2)| \le A\,|t_1 - t_2|^\alpha, \quad 0 < \alpha \le 1, \qquad (1.4.6)$$

for arbitrary points $t_1, t_2 \in \Gamma$ ($t_1 \ne t_2$), where $A > 0$ and α are real constants, then $f(t)$ is said to satisfy the *Hölder condition of order* α, or simply the condition H^α, denoted by $f(t) \in H^\alpha$. The condition H^1 is known as the *Lipschitz condition*. If $f(t) \in C(\Gamma)$ and $f(t) \in H^\alpha$, then we say that $f(t)$ is *Hölder-continuous* on Γ. If $f \in C(\Gamma)$ and $f \in H^1$, then $f(t)$ is said to be *Lipschitz-continuous*.

Theorem 1.4.1. (Fatou's Lemma) *Let $f_n : I \mapsto \mathbb{R}$ be nonnegative, extended real-valued, measurable functions defined in an interval I and such that the sequence $\{f_n\}$ converges pointwise to the function $f : I \mapsto \mathbb{R}$. If*

$\varinjlim_{n\to\infty} \int_\Gamma f_n < \infty$, *then f is integrable and*

$$\int_I f \le \varinjlim_{n\to\infty} \int_\Gamma f_n. \tag{1.4.7}$$

PROOF. For every $k \in \mathbb{N}$, let $g_k = \inf_{n\ge k} f_n$ define a sequence of pointwise convergent functions. Then the sequence $\{g_k\}$ is increasing, i.e., $g_k \le g_{k+1}$ for all k, and it converges pointwise to a limit function f. For all $k \le n$ we have $g_k \le f_n$, which implies that $\int_I g_k\, ds \le \int_I f_n\, ds$. Hence $\int_I g_k\, ds \le \inf_{n\ge k} \int_I f_n\, ds$. By using the monotone uniform convergence theorem, we get

$$\int_I f\, ds = \lim_{k\to\infty} \int_I g_k\, ds \le \lim_{k\to\infty} \inf_{n\ge k} \int_\Gamma f_n\, ds = \varinjlim_{n\to\infty} \int_\Gamma f_n\, ds. \blacksquare$$

1.5 Cauchy's Theorems

Some basic theorems from the theory of functions of a complex variable are presented without proofs as they can be found in standard textbooks on the subject.

Theorem 1.5.1. (Cauchy's theorem) *Let f be analytic on a domain D, and let Γ be a closed contour which is homotopic to a point in D. Then*

$$\int_\Gamma f = 0.$$

Note that a set is said to be simply connected if every closed contour $\Gamma \subset D$ is homotopic (as a closed curve) to a point in D, i.e., to some constant curve. Some local versions of Cauchy's theorem are as follows:

Theorem 1.5.2. (Cauchy-Goursat theorem for a disk) *Let $f : B \mapsto \mathbb{C}$ be analytic on a disk $B = B(z_0, r) \subset D$. Then (i) there exists a function $F : B \mapsto \mathbb{C}$ which is analytic on B and is such that $F'(z) = f(z)$ for all $z \in B$ (i.e., f has an antiderivative on B); and (ii) If Γ is a closed contour in B, then* $\int_\Gamma f = 0.$

This theorem also holds if f is continuous on B and analytic on $D\backslash\{z_1\}$ for some fixed $z_1 \in D$.

Theorem 1.5.3. (Cauchy-Goursat theorem for a rectangle) *Let R denote a rectangle with sides parallel to the coordinate axes, and let f be a function defined and analytic on an open set D containing R. Then* $\int_\Gamma f = 0.$

Even if the function f is analytic on D except at some fixed point $z_1 \in D$ which does not lie on the contour R, and if $\lim_{z\to z_1} (z - z_1)f(z) = 0$, then also

$\int_\Gamma f = 0$. The two theorems 1.5.2 and 1.5.3 hold (a) if f is bounded on a deleted neighborhood of z_1, or (b) if f is continuous on D, or (c) if $\lim\limits_{z \to z_1} f(z)$ exists.

The *index* of a curve Γ with respect to a point $z_0 \in \mathbb{C}$ is the integer n that expresses how many times Γ winds around z_0. This index is denoted by $I(\Gamma, z_0)$ and is called the *winding number* of Γ with respect to z_0. Thus,

$$I(\Gamma, z_0) = \frac{1}{2i\pi} \int_\Gamma \frac{dz}{z - z_0}. \tag{1.5.1}$$

In fact,

$$I(\Gamma, z_0) = \begin{cases} \pm n & \text{if } z_0 \in \text{Int}(\Gamma), \\ 0 & \text{if } z_0 \in \text{Ext}(\Gamma). \end{cases}$$

Theorem 1.5.4. (Cauchy's integral formula) *Let f be analytic on a domain D, and let Γ be a simple closed contour in D that is homotopic to a point in D. Let $z_0 \in D$ be a point not on Γ. Then*

$$f(z_0) \cdot I(\Gamma, z_0) = \frac{1}{2i\pi} \int_\Gamma \frac{f(\zeta)}{\zeta - z_0} \, d\zeta. \tag{1.5.2}$$

The integrand in (1.5.2) is known as the *Cauchy kernel* defined by

$$H(z, z_0) = \frac{1}{2i\pi} \frac{f(z)}{z - z_0}. \tag{1.5.3}$$

The formula (1.5.2) is a special case of integrals of the Cauchy type. If we set

$$F(z) = \frac{1}{2i\pi} \int_\Gamma \frac{g(\zeta)}{\zeta - z} \, d\zeta, \tag{1.5.4}$$

where $\Gamma : [a, b] \mapsto \mathbb{C}$ is a simple contour and g a C-function defined on the image $\Gamma([a, b])$, then F is analytic on $\mathbb{C} \backslash \Gamma([a, b])$ and is infinitely differentiable, such that its kth derivative is given by

$$F^{(k)}(z) = \frac{k!}{2i\pi} \int_\Gamma \frac{g(\zeta)}{(\zeta - z)^{k+1}} \, d\zeta, \quad k = 1, 2, \dots. \tag{1.5.5}$$

Then Cauchy's integral formula for the derivatives is

$$f^{(k)}(z) \cdot I(\Gamma, z_0) = \frac{k!}{2i\pi} \int_\Gamma \frac{f(\zeta)}{(\zeta - z)^{k+1}} \, d\zeta, \quad k = 1, 2, \dots. \tag{1.5.6}$$

Let f be analytic on a domain D and let $\Gamma = \partial B(z_0, R)$ be a circle lying in D. If $|f| \le M$ for all $z \in \Gamma$, then for $k = 0, 1, 2, \ldots$

$$|f^{(k)}(z_0)| \le \frac{k!}{R^k} M. \tag{1.5.7}$$

This result is known as *Cauchy's inequality*. A corollary is the *Liouville theorem* which states that the only bounded entire functions are constants. A partial converse of Cauchy's theorem is known as *Morera's theorem* which states that if f is continuous on a domain D and if $\int_\Gamma f \, dz = 0$ for every closed contour Γ in D, then f is analytic on D, and $f = F'$ for some analytic function F on D.

Two very useful corollaries of Cauchy's integral formula (1.5.2) are:

(i) The *maximum modulus theorem* which states that if f is a nonconstant analytic function on a domain D with a simple boundary Γ, then $|f|$ cannot have a local maximum anywhere in Int (Γ).

(ii) The *mean value property* of an analytic function f defined on the circle $\partial B(z_0, r)$ states that

$$f(z_0) = \frac{1}{2\pi} \int_0^{2\pi} f\left(z_0 + re^{i\theta}\right) d\theta. \tag{1.5.8}$$

An application of the maximum modulus theorem is Schwarz's lemma (see Theorem 1.5.8).

A generalization of Cauchy's integral theorem and Cauchy's integral formula is Cauchy's residue theorem.

Theorem 1.5.5. (Cauchy's Residue theorem) *Let D be a simply connected subset of $\mathbb{C}$, and let $a_1, a_2, \ldots, a_n$ be finitely many isolated points in D, and let a function f be defined and analytic on $D \backslash \{a_1, \ldots, a_n\}$. If γ is a rectifiable closed curve in D which does not meet any of the points a_k, $k = 1, \ldots, n$, then*

$$\oint_\gamma f(z) \, dz = 2\pi i \sum_{k=1}^n I\left(\gamma, a_k\right) \text{Res}\left(f, a_k\right), \tag{1.5.9}$$

where Res (f, a_k) denotes the residue of f at a_k.

Since

$$I\left(\gamma, a_k\right) = \begin{cases} 1 & \text{if } \gamma \text{ is a positively oriented simple closed curve,} \\ 0 & \text{otherwise,} \end{cases}$$

we have

$$\oint_\gamma f(z) \, dz = 2\pi i \sum_{k=1}^n \text{Res}\left(f, a_k\right), \tag{1.5.10}$$

where the sum is taken over those k for which a_k is inside γ. Proof of this theorem can be found in a textbook on complex analysis.

Theorem 1.5.6. (Cauchy's Argument Principle) *Let the number of zeros and poles, counting their multiplicity, of a meromorphic function $f(z)$ inside and on a simple (i.e., without double points and oriented counter-clockwise) closed contour Γ be N and P. Then*

$$\oint_\Gamma \frac{f'(z)}{f(z)}\, dz = 2\pi i(N - P). \tag{1.5.11}$$

A general form of (1.5.11) is

$$\oint_\Gamma \frac{f'(z)}{f(z)}\, dz = 2\pi i \left(\sum_N I(\Gamma, a) - \sum_P I(\Gamma, b) \right), \tag{1.5.12}$$

where the first summation is over all N and the second over all P.

PROOF. Let z_N be a zero of f. Then $f(z) = (z - z_N)^k\, g'(z)$, where k is the multiplicity of the zero z_N, and $g(z_N) \neq 0$. Then

$$f'(z) = k(z - z_N)^{k-1} g(z) + (z - z_N)^k\, g'(z),$$

and

$$\frac{f'(z)}{f(z)} = \frac{k}{z - z_N} + \frac{g'(z)}{g(z)}.$$

Thus, $\dfrac{g'(z)}{g(z)}$ has no singularity at z_N, and is, therefore, analytic at z_N, which implies that the residue of $\dfrac{f'(z)}{f(z)}$ at z_N is k. Similarly, let z_P be a pole of f. Then $f(z) = (z - z_P)^{-m}\, h(z)$, where m is the order of the pole, and $h(z_P) \neq 0$. Then

$$f'(z) = -m(z - z_P)^{-m-1} h(z) + (z - z_P)^{-m} h'(z),$$

and

$$\frac{f'(z)}{f(z)} = \frac{-m}{z - z_P} + \frac{h'(z)}{h(z)}.$$

Thus, $\dfrac{h'(z)}{h(z)}$ has no singularity at z_P, and therefore, it is analytic at z_P, which implies that the residue of $\dfrac{f'(z)}{f(z)}$ at z_P is $(-m)$. Hence, we see that each zero z_N of multiplicity k of F gives a simple pole with residue k, while each pole z_P of order m of f gives a simple pole for $\dfrac{f'(z)}{f(z)}$ with the residue $-m$. There are no other poles of $\dfrac{f'(z)}{f(z)}$, and hence, no other residues. Using the residue theorem, the result follows. ∎

The name of this theorem follows from the fact that the contour integral $\oint \dfrac{f'(z)}{f(z)}\,dz$ is the total change in the argument of $f(z)$ as z travels around Γ; this follows from $\dfrac{d}{dz}\log f(z) = \dfrac{f'(z)}{f(z)}$ and the relation between arguments and logarithms.

Example 1.5.1. A function f is said to be complex differentiable (i.e., holomorphic) at a point $z_0 \in D$ if the limit (1.3.2) exists, i.e., if $f'(z_0) = \lim\limits_{\Delta z_0 \to 0} \dfrac{f(z_0 + \Delta z_0) - f(z_0)}{\Delta z_0}$ exists, where the quantity Δz_0 is a complex number that may approach zero from any direction. If f is complex differentiable at all points of D, then f is said to be holomorphic on D. A holomorphic function is infinitely many times differentiable, i.e., the existence of f' guarantees the existence of derivatives of any order. Let f be holomorphic on D and $\bar{C}(z_0, r) \subset D$. Then the function f has a power series expansion in $C(z_0, r)$ expressed by

$$f(z) = \sum_{n=0}^{\infty} a_n(z - z_0)^n \quad \text{for all } z \in C(z_0, r), \tag{1.5.13}$$

where

$$a_n = \frac{f^{(n)}(z_0)}{n!} = \frac{1}{2\pi i} \int_{C_r} \frac{f(w)}{(w - z_0)^{n+1}}\,dw. \tag{1.5.14}$$

To prove, assume without loss of generality that $z_0 = 0$ and $\rho < r$. Let $S_N(z) = 1 + z + z^2 + \cdots + z^N$, and $R_N(z) = \dfrac{z^{N+1}}{1 - z}$. Then $\dfrac{1}{1 - z} = S_N(z) + R_N(z)$, and denote $\partial C(z_0, r)$ by C_r. Then, by Cauchy's formula (1.5.4),

$$\begin{aligned}
f(z) &= \frac{1}{2\pi i} \int_{C_r} \frac{f(w)}{(w - z)}\,dw = \frac{1}{2\pi i} \int_{C_r} \frac{f(w)}{w} \frac{1}{1 - z/w}\,dw \\
&= \frac{1}{2\pi i} \int_{C_r} \frac{f(w)}{w} S_N(z/w)\,dw + \frac{1}{2\pi i} \int_{C_r} \frac{f(w)}{w} R_N(z/w)\,dw \\
&= \sum_{n=0}^{N} \left(\frac{1}{2\pi i} \int_{C_r} \frac{f(w)}{w}\,dw \right) + \varepsilon_N(z),
\end{aligned}$$

where

$$|\varepsilon_N(z)| \le \frac{1}{2\pi} \frac{\sup_{C_r} |f|}{r} \frac{(\rho/r)^{N+1}}{1 - \rho/r}(2\pi r) \to 0 \quad \text{as } N \to \infty.$$

Also,

$$a_n = \frac{1}{2\pi i} \int_{C_r} \frac{f(w)}{w}\,dw = \frac{f^{(n)}(0)}{n!}.$$

Hence, this result takes a function $f \in C^{(k)}(D)$ from holomorphic to analytic. ∎

Example 1.5.2. An analytic function $w = f(z)$, $w = u(x, y) + iv(x, y)$, is said to be *smooth* if (i) the first derivatives u_x, u_y, v_x and v_y exist and are continuous, and (ii) the second derivatives u_{xx}, u_{xy}, u_y, x and u_{yy} exist and are continuous. Note that for functions of a real variable $f(x)$, the existence of the first derivative f_x does nor guarantee its continuity, and even if the derivative is continuous, there is no guarantee that higher derivatives will exist. Similar issues are also valid for functions of several real variables. For example, consider $f(x) = x^2 \sin(1/x)$ if $x \neq 0$; $f(0) = 0$; then f is differentiable at and near $x = 0$ but the derivative $f'(x)$ is discontinuous at $x = 0$ since $\lim_{x \to 0} f'(x)$ does not exist. Does f'' exist at $x = 0$? ∎

Theorem 1.5.7. (Maximum modulus principle) *If f is analytic on the unit disk E and $r < 1$, then*

$$\max_{z \leq r} |f(z)| \leq \max_{|z|=r} |f(z)|.$$

PROOF. By Cauchy's integral formula

$$|f(z)| = \left| \frac{1}{2i\pi} \int_\gamma \frac{f(t)}{t - z} \, dt \right| \leq \frac{1}{2\pi r} \int_0^{2\pi} |f(r\, e^{i\theta})| \, d\theta \leq \max_{|z|=r} |f(z)|. \quad \blacksquare$$

Thus if f is analytic on E and $\limsup_{|z| \uparrow 1} |f(z)| \leq M$, then $|f(z)| \leq M$ on the entire unit disc E. Note that the analytic map $f : E \mapsto E$ is a contraction of the hyperbolic metric on E, defined by $d\rho = 2|dz|/\left(1 - z^2\right)$, with equality if and only if f is a Möbius transformation (see §2.6). This is generally known as *Schwarz's lemma*. Recall that Möbius transformations map circles into circles, using the convention that straight lines are also circles passing through the point at infinity.

Theorem 1.5.8. (Schwarz's lemma) *If $f : E \mapsto E$ is analytic and $f(0) = 0$, then $|f'(0)| \leq 1$ with equality if and only if f is a rotation.*

PROOF. Apply the maximum modulus principle to $f(z)/z$. ∎

1.6 Harmonic functions

The functions whose Laplacian is zero are known as harmonic functions. Thus, a real-valued function $u(x, y) \in C^2(D)$ is said to be *harmonic* in a domain D if $\nabla^2 u = 0$. Some properties of harmonic functions in $\mathbb{R}^2$ are as follows:

(i) The function

$$\frac{1}{r} = \frac{1}{\sqrt{(x - x_0)^2 + (y - y_0)^2}} \tag{1.6.1}$$

is harmonic in a domain that does not contain the point (x_0, y_0).

(ii) If $u(x, y)$ is a harmonic function in a simply connected domain D, then there exists a conjugate harmonic function $v(x, y)$ in D such that $u(x, y) + i\,v(x, y)$ is an analytic function of $z = x + iy = (x, y)$ in D. In view of the Cauchy-Riemann equations (1.3.4),

$$v(x, y) - v(x_0, y_0) = \int_{x_0, y_0}^{x, y} (-u_y \, dx + u_x \, dy), \qquad (1.6.2)$$

where $(x_0, y_0) = z_0$ is a given point in D. This property is also true if D is multiply connected. However, in that case the conjugate function $v(x, y)$ can be multiple-valued, as we see by considering $u(x, y) = \log r = \log \sqrt{x^2 + y^2}$ defined on a domain D containing the origin which has been indented by a small circle centered at the origin. Then, in view of (1.6.2), we have

$$v(x, y) - v(x_0, y_0) = \tan^{-1} \frac{y}{x} \pm 2n\pi + \text{const}, \quad n = 1, 2, \ldots,$$

which is multiple-valued.

(iii) Since derivatives of all orders of an analytic function exist and are themselves analytic, any harmonic function will have continuous partial derivatives of all orders, i.e., a harmonic function belongs to the class $C^\infty(D)$, and a partial derivative of any order is again harmonic.

(iv) A harmonic function must satisfy the *mean-value theorem*, where the mean value at a point is evaluated for the circumference or the area of the circle around that point. If u is harmonic on a domain containing the closed disk $\bar{C}(z_0, r)$, where $z_0 = x_0 + iy_0$, then

$$u(x_0, y_0) = \frac{1}{2\pi} \int_0^{2\pi} u\left(z_0 + r\,e^{i\theta}\right) d\theta. \qquad (1.6.3)$$

(v) In view of the maximum modulus principle (Theorem 1.5.7), the maximum (and also the minimum) of a harmonic function u in a domain D occurs only on the boundary of D. This result is known as

Theorem 1.6.1. (Maximum principle for harmonic functions) *A nonconstant function which is harmonic inside a bounded domain D with boundary Γ and continuous in the closed domain $\bar{D} = D \cup \Gamma$ attains its maximum and minimum values only on the boundary of the domain.*

Thus, u has a maximum (or minimum) at $z_0 \in D$, i.e., if $u(z) \leq u(z_0)$ (or $u(z) \geq u(z_0)$) for z in a neighborhood $C(z_0, \varepsilon)$ of z_0, then $u = \text{const}$ in $C(z_0, \varepsilon)$.

(vi) The value of a harmonic function u at an interior point z_0 in terms of the boundary values of u and $\dfrac{\partial u}{\partial n}$ is given by Green's third identity (C.1.8)

$$2\pi u(z_0) = \int_\Gamma \left[\frac{\partial}{\partial n}(\log r) - (\log r)\frac{\partial u}{\partial n} \right] ds.$$

(vii) If u and U are continuous in $\bar{D}$ and harmonic in D such that $u \leq U$ on Γ, then $u \leq U$ also at all points inside D. In fact, the function $U - u$ is continuous and harmonic in D. Hence $U - u \geq 0$ on Γ. Then, in view of the maximum modulus principle (Theorem 1.6.1), we require that $U - u \geq 0$ at all points inside D.

(viii) If u and U are continuous in $\bar{D}$ and harmonic in D for which $|u| \leq U$ on Γ, then $|u| \leq U$ also at all points inside D. In fact, the three harmonic functions $-U$, u, and U satisfy the relation $-U \leq u \leq U$ on Γ. Then, by (vii), $-U \leq u \leq U$ at all points inside D, or $|u| \leq U$ inside D.

Theorem 1.6.2. (Harnack's theorem) *Suppose that* $\{u_n(z)\}$ *is a monotone increasing sequence of harmonic functions on a domain* D, *which is convergent at a point* $z_0 \in D$. *Then* $\{u_n(z)\}$ *is uniformly convergent on closed sets in* D.

PROOF. Since the sequence $\{f_n(z)\}$ is harmonic in the domain $D \subset \mathbb{C}$, and $u_1(z) \leq u_2(z) \leq \cdots$ at every point of D, then the limit $\lim_{n \to \infty} u_n(z)$ is either infinite at every point of D or is finite at every point of D, but in each case uniform in every compact subset of D. In the latter case, the function $u(z) = \lim_{n \to \infty} u_n(z)$ is harmonic in D. ∎

This theorem follows from the *Harnack inequality* in $\mathbb{R}^n$, which is stated as follows: *Let* f *be a non-negative function defined on a closed ball* $\bar{B}(x_0, R) \in \mathbb{R}^n$ *with center at* x_0 *and radius* R. *If* f *is continuous on* $\bar{B}(x_0, R)$ *and harmonic on the boundary* $\partial B(x_0, R)$, *then for any point* x *such that* $|x - x_0| = r < R$,

$$\frac{1 - (r/R)}{(1 + (r/R))^{n-1}} f(x_0) \leq f(x) \leq \frac{1 + (r/R)}{(1 - (r/R))^{n-1}} f(x_0). \tag{1.6.4}$$

For $n = 2$, the inequality (1.6.4) becomes

$$\frac{R - r}{R + r} f(x_0) \leq f(x) \leq \frac{R + r}{R - r} f(x_0), \tag{1.6.5}$$

which holds in $C(x_0, R)$. For a proof, see Moser [1961].

Let $z_0 \neq \infty$ be any point inside D, and let K denote the closed disk $\bar{C}(z_0, R)$ such that $K \subset D$. Then, if $r = |z - z_0| \leq R$, the Harnack inequality

$$u_n(z_0) \frac{R - r}{R + r} \leq u_n(z) \leq u_n(z_0) \frac{R + r}{R - r} \tag{1.6.6}$$

holds for any annulus $0 < r < R$ with center at z_0, provided that $u_n(z)$ are harmonic and nonnegative on the disk $C(z_0, R)$.

Theorem 1.6.3. (Identity Theorem) *Let* $D \subset \mathbb{C}$, *and let* $f, g : D \mapsto \mathbb{C}$ *be analytic. Then the following three statements are equivalent:*

(i) $f = g$;

(ii) there exists a non-discrete subset $G \subset D$ such that $f(z) = g(z)$ for all $z \in G$;

(iii) there exists a point $z_0 \in D$ such that $f^{(n)}(z_0) = g^{(n)}(z_0)$ for all $n \in \mathbb{N}$. The third statement implies that f and g have the same Taylor series expansion at z_0, and therefore, they are equal on some neighborhood of z_0.

PROOF. (i) implies (iii) is obvious; (iii) implies (ii) is clear; (ii) implies (i) follows from Theorem 1.6.1. In fact, let $f(z) = \sum_{n=0}^{\infty} a_n(z - z_0)^n$ is a non-zero power series with radius of convergence $\rho > 0$, and let $m = \inf\{n \in \mathbb{N}; a_n \neq 0\}$ at z_0. Then $f(z) = a_m(z - z_0)^m + a_{m+1}(1 + h(z))$, where $h(z)$ is a power series of the form $h(z) = b_1(z - z_0) + b_2(z - z_0)^2 + \cdots$. Then

$$f(z) = a_m(z - z_0)^m + a_{m+1}(z - z_0)^{m+1} + \cdots$$
$$= a_m(z - z_0)^m \left[b_1(z - z_0) + b_2(z - z_0)^2 + \cdots \right],$$

where $b_k = a_{m+k}/a_m$ for $k \in \mathbb{N}$. Thus, $\limsup_k |b_k|^{1/k} = \limsup_k |a_n|^{1/n}$ since $\lim_k |a_m|^{1/k} = 1$. ∎

This theorem is useful in establishing certain identities, including the concept of analytic continuation.

Example 1.6.1. (Argument Principle) This principle implies that every zero of a meromorphic function winds around the origin *anti-clockwise*, whereas every pole does so *clockwise*. The total winding number (index) of a curve around the origin is equal to the number of zeros *minus* the number of poles inside γ, each zero and pole winding as many time as its multiplicity.

Consider the polynomial $f(z) = z^5$. Let γ be a closed curve containing the origin defined by $\gamma(t) = e^{it}$, $0 \leq t \leq 2\pi$. Thus, $f(z)$ has a zero at 0 of multiplicity 5, i.e., it has 5 zeros at 0 and no poles inside γ. The curve $f(z) = f(\gamma(t)) = e^{5it}$, $0 \leq t \leq 2\pi$, winds 5 times anti-clockwise around the origin. Since $5 - 0 = 5$, it is consistent with the argument principle. Now, consider another function $g(z) = z^{-5}$ which has 5 poles inside γ and no zeros. The curve $g(z) = g(\gamma(t)) = e^{-5it}$, $0 \leq t \leq 2\pi$, winds 5 times clockwise around the origin. Sine $0 - 5 = -5$, it is consistent with the argument principle. ∎

Example 1.6.2. Use the argument principle to find the number of zeros of the polynomial $p(z) = z^{2010} + z + 2015$ in the first quadrant. A solution is as follows: choose a closed contour $\gamma = L_0^R + C_R^I + R_{iR}^0$, where C_R^I is the circular arc of very large radius R in the first quadrant, L_0^R the real-axis side, and L_{iR}^0 the imaginary-axis side of the domain in the first quadrant. Note that $\Delta_{L_0^R} \arg\{p(z)\} = 0$, and $\Delta_{L_{iR}^0} \arg\{p(z)\} = -\pi$. Now on C_R^I,

$$\int_{C_R^I} \frac{p'(w)}{p(w} \, dw == \int_{C_R^I} \frac{2010w^{2009} + 1}{w^{2010} + w + 2015} \, dw$$

$$\approx \int_{C_R^I} \frac{2010}{w} \, dw = 2010\frac{2\pi i}{4} = 1005\pi i.$$

Thus, the number of zeros of $p(z)$ in the first quadrant is $\dfrac{1005\pi - \pi}{2\pi} = 502$. ∎

1.7 Piecewise Bounded Functions

A family $\mathcal{B}$ of functions is said to be *piecewise bounded* if for every x, the set $\{f(x) : f \in \mathcal{B}\}$ is a bounded set, where different bounds for different x are permitted.

Theorem 1.7.1. *Let $\{f_n(z)\}$ be a sequence of functions, each continuous on a piecewise smooth path Γ, and suppose $\{f_n(z)\}$ converges uniformly on Γ to a limit function $f(z)$. Then*

$$\lim_{n \to \infty} f_n(z) \, dz = \int_\Gamma f(z) \, dz. \tag{1.7.1}$$

PROOF. The continuity of f implies its integrability, which in turn follows from the uniform limit of a sequence of continuous functions. To prove (1.7.1), we have

$$\left| \int_\Gamma [f(z) - f_n(z)] \, dz \right| \leq l \sum_{z \in \Gamma} |f(z) - f_n(z)|, \tag{1.7.2}$$

where l is the length of Γ. But the right-hand side approaches 0 as $n \to \infty$ by the assumed convergence of $\{f_n(z)\}$. ∎

Theorem 1.7.2. *Let $f_n(z)$ be a sequence of functions, each continuous on a piecewise smooth path Γ and dependent on a parameter λ varying over some set Λ. Further, suppose that $f_n(z, \lambda)$ converges uniformly on the direct product $G \times \Lambda$ to a limit function $f(z, \lambda)$. Then the sequence of functions $\{F_n(\lambda)\} = \int_\Gamma f_n(z, \lambda, ds$ converges uniformly on Λ to the function $f(z) = \int_\Gamma f(z, \lambda) \, dz$.*

PROOF. Instead of (1.7.2) we have

$$\left| \int_\Gamma [f(z, \lambda) - f_n(z, \lambda)] \, dz \right| \leq l \sum_{z \in \Gamma} |f(z, \lambda) - f_n(z, \lambda)|, \tag{1.7.3}$$

where the right-hand side approaches 0 as $n \to \infty$, by the assumed convergence of $\{f_n(z, \lambda)\}$. ∎

Theorem 1.7.3. *Let $f(z, \zeta)$ be a function of two complex variables z and ζ, where $z = x + iy$ varies over a domain D in the z-plane and $\zeta = \xi + i\eta$ varies over a piecewise smooth path γ in the ζ-plane. Suppose $f(z, \zeta)$ is analytic on D for every fixed $\zeta \in \Gamma$, while both $f(z, \zeta)$ and its partial derivatives $f_z(z, \zeta)$ with respect to z are continuous on $D \times \gamma$. Then the function $\Phi(z) = \int_\Gamma f(z, \zeta)\, d\zeta$ is analytic on D with derivatives $\phi(z) = \int_\Gamma f_z(z, \zeta)\, d\zeta$.*

PROOF. Obviously, $\oint_\Gamma \phi(z)\, dz = 0$ for every piecewise smooth path $\Gamma \subset D$. In fact,

$$\oint_\Gamma \phi(z)\, dz = \oint_\Gamma \left\{ \int_\gamma f_z(z, \zeta)\, d\zeta \right\} dz = \int_\gamma \left\{ \oint_\Gamma f_z(z, \zeta)\, dz \right\} d\zeta = 0.$$

Hence, there exists a function $\Phi_0(z)$ analytic on D, with derivative $\Phi'(z) = \phi(z)$, where

$$\Phi_0(z) = \int_{z_0}^z \phi(t)\, dt = \int_{z_0}^z \left\{ \int_\gamma f_z(t, \zeta)\, d\zeta \right\}$$

$$= \int_\gamma \left\{ \int_{z_0}^z f_z(t, \zeta(\, dt \right\} d\zeta = \int_\gamma f(z, \zeta)\, d\zeta - \int_{z_0}^z f(z_0, \zeta)\, d\zeta.$$

Thus, $\Phi_0(z)$ differs from $\Phi(z)$ by only a constant, and hence, $\Phi(z)$ is analytic on D. ∎

DEFINITION 1.7.1. By an *integral of the Cauchy type* we mean an expression of the form

$$F(z) = \int_\Gamma \frac{f(\zeta)}{\zeta - z}\, d\zeta, \quad z \notin \Gamma, \tag{1.7.4}$$

where $f(\zeta)$ is analytic on a piecewise smooth path $\Gamma \subset \mathbb{C}$. Compare this definition with Theorem 1.5.4. Clearly, $F(z)$ is defined at every point of the open set $D = \mathbb{C} \backslash \Gamma$ (if D is not connected, then it may not be a set).

Theorem 1.7.4. *The function $F(z)$ defined by (1.7.4) is analytic on D, with derivatives of all orders on D (they are analytic, too), given by*

$$F^{(n)}(z) = n! \int_\Gamma \frac{f(\zeta)}{(\zeta - z)^{n+1}}\, d\zeta, \quad n = 1, 2, \dots .$$

PROOF. For a fixed $\zeta \in \Gamma$, the function $f(z, \zeta) = \dfrac{f(\zeta)}{\zeta - z}$ and its partial derivatives

$$f_z^{(n)}(z, \zeta) = n! \frac{f(\zeta)}{(\zeta - z)^{n+1}}, \quad n = 1, 2, \dots ,$$

are analytic on D. Moreover, $f(z, \zeta)$ and all its derivatives $f_z^{(n)}(z, \zeta)$ are continuous on $D \times \Gamma$. The result then follows by Theorem 1.7.3. ∎

Theorem 1.7.5. (Weierstrass' Theorem) *If $\{f_n(z)\}$ is a sequence of analytic functions converging uniformly inside a domain D to a limit function $f(z)$, then $f(z)$ is analytic in D. Moreover, the sequence of derivatives $\{f_n^{(m)}(z)\}$ of any order m converges uniformly inside D to the corresponding derivative $f^{(m)}(z)$ of $f(z)$.*

PROOF. By Cauchy's formula (1.5.2)

$$f_n(z) = \frac{1}{2\pi i} \oint_\Gamma \frac{f_n(z)}{\zeta - z} \, dz$$

for any piecewise smooth path $\Gamma \subset D$ surrounding a given point $z \in D$. However, $\sup_{\zeta \in \Gamma} \left| \frac{1}{\zeta - z} \right| \le \frac{1}{\delta}$, where $\delta = \inf_{\zeta \in \Gamma} |\zeta - z|$, while the sequence $\{f_n(z)\}$ converges uniformly to $f(\zeta)$ on Γ as $n \to \infty$, since Γ is a compact set. It implies that $\frac{f_n(\zeta)}{\zeta - z}$ converges uniformly to $\frac{f(\zeta)}{\zeta - z}$, and hence, by Theorem 1.7.1, we have

$$f(z) = \frac{1}{2\pi i} \oint_\Gamma \frac{f_n(z)}{\zeta - z} \, dz.$$

Thus, $f(z)$ can be represented on D as an integral of the Cauchy type. Therefore, $f(z)$ is analytic on D by Theorem 1.7.3.

To prove the second part, by Theorem 1.7.2 we have

$$f^{(m)}(z) = \frac{m!}{2\pi i} \oint_\Gamma \frac{f(\zeta)}{(\zeta - z)^{m+1}} \, d\zeta,$$

$$f_n^{(m)}(z) = \frac{m!}{2\pi i} \oint_\Gamma \frac{f_n(\zeta)}{(\zeta - z)^{m+1}} \, d\zeta,$$

where the integrals are taken along a piecewise smooth closed path $\Gamma \subset D$ surrounding a given compact set $G \subset D$. Then in this case $\sum_{\zeta \in \Gamma, z \in G} \left| \frac{1}{(\zeta - z)^{m+1}} \right| \le \frac{1}{\delta^{m+1}}$, where $\delta = \inf_{\zeta \in \Gamma, z \in G} |\zeta - z|$, while the sequence $\frac{f_n(\zeta)}{(\zeta - z)^{m+1}}$ converges uniformly on $\Gamma \times G$ to $\frac{f(\zeta)}{(\zeta - z)^{m+1}}$. Hence, by Theorem 1.7.2, $f^{(m)})z = \lim_{n \to \infty} f_n^{(m)}(z)$, where the convergence is uniform on every compact set $G \subset D$. ∎

1.8 Metric Spaces

Let S denote the set of all real-valued sequences. Then S is a vector space if addition and scalar multiplication of vectors s_i, $t_i \in S$ for $i = 1, 2, \ldots$ are defined coordinatewise, i.e., $\{s_i\} + \{t_i\} = \{s_i + t_i\}$, and $\lambda \{s_i\} = \{\lambda s_i\}$ for a scalar λ. The *Fréchet metric* (distance) $d\left(\{s_i\}, \{t_i\}\right)$ is defined for $\{s_i\}, \{t_i\} \in S$ by

$$d\left(\{s_i\}, \{t_i\}\right) = \sum_{i=1}^{\infty} 2^{-i} \frac{|s_i - t_i|}{(1 + |s_i - t_i|)}. \tag{1.8.1}$$

The space ℓ^p, $1 \le p < \infty$, is defined by $\ell^p = \left\{ \{s_i\} : \sum_{i=1}^{\infty} |s_i|^p < \infty \right\}$. Since the sum of any two elements in ℓ^p is also in ℓ^p, then ℓ^p is a vector subspace of S. Define a norm $\| \cdot \|_p$ on ℓ^p by

$$\|\{s_i\}\|_p = \left(\sum_{i=1}^{\infty} |s_i|^p \right)^{1/p}. \tag{1.8.2}$$

Then it can be shown that ℓ^p is closed under vector addition and $\| \cdot \|_p$ satisfies the triangle inequality (see below). Let $\{s_i\}$, $\{t_i\} \in \ell^p$ be such that $\sum |s_i|^p = 1$ and $\sum |t_i|^q = 1$, where p and q are called conjugate exponents such that $p > 1$ and $\dfrac{1}{p} + \dfrac{1}{q} = 1$. Then we have

Hölder inequality: $\displaystyle\sum_{i=1}^{\infty} |s_i\, t_i| \le \left(\sum_{i=1}^{\infty} |s_i|^p \right)^{1/p} \left(\sum_{i=1}^{\infty} |t_i|^q \right)^{1/q}. \tag{1.8.3}$

Cauchy-Schwarz inequality: $\displaystyle\sum_{i=1}^{\infty} |s_i\, t_i| \le \sqrt{\sum_{i=1}^{\infty} |s_i|^2} \sqrt{\sum_{i=1}^{\infty} |t_i|^2}. \tag{1.8.4}$

Minkowsky's inequality:

$$\left(\sum_{i=1}^{\infty} |s_i + t_i|^p \right)^{1/p} \le \left(\sum_{i=1}^{\infty} |s_i|^p \right)^{1/p} + \left(\sum_{i=1}^{\infty} |t_i|^p \right)^{1/p}, \quad p \ge 1. \tag{1.8.5}$$

Note that $p = 1$ gives the triangle inequality.

1.8.1. Poincaré Metric. The Poincaré lemma states that a closed form on a starshaped set is exact. This lemma is a generalization of the fundamental theorem of calculus. The vector version of this lemma is as follows: If $\mathbf{v}$ is a smooth vector field and f a smooth scalar field function in a ball in $\mathbb{R}^3$ centered at the origin, then

$$\mathbf{v}(\mathbf{r}) = \nabla \left(\int_0^1 \mathbf{v}(t\mathbf{r}) \right) + \int_0^1 \operatorname{curl} \mathbf{v}(t\mathbf{r}) \times t\mathbf{r}\, dt,$$

$$\mathbf{v}(\mathbf{r}) = \operatorname{curl} \left(\int_0^1 (\mathbf{v}(t\mathbf{r}) \times t\mathbf{r})\, dt \right) + \int_0^1 t^2 \mathbf{r}\, \operatorname{div} \mathbf{v}(t\mathbf{r})\, dt, \tag{1.8.6}$$

$$f(\mathbf{r}) = \operatorname{div} \left(\int_0^1 t^2 \mathbf{r} \cdot f(t\mathbf{r})\, dt \right),$$

where $\mathbf{r} = x\mathbf{i} + y\mathbf{j} + x\mathbf{k}$ is a point in $\mathbb{R}^3$. For a proof, see Spivak [1965].

1.9 Equicontinuity

Let $\mathcal{C}[a, b]$ denote the set of all continuous real-valued functions defined on the interval $[a, b]$ of the real line $\mathbb{R}$. A subset $Y \subset \mathcal{C}[a, b]$ is said to be *equicontinuous* on the interval $[a, b]$ if for every $\varepsilon > 0$ there exists a $\delta > 0$ such that $|x(t) - x(t_0)| < \varepsilon$ for all $x \in Y$ and all t, t_0 such that $|t - t_0| < \delta$.

Theorem 1.9.1. *For* $x, y \in \mathcal{C}[a, b]$, *let* $\rho(x, y) = \sup_{a \leq t \leq b} |x(t) - y(t)|$. *Then* $\{\mathcal{C}[a, b]; \rho\}$ *is a metric space.*

PROOF. Note that (i) $\rho(x, y) = \rho(y, x)$; (ii) $\rho(x, y) \geq 0$ for all x and y; and (iii) $\rho(x, y) = 0$ if and only if $x(t) = y(t)$ for all $t \in [a, b]$. Next,

$$
\begin{aligned}
\rho(x, y) &= \sup_{a \leq t \leq b} |x(t) - y(t)| \\
&= \sup_{a \leq t \leq b} |x(t) - z(t) + z(t) - y(t)| \\
&\leq \sup_{a \leq t \leq b} \{|x(t) - z(t)| + |z(t) - y(t)|\} \\
&\leq \sup_{a \leq t \leq b} |x(t) - z(t)| + \sup_{a \leq t \leq b} |z(t) - y(t)| \\
&= \rho(x, z) + \rho(y, z),
\end{aligned}
$$

which shows that ρ satisfies the triangle inequality. Hence, $\{\mathcal{C}[a, b]; \rho\}$ is a metric space. ∎

Theorem 1.9.2. (Arzela-Ascoli theorem) *Let* $\{\mathcal{C}[a, b]; \rho\}$ *be the metric space defined in Theorem 1.9.1. Le* Y *be a bounded subset of* $\mathcal{C}[a, b]$. *If* Y *is equicontinuous on* $[a, b]$, *then* Y *is relatively compact in* $\mathcal{C}[a, b]$.

This theorem provides necessary and sufficient conditions for determining whether every sequence of a given family of real-valued continuous functions defined on a compact interval has a uniformly convergent subsequence. The main condition required for this determination is the equicontinuity of the family of functions.

The sufficient condition for compactness was established by Ascoli in 1983–1884, while the necessary condition was established by Arzelà in 1895. A generalization of this theorem to sets of real-valued continuous functions with a compact metric space as domain was given by Fréchet in 1906. The details are available in Dunford and Schwartz [1958: 382].

Theorem 1.9.2a. (Arzela-Ascoli theorem in $\mathbb{R}$) *Consider a sequence* $\{f_n\}_{n \in \mathbb{N}}$ *of real-valued continuous functions defined on a compact interval* $I = [a, b]$. *If the sequence is uniformly bounded and equicontinuous, then there exists a subsequence* $\{f_{n_k}\}$ *that converges uniformly.*

PROOF. The sequence $\{f_n\}$ is uniformly bounded if there exists an $M > 0$

such that $|f_n(x)| \leq M$ for every function f_n in the sequence and for every $x \in [a, b]$. The sequence is equicontinous if for every $\varepsilon > 0$ there exists a $\delta > 0$ such that $|f_n(x) - f_n(y)| \leq \varepsilon$ whenever $|x - y| < \delta$ for all functions in the sequence. The converse also holds in the sense that if every subsequence of $\{f_n\}$ has a uniformly convergent subsequence, then the sequence $\{f_n\}$ is uniformly bounded and equicontinous.

Let F be an infinite set of functions $f : I \mapsto \mathbb{R}$ which is uniformly bounded and equicontinuous, then there exists a sequence of functions $f_n \in F$ that converges uniformly on I. Let x_{n_k} be a fixed enumeration of rational points on I. Since F is uniformly bounded, so is the set of points $\{f_{x_n}\}_{f \in F}$, and hence, by the Bolzano-Weierstrass theorem, there is a subsequence $\{f_{n_1}\}$ of distinct functions in F such that $\{f_{n_1}(x_1)\}$ converges. Similarly there exists a subsequence of distinct functions sequence $\{f_{n_2}(x_2)\}$ such that $\{f_{n_1}\} \supseteq \{f_{n_2}\}$. Continuing this process, we obtain by induction a chain of subsequences $\{f_{n_2}(x_2)\}$ such that $\{f_{n_1}\} \supseteq \{f_{n_2}\}, \ldots$ such that for each $k = 1, 2 \ldots$ the subsequences $\{f_{n_k}\}$ converge at the points $x_1, x_2, \ldots$.

Next, by forming the diagonal subsequences $\{f_m\}$, where f_m is the mth term in the mth subsequence $\{f_{n_m}\}$. This guarantees that $\{f_m\}$ converges at every rational point of I. Thus, for any $\varepsilon > 0$ and rational $x_k \in I$, there is an integer $N = N(\varepsilon, x_k)$ such that $|f_n(x_k) - f_m(x_k)| < \varepsilon/3$ for $n, m \geq N$. The equicontinuity property of the set F implies that for a fixed ε and for every $x \in I$, there is an open interval T_k containing x such that $|f(s) - f(t)| < \varepsilon/3$ for all $f \in F$ and all $s, t \in I$ such that $s, t \in T_k$. Thus, the collection of all intervals T_k, $x \in I$, forms an open cover of I. Since I is compact, this cover admits a finite subcover $T_1, \ldots, T_J$. There exists an integer K such that each open interval T_j, $1 \leq j \leq J$, contains a rational x_k for $1 \leq k \leq K$. Finally, for any $t \in I$, there are j and k so that t and x_k belong to the same interval T_j. Then for such a choice of k, we get $|f_n(t) - f_m(t)| \leq |f_n(t) - f_n(x_k)| + |f_n(x_k) - f_m(x_k)| + |f_m(x_k) - f_m(t)| < \varepsilon/3 + \varepsilon/3 + \varepsilon/3 = \varepsilon$ for all $n, m > N = \max\{N(\varepsilon, x_1), \ldots, N(\varepsilon, x_K)\}$. Hence, the sequence $\{f_n\}$ is a Cauchy sequence, and therefore, converges uniformly to a continuous function. $\blacksquare$

Corollary 1.9.1. *Let $\{\phi_n\}$ be a sequence of functions in $\{\mathcal{C}[a, b]; \rho\}$. If $\{\phi_n\}$ is equicontinuous on $[a, b]$ and uniformly bounded on $[a, b]$ (i.e., there exist an $M > 0$ such that $\sup_{a \leq t \leq b} |\phi_n(t)| \leq M$ for all n) , then there exists a $\phi \in \mathcal{C}[a, b]$ and a subsequence $\{\phi_{n_1}\}$ of $\{\phi_n\}$ such that $\{\phi_{n_1}\}$ converges to ϕ uniformly on $[a, b]$.*

Corollary 1.9.2. *Let Y be a subset of $\mathcal{C}[a, b]$ which is relatively compact in the metric space $\{\mathcal{C}[a, b]; \rho\}$. Then Y is a bounded set and is equicontinuous on $[a, b]$.*

1.10 Exercises

1.10.1. Show that the series (1.3.15) converges uniformly on the circle γ_r, as stated in the proof of Theorem 1.3.3. HINT. Using the Weierstrass test, since f is continuous on γ_r, it is bounded, so that $|f(z)| \le M$ for all z on Γ_r. The terms in series (1.3.15) have the bounds

$$\left| f(z) \frac{(\zeta - a)^n}{(z-a)^{n+1}} \right| \le M \left| \frac{(\zeta - a)^n}{(z-a)^{n+1}} \right| = \frac{M}{r^{n+1}} |\zeta - a|^n = M_n$$

for all z on γ_r, and the sum of these bounds is finite,

$$\sum_{n=0}^{\infty} M_n = \frac{M}{r} \sum_{n=0}^{\infty} \left(\frac{|\zeta - a|}{r} \right)^n < +\infty,$$

because $|\zeta - a| < r$. ∎

1.10.2. Consider $f(z) = z^i$, which is defined and analytic on the disk $D = \{z : |z - 1| < \frac{1}{4}\}$ with center at $z = 1 + i0$. Other branches of this function on the complex plane with slit along the negative real axis are $f_k(z) = e^{i(\log z + 2\pi k i)} = e^{-2\pi k} z^i$ for $k = 0, \pm 1, \pm 2, \ldots$, where $\log z$ denotes its principal brach. Determine the limit values of the branch $f_0(z) = z^i$. HINT: The limit values as z approaches a point $z_0 = -R + i0$ on the slit negative real axis from the upper or lower half-plane are

$$\lim_{\varepsilon \to 0^+} (-R + i\varepsilon)^i = e^{i(\log R + i\varepsilon + i\pi)} = e^{-\pi + i \log R},$$

or

$$\lim_{\varepsilon \to 0^+} (-R - i\varepsilon)^i = e^{i(\log R + i\varepsilon - i\pi)} = e^{\pi + i \log R} \quad ∎$$

1.10.3. Let $f : D \mapsto \mathbb{C}$ be analytic in a domain $D \subset \mathbb{C}$ such that $f \ne 0$. Prove that $\{z \in D; f(z) = 0\}$ is a discrete subset of D. In fact, let $z_0 \in D$ with $f(z_0) \ne 0$. Since f is analytic , there exists either $r > 0$ such that $f(z) \ne 0$ for some $in B(z_0, r) \backslash \{z_0\}$, or $f = 0$ in some neighborhood of z_0. In the latter case, $z_0 \notin \text{supp}(f)$. Since D is connected, we have $\text{supp}(f) = \emptyset$, and hence, $f = 0$, which is a contradiction. ∎

1.10.4. For a power series in negative powers of z, what is the domain of convergence and on which sets in the complex plane is the series uniformly convergent, and are the sums analytic functions on the domain of convergence? In fact, consider the series

$$\sum_{n=0}^{\infty} a_n (z - z_0)^{-n} = a_0 + \frac{a_1}{z - z_0} + \frac{a_2}{(z - z_0)^2} + \cdots, \qquad (1.10.1)$$

Let $f_n(z) = a_n (z - z_0)^{-n}$. This function which is a general term of the series (1.10.1) is well defined except at the point $z = z_0$. Let us compare the series

(1.10.1) with the power series

$$\sum_{n=0}^{\infty} a_n w^n = a_0 + a_1 w + a_2 w^2 + \cdots , \tag{1.10.2}$$

which has the same coefficients a_n. The series (1.10.2) has a definite radius of convergence $0 \leq r \leq \pm \infty$; it is absolutely convergent at points in any open disk bounded by the circle $|w| = r$, and it diverges at all points outside this circle. Moreover, this series this uniformly convergent on any closed disk whose radius is smaller than r, and the sum $g(w) = \sum_{n=0}^{\infty} a_n w^n$ is analytic on the open disk $|w| < r$. If we substitute $w = 1/(z - z_0)$ in the series (1.10.2), we get (i) $|w| < r$ if and only if $|z - z_0| > 1/r$; (ii) $|w| = r$ if and only if $|z - z_0| = 1/r$; and (iii) $|w| > r$ if and only if $|z - z_0| < 1/r$. Hence, the series (1.10.1) diverges if $|z - z_0| < 1/r$; it converges absolutely if $|z - z_0| > 1/r$; but its behavior on the circle $|z - z_0| = 1/r$ is undecided. Note that the mapping $w = \phi(z) = 1/(z - z_0)$ maps the circle $|z - z_0| = 1/r$ onto the domain $\{w : 0 \leq |w| < r\}$, as shown in Figure 1.10.1. Obviously, the circle of radius $R = 1/r$ divides the z-plane into separate domains on which the series (1.10.1) converges or diverges, but the set of points lying outside this circle of radius R is the domain of convergence. If we write $f(z)$ for the sum of negative powers (1.10.1), it is well defined on $D = \{z : |z - z_0| > 1/r = R\}$, and thus,

$$f(z) = g(w)\Big|_{w=1/(z-z_0)} = g\left(\frac{1}{z - z_0}\right) = g(\phi(z)) \quad \text{for all } z \in D.$$

Since $\phi(z)$ is analytic and maps D into the domain $G = \{w : w < r\}$ on which $g(w)$ is analytic. Therefore, the composite function $f(z) = (g \circ \phi)(z)$, which is the sum of the series (1.10.1) of negative powers, is analytic throughout the domain of convergence D. Thus, the series (1.10.1) converges uniformly to its limit on any set of the form $D^* = \{z : |z - z_0| \geq R'\}$ as long as R' is greater than the radius of convergence $R = 1/r$ of the series (1.10.1). Note that sets like D^* are unbounded; however, the series is uniformly convergent to its limit on any set of this form. ∎

1.10.5. Find the radius of convergence of the series

$$1 + \frac{1}{1!z} + \frac{1}{2!z^2} + \cdots , \tag{1.10.3}$$

by comparing this series with the exponential series $e^w = 1 + w + \frac{w^2}{2!} + \cdots$. Note that the series (1.10.3) converges absolutely for all points $z : |z| > 0$, since $R = 1/r = 0$, i.e., it converges to an analytic function throughout the punctured plane $\mathbb{C} \setminus \{0\}$; it diverges because the functions appearing in the series are undefined at the point $z = 0$. If we substitute $w = 1/z$ into the series (1.10.3), the sum of the series (1.10.3) is $e^{1/z}$. ∎

1.10.6. Discuss the power series $\sum\limits_{n=0}^{\infty} z^{-n} = 1 + \dfrac{1}{z} + \dfrac{1}{z^2} + \cdots$. A solution is as follows: compare this series with the geometric series $\sum\limits_{n=0}^{\infty} w^n$ which converges to $\dfrac{1}{1-w}$ with radius of convergence $r = 1$. Since $R = 1/r = 1$, the given series converges absolutely for all points lying outside the circle $|z| = 1$. Substituting $w = 1/z$ in the geometric series, we get

$$\sum_{n=0}^{\infty} z^{-n} = \frac{1}{1 - 1/z} = \frac{z}{z-1} \text{ for all } \{z : |z| > 1\}.$$

Note that

$$\sum_{n=1}^{\infty} z^{-n} = \frac{1}{z} + \frac{1}{z^2} + \cdots = \frac{1}{z-1} = -\frac{1}{1-z}. \ ∎$$

1.10.7. The *Parseval's identity* states: If $f(z) = \sum\limits_{n=0}^{\infty} a_n\, z^n$ converges for $|z| < R$, then for $0 < r < R$ (with $z = r\, e^{i\theta}$, $0 \le \theta < 2\pi$)

$$\frac{1}{2\pi} \int_0^{2\pi} \left| f(r\, e^{i\theta}) \right|^2 d\theta = \sum_{n=0}^{\infty} |a_n|^2\, r^{2n}.$$

For proof, see, e.g., Marsden and Hoffman [1987:184]. ∎

1.10.8. Prove that if $f(z) = z + a_2 z^2 + \cdots + a_n z^n$ is a univalent polynomial in D, then $|a_n| \le 1/n$. HINT. Since $f'(z) = 1 + 2a_2 z + \cdots + n a_n z^{n-1} = n a_n \left(\dfrac{1}{n a_n} + \cdots + z^{n-1} \right)$, then by the fundamental theorem of algebra the polynomial with the parentheses has $n - 1$ complex roots $c_1, \ldots, c_{n-1}$ with their proper multiplicity. Thus, $f'(z) = n a_n (z - c_1)(z - c_2) \cdots (z - c_{n-1})$. Since f is univalent in D, f' has no zeros in D. Hence, each c_i lies outside D, i.e., $|c_i| \ge 1$ for each i. Since f is in $\mathcal{S}$, we have $1 = |f'(0)| = |n A_n||c_1||c_2| \cdots |c_{n-1}| \ge |n a_n|$. ∎

1.10.9. Prove that $f(z) = z + a_n z^n$ is a univalent polynomial in D if and only if $|a_n| \le 1/n$. The proof follows from the above exercise. ∎

2

Univalent Functions

The subject of conformal mapping and the Riemann mapping theorem provides an important beginning for the central theme of this book. Most of the topics presented in this chapter are available in books and treatises on the subject. However, they are presented for a complete description of the broad topic of conformal inequalities.

2.1 Conformal Mapping

The concept of conformal mapping applies when a small neighborhood of a point z_0 in the z-plane is mapped onto the w-plane by the mapping $w = f(z)$. Suppose that $f(z)$ is analytic at z_0 and that $f'(z_0) \neq 0$. Then at points z in the neighborhood of z_0 we have $f(z) = f(z_0) + (z - z_0)f'(z_0)$. Let $f'(z_0) = m\,e^{i\,\alpha}$, and $z - z_0 = \Delta r_z\, e^{i\,\theta_z}$. Assuming that the points z and z_0 are mapped onto points w and w_0 in the w-plane, let $w - w_0 = \Delta r_w\, e^{i\,\Delta\theta_w}$. Then a small movement along a path from z_0 to z generates a path from w_0 to w, where $w - w_0 = (z - z_0)f'(z_0)$, or $\Delta r_w e^{i\,\Delta\theta_w} = \Delta r_z\, e^{i\,\Delta\theta_z} = m\Delta r_z e^{i\,(\alpha+\Delta\theta_z)}$. This implies that

$$\Delta r_w = m\Delta r_z, \quad \Delta\theta_w = \alpha + \Delta\theta_z, \text{where } m = |f'(z_0)|, \ \alpha = \arg\{f'(z_0)\}.$$
$$(2.1.1)$$

Note that α depends only on z_0, not on z. A change occurs when a small segment γ_z in the z-plane is mapped onto the w-plane where γ_z becomes γ_w which is magnified (changed in size) by the scale factor m and rotated by an angle α. This change leads to a limiting process in terms of the infinitesimally increments of areas ΔA_z and ΔA_w, which are related by the ratio

$$\lim_{\Delta z \to 0} \frac{\Delta A_w}{\Delta A_z} = |f'(z)|^2 = \left(\frac{\partial u}{\partial x}\right)^2 + \left(\frac{\partial u}{\partial y}\right)^2 = \frac{\partial u}{\partial x}\frac{\partial v}{\partial y} - \frac{\partial u}{\partial y}\frac{\partial v}{\partial x} = \begin{vmatrix} \dfrac{\partial u}{\partial x} & \dfrac{\partial u}{\partial y} \\ \dfrac{\partial v}{\partial x} & \dfrac{\partial v}{\partial y} \end{vmatrix} = J,$$

where J is the Jacobian of the transformation $w = f(z)$. This transformation is one-to-one when $f'(z) \neq 0$ or $J \neq 0$.

Note that since the transformation of the incremental length is given by $|dw| = |f'(z)|\,|dz|$, and that of incremental area $\Delta A_z \to \Delta A_w$ is given by $dw\,d\bar{w} = f'(z)\overline{f(z)}\,dz\,d\bar{z}$, or $du\,dv = |f'(z)|^2\,dx\,dy$, the bounded area is given by

$$A = \iint_{B_z} |f'(z)|^2\,dx\,dy, \tag{2.1.2}$$

where $|f'(z)|^2$ is defined by (1.3.3).

If we place a sphere such that the complex plane is tangent to it and the origin coincides with the south pole, then we can transfer all points of $\mathbb{C}_\infty$ to the sphere by projection from the north pole. This is called a *stereographic projection* which is a one-to-one map of $\mathbb{C}_\infty$ onto the sphere such that its image is the whole sphere except the north pole that corresponds to the point at infinity $z = \infty$ (see Silverman [1967]).

A mapping f of a region D onto a region G is called *analytic* if and only if it is differentiable. The mapping f is called *conformal* if it is bijective (i.e., one-to-one and onto) and analytic (see Appendix A for definitions). The *conformal mapping theorem* states that if a mapping $f : D \mapsto G$ is analytic and $f'(z_0) \neq 0$ for each $z \in D$, then f is conformal. Thus, f rotates tangent vectors to curves through z_0 by a definite angle θ and magnifies (or contracts) them by a factor r. The mapping f is *conformal* if it is analytic with a nonzero derivative. Two important properties are the following:
(i) If $f : D \mapsto G$ is conformal and bijective, then $f^{-1} : G \mapsto D$ is also conformal and bijective.
(ii) If $f : D \mapsto G$ and $g : G \mapsto E$ are conformal, then the composition $f \circ g : D \mapsto E$ is conformal and bijective.

Property (i) is useful in solving certain boundary value problems (e.g., the Dirichlet problem) for a region D. The method involves finding a map $f : D \mapsto G$ such that G is a simple region on which the problem can be first solved, and then the result for the original problem is provided by f^{-1}. Since the Dirichlet problem involves harmonic functions, the following result on the composition of a harmonic function with a conformal map is useful: If u is harmonic on a region G and if $f : D \mapsto G$ is conformal, then $u \circ f$ is harmonic on D. In fact, let $z \in D$ and $w = f(z) \in G$. Then there is an analytic function g on the open disk $C(w, \rho) \subset G$ such that $u = \Re\{g\}$. Thus, $u \circ g = \Re\{g \circ f\}$, which is harmonic since $g \circ f$ is analytic.

A mapping that preserves both the magnitude and the sense of the angles between the curves and their images is said to be a *conformal mapping of the first kind*, but if the sense of the angles is reversed, then it is called a *conformal mapping of the second kind*.

Theorem 2.1.1. (Riemann mapping theorem) *Let $D \subset \mathbb{C}$ be a simply connected region. Then there exists a bijective conformal map $f : D \mapsto E$,* *where E is the open unit disk. Moreover, the map f is unique provided that $f(z_0) = 0$ and $f'(z_0) > 0$ for $z_0 \in D$.*

Ahlfors [1953: 172] has hailed this theorem as "one of the most important theorems of complex analysis." We will first provide a short historical development of this theorem before proving it. Riemann [1851: 40] stated that "two simply connected plane surfaces can always be mapped onto each other such that each point of the one corresponds to a unique point of the other in a continuous manner and the correspondence is conformal; moreover, the correspondence between an arbitrary interior point and an arbitrary boundary point of the one and the other may be given arbitrarily, but when this is done the correspondence is determined completely." This assertion came to be known as the *Riemann mapping theorem* which eventually came to be stated as Theorem 2.1.1, in which the explicit requirement that the boundary has at least two points is necessary to rule out the cases where the domain is either the complex plane or the sphere; these are the counter-examples provided by Riemann himself. The uniqueness of the mapping f was first specified by conditions on it only at an interior point as it was made explicit by Carathéodory [1913a,b; 1914] and Koebe [1913].

As Gray [1994] has narrated, Riemann's own proof used the Dirichlet principle, and for this reason it was not accepted, specially in view of Hilbert's first paper [1905] on Dirichlet principle, in which he stated the problem as follows: Suppose a boundary curve and a function on this curve are given. Let S be the part of the plane bounded by this curve. Then the function $f(x, y)$ is taken for which the value of the integral

$$L(\alpha, \beta) = \int_S \left[\left(\frac{\partial \alpha}{\partial x} - \frac{\partial \beta}{\partial y} \right)^2 + \left(\frac{\partial \alpha}{\partial y} + f \frac{\partial \beta}{\partial x} \right)^2 \right] dS \qquad (2.1.3)$$

is finite, where α and β are two arbitrary real functions of x and y, and S is the surface over which the integral is taken. The function f is necessarily harmonic. Hilbert claimed that such considerations had led Riemann to his proof of the existence of functions with the given boundary values. However, Weierstrass was first to show that this approach was not reliable. According to Hilbert, the Dirichlet principle had 'fallen into disrepute, and only Brill and Noether [1894] continued to hope that it could be resurrected, perhaps in a modified form.'

If we vary α in the integral (2.1.3) by a continuous function, or by one discontinuous only at a single point, the integral $L(\alpha, \beta)$ attains a minimal value, and this minimum is attained by a unique function if we exclude the points of discontinuity. This unique minimizing function is harmonic. In Riemann's proof, a function λ is constructed such that it vanishes on the boundary, may be discontinuous at isolated points, and for which the integral

$$L(\lambda) = \int \left[\left(\frac{\partial \alpha}{\partial x} \right)^2 + \left(\frac{\partial \lambda}{\partial y} \right)^2 \right] dT \qquad (2.1.4)$$

is finite. The integral $L(\lambda)$ was called the *Dirichlet integral* later by Hilbert. Let $\alpha + \lambda = w$. Then Riemann considered the integral

$$\Omega = \int \left[\left(\frac{\partial \alpha}{\partial w} - \frac{\partial \beta}{\partial y} \right)^2 + \left(\frac{\partial w}{\partial y} + \frac{\partial \beta}{\partial x} \right)^2 \right] dT, \tag{2.1.5}$$

and wrote: "The totality of these functions λ represents a connected domain closed in itself, in which each function can be transformed continuously into every other, and a function cannot approach indefinitely close to one which is discontinuous along a curve with $L(\lambda)$ becoming infinite." Thus, for each λ, the integral Ω only becomes infinite with L, which depends continuously on λ and can never be less than zero. Hence, Ω has at least one minimum. The uniqueness follows directly from functions of the form $u + h\lambda$ near to a minimum u.

After this brief historical background we will provide a complete proof of the Riemann mapping theorem, which requires a prior knowledge of the maximum principle (Theorem 1.5.7) and Schwarz's lemma (Theorem 1.5.8). There are other proofs of the Riemann mapping theorem, available in Gray [1994], and Rudin [1976].

PROOF. This proof takes into account Riemann's own approach and is based on the existence of a solution to the Dirichlet problem in any domain. We will use real variables, so that $z = x + iy$ is written as (x, y), and $f(z) = u + iv$ is written as (u, v). Consider the boundary value problem in a domain $D \subset \mathbb{C}$:

$$\frac{\partial^2 u}{\partial x^2} + \frac{\partial^2 u}{\partial y^2} = 0,$$
$$u(x, y) = \log |x + iy - a|, \tag{2.1.6}$$

where $x + iy \in \partial D$, and $a \in D$. Thus, $\log |x + iy - a|$ is bounded on ∂D. In view of the existence theorem, there exists a unique function u satisfying the boundary value problem (2.1.6). Since the boundary value of u is bounded, u remains bounded on the interior of D. The function u satisfies the Laplace equation in (2.1.6). By regularity theorem for the Laplace equation, u is differentiable and harmonic, thus analytic (synonymously, holomorphic or regular) on D, and the curl of this vector field is given by $(-u_y, u_x)$. Since D is simply connected, by Poincaré's lemma (see §1.8.1), there exists a function v such that this vector field is $\mathrm{grad}\, v = v_x + iv_y$. Since v is only determined up to an additive constant, we impose the condition $v(a) = 0$. This leads to the Cauchy-Riemann equations (1.3.4), and $u + iv$ is analytic on D.

Now, define three functions p, q, and f as

$$\begin{aligned}
p(z) &= u(z) - \log |z - a|, \\
q(z) &= v(z) - \arg |z - a|, \\
f(z) &= e^{-p(z) - iq(z)} = (z - a)\, e^{-u(z) - iv(z)}.
\end{aligned} \tag{2.1.7}$$

Note that the function p is single-valued while q is multiple-valued with branch point at a, yet both satisfy the Laplace equation in $D\backslash\{a\}$. Also, the value of q increases by $2\pi i$ after one winding around the point a. However, the function f is single-valued since the exponentiation operation cancels the multiple-valued property of q. Moreover, f is analytic on D.

Notice that $p(z) = 0$ for $z \in \partial D$. We will show that $p(z) \geq 0$ whenever $z \in D\backslash\{a\}$. Since $\log|z - a| \to -\infty$ as $z \to a$, and $p(a)$ is finite, there exists an $\varepsilon > 0$ such that $p(z) > 0$ when $0 < |z - a| \leq \varepsilon$. Consider the region $G = \{z \in D : |z| > \varepsilon\}$. For a point $z \in \partial G$, either $|z - a| = \varepsilon$ or z must lie on ∂D. In either case we get $p(z) \geq 0$, which, in view of the maximum principle (Theorem 1.3.1), implies that $p(z) \geq 0$ for all $z \in G$. We already have $p(z) \geq 0$ whenever $0 < |z - a| < \varepsilon$, so $p(z) \geq 0$ whenever $z \in D\backslash\{a\}$. Since $|f(z)| = e^{-p(z)} \leq 1$, we have $|f(z)| \leq 1$ when $z \in D$. Also, f(a)=0 and $f'(a)$ is real and positive because $f'(a) = e^{-u(a)-i\,v(a)} = e^{-u(a)} \in \mathbb{R}$.

To show that f is bijective, consider the level sets of p. For the sake of simplicity, we exclude those points where $f' = 0$. Then for every real number $r > 0$, define $A(r) = \{z \in D : |f(z)| \geq e^{-r} \text{ and } f'(z) = 0\}$. We must show that $A(r)$ is finite. Note that the set $\{z \in D : |f(z)| \geq e^{-r}\}$ is compact. Hence, if $A(r)$ were infinite, it would have an accumulation point. But since $f(z) = 0$ whenever $z \in A(r)$ and f is analytic, this would imply that $f(z) \equiv 0$, which is a contradiction. Hence, $A(r)$ is finite.

Next, choose r such that $f'(z) \neq 0$ whenever $|z| = r$. Let $C(r)$ denote the level set $C(r) = \{z : p(z) = r\}$. We will show that $C(r)$ is smooth and homeomorphic[1] to a circle. Since $C(r)$ is a level set of a continuous function on a compact set, $C(r)$ is compact. Let w be a point on $C(r)$. By assumption, $f'(w) \neq 0$ Thus, by the inverse function theorem (Theorem 1.3.1), there exists a neighborhood E of w on which f is invertible, i.e., f^{-1} exists, and is an analytic function. Since $z \in C(r)$ if and only if $|f(z)| = e^{-r}$, it follows that $C(r) \cap E$ is the image of an arc of the circle $\{z : |z| = e^{-r}\}$ under the mapping f^{-1}. Also, f^{-1} is an analytic function which is differentiable also. Hence, $C(r) \cap E$ is bijectively mapped onto a line segment. Since this is true for every point $z \in C(r)$, $C(r)$ is a compact one-dimensional manifold; as such it must be either a circle or a finite union of circles. Suppose $C(r)$ is a union of more than one circle. Then by Jordan theorem (§1.2), each of these circles divides the complex plane into two components, one the interior and the other the exterior, with $C(r)$ as a common boundary. If there were two circles which together comprise $C(r)$, one of these circles would have to lie inside the other and there would be an open set Q which would have the two circles as its boundary. However, since $p(z)$ is assumed to be constant on both circles and assumes the same value on both, the maximum principle would imply that p is constant in the region Q, which in turn would imply that $f = \text{const}$ on D.

---

[1] Two curves are homeomorphic if they can be deformed into each other by a continuous, bijective and invertible mapping.

But this is impossible. Hence $C(r)$ consists of only one circle.

Next, we must show that a lies on $\text{Int}(C(r))$. Since the winding number of $C(r)$ about the point a is 1, the phase of $q(z)$ will increase by 2π after traversing $C(r)$ once. Since f^{-1} is analytic, $C(r)$ is not only homeomorphic to a circle; however, it is a smooth curve, and therefore, it has a tangent and a normal. The normal and tangential derivatives are obtained from the Cauchy-Riemann equations which in this case are $\dfrac{\partial p}{\partial t} = \dfrac{\partial q}{\partial n}$ and $\dfrac{\partial p}{\partial n} = -\dfrac{\partial q}{\partial t}$. Since $p(x) = r$ for $z \in C(r)$ and $p(x) \geq r$ for $x \in \text{Int}(C(r))$, so by the Cauchy-Riemann equations, both $\dfrac{\partial p}{\partial t} > 0$ and $\dfrac{\partial p}{\partial n} > 0$. This implies that $\dfrac{\partial q}{\partial t} \leq 0$, which is impossible because otherwise all the derivatives of p and q would vanish, i.e., f' would be zero, contrary to the hypothesis. Hence, q is a monotonically decreasing function on $C(r)$, and $\arg\{q\}$ decreases by 2π upon traversing $C(r)$, which imply that the function e^{-iq} is a bijection from $C(r)$ onto the unit circle E. Hence, f is a bijection from $C(r)$ onto the circle of radius e^{-r}.

Finally, we must show that f' cannot have any zeros inside $C(r)$. Consider the points z for which $f'(z) = 0$, and choose r such that $f'(z) \neq 0$ whenever $|z| = r$. Note that $C(r)$ is a smooth closed curve, and since $f'(z) = -f(z)(u'(z)+i\,v'(z))$. We use the fact that the argument of a product is the sum of the arguments of the factors. Consider f' and $u'+i\,v'$ separately. By argument principle (Example 1.6.1), f has only one simple zero, located at a, inside $C(r)$, and so $\arg\{f\}$ will increase by 2π after traversing $C(r)$ once. On the other hand, $\arg\{u' + i\,v'\}$ is computed by using the fact that the argument of the derivative of a an analytic function is the same along any direction chosen to compute it. So we choose the normal direction, and find that $\arg\{u'+i\,v'\}$ changes by -2π after traversing $C(r)$ once. Hence, $\arg\{f'\}$ remains the same after traversing $C(r)$ once. Since f' is analytic inside $C(r)$, by the argument principle, f' cannot have any zero inside $C(r)$. ∎

Example 2.1.1. Let K be a subset of the upper half-plane $H = \{z : |z| > 0\}$ such that a domain $D = H \backslash K$ is simply connected. By the Riemann mapping theorem, there is a bijection $g : D \mapsto H$. Then for any such map g and $z \in D$, we have $R = \dfrac{2\Im\{g(z)\}}{|g'(z)|}$. For example, let $K = \emptyset$ and $z = i$. Then g is the identity map, i.e., $g(z) = z$, and $R = 2$. ∎

Example 2.1.2. The mapping $w = e^z$ is not univalent on any domain $D \in \mathbb{C}$ due to the periodicity of this function: $e^z = e^{z+2n\pi i}$. But this function becomes univalent if its domain of definition is restricted to the horizontal strip $M = \{z : -\pi < \Im\{z\} < +\pi\}$ or any other horizontal strip of width 2π. Then the exponential function maps the strip M one-to-one onto the w-plane with a slit along the negative real axis. The inverse is given by principal branch $z = \log w$. ∎

Example 2.1.3. To determine all invertible conformal mappings of the unit disk E onto itself, note that by Schwarz lemma, if S and T are bilinear transformation of E onto E, and if $S(0) = T(0)$, then S and T can differ by a rotation $R_\alpha = \alpha z = e^{i\theta} z$, $\alpha = e^{i\theta}$, and $|\alpha| = 1$, such that $S(z) = T(\alpha z) = (T \circ R_\alpha)(z)$. If the derivatives at the origin have the same values, or even the same argument, then $\arg\{dS/dz(0)\} = \arg\{dT/dz(0)\}$, $\mathrm{mod} 2\pi$, then $S = T$. Thus, the required mapping T must be bilinear. In fact, if $T(0) = w_0$, then there is a bilinear transformation $S : E \mapsto E$ such that $(S \circ T)(0) = 0$, i.e., S takes w_0 back to the origin. Thus, $S \circ T$ is an invertible conformal mapping of E onto E such that $(S \circ T)(z) = e^{i\theta} z$ and $T = \tilde{S}\left(e^{i\theta} z\right) = (\tilde{S} \circ R_\alpha)(z)$. Since $\tilde{S}$ and $\tilde{S} \circ R_\alpha$ are bilinear transformations, so is T. ∎

2.2 Some Theorems

Theorem 2.2.1. *Let f be a nonzero analytic function defined on a simply connected domain D. Then for any $N \in \mathbb{Z}^+$, there exists an analytic function g such that $g^N = f$.*

PROOF. Since $f \neq 0$, the function f'/f is analytic. Fix a point A in D and let $z \in D$. Since D is path-connected, there is at least one path $\gamma_A(z)$ in D. Define $F(z) = \int_{\gamma_A(z)} \dfrac{f'(t)}{f(t)} \, dt$. Since D is simply connected, the function F is independent of paths and thus well-defined. Now, we show that F is analytic. Since D is open, the open disk $C(z_0, r)$ is contained in D. Let $z_0 \in C(z_0, r)$. Then for any $z \in C(z_0, r)$, the straight line L joining z and z_0 lies in $C(z_0, r)$ and thus in D. Using simple connectedness of the domain D we have

$$F(z) - F(z_0) = \int_{\gamma_A(z) - \gamma_A(z_0)} \frac{f'(t)}{f(t)} \, dt = \int_L \frac{f'(t)}{f(t)} \, dt. \tag{2.2.1}$$

Thus, since f'/f is continuous, we get

$$\lim_{z \to z_0} \left| \frac{F(z) - F(z_0)}{z - z_0} - \frac{f'(z_0)}{f(z_0)} \right| = \lim_{z \to z_0} \left| \frac{\int_L \dfrac{f'(t)}{f(t)} - \dfrac{f'(z_0)}{f(z_0)}}{z - z_0} \, dt \right|$$
$$\leq \lim_{z \to z_0} \frac{1}{z - z_0} \int_L \left| \frac{f'(t)}{f(t)} - \frac{f'(z_0)}{f(z_0)} \right| dt \leq \lim_{z \to z_0} \max_{t \in L} \left| \frac{f'(t)}{f(t)} - \frac{f'(z_0)}{f(z_0)} \right| dt = 0. \tag{2.2.2}$$

This shows that F is analytic. Since $\left(f\, e^{-F}\right)' = f'\, e^{-F} - f F'\, e^{-F} = f'\, e^{-F} - f(f'/f)\, e^{-F} = 0$, the function $f\, e^{-F} = c$ (const). Thus, $f = c\, e^F = e^{\ln c + F}$. Set $h = \ln c + F$. Then h is analytic and $f = e^h$. Hence, $g = e^{h/N}$ is analytic, and therefore, $g^N = f$. ∎

Theorem 2.2.2. *Let f be analytic. If f is one-to-one, then f' is nonzero.*

PROOF. Let z_0 be in the domain D of f, and let $w_0 = f(z_0)$. Then $f - w_0$ has a zero at z_0. Also, $f - w_0 \neq 0$ since f is one-to-one. Also, since f is analytic, z_0 is an isolated zero of $f(z) - w_0$. Hence, $f(z) - w_0 = (z - a)^N G(z)$ for some $N \in \mathbb{Z}^+$, where G is analytic and $G \neq 0$ on some disk $C(z_0)$. Then by Theorem 2.1.1, there is an analytic function g such that $G = g^N$ and $g(z_0) \neq 0$. Thus, $f(z) - w_0 = [(z - z_0) g(z)]^N$. Let $h(z) = (z - z_0)g(z)$. Then $f(z) - w_0 = h^N(z)$. Since $h'(z_0) = g(z_0) \neq 0$, by complex inverse function theorem, h is locally one-to-one. Thus, f is locally N-to-one. But since f is one-to-one, we must have $N = 1$. Therefore, $f(z) - w_0 = h(z)$, which gives $f'(z_0) = h'(z_0) \neq 0$. ∎

Theorem 2.2.3. (Rouché's Theorem)[1] *Suppose f and g are analytic on $C(a, R)$ and let $0 < r < R$. If $|f(z) - g(z)| < |g(z)|$ on $\partial C(a, r)$, then f and g must have the same number of zeros inside $\partial C(a, r)$.*

PROOF. Obviously, both f and g are nonzero on $\partial C(a, r)$, since $|f(z) - g(z)| < |g(z)|$ on $\partial C(a, r)$. Then, by the argument principle, the number of zeros of f and g inside $\partial C(a, r)$ are given by

$$Z_f = \frac{1}{2\pi i} \int_{\partial C(a,r)} \frac{f'(z)}{f(z)} \, dz, \quad \text{and} \quad Z_g = \frac{1}{2\pi i} \int_{\partial C(a,r)} \frac{f'(z)}{f(z)} \, dz, \qquad (2.2.3)$$

respectively. Then the difference between Z_f and Zg inside $\partial C(a, r)$ is given by

$$\frac{1}{2\pi i} \int_{\partial C(a,r)} \frac{\left(f'(z)/g(z)\right)'}{f(z)/g(z)} \, dz = \frac{1}{2\pi i} \int_{\partial C(a,r)} \left(\frac{f'(z)}{f(z)} - \frac{g'(z)}{g(z)}\right) \, dz. \qquad (2.2.4)$$

Since $|f(z) - g(z)| < |g(z)|$ on $\partial C(a, r)$, which implies that $\left|\frac{f(z)}{g(z)} - 1\right| < 1$ on $\partial C(a, r)$, i.e., the negative real axis $\{z \in \mathbb{C} : \Re\{z\} \leq 0, \Im\{z\} = 0\}$ is a branch cut for the function $\log(f/g)$. Thus,

$$\frac{1}{2\pi i} \int_{\partial C(a,r)} \frac{\left(f'(z)/g(z)\right)'}{f(z)/g(z)} \, dz = \frac{1}{2\pi i} \int_{\partial C(a,r)} \frac{d}{dz} \log \frac{f(z)}{g(z)} \, dz = 0. \qquad (2.2.5)$$

Hence, $Z_f = Z_g$, i.e., f and g have the same number of zeros inside $\partial C(a, r)$. ∎

Rouché's theorem is used to determine the number of zeros of an analytic functions inside a given closed curve, and also to prove the fundamental theorem of algebra. Some applications of Rouché's Theorem are as follows:

(i) There are no such sequence of polynomials that uniformly converges to $f(z) = \frac{1}{4}$ on the unit circle $\partial E = \{z : |z| < 1\}$. To prove, suppose there exists

---

[1] Eugene Rouché published this result in the *Journal of the École Polytechnique* in 1862.

a polynomial $p_n(z)$ of degree n that converges uniformly to $1/4$ on ∂E. Then $|zp_n(z) - 1| = \left|p_n(z) - \frac{1}{z}\right| \to 0$ for all $z \in \partial E$. There exists a sufficiently large N such that $|zp_n(z) - 1| < 1$ for all $n > N$ and all $z \in \partial E$. By Rouché's theorem, the number of zeros of $zp_n(z)$ is N_1. But $zp_n(z)$ has at least one zero, while 1 has no zero, which is a contradiction. ∎

(ii) There are no such sequence of polynomials that uniformly converges to $f(z) = (\bar{z})^2$ on the unit circle ∂E. To prove, suppose there exits a polynomial $p_n(z)$ of degree n such that $p_n(z) \to (\bar{z})^2$ uniformly on ∂E. Then $|z^2 p_n(z)| \to |z^2(\bar{z})^2| = |z|^4 = 1$ for all $z \in \partial E$. Following the argument as in (i), this leads to a contradiction. ∎

Theorem 2.2.4. (Hurwitz theorem [1897]) *(i) Let $\{f_n(z)\}$ be a sequence of nonzero analytic functions on a domain D. If $f_n \to f$ uniformly on compact sets in D, then either f is nonzero on D, or $f \equiv 0$ on D. (ii) If a sequence of univalent functions f_n defined on a domain D converges uniformly on compact sets to f, then either f is univalent or f is constant on D.*

PROOF. (Both parts by contradiction) (i) If $f \equiv 0$, there is nothing to prove. Suppose $f \not\equiv 0$. Then there is a $z_0 \in D$ such that $f(z_0) = 0$. Since f_n are analytic and converge uniformly on compact sets, f is analytic, by Morera's theorem and uniform convergence. Thus, z_0 is an isolated zero of f, and $f(z) = (z - z_0)^N g(z)$, $g(z_0) \neq 0$. This means that there is an $r > 0$ such that f is nonzero on $\bar{C}(z_0, r) \backslash \{z_0\}$, where $\bar{C}(z_0, r) \in D$. Now, analyticity of f implies its continuity, so that $|f|$ is continuous. Since $\bar{C}(z_0, r)$ is compact and f does not vanish on this disk, then $\min_{C(z_0,r)} |f|$ exists and is nonzero. Also, since f_n converges uniformly on compact sets, there is an $n \in \mathbb{Z}^+$ such that for all $n > N$, $|f_n(z) - f(z)| < \min_{C(z_0,r)} |f|$. Thus, $|f_n(z) - f(z)| < |f(z)|$ on $C(z_0, r)$. Then, by lemma 2.1.3, both f_n and f have the same number of zeros inside $C(z_0, r)$. But f_n has no zeros there and f has only one zero there, which is a contradiction. Hence f is nonzero on D. ∎

(ii) If $f \equiv c$, there is nothing to prove. Suppose $f \neq c$. For the sake of contradiction, assume that there are distinct points $z_1, z_2 \in D$ with $z_1 \neq z_2$ such that $f(z_1) = f(z_2) = w$. Define $F_n = f_n - w$ and $F = f - w$. Then $F(z_1) = f(z_2) = 0$ and F is a non-constant analytic function. Following the argument in the proof of part (i), there are two disjoint disks $C(z_1, r_1)$ and $C(z_2, r_2)$ and there is an $n \in \mathbb{N}$ such that for n greater than N_1 and N_2; thus, both F_1 and F_2 have the same number of zeros in $C(z_1, r_1)$ and $C(z_2, r_2)$, respectively. Hence, there are two distinct points that make F_n zero. Therefore, f_n is not one-to-one, which is a contradiction. ∎

An alternate version of Hurwitz theorem is as follows.

Theorem 2.2.5. (Hurwitz's theorem, second version) *Let $\{f_n\}$ be a sequence of analytic functions on a simply connected domain D, which converges* *uniformly to an analytic function f. If f has a zero of order m at z_0, then for every sufficiently small $r > 0$ and for sufficiently large $n \in \mathbb{N}$ dependent on r, f_n has precisely m zeros, including multiplicity, in the disk $|z - z_0| < r$. Also, these zeros converge to z_0 as $n \to \infty$.*

PROOF. Let f be an analytic function on D with zeros of order m at z_0, and suppose the $\{f_n\}$ is a sequence of functions converging uniformly to f on compact subsets of D. For fixed $r > 0$ such that $f(z) \neq 0$ in $0 < |z - z_0| < r$, choose a δ such that $|f_k(z)| \geq \delta/2$ for every $k \geq N$ and every z on the circle $C : |z - z_0| = r$, thus ensuring that the quotient $f_k'(z)/f_k(z)$ is well defined for all z on C. Then by Morera's theorem (§1.5), the quotient $f_k'(z)/f_k(z)$ converges uniformly to $f'(z)/f(z)$. Let the number of zeros of $f_k(z)$ in the disk $|z - z_0| < r$ be N_k, counting multiplicity. Then using the argument principle (Example 1.6.1)

$$
m = \frac{1}{2\pi i} \int_C \frac{f'(z)}{f(z)}\, dz = \frac{1}{2\pi i} \int_C \lim_{k\to\infty} \frac{f_k'(z)}{f_k(z)}\, dz = \lim_{k\to\infty} \frac{1}{2\pi i} \int_C \frac{f_k'(z)}{f_k(z)}\, dz
$$
$$
= \lim_{k\to\infty} N_k, \tag{2.2.6}
$$

where the integral and the limit are interchanged because of uniform convergence of the integrand. Since N_k are integers, N_k must equal to m for sufficiently large k. ■

Note that this theorem may not hold for arbitrary disks. For example, the theorem fails if the disk is such that f has all zeros on its boundary. Consider the unit disk and the sequence defined by $f_k(z) = z - 1 + 1/k$, $z \in \mathbb{C}$, which converges uniformly to $f(z) = z - 1$. The function $f(z)$ has no zeros in the disk, but each f_k has exactly one zero in the disk of value $1 - 1/k$ for each k.

Hurwitz's theorem is used in the proof of the Riemann mapping theorem. This theorem has the following two corollaries.

Corollary 2.2.1. *Let $\{f_n\}$ be a sequence of analytic functions defined on a domain $D \in \mathbb{C}$, which converge uniformly on compact subsets of D to an analytic function f. If each f_n is nonzero everywhere in D, then f is either identically zero or nowhere zero.*

Corollary 2.2.2. *If $\{f_n\}$ is a sequence of univalent functions on a domain D that converge uniformly on compact subsets of D to an analytic function f, then f is univalent or constant.*

Theorem 2.2.6. (Montel theorem [1907]) *Let a set of analytic function $\{f_n\}$ defined on a domain D be uniformly bounded on compact sets of D. There is a subsequence $\{f_{n_k}\}$ which converges uniformly on compact subsets to an analytic function f.*

PROOF. We will show that the set $\{f_n\}$ is equicontinuous on D. Let $z_0 \in D$, and let $C(z_0, 2r) \subset D$, as D is open. Since the functions f_n are uniformly bounded on compact sets, let $M = \max\limits_{C(z_0,2r)} |f_n|$. Also,

$$f(z) = \frac{1}{2\pi i} \int_{C(z_0,2r)} \frac{f(t)}{t-z}\,dt. \tag{2.2.7}$$

If $z \in C(z_0,r)$, then

$$f_n(z) - f_n(z_0) = \frac{1}{2\pi}|z-z_0|\left|\int_{C(z_0,2r)} \frac{f_n(t)}{(t-z)(t-z_0)}\,dt\right|$$

$$< \frac{1}{2\pi}|z-z_0|\frac{M}{(2r-r)^2}2\pi(2r) = |z-z_0|\frac{2M}{r}, \tag{2.2.8}$$

which shows that f_n are equicontinuous. Next, since $\mathbb{C}$ is separable and at each point (recall that a one-point set is compact) the set $\{f_n\}$ is bounded. So by Arzela-Ascoli theorem (Theorem 1.9.2a) there is a subsequence $\{f_{n_k}\}$ which converges uniformly on compact subsets to a function f which is obviously an analytic function. ∎

Note that Montel's theorem simply states that a uniformly bounded families of functions $\{f_n(z)\}$ is *normal*. A stronger version of Montel's theorem states that a family of analytic functions, all of which omit the same two values $a, b \in \mathbb{C}$, is normal. This is known as the *fundamental normality test*. In fact, there exists an analytic universal covering from the unit disk to the twice punctured plane $\mathbb{C}\backslash\{a,b\}$. For example, such a covering is obtained by the elliptic modular function.

2.3 Implications

The Riemann mapping theorem implies that if D and G, both contained in $\mathbb{C}$, are any two simply connected regions, then there exists a bijective conformal map $g : D \mapsto G$. If $f : D \mapsto E$ and $h : G \mapsto E$, then $g = h^{-1} \circ f$ is bijective conformal (Figure 2.3.1). Thus, the two regions D and G are said to be *conformal* if there exists a bijective conformal map between them.

A bijective conformal map is also called a *univalent* map. A function $w = f(z)$ defining a univalent mapping is called a *univalent (or schlicht) function*. Its inverse image is also a univalent function defined on the image region. Thus, a function $f(z)$ is said to be univalent in a domain D if $f(z_1) \neq f(z_2)$ if $z_1 \neq z_2$, $(z_1, z_2 \in D)$, i.e., $f(z)$ never takes the same value twice. In the study of univalent mappings the first question asked is whether a given region can be mapped univalently onto another region. In the case of simply connected regions it is necessary that the two regions have the same connectivity. Once this condition is met, we can inquire about the possibility of univalent conformal mapping of various regions onto a given simply connected region.

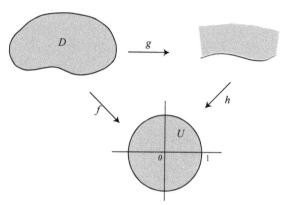

Figure 2.3.1 Conformal Mapping onto the Unit Disk.

To determine how a simply connected region is mapped onto a simply connected region, we must know their mappings onto a standard region, such as the unit disk. This enables us to obtain the required mapping first by mapping the given region onto the unit disk and then mapping the unit disk onto the other region.

Now the question arises whether an arbitrary simply connected region can be mapped onto the unit disk. It turns out that this is always possible except for two cases, namely, when the region is the entire plane, and when it has a single boundary point. Note that the requirement that the boundary orientation be preserved even in the mapping of regions with non-Jordan boundaries is important. Consider, for example, the map $w = \sin z$ which is analytic on the strip $-\pi/4 < x \leq 3\pi/4$ and maps bijectively the two boundaries of this strip onto the two boundaries of the hyperbola $u^2 - v^2 = 1/2$, which make the boundary of a simply connected region. But the boundary of this curved strip is not traversed in the positive sense. Other simple examples of functions that are not univalent in the lower or the upper half-plane are $w = z^2$, and $w = \sinh z$, although they are analytic in $\mathbb{C}$ and map bijectively the real axis $\Im\{z\} = 0$ onto the real axis $\Im\{w\} = 0$. Thus, the function $w = f(z)$, analytic on the half-plane $\Im\{z\} > 0$ and continuous on the closed region $\Im\{z\} \geq 0$, grows as $|z| \to \infty$ at most as fast as cz^2, i.e., the ratio $\dfrac{f(z)}{z^2}$ is bounded for $0 \leq \arg\{z\} \leq \pi$ and a sufficiently large $|z|$.

The Riemann mapping theorem also implies that there exists a unique function $w = F(z)$ that is regular in D, is normalized at a finite point $z_0 \in D$ by the conditions $F(z_0) = 0$ and $F'(z_0) = 1$, and maps the region D univalently onto the disk $|w| < 1$. In fact, the function $F(z) = \dfrac{f(z)}{f(z_0)}$ is such a function, where $f(z)$, with $f(z_0) = 0$ and $f'(z_0) > 0$, is the function mentioned in the Riemann mapping theorem, and the radius of the disk onto which the function $w = F(z)$ maps the region D is $R = \dfrac{1}{f'(z_0)}$. If there exists another function $w = F_1(z)$, with $F_1(z_0) = 0$ and $F_1'(z_0) = 1$, that maps D onto a disk $|w| < R_1$, then, by the Riemann mapping theorem, we could have $\dfrac{F_1(z)}{R} = f(z)$, and hence, $\dfrac{1}{R_1} = f'(z_0)$, i.e., $F_1(z) = \dfrac{f(z)}{f'(z_0)} = F(z)$, which proves the uniqueness of the mapping function $w = F(z)$. The quantity $R = \dfrac{1}{f'(z_0)}$ is called the *conformal radius* of the region D at the point $z_0 \in D$ (see Exercise 2.8.9).

Since $D^* = \mathrm{Ext}\,(\Gamma)$ is simply connected, Green's function $\mathcal{G}(z, \infty)$ for D^* coincides with $\log |f(z)|$, where the function $w = f(z)$ maps D^* univalently onto $|w| > 1$ such that $f(\infty) = \infty$. Then, Robin's constant γ for the region D^* is equal to $\log |f'(\infty)|$, and the capacity $\mathrm{cap}\,(D^*) = \dfrac{1}{|f'(\infty)|} = R$, where R is the conformal radius of the region D^* (with respect to ∞), i.e., the number R is such that the region D^* is mapped univalently onto $|w| > R$ by a normalized function $w = F(z)$ with $F(\infty) = \infty$ and $F'(\infty) = 1$. Thus, since the capacity of an arbitrary compact domain D is equal to its transfinite diameter, we have, in view of the maximum principle (Theorem 1.3.1), the following result.

Theorem 2.3.1. *The capacity, and hence the transfinite diameter of a bounded simply connected region D, is equal to the conformal radius of the region D^* which is the complement of the region D in $\mathbb{C}_\infty$ and contains the point at infinity.*

2.3.1. Chain Property: Let $D_0, \ldots, D_n$ be regions in $\mathbb{C}$, and let $f_k : D_{k-1} \mapsto D_k$ denote conformal mappings for $k = 1, \ldots, n$. Then the mapping $g = f_n \circ \cdots \circ f_1$, defined by

$$g(z) = f_n \left(f_{n-1} \left(\cdots f_2 \left(f_1(z) \right) \cdots \right) \right) \tag{2.3.1}$$

is a conformal mapping of D_0 onto D_n. The mapping g is said to be composed of a *chain* of mappings $f_1, \ldots, f_n$ and is represented by the scheme

$$D_0 \xrightarrow{f_1} D_1 \xrightarrow{f_2} D_2 \xrightarrow{f_2} \cdots \xrightarrow{f_{n-1}} D_{n-1} \xrightarrow{f_n} D_n. \tag{2.3.2}$$

Thus, the set of regions on $\mathbb{C}$ can be divided into *mapping classes* such that two regions can be mapped conformally onto each other if and only if they belong to the same mapping class. The chain property is also known as *composite mapping*.

The practical applications of conformal mappings are related to the problem of constructing a function which maps a given region onto another given region. Often we find an explicit expression for the mapping function and determine it by applying the chain property; see Kythe [1998: 34].

Note that if a function $w = f(z)$ maps a region D conformally onto another region G, and if z_0 is an isolated boundary point of D, then z_0 is a removable singularity or a simple pole of $f(z)$ (Goluzin [1969:205]; Wen [1992:95]). That is why we assume the regions to be without isolated boundary points.

Example 2.3.1. The chain property of conformal mappings is illustrated in Figure 2.3.2.

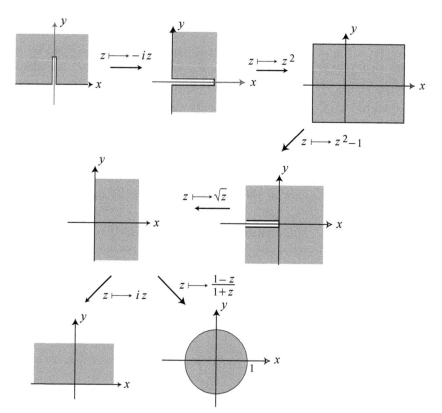

Figure 2.3.2 Chain property of conformal mappings. ∎

2.4 Analytic Continuation

If A and B are two domains that overlap, i.e., $A \cap B \neq \emptyset$, and if f is analytic on A, g is analytic on B, and $f(z) \equiv g(z)$ on $A \cap B$, then g is called the analytic continuation of f from A into B. This analytic continuation is unique. In fact, suppose that g_1 and g_2 are analytic continuations of f from A into B. Then $g_1(z) \equiv g_2(z)$ on $A \cap B$, and thus, by identity theorem (Theorem 1.5.3), $g_1 = g_2$ throughout B. Thus, in practice we shall determine a single function $F(z)$ that is analytic on $A \cup B$ and is given by $F(z) = f(z)$ on A, and $F(z) = g(z)$ on B. For example, let $f(z) = \int_0^\infty e^{-zt} \, dt$. Notice that $f(z)$ is analytic on $\Re\{z\} > 0$ because after evaluating the improper integral we find that $f(z) = 1/z$ is analytic on $\Re\{z\} > 0$. So we take $g(z) = 1/z$. Since $g(z)$ is analytic on $\mathbb{C}\backslash\{0\}$, $g(z)$ becomes the analytic continuation of $f(z)$ from the right half-plane into the whole plane indented at the origin. Similarly, the Laplace transform of $\cos at$ is analytic for $\Re\{z\} > 0$; but the function $\dfrac{z}{z^2 + a^2}$ is its analytic continuation from the right half-plane into the whole plane indented at the points $\pm ia$.

Analytic continuation is closely related to the Taylor series. Consider a function $f(z)$ that has its series expansion $f_1(z)$ about $z = z_1$ in the circle of convergence D_1 inside a domain D. If there exists a singularity at $z_s \in D$, then the circle of convergence D_1 within which $f_1(z)$ is analytic has radius $|z_s - z_1|$ and is centered at z_1. Within D_1 there exists another point $z_2 \neq z_1$ for which the series expansion of $f(z)$ is $f_2(z)$ with the circle of convergence D_2 that does not include the singular point at z_s. If z_2 is close to the circle D_1 and id D_1 and D_2 overlap, the $f_2(z)$ is the analytic continuation of $f_1(z)$ into D_2, while $f_1 = f_2$ in $D_1 \cap D_2$, as shown in Figure 2.4.1.

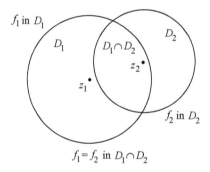

Figure 2.4.1 Analytic Continuation.

Example 2.4.1. Let the points p and q, not equal to $+i$ or $-i$, be given. Consider the function $f(z) = 1/(1 + z^2)$ on a small disc $C(p, \rho)$, where $\rho < |p \pm i|$. Then every analytic continuation f^* of f to a domain D that includes the disc $C(p, \rho)$ and the point q reproduces the function $f(z)$ not only near q but also throughout D. Note that no continuation can have the points $\pm i$ within its domain D. Note that the product $(1 + z^2)f^*(z) = (1 = z^2)f(z) = 1$ near p, and this identity holds throughout D. Thus $f^*(z) = 1/(1 + z^2)$ for all $z \in D$. $\blacksquare$

2.5 Schwarz Reflection Principle

Analytic continuation plays an important role in the Schwarz reflection principle defined as follows: If $f(z)$ is analytic on a domain containing a segment of the real axis and is real-valued on this segment, then $\overline{f(z)} = f(\bar{z})$. This leads to the concept of analytic continuation across the real axis when the function is known only on one side. Let I denote a segment $a < x < b$, and let Γ be a Jordan arc joining a and b and lying in $\Re\{z\} > 0$. Then $C = I \cup \Gamma$ is a contour that encloses a domain D lying entirely in $\Re\{z\} > 0$. If we reflect this domain into the real axis, we get a domain D' that lies in the lower half-plane and is bounded by I and Γ' which is the reflection of Γ into the real axis. Then the domain $I \cup D'$ is symmetric about the real axis. If $f(z)$ is continuous on $D \cup I$, analytic on D, and real-valued on I, then the function $F(z)$ defined by

$$F(z) = \begin{cases} f(z), & \text{if } z \in D \cup I, \\ \overline{f(z)}, & \text{if } z \in D' \end{cases} \qquad (2.5.1)$$

is analytic on $D \cup I \cup D'$.

If f is analytic on one side of a curve Γ and cannot be continued analytically across Γ, then Γ is called a *natural boundary* for f. For example, the unit circle $|z| = 1$ is a natural boundary for the function $f(z) = \sum_{n=0}^{\infty} z^{2^k}$.

Let S^+ (S^-) denote the upper (lower) half-plane $\Re\{z\} > 0$ ($\Re\{z\} < 0$), respectively, or vice-versa, with L as their common boundary (i.e., the real axis). Let $f(z)$ be a function defined at $z \in S^+$, and let it be connected with the function $f^*(z)$ defined in S^- by the relation

$$f^*(z) = \overline{f(\bar{z})}, \qquad (2.5.2)$$

i.e., $f(z)$ and $f^*(z)$ take conjugate complex values at points symmetric with respect to the real axis, since the points z and $\bar{z}$ are reflections of each other in L, and in the case of a mirror image about the real axis, we have $f^*(z) = f(\bar{z})$. We can rewrite the relation (2.5.2) as $f^*(z) = \bar{f}(z)$, i.e., if $f(z) = u(x, y) + iv(x, y)$, then $\bar{f}(z) = u(x, -y) - iv(x, -y)$. If $f(z)$ is regular (or meromorphic) in S^+, then $f^*(z) = \bar{f}(z)$ is regular (or meromorphic) in S^-, and the relation (2.5.2) is symmetric as regards f and f^*, i.e.,

$$f(z) = \overline{f^*(\bar{z})}, \quad (f^*(z))^* = f(z). \qquad (2.5.3)$$

If $f(z)$ is a rational function

$$f(z) = \frac{a_n z^n + a_{n-1} z^{n-1} + \cdots + a_0}{b_m z^m + b_{m-1} z^{m-1} + \cdots + b_0}, \qquad (2.5.4)$$

then $f^*(z) = \bar{f}(z)$ is obtained by simply replacing the coefficients by their conjugate complex values. Let t be a real number, and assume that $f(z)$ takes a definite limit value $f^+(t)$ as $z \to t$ from S^+. Then $f^{*-}(t)$ exists and

$$f^{*-}(t) = \bar{f}^-(t) = \overline{f^+(t)}, \qquad (2.5.5)$$

because $\bar{z} \to t$ from S^- as $z \to t$ from S^+, and hence, $f^*(z) = \overline{f(\bar{z})} \to \overline{f^+(t)}$.

We shall assume that $f(z)$ is regular on S^+, except possibly at infinity, and continuous on L from the left. Define a sectionally regular function $F(z)$ by

$$F(z) = \begin{cases} f(z) & \text{for } z \in S^+, \\ f^*(z) & \text{for } z \in S^-. \end{cases} \tag{2.5.6}$$

Then, in view of (2.5.5)

$$F^-(t) = \overline{F^+(t)}, \quad F^+(t) = \overline{F^-(t)}. \tag{2.5.7}$$

These relations are useful when transforming the boundary conditions in any boundary problem containing $f^+(t)$ and $\overline{f^+(t)}$, or $f^-(t)$ and $\overline{f^-(t)}$ into those involving $F^+(t)$ and $F^-(t)$.

In view of the Schwarz reflection principle, another property is that of extending $f(z)$. If $\Im\{f^*(t)\} = 0$ in any interval I of the real axis, then the function $f^*(z)$ is the analytic continuation of $f(z)$ through the interval I because $f^{*-}(t) = f^+(t)$ on this interval.

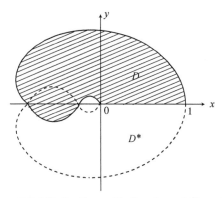

Figure 2.5.1 Reflection with Overlap of D and D^*.

The reflection principle is an important application of the Riemann mapping theorem for removability of isolated singularities. This principle can also be applied to certain cases where a simply connected domain D does not lie entirely in the upper half-plane. But there are cases when this principle does not hold. For example, consider the domain D in Figure 2.5.1, and let $f(z)$ be determined by $f(re^{i\theta}) = r^{1/2}e^{i\theta/2}$, normalized by $-\pi/2 < \theta < 3\pi/2$. Then the function $f(z)$ on D and its reflection $\tilde{f}(z) = \overline{f(\bar{z})}$ on D^* agree along the segment $(0, 1)$, but $\tilde{f}$ is determined as square root function on the intersection $D \cap D^*$.

2.6 Bilinear Transformations

The *bilinear (linear-fractional, or Möbius) transformation* has the form

$$w = f(z) = \frac{az + b}{cz + d},$$
(2.6.1)

where a, b, c, d are complex constants such that $ad - bc \neq 0$ (otherwise the function $f(z)$ would be identically constant). If $c = 0$ and $d = 1$, or if $a = 0$, $d = 0$ and $b = c$, then the function (2.6.1) reduces to a linear transformation $w = az + b$, or an inversion $w = \frac{1}{z}$, respectively. The transformation (2.6.1) can also be written as

$$w = \frac{a}{c} + \frac{bc - ad}{c(cz + d)},$$
(2.6.2)

that can be viewed as composed of the following three successive functions:

$$z_1 = cz + d, \quad z_2 = \frac{1}{z_1}, \quad w = \frac{a}{c} + \frac{bc - ad}{c} z_2.$$

This shows that the mapping (2.6.1) is a linear transformation, followed by an inversion followed by another linear transformation. The bilinear transformation (2.6.1) maps the extended z-plane conformally onto the extended w-plane such that the pole at $z = -d/c$ is mapped into the point $w = \infty$. The inverse transformation

$$z = f^{-1}(w) = \frac{b - dw}{-a + cw}$$
(2.6.3)

is also bilinear defined on the extended w-plane, and maps it conformally onto the extended z-plane such that the pole at $w = a/c$ is mapped into the point $z = \infty$. Note that $f'(z) = \frac{ad - bc}{(cz + d)^2} \neq 0$; also $[f^{-1}(w)]' = \frac{-ad + bc}{(cw - a)^2} \neq 0$. A bilinear transformation carries circles into circles (in the extended sense, i.e., there is no distinction between circles and straight lines in the theory of bilinear transformations).

The *cross-ratio* between four distinct finite points z_1, z_2, z_3, z_4 is defined by

$$(z_1, z_2, z_3, z_4) = \frac{z_1 - z_2}{z_1 - z_4} \cdot \frac{z_3 - z_4}{z_3 - z_2}.$$
(2.6.4)

If z_2, z_3, or z_4 is a point at infinity, then (2.6.4) reduces to

$$\frac{z_3 - z_4}{z_1 - z_4}, \quad \frac{z_1 - z_2}{z_1 - z_4}, \quad \text{or} \quad \frac{z_1 - z_2}{z_3 - z_2},$$

respectively. The cross-ratio (z, z_1, z_2, z_3) is invariant under bilinear transformations.

Theorem 2.6.1. *A bilinear transformation is uniquely defined by a correspondence of the cross-ratios*

$$(w, w_1, w_2, w_3) = (z, z_1, z_2, z_3),$$ (2.6.5)

which maps any three distinct points z_1, z_2, z_3 in the extended z-plane into three prescribed points w_1, w_2, w_3 in the extended w-plane.

The cross-ratio (z, z_1, z_2, z_3) is the image of z under a bilinear transformation that maps three distinct points z_1, z_2, z_3 into $0, 1, \infty$.

The points z and z^* are said to be *symmetric* with respect to a circle C (in the extended sense) through three distinct points z_1, z_2, z_3 if and only if

$$(z^*, z_1, z_2, z_3) = \overline{(z, z_1, z_2, z_3)}.$$ (2.6.6)

The mapping that carries z into z^* is called a *reflection* with respect to C. Two reflections obviously yield a bilinear transformation. For an example of symmetric points, see Exercise 2.8.1.

The *symmetry principle* states that if a bilinear transformation maps a circle C_1 onto a circle C_2, then it maps any pair of symmetric points with respect to C_1 into a pair of symmetric points with respect to C_2. This means that bilinear transformations preserve symmetry. A practical application of the symmetry principle is given in Exercise 2.8.4.

Example 2.6.1. The bilinear transformation that maps the three points $\{-1, 0, +1\}$ on the real axis to the points $\{-1, -i, +1\}$ on the unit circle is $w = f(z) = i\left(\dfrac{z-i}{z+i}\right)$, and this functions maps the upper half-plane $\Im\{z\} > 0$ conformally onto the unit disk $|w| < 1$. ∎

2.7 Poisson's Formula

Theorem 2.7.1. *If u is defined and continuous on the disk $D(0, r) = \{z : |z| < r\}$, then for $\rho < r$*

$$u\left(\rho\, e^{i\phi}\right) = \frac{r^2 - \rho^2}{2\pi} \int_0^{2\pi} \frac{u\left(re^{i\theta}\right)}{r^2 - 2r\rho\cos(\theta - \phi) + \rho^2} \, d\theta.$$ (2.7.1)

PROOF. $D(0, r)$ is simply connected; u is harmonic on $D(0, r)$. Then there exists an analytic function $f(z)$ in $D(0, r)$ such that

$$u(z) = \Re\{f(z)\}, \quad |z| < r.$$

By Cauchy's integral formula

$$f(z) = \frac{1}{2i\pi} \int_{\Gamma_s : |z| = s|r} \frac{f(\zeta)}{\zeta - z} \, d\zeta$$ (2.7.2)

for all z within Γ_s, where $\rho \in \Gamma_s$. (Recall that the symmetric point of z with respect to the circle $|z - a| + R$ is $z^* = \dfrac{R^2}{\bar{z} - \bar{a}} + a$.) Then the symmetric point z^* of z with respect to the circle Γ_s is given by $z^* = \dfrac{s^2}{\bar{z}}$. ($z^*$ is also called the reflection of z in the circle $z| = s$.) Note that if z is within Γ_s, then z^* is given by

$$\frac{1}{2i\pi} \int_{\Gamma_s : |z| = s|r} \frac{f(\zeta)}{\zeta - z^*} \, d\zeta \qquad (2.7.3)$$

Subtracting (2.7.2) and (2.7.3) we get

$$f(z) = \frac{1}{2i\pi} \int_{\Gamma_s : |z| = s|r} f(\zeta) \left\{ \frac{1}{\zeta - z} - \frac{1}{\zeta - z^*} \right\} d\zeta.$$

Simplifying and noting that $s^2 = |\zeta|^2 = \zeta\bar{\zeta}$, we have

$$\frac{1}{\zeta - z} - \frac{1}{\zeta - z^*} = \frac{1}{\zeta - z} - \frac{1}{\zeta - \dfrac{s^2}{z}} = \frac{1}{\zeta - z} - \frac{\bar{z}}{\zeta\bar{z} - \zeta\bar{\zeta}}$$

$$= \frac{1}{\zeta - z} - \frac{\bar{z}}{\zeta(\bar{z} - \bar{\zeta})} = \frac{\zeta(\bar{z} - \bar{\zeta}) + \bar{z}(\zeta - z)}{\zeta(\zeta - z)(\bar{z} - \bar{\zeta})}$$

$$= \frac{|\zeta|^2 - |z|^2}{\zeta|\zeta - z|^2}.$$

Hence,

$$f(z) = \frac{1}{2i\pi} \int_{\Gamma_s : |z| = s|r} f(\zeta) \frac{|\zeta|^2 - |z|^2}{\zeta|\zeta - z|^2} \, d\zeta.$$

Now, let $z = \rho e^{i\phi}$, $\rho < s < r$ and $\zeta = se^{i\theta}$. Then

$$f(z) = \frac{1}{2i\pi} \int_0^{2\pi} \frac{f(se^{i\theta})(s^2 - \rho^2)}{se^{i\theta} \left| se^{i\theta} - \rho e^{i\phi} \right|^2} \, ise^{i\theta} \, d\zeta.$$

But

$$\left| se^{i\theta} - \rho e^{i\phi} \right|^2 = \left| s\cos\theta - \rho\cos\phi + i(s\sin\theta - \rho\sin\phi) \right|^2$$

$$= (s\cos\theta - \rho\cos\phi)^2 + (s\sin\theta - \rho\sin\phi)^2$$

$$= s^2 - 2s\rho\sin\theta\cos\phi + \rho^2.$$

Then

$$u(\rho e^{i\phi}) = \Re\{f(\rho e^{i\phi})\} = \frac{1}{2\pi} \int_0^{2\pi} \frac{u(se^{i\theta})(s^2 - \rho^2)}{s^2 - 2s\rho\sin\theta\cos\phi + \rho^2} \, d\theta.$$

Let $s \to r$ through a sequence of radii $\{\rho_n\} \to r$. Then

$$u(\rho e^{i\phi}) = \frac{1}{2\pi} \int_0^{2\pi} \frac{u(re^{i\theta})(r^2 - \rho^2)}{r^2 - 2r\rho \sin\theta \cos\phi + \rho^2} \, d\theta, \quad \rho < r. \blacksquare \qquad (2.7.4)$$

Theorem 2.7.2. (Schwarz integral formula for the unit disk) *Let $f(z)$ be an analytic function defined on the unit disk $E : |z| < 1$. Then*

$$f(z) = \frac{1}{2\pi i} \oint_{|\zeta|=1} \frac{\zeta + z}{\zeta - z} \Re\{f(\zeta)\} \frac{d\zeta}{\zeta} + i\Im\{f(0)\} \qquad (2.7.5)$$

for all $|\zeta| < 1$.

This formula follows from the Poisson integral formula (Theorem 2.7.1).

2.7.1 Herglotz Representation. A holomorphic function $f \in E$ with $f(0) = 1$ is harmonic if and only if there is a probability measure μ on E such that

$$f(re^{i\theta}) = \int_0^{2\pi} \frac{1 - r^2}{1 - 2r\cos(\theta - \phi) + r^2} \, d\mu(\theta), \quad z = e^{i\theta}. \qquad (2.7.6)$$

This formula is known as the *Herglotz representation for harmonic functions* on E; it defines a positive harmonic function with $f(0) = 1$. Conversely, if f is positive and harmonic and r_n increases to 1, define $f_n(z) = f(r_n z)$. Then

$$\begin{aligned} f_n(re^{i\theta}) &= \frac{1}{2\pi} \int_0^{2\pi} \frac{1 - r^2}{1 - 2r\cos(\theta - \phi) + r^2} \, d\mu(\theta) f_n(\phi) \, d\phi \\ &= \int_0^{2\pi} \frac{1 - r^2}{1 - 2r\cos(\theta - \phi) + r^2} \, d\mu(\phi), \qquad (2.7.7) \end{aligned}$$

where $d\mu(\phi) = \frac{1}{2\pi} f(r_n e^{i\phi}) \, d\phi$. Since $r_n \to 1$, so that $f_n(a) \to f(z)$, and formula (2.7.6) follows.

The Herglotz representation for a holomorphic function f on E with $f(0) = 1$ has positive real part if and only if there is a probability measure μ on E such that

$$f(z) = \int_0^{2\pi} \frac{1 + e^{i\theta} z}{1 - e^{i\theta} z} \, d\mu(\theta), \qquad (2.7.8)$$

This formula follows from (2.7.6).

Carathéodory's positivity criterion for holomorphic functions (Carathéodory [1907]) is as follows.

Theorem 2.7.3. *Let* $f(z) = 1 + \sum\limits_{n=1}^{\infty} a_n z^n$ *be holomorphic on* E. *Then* $f(z)$ *has positive real part on* E *if and only if*

$$\sum_m \sum_n a_{m-n} \lambda_m \overline{\lambda_n} \geq 0 \qquad (2.7.9)$$

for any complex numbers $\lambda_0, \lambda_1, \ldots, \lambda_N$, *where* $a_0 = 2$, $a_{-m} = \overline{a_m}$ *for* $m > 0$.

PROOF. From the Herglotz representation (2.7.7) for $n > 0$, we have $a_n = 2 \int_0^{2\pi} e^{-in\theta} d\mu(\theta)$. Hence

$$\sum_m \sum_n a_{m-n} \lambda_m \overline{\lambda_n} = \int_0^{2\pi} \left| \sum_n \lambda_n e^{in\theta} \right|^2 d\mu(\theta) \geq 0. \qquad (2.7.10)$$

Conversely, setting $\lambda_n = z^n$,

$$\sum_{m=0}^{\infty} \sum_{n=0}^{\infty} a_{m-n} \lambda_m \overline{\lambda_n} = 2(1 - |z|^2) \Re\{f(z)\}. \ \blacksquare$$

2.8 Exercises

Some selected problems related to conformal mappings are presented in this section. More examples can be found in Kythe [1998: 65ff; 293; 398].

2.8.1. If C is a straight line, then we choose $z_3 = \infty$, and the condition for symmetry (2.6.6) gives

$$\frac{z^* - z_1}{z_1 - z_2} = \frac{\bar{z} - \bar{z}_1}{\bar{z}_1 - \bar{z}_2}. \qquad (2.8.1)$$

Let z_2 be any finite point on the line C. Then, since $|z^* - z_1| = |z - z_1|$, the points z and z^* are equidistant from the line C. Moreover, since $\Im\left\{\dfrac{z^* - z_1}{z_1 - z_2}\right\} = -\Im\left\{\dfrac{z - z_1}{z_1 - z_2}\right\}$, the line C is the perpendicular bisector of the line segment joining z and z^*. If C is the circle $|z - a| = R$, then

$$\overline{(z, z_1, z_2, z_3)} = \overline{(z - a, z_1 - a, z_2 - a, z_3 - a)}$$

$$= \left(\bar{z} - \bar{a}, \frac{R^2}{z_1 - a}, \frac{R^2}{z_2 - a}, \frac{R^2}{z_3 - a} \right)$$

$$= \left(\frac{R^2}{\bar{z} - \bar{a}}, z_1 - a, z_2 - a, z_3 - a \right)$$

$$= \left(\frac{R^2}{\bar{z} - \bar{a}} + a, z_1, z_2, z_3 \right).$$

Hence, in view of (2.6.6), we find that the points z and $z^* = \dfrac{R^2}{\bar{z} - \bar{a}} + a$ are symmetric, i.e.,

$$(z^* - a)(\bar{z} - \bar{a}) = R^2. \tag{2.8.2}$$

Note that $|z^* - a||z - a| = R^2$; also, since $\dfrac{z^* - a}{z - a} > 0$, the points z and z^* are on the same ray from the point a (Fig 2.8.1); moreover, the point symmetric to a is ∞.

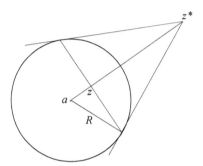

Fig. 2.8.1 Symmetry with respect to a circle.

A generalization of this result is as follows: If Γ denotes an analytic Jordan curve with parametric equation $z = \gamma(s)$, $s_1 < s < s_2$, then for any point z sufficiently close to Γ, the point

$$z^* = \gamma\overline{\left(\gamma^{[-1]}(z)\right)} \tag{2.8.3}$$

defines a symmetric point of z with respect to Γ (Sansone and Gerretsen [1969:103], and Kythe [1998:48]). Some examples are as follows:
(i) If Γ is the circle $x^2 + y^2 = a^2/9$, then $z^* = a^2/9\bar{z} = z^*$.
(ii) If Γ is the ellipse $\dfrac{(x + a/2)^2}{a^2} + y^2 = 1$, then

$$z^* = -\frac{a}{2} + \frac{(a^2 + 1)(\bar{z} + a/2) + 2ia\sqrt{a^1 - 1 - (\bar{z} + a/2)}}{a^2 - a}.$$

(iii) If Γ_1 is a cardioid defined by $z = \gamma(s) = \left(\dfrac{1}{2} + \cos\dfrac{s}{2}\right) e^{is}$, $-\pi < s \le \pi$, then from (2.8.3) we cannot write an explicit expression for symmetric points with respect to Γ_1. However, for any real t,

$$\gamma(\pm it) = \left(\frac{1}{2} + \cosh\frac{t}{2}\right) e^{\mp t}$$

defines two real symmetric points with respect to Γ_1, provided the parameter t satisfies the equation $\gamma(it)\,\gamma(-it) = a^2$, i.e., $\dfrac{1}{2} + \cosh\dfrac{t}{2} = a$ which has the roots

$$t = 2\,\cosh^{-1}(a - 1/2) = \pm 2\,\log\rho, \qquad (2.8.4)$$

where $\rho = a - \dfrac{1}{2} + \sqrt{\left(a - \dfrac{1}{2}\right)^2 - 1}$.

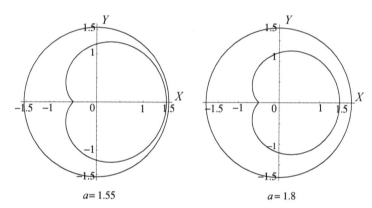

$a = 1.55$ $a = 1.8$

Fig. 2.8.2 Cardioid inside a Circle.

(iv) If a doubly connected region Ω is bounded outside by the circle $\Gamma_2 = \{z : |z| = a,\ a > 1.5\}$ and inside by the cardioid Γ_1 defined above in (iii) (see Fig. 2.8.2), then it follows from (2.8.4) that there is one pair of real common symmetric points $\zeta_1 \in \mathrm{Int}\,(\Gamma_1)$ and $\zeta_2 \in \mathrm{Ext}\,(\Gamma_2)$ such that $\zeta_1 = a/\rho^2$ and $\zeta_2 = a\rho^2$, where ρ is defined in (2.8.4). ∎

2.8.2. The bilinear transformation that maps the three points $\{-1, 0, +1\}$ on the real axis to the points $\{-1, -i, +1\}$ on the unit circle is $w = f(z) = i\left(\dfrac{z - i}{z + i}\right)$, and this functions maps the upper half-plane $\Im\{z\} > 0$ conformally onto the unit disk $|w| < 1$. ∎

2.8.3. Use Rouché's theorem to find the roots of the transcendental equation $\tan z - \lambda z = 0$, where $\lambda > 0$ is a real parameter. This equation arises in certain nonlinear problems of heat conduction. SOLUTION: Let $h(z) = \tan z - \lambda z$. Since $h(-z) = -h(z)$, the zeros of $h(z)$ are located symmetrically about the origin, so that the real roots can be located by graphical method; for example, there are three real zeros in the interval $(-\pi/2, \pi/2)$ when $1 < \lambda < \infty$ and one real zero when $0 < \lambda < 1$; in general, for all values of λ there is a positive real zero in each of the intervals $(n\pi, n\pi + \pi/2)$ and a negative zero in each of the intervals $(-n\pi - \pi/2, -n\pi)$, $n = 1, 2, \ldots$ (see

Figure 2.8.3).

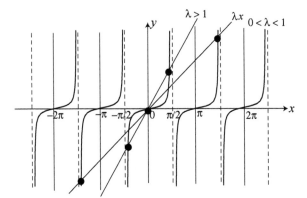

Figure 2.8.3 Graph of $h(x) = \tan x - \lambda x$. ∎

2.8.4. Use the symmetry principle and find bilinear transformations which map a circle C_1 onto a circle C_2. SOLUTION: We already know that the transformation (2.6.5) can always be determined by requiring that the three points $z_1, z_2, z_3 \in C_1$ map onto three points $w_1, w_2, w_3 \in C_2$. But a bilinear transformation is also determined if a point $z_1 \in C_1$ should map into a point $w_1 \in C_2$ and a point $z_2 \notin C_1$ should map into a point $w_2 \notin C_2$. Then, by the symmetry principle, the point z_2^* which is symmetric to z_2 with respect to C_1 is mapped into the point w_2^* which is symmetric to w_2 with respect to C_2, and then the desired bilinear transformation is given by

$$(w, w_1, w_2, w_2^*) = (z, z_1, z_2, z_2^*). \tag{2.8.5}$$

Let E denote the unit disk $|z| < 1$, E^* the region $|z| > 1$, and let $C = \{|z| = 1\}$ be their common boundary. Let $f(z)$ be a function defined on E. Then this function can be related to the function $g(z)$ defined in E^* as $g(z) = \overline{f(\bar z)}$ for half-planes, except that now the conjugate complex points are replaced by points inverse with respect to the circle C according to the relation (2.8.2). Thus,

$$g(z) = \overline{f\left(\frac{1}{\bar z}\right)} = \bar f\left(\frac{1}{z}\right). \tag{2.8.6}$$

This relation is symmetrical, i.e.,

$$f(z) = \overline{g\left(\frac{1}{\bar z}\right)}, \quad (g(z))^* = f(z). \tag{2.8.7}$$

If $f(z)$ is regular or meromorphic in E, then $g(z)$ is regular or meromorphic in E^*. Also, if $f(z)$ is a rational function defined by (2.5.4), then

$$g(z) = \frac{\bar a_n\, z^{-n} + \bar a_{n-1}\, z^{-n+1} + \cdots + \bar a_0}{\bar b_m\, z^{-m} + \bar b_{-m+1}\, z^{m-1} + \cdots + \bar b_0}. \tag{2.8.8}$$

Moreover, if $f(z)$ has the power series expansion $f(z) = \sum\limits_{k=-\infty}^{\infty} a_k z^k$ $z \in E$, then

$$g(z) = \sum_{k=-\infty}^{\infty} \bar{a}_k z^{-k}, \quad z \in E^*. \tag{2.8.9}$$

If $f(z)$ has a zero (pole) of order k at $z = \infty$ ($z = 0$), so does $g(z)$. Let us assume that $f(z)$ approaches a definite limit value $f^+(t)$ as $z \to t \in C$ from E. Then $f^{*-}(t)$ exists and

$$f^{*-}(t) = \bar{f}^{-}\left(\frac{1}{t}\right) = \overline{f^+(t)}, \tag{2.8.10}$$

because $1/\bar{z} \to t$ from E as $z \to t$ from E^*, and hence $g(z) = \bar{f}\left(\dfrac{1}{z}\right) = \overline{f\left(\dfrac{1}{\bar{z}}\right)} \to \overline{f^+(t)}$. If $f(z)$ is regular in E except possibly at infinity and continuous on C from the left, then let $F(z)$ be sectionally regular and be defined by

$$F(z) = \begin{cases} f(z) & \text{for } z \in E, \\ g(z) & \text{for } z \in E^*. \end{cases} \tag{2.8.11}$$

Then, $F^*(z) = F(z)$, and, as in (2.5.7),

$$F^-(t) = \overline{F^+(t)}, \quad F^+(t) = \overline{F^-(t)}. \tag{2.8.12}$$

Moreover, in view of the Schwarz reflection principle, if $\Im\{f^+(t)\} = 0$ on some part of the circle C, then $g(z)$ is the analytic continuation of $f(z)$ through this part of C. ∎

2.8.5. The argument principle (Example 1.6.1) implies the Rouché's theorem, which states that if γ is a closed curve enclosing a domain D, and if f, g are meromorphic functions on $\gamma \cup D$ with finitely many zeros, no poles, and no removable singularities (i.e., they are all inside γ), and if $|f(z) - g(z)| < |f(z)|$ for all $z \in \gamma$, i.e., at every point $z \in \gamma$, $f(z)$ is closer to $g(z)$ than the origin is, then both $f(\gamma)$ and $g(\gamma)$ have the same winding number around the origin. In view of the argument principle, the number of zeros minus the number of poles of f in D is equal to the number of zeros minus that of the poles of g. The Rouché's theorem is sometimes referred as 'walking dog theorem' in the sense that if you (at $f(z)$) are walking a dog (at $g(z)$) around a lamp post (at 0) and you always maintain the length of the leash ($|f(z) - g(z)|$) between yourself and the dog shorter than the length ($|f(z)|$) to the lamp post, then you and the dog will have the same winding number, i.e., the leash cannot be tangled up around the lamp post. Prove Rouché's theorem.

PROOF. By hypothesis, we have $|f(z)| > |g(z) - f(z)|$, and $|f(z)| > 0$ for any $z \in \gamma$, i.e., $f(z)$ is non-zero. Also, $g(z)$ is non-zero, since $g(z) = 0$ would give $|f(z)| = |g(z) - f(z)|$, which is a contradiction. Thus, define a function $h(z) = g(z)/f(z)$ on γ, $f \neq 0$. The function h is analytic everywhere on γ and also meromorphic on D. Then $\dfrac{|g(z) - f(z)|}{|f(z)|} \leq \dfrac{|g(z)|}{|f(z)|} - 1 = |h(z)| - 1 < 1$, which means that for every $z \in \gamma$, $h(z)$ is contained in the disk $C(1,1)$, i.e., $h(z)$ is never on the negative real axis. Thus, taking the principal branch of $\log h(z)$, this function is analytic on γ, and $\dfrac{d}{dz} \log h(z) = \dfrac{h'(z)}{h(z}$, and by the fundamental theorem of calculus, $\int_\gamma \dfrac{h'(z)}{h(z)} dz = 0$ since γ is a closed curve. Hence, $\dfrac{h'(z)}{h(z)} = \dfrac{f'(z)}{f(z)} - \dfrac{g'(z)}{g(z)}$, and

$$\int_\gamma \frac{f'(z)}{f(z)} dz = \int_\gamma \frac{g'(z)}{g(z)} dz, \tag{2.8.13}$$

which implies, by the argument principle, that $f(z)$ and $g(z)$ have the same winding number around the origin. ∎

2.8.6. Application of Rouché's theorem: Let $p(z) = z^7 - 5z^3 + 12$. Show that $p(z)$ has no zeros in $|z| < 1$. Take $g(z = z^7 - 5z^3 + 12$, and $f(z) = 12$. Then for $|z| < 1$, we have $|f(z) - g(z)| = |z^7 - 5z^3| \leq |z|^7 + 5|z|^3 < 1 + 5 = 6 < 12 = |f(z)|$. Thus, $p(z)$ has 7 zeros in $|z| < 2$. ∎

2.8.7. Consider the functions $f(z) = z^5$ and $g(z) = z^5 + z + 1$. Then show that $g(z)$ has five zeros inside the circle $|z| = 2$.

SOLUTION. $f(z)$ has 5 zeros at the origin. Let the circle C be defined by $\gamma(t) = 2e^{it}$, $0 \leq t \leq 2\pi$. Then on C, we have $|f(z)| = |z|^5 = 2^5 = 42$, and by triangle inequality, $|g(z) - f(z)| = |z + 1| \leq |z| + 1 = 2 + 1 = 3$. Thus, by Rouché's theorem, the number of zeros of g minus the number of poles inside C must be equal to that of f, i.e., equal to 5. Since g is a polynomial, it has no poles, and so g has 5 zeros inside C. ∎

2.8.8. With the same function g as in Exercise 2.8.7, and $f(z) = 1$, show that g has all five zeros inside the annulus $\Gamma = \{z : \frac{1}{2} < |z| < 2\}$. SOLUTION. Take $\gamma'(t) = \frac{1}{2} e^{it}$, $0 \leq t \leq 2\pi$, which traverses the circle $C' = \{z : |z| = \frac{1}{2}\}$. On the curve C' we have $|f(z)| = 1$, and $|g(z) - f(z)| = |z^5 + z| \leq |z|^5 + |z| = \frac{1}{2^5} + \frac{1}{2} = \frac{17}{32}$. Then by Rouché's theorem, both f and g has the same number of zeros inside C', which is 0, since $1 \neq 0$. Thus, g has no zeros in the disk $\{z : |z| < \frac{1}{2}\}$. Hence, in view of the Exercise 2.8.7, all zeros must lie in the annulus Γ. ∎

2.8.9 The *conformal radius* of a simply connected domain $D \subset \mathbb{C}$ with respect to a point $z \in D$ is defined in §2.3 as the radius R of the largest disk $C(z, R)$ that contains D. By the Riemann mapping theorem there is a conformal bijection $f : D \mapsto E$, $f(z) = 0 \in D$ and $f'(z) \in \mathbb{R}$. The conformal radius, which measures the size of a simply connected domain, for D is defined by $\mathrm{rad}(z, D) = R = \dfrac{1}{|f'(z)|}$. Determine $\mathrm{rad}(i, H)$, where H is the upper half-plane and $D \subset H$. SOLUTION. By the Riemann mapping theorem there is a conformal bijection $f : D \mapsto H$. Since the conformal map is $f(z) = i\dfrac{z - i}{z + i}$, we find that $f'(z) = \dfrac{-2}{(z + i)^2}$, and thus, $\mathrm{rad}(i, H) = 2$. ∎

2.8.10. A Schwarz function $S(z)$ defined on the boundary ∂D of a domain in $\mathbb{C}$ is an analytic function such that $S(z) = \bar{z}$ on ∂D. The Schwarz function $S(z)$ on a curve $\Gamma(z) = 0$ is obtained by substituting $x = \frac{1}{2}(z + \bar{z})$ and $y = \frac{1}{2i}(z - \bar{z})$, and solving for $\bar{z}$. Some simple examples are: (i) $S(z) = 1/z$ on ∂E; and (ii) $S(z) = z$ on the real axis. Let $z = f(\zeta)$ be a conformal map from $E_\zeta = \{|\zeta| < 1\}$ onto some domain G in the z-plane. Let $\zeta \in E_\zeta$. Since $\bar{\zeta} = \zeta^{-1}$ on ∂E_ζ, the Schwarz function on ∂E_ζ is $S(z) = \bar{z} = \overline{f(\zeta)} = \bar{f}\left(\zeta^{-1}\right)$. Some properties of the Schwarz function are:

(i) *unitary property*: $\bar{z} = S(z)$ implies $z = \overline{S(a)} = \bar{S}(\bar{z}) = \bar{S}\left(S(z)\right)$. By differentiating with respect to z we get $S' = 1/\bar{S}'$.

(ii) *Geometric property*: Since $ds^2 = dx^2 + dy^2 = dz\, d\bar{z}$, we get

$$\frac{dz}{ds} = \frac{dz}{\sqrt{dz\, d\bar{z}}} = \sqrt{\frac{dz}{d\bar{z}}} = \frac{1}{(S')^{1/2}}, \quad \text{by using unitary property.}$$

Schwarz functions are used in Laplacian growth problem (using the Schwarz potential), coastal current near a gap, industrial dip-coating of substrates and periodic substrates, and coating of thin rods; see Davis [1974], Richtmyer [1978], and Gustafsson and Vasil'ev [2006].

(a) Using $x = \frac{1}{2}(z + \bar{z})$ and $y = \frac{1}{2i}(z - \bar{z})$, and solving for $\bar{z}$. show that the Schwarz function on the ellipse $\dfrac{x^2}{a^2} + \dfrac{y^2}{b^2} = 1$ is

$$S(z) = \frac{a^2 + b^2}{a^2 - b^2}\, z + \frac{2ab}{b^2 - a^2}\,\sqrt{z^2 + b^2 - a^2}. \tag{2.8.14}$$

Also, show that the Schwarz function is $S(z) = 1/z$ on ∂E in the limit as $a, b \to 1$ in (2.8.14).

(b) The function $z = i\dfrac{1 - \zeta}{1 + \zeta}$ maps the unit disk E_ζ onto the upper half of the z-plane. Then the Schwarz function on the boundary of the image domain is

$$S(z) = \bar{z} = -i\left(\frac{1 - 1/\zeta}{1 + 1/\zeta}\right) = z. ∎$$

3

Area Principle

The area principle started with the work of Gronwall [1914]. Although classical, it is important in that it developed into other area theorems which played an important role in the development the subject of geometric function theory. Another important topic to discuss involves the Grunsky inequalities, especially the strong and the weak forms of these inequalities.

3.1 Area Theorems

The first result is due to Gronwall [1914], known as the *area principle*, which states that the area of the complement of the image of a domain D under a mapping by a function is non-negative. Let $f(\zeta) \in \Sigma$ have the series expansion of the form $f(\zeta) = \zeta + b_0 + \dfrac{b_1}{\zeta} + \dfrac{b_2}{\zeta^2} + \cdots$, $|\zeta| > 1$. The functions in the class Σ are normalized by $f(\infty) = \infty$ and $f'(\infty) = 0$. The coefficient b_0 is given by

$$b_0 = \frac{1}{2\pi} \int_0^{2\pi} f\left(\rho e^{i\theta}\right) d\theta, \quad \rho > 1.$$

Let Σ_0 denotes the class of functions $f(z)$ such that $f \in \Sigma$, $f(\zeta) \neq 0$ for $\zeta \in E^* = \{|\zeta| > 1\}$.[1] Let $f \in S$ with the series expansion $f(z) = z + a_2 z^2 + \cdots$. Then the function

$$g(z) = f(1/z) = \frac{1}{z^{-1} + a_2 z^{-2} + \cdots} = \frac{z}{1 + a_2 z^{-1} + \cdots} = z - a_2 + \frac{a_3}{z} + \cdots \in \Sigma,$$

where $g(z) \neq 0$ for all $z \in E^*$, since $f \in S$ has no poles. Conversely, if $g \in \Sigma$, with series expansion $g(z) = z + b_0 + \dfrac{b_1}{z} + \cdots$, and if $c \in \mathbb{C} \backslash g(E^*)$, then the function

$$f(z) = \frac{1}{g(1/z) - c} = \frac{z}{1 + (b_0 - c)z + \cdots} = z + (c - b_0)z^2 + \cdots \in S.$$

---

[1]Note that $\mathbb{C} = E \cup \partial E \cup E^*$.

Theorem 3.1.1. (Gronwall's area theorem) *Let* $f(\zeta) = \zeta + \dfrac{b_1}{\zeta} + \cdots \in \Sigma$.
Let $E^* = \{\zeta : 1 < |\zeta| < \infty\}$. *Then the area of the domain* $\{f(E^*)\backslash \mathbb{C}\}$ *of the image* $f(E^*))$ *of* E^* *under the mapping* $w = z(z) \in \Sigma$ *is determined by*

$$\text{Area}\{f(E^*)\backslash\mathbb{C}\} = \pi\left(1 - \sum_{n=1}^{\infty} n|b_n|^2\right) \geq 0. \tag{3.1.1}$$

PROOF. This principle gets its name from the geometric fact that the map of the region $|\zeta| = \rho > 1$ by the function $w = f(\zeta) \in \Sigma$ covers the entire w-plane except for a region of positive area. We will use this fact to prove this theorem. Let $\zeta = \rho\, e^{i\theta}$, $0 \leq \theta \leq 2\pi$. Then the simple closed curve $w = f(\rho\, e^{i\theta})$, $w = u + iv$, encloses a region whose area is given by

$$\int_0^{2\pi} u\, dv = \int_0^{2\pi} u \frac{\partial v}{\partial \theta}\, d\theta = \int_0^{2\pi} \left(\frac{w + \bar{w}}{2}\right) \frac{\partial}{\partial \theta}\left(\frac{w - \bar{w}}{2i}\right) d\theta$$

$$= \int_0^{2\pi} \left[\frac{\rho\, e^{i\theta} + \rho\, e^{-i\theta}}{2} + \sum_{n=1}^{\infty} \frac{b_n\, e^{-in\theta} + \bar{b}_n\, e^{in\theta}}{2\rho^n}\right] \times$$

$$\left[\frac{\rho\, e^{i\theta} + \rho\, e^{-i\theta}}{2} - \sum_{n=1}^{\infty} \frac{nb_n\, e^{-in\theta} + n\bar{b}_n\, e^{in\theta}}{2\rho^n}\right] d\theta = \pi\left[\rho^2 - \sum_{n=1}^{\infty} \frac{n|b_n|^2}{\rho^{2n}}\right] > 0.$$

Since the difference in the square brackets is positive for any $\rho > 1$, taking $\rho \downarrow 1$, we obtain (3.1.1). ∎

From this theorem we obtain

$$\sum_{n=1}^{\infty} n|b_n|^2 \leq 1, \tag{3.1.2}$$

where equality holds for the function $g_\theta(z) = z + \dfrac{e^{i\theta}}{z}$, $\theta \in \mathbb{R}$. The following conclusions from (3.1.2) are obvious:

(i) The inequality $|b_1| \leq 1$ holds for $b_2 = b_3 = \cdots = 0$, where the equality is true for $b_1 = e^{i\theta}$, i.e., if $f(\zeta) = \zeta + e^{i\theta}/\zeta$. This function is in the class Σ since it maps $|\zeta| > 1$ onto the w-plane minus a rectilinear cut from $w = 2e^{i\theta/2}$ to $w = -2e^{i\theta/2}$. Goluzin [1936] has shown that $|b_2| \leq 2/3$ for $b_1 = b_3 = \cdots = 0$, where equality holds for $w = f(\zeta) = \zeta\left(1 + e^{i\theta}/\zeta^3\right)^{2/3} \in \Sigma$ (see also Schiffer [1938b]).

(ii) Let $f(z) = z + \sum_{n=1}^{\infty} a_n z^n$ be any function in the class $\mathcal{S}$. Then the function $f_1(z) = \sqrt{f(z^2)} = z + \frac{1}{2}a_2 z^2 + \cdots$ is also in the class $\mathcal{S}$, because $z' \mapsto z^2$ transforms the unit circle $E = \{z : |z| < 1\}$ into a two-sheeted circle with a branch at $z' = 0$, and the function $w = f(z')$ maps this domain onto a region with a branch at $w = 0$. Since $\sqrt{w}$ reverses this branching, we have

$$w = f(\zeta) = \frac{1}{f_1(1/\zeta)} = \left[\frac{1}{\zeta}\left(1 + \frac{a_2}{2}\frac{1}{\zeta^2} + \cdots\right)\right]^{-1}$$

$$= \zeta\left(1 - \frac{a_2}{2}\frac{1}{\zeta^2} + \cdots\right) = \zeta - \frac{a_2}{2}\frac{1}{\zeta} + \cdots \in \Sigma,$$

which, in view of (3.1.2) yields $|a_2/2| \leq 1$, or

$$|a_2| \leq 2. \tag{3.1.3}$$

This result is due to Bieberbach [1916], and is known as the *Bieberbach inequality*. Equality holds in (3.1.3) for $f(\zeta) = \zeta - \dfrac{e^{i\theta}}{\zeta}$, i.e., for

$$w = f(z) = \frac{z}{(1 - e^{i\theta}z)^2} = z + 2\,e^{i\theta}\,z^2 + \cdots \in \mathcal{S}. \tag{3.1.4}$$

This functions maps the unit circle E onto the w-plane minus the half-line from the point $w = -\frac{1}{4}e^{i\theta}$ to the origin $w = 0$ inclusive, i.e., the slit $(-\frac{1}{4}e^{i\theta}, 0]$.

(iii) If w is not in $f(E)$, then let $h(z) = \dfrac{wf(z)}{w - f(z)}$. Then $h(z)$ is injective on the unit disk and $h(z) = z + (a_2 + w^{-1})z^2 + \cdots$. Then the Bieberbach's inequality (3.1.3) gives $|a_2 + w^{-1}| \leq 2$, which implies that $|w| \geq \frac{1}{4}$ and Koebe's 1/4-theorem follows, which is proved below (Theorem 3.4.1).

(iv) Let $f \in \mathcal{S}$. Then $|a_3 - a_2^2| \leq 1$. Also, if f is an *odd function*, then $|a_3| \leq 1$, with equality if and only if $f(z) = \dfrac{z}{1 + e^{2i\theta}z^2}$, $\theta \in \mathbb{R}$. For more on odd functions, see §3.1.2.

3.1.1. Gronwall's Inequality. We will present three versions, as follows.

Theorem 3.1.2. *Let $f : [0, T] \mapsto \mathbb{R}$ be differentiable on $\mathbb{R}$, and suppose f satisfies the inequality $f'(t) \leq f(t)g(t) + h(t)$, where g is continuous and h is locally integrable. If $G(t) = \int_0^t g(x)\,dx$, then*

$$f(t) \leq f(0)\,e^{G(t)} + \int_0^t e^{G(t)-G(s)}\,h(s)\,ds. \tag{3.1.5}$$

PROOF. Since $\dfrac{d}{dt}e^{-G(t)}f(t) = e^{-G(t)}\dfrac{d}{dt}f(t) + f(t)\dfrac{d}{dt}e^{-G(t)}$, and since $f(t)\dfrac{d}{dt}e^{-G(t)} = -f(t)\,e^{-G(t)}\dfrac{d}{dt}G(t) = -f(t)\,e^{-G(t)}g(t)$, we have

$$\frac{d}{dt}e^{-G(t)}f(t) \leq e^{-G(t)}[g(t)f(t) + h(t)] - g(t)f(t)\,e^{-G(t)} = e^{-G(t)}h(t).$$

Then on integration we get

$$\int_0^t \frac{d}{ds} e^{-G(s)} f(s)\, ds \leq e^{-G(t)} f(t) - f(0)\, e^{-G(0)} \leq \int_0^t e^{-G(s)} h(s)\, ds.$$

Since $G(0) = 0$, the right part of the above inequality yields (3.1.5), which is the *first version* of Gronwall's inequality. ∎

Theorem 3.1.3. *Let $f, g, h : [0, T] \mapsto \mathbb{R}$, where f and h are continuous and g is locally integrable and $h \geq 0$. Suppose f satisfies the integral inequality $f(t) \leq g(t) + \int_0^t h(s)f(s)\, ds$, $H(t) = \int_0^t h(s)\, ds$. Then*

$$f(t) \leq g(t) + \int_0^t g(s)h(s)\, e^{H(t)-H(s)}\, ds. \tag{3.1.6}$$

PROOF. Define a function $y(t) = e^{-H(t)} \int_0^t h(s)f(s)\, ds$. Then

$$\frac{d}{dt} y(t) = e^{-H(t)} h(t)f(t) - h(t)\, e^{-H(t)} \int_0^t h(s)f(s)\, ds$$

$$= h(t)\, e^{-H(t)} \left(f(t) - \int_0^t h(s)f(s)\, ds \right).$$

Since $h(t) \geq 0$, and $\frac{d}{dt} y(t) \leq g(t)h(t)\, e^{-H(t)}$, and $y(0) = 0$, integrating both sides from 0 to t we get

$$y(t) = e^{-H(t)} \int_0^t h(s)f(s)\, ds \leq \int_0^t g(s)h(s)\, e^{-H(s)}\, ds,$$

which, after rearranging the terms, gives (3.1.6). ∎

Theorem 3.1.4. *Let $f : [0, T] \mapsto \mathbb{R}$ be integrable and differentiable on $[0, T]$. Suppose $g, h : [0, T] \mapsto \mathbb{R}$ are integrable so that $\frac{d}{dt} f(t) \leq g(t)f(t) + h(t)$ for almost every $t \in [0, T]$. Let $G(t) = \int_0^t g(s)\, ds$. Then for almost every $t \in [0, T]$*

$$f(t) \leq f(0)\, e^{G(t)} + \int_0^t e^{G(t)-G(s)} h(s)\, ds. \tag{3.1.7}$$

PROOF. The inequality (3.1.7) follows from the two lemmas given below.

Lemma 3.1.1. *(Lebesgue differentiation theorem) Let $f : \mathbb{R} \mapsto \mathbb{R}$ be Lebesgue integrable. Then for almost every $x \in \mathbb{R}$, $F(x) = \int_a^x f(t)\, dt$ is differentiable. Also, for such x, we have $F'(x) = f(x)$.*

Lemma 3.1.2. *(Generalization of the second fundamental theorem of calculus) If $f : \mathbb{R} \mapsto \mathbb{R}$ is Lebesgue integrable, and for almost every $x \in \mathbb{R}$,* *there exists a function F such that $F'(x) = f(x)$, then $\int_a^b f(t)\,dt = F(b) - F(a)$.*

3.1.2 Odd Functions. For $f = z + \sum\limits_{n=2}^{\infty} a_n z^n \in \mathcal{S}$, consider the function

$$f_1(z) = \sqrt{f(z^2)} = \sum_{n=1}^{\infty} c_{2n-1} z^{2n-1}, \quad c_1 = 1, \tag{3.1.8}$$

This is an odd function in $\mathcal{S}$, and we find that

$$a_n = c_1 c_{2n-1} + \cdots + c_{2n-1} c_1. \tag{3.1.9}$$

If f is the Koebe function $z/(1-z)^2$, then

$$f_1(z) = \frac{z}{1-z^2} = z + \sum_{n=2}^{\infty} z^{2n-1}.$$

Littlewood and Paley [1932] proved that there is a constant $C \leq 14$ such that for all odd functions in $\mathcal{S}$, $|c_{2n-1}| \leq C$ for $n = 1, 2, \ldots$. However, in a footnote they remarked, "No doubt the true bound is given by $C = 1$." If one looks at Eq (3.1.9), it is obvious that their conjecture would imply the Bieberbach conjecture! However, Hayman proved that $\lim\limits_{n \to \infty} |c_{2n-1}|$ exists for every odd function $f_1 \in \mathcal{S}$, and this limit is smaller than 1 unless f_1 is the root transform of a Koebe function; later Milin [1965] proved that $C = 1.14$. These results seem to support the Littlewood-Paley conjecture. But Fekete and Szegö [1933] disproved this conjecture, by proving that there exist odd functions in $\mathcal{S}$ for which $|c_5| > 1$.

It was observed by Robertson [1936] that using (3.1.9) and the Cauchy-Schwarz inequality, the Bieberbach conjecture would already follow from the inequality

$$\sum_{k=1}^{n} |c_{2k-1}|^2 \leq n. \tag{3.1.10}$$

This inequality is known as the *Robertson conjecture for odd functions in $\mathcal{S}$*.

3.2 Bieberbach Conjecture

Let $f(z) = z + a_2 z^2 + \cdots$ be defined injectively on the unit disk. Since the function $f(z)/z$ nowhere vanishes on the unit disk, we take its square root function $\psi(z) = \left\{ \dfrac{f(z)}{z} \right\}^{1/2}$, and choose its branch such that $\psi(0) = 1$. Let $h(z) = z\psi(z)^2$. Then $h(z)^2 = f(z^2)$, and so h is injective. For the

Laurent series expansion of $f(z) = \dfrac{1}{h(z)} = z^{-1} - \frac{1}{2}a_2 z + \cdots$, the Gronwall's area theorem (3.1.2) yields the *Bieberbach's inequality* (3.1.3). If w is not in $f(D)$, then let $h(z) = \dfrac{wf(z)}{w - f(z)}$. Then $h(z)$ is injective on the unit disk and $h(z) = z + (a_2 + w^{-1})z^2 + \cdots$. Then the Bieberbach's inequality (3.1.3) gives $|a_2 + w^{-1}| \leq 2$, which implies that $|w| \geq \frac{1}{4}$ and Koebe's 1/4-theorem follows, which is stated and proved as Theorem 3.4.1 below.

In fact, Bieberbach [1914] proved that of all the functions $f(z)$, analytic on D with $f(0) = 0$ and $f'(0) = 1$, the one which minimizes the integral $I_A = \displaystyle\iint_{\partial D} |f'(z)|^2\, dx\, dy$, maps D onto the unit disk. Gronwall [1914] proved the *area theorem* (3.1.2) for functions $g \in \Sigma$. This was the first conformal inequality, obtained at the beginning of the last century, which shows that the area of the complement of the range of the function $g \in \Sigma$ is nonnegative.

The footnote on page 940 in Bieberbach [1916] reads: "Daß $k_n \geq n$ zeigt das Beispiel $\sum nz^n$. Vielleicht ist überhaupt $k_n = n$." (Translation: "The example $\sum nz^n$ shows that $k_n \geq n$. Perhaps $k_n = n$ is generally valid.") This led to the *Bieberbach conjecture*:

$$|a_n| \leq n. \tag{3.2.1}$$

To understand this famous conjecture that occupied mathematicians for the next sixty-eight years until it was proof by de Branges [1984], we follow Zorn [1986] and explain this conjecture in very simple mathematical terms. The identity map $f(z) = z$ is the most basic, yet trivial, mapping in the class $\mathcal{S}$, since the normalization implies that every function in $\mathcal{S}$ agrees with the identity up to order one at the origin. The Koebe function $K(z) = z/(1-z)^2$ is the most important nontrivial function in $\mathcal{S}$. Since $K(z) = z\dfrac{d}{dz}\left[\dfrac{1}{1-z}\right]$, we can write $K(z)$ as the power series $K(z) = z + 2z^2 + 3z^3 + \cdots$, which converges for every z in a domain D in the complex plane. The Bieberbach conjecture assumes that $K(z)$ is the other extreme from the identity functions: every coefficient of $K(z)$ is *as large as possible*. The Koebe mapping

$$K(z) = \frac{z}{(1-z)^2} = \frac{1}{4}\left[\left(\frac{1+z}{1-z}\right)^2 - 1\right] = \sum_{n=1}^{\infty} k_n z^n. \tag{3.2.2}$$

is composed of the mappings $\zeta = (1+z)/(1-z)$; $t = s^2$; and $w = \dfrac{1}{4}(t-1)$, in that order, where the first one is a linear fractional transformation that maps D univalently on the right-half of the ζ-plane; the second is one-to-one (injective) when restricted to the right half-plane, its image being the entire t-plane minus the non-positive real axis; and the third is a translation one unit to the left followed by a dilation factor $\frac{1}{4}$.

The question arises: is the Koebe function $K(z)$ unique among the class

S in the sense of having the largest coefficients? The answer is 'essentially, but not quite'. For example, if $f \in S$ and θ is any real number, consider the functions $f_\theta(z) = e^{-i\theta} f\left(e^{i\theta} z\right)$, which are (counterclockwise) rotations of f about $z = 0$ by θ radians. If f has the power series expansion $f(z) = z + \sum_{n=2}^{\infty} a_n z^n$, then $f_\theta(z) = z + \sum_{n=2}^{\infty} b_n z^n$, where $b_n = a_n e^{i(n-1)\theta}$. Since $|e^{i\theta}| = 1$, we have $|a_n| = |b_n|$. Thus, the Bieberbach conjecture states precisely that the extreme function $K(z)$ is unique up to rotations, i.e., if $f(z) = z + \sum_{n=2}^{\infty} a_n z^n \in S$, then $|a_n| \leq n$ for every $n \geq 2$, and if , for any n, $|a_n| = n$, then f is a rotation of the Koebe function.

Next, consider the optimization problem: for each $n \geq 2$, find a function $f \in S$ for which $|a_n|$ is the largest. We may ask whether there is, for each n, an absolute bound on $|a_n|$ as f ranges over S, and whether, if the nth coefficients of $f \in S$ are bounded, there is a function $f \in S$ whose nth coefficient attains this bound. Although the answer to both these questions is in the affirmative, yet the question is why so, without assuming the Bieberbach conjecture. Why are the nth coefficients of $f \in S$ bounded? By the Cauchy formula for the radius of convergence of the power series, we have $\lim\limits_{n \to \infty} \dfrac{a_n}{r^n} = 0$ for every $r > 1$, i.e., $|a_n| < r^n$ asymptotically as $n \to \infty$. Since analyticity does not imply the boundedness of the coefficients of $f \in S$, we will consider a relationship between univalence of f and the size of a_n. Firstly, note that the complex monomial z^n is an n-to-one function in D. In fact, if $f(z) = z + a_2 z^2 + \cdots + a_n z^n$ is a univalent polynomial in D, then $|a_n| \leq 1/n$. However, the converse is false, although $f(z) = z + a_n z^n$ is a univalent polynomial in D if and only if $|a_n| \leq 1/n$. (For a proof see Exercise 1.10.8). Since S is a compact (and dense) subset of the space of all analytic functions on D, the uniform convergence on compact subsets of D justifies the boundedness of every coefficient in the power series expansion of f. The functional T which associates to an analytic function on D its nth coefficient is continuous, i.e., $T(f) = T\left(\sum_n a_n z^n\right) = a_n$. Thus, T attains a maximum modulus somewhere on S.

Unbeknownst to him, Bieberbach [1916] proved Gronwall's area theorem and obtained the first coefficient bound (3.1.3) in the class S. He derived this result from (3.1.2), which implies $|b_1| \leq 1$ by a simple transformation to S. Since by (3.1.2), $|b_1| = 1$ yields $b_2 = b_3 = \cdots = 0$, it follows that $a_2 = 2$ occurs if and only if f is the Koebe function $K(z)$ defined by (3.2.2). As shown above, the image domain of the function $K(z)$ is a plane that is radially slit on the negative real axis from $-1/4$ to the point at infinity.

3.3 Lebedev-Milin's Area Theorem

Theorem 3.3.1. (area theorem, Lebedev and Milin [1951]) *Let $\bar{E}^*$ denote the domain $E^* \cup \partial E^*$, where $\bar{E}^* = \{g(E^*)\backslash \mathbb{C}\}$, $g \in \Sigma$, and let $h(w)$ be a regular* *function in $\bar{E}^*$. If $h(g(z)) = \sum\limits_{k=-\infty}^{\infty} c_k z^{-k}$ is a non-constant regular function defined in $\bar{E}^*$, then*

$$\sum_{k=-\infty}^{\infty} k|c_k|^2 \le 1, \tag{3.3.1}$$

where the equality holds only if the area of $\bar{E}^*$ is zero.

This is a general area theorem in the class Σ. Let $g \in \Sigma$ and let

$$\log \frac{g(z) - g(\zeta)}{z - \zeta} = - \sum_{m,n=1}^{\infty} \gamma_{mn} z^{-m} \zeta^{-n}, \quad z, \zeta \in g(D). \tag{3.3.2}$$

Then by a proper choice of a regular function in $\bar{E}^*$, the inequality (3.3.1) can be written as

$$\sum_{n=1}^{\infty} n \left| \sum_{m=1}^{\infty} \gamma_{mn} x_m \right|^2 \le \sum_{m=1}^{\infty} \frac{1}{m} |x_m|^2, \tag{3.3.3}$$

where x_m are any numbers not all zero such that $\limsup\limits_{m \to \infty} |x_m|^{1/m} < 1$. For more general area theorems, see Nehari [1969]. For example, using Cauchy's inequality (1.6.7), we obtain from (3.3.3)

$$\left| \sum_{m,n=1}^{\infty} n \gamma_{mn} x_m x'_n \right|^2 \le \sum_{m=1}^{\infty} \frac{1}{m} |x_m|^2 \sum_{n=1}^{\infty} \frac{1}{n} |x'_n|^2, \tag{3.3.4}$$

where x_m and x'_n are such that both series on the right side of (3.3.4) converge. For example, if we take $x_m = z^{-m}, x'_n = \zeta^{-n}, |z| > 1, |\zeta| > 1$ in (3.3.4), we obtain the *chord-distortion theorem*:

$$\left| \log \frac{g(z) - g(\zeta)}{z - \zeta} \right|^2 \le \log \frac{|z|^2}{|z|^2 - 1} \log \frac{|\zeta|^2}{|\zeta|^2 - 1}. \tag{3.3.5}$$

Note that the inequalities (3.3.3) and (3.3.4) are of the form of Grunsky inequalities, which are discussed in §3.5.

Theorem 3.3.2. (Goluzin's theorem) *If $g(\zeta) \in \Sigma$, then for any two points ζ_1, ζ_2 on the circle $|\zeta| = \rho > 1$ the inequality*

$$\left| \frac{g(\zeta_1) - g(\zeta_2)}{\zeta_1 - \zeta_2} \right| \ge 1 - \frac{1}{\rho^2} \tag{3.3.6}$$

holds, such that the equality is attained only for the functions $g(\zeta) = \zeta + c + \dfrac{e^{2i\phi}}{\zeta}$, where c is a constant and $\phi = \frac{1}{2} (\arg\{\zeta_1\} + \arg\{\zeta_2\})$.

There are various generalizations of the inequality (3.3.6), which are available in Goluzin [1969].

3.4 Koebe's Theorem

For the class S of functions $f(z) = z + a_2 z^2 + \cdots$, which are analytic and univalent in the unit disk $E = \{z : |z| < 1\}$, the following distortion inequalities hold for all z, $0 < |z| < 1$:

$$\frac{1-|z|}{(1+|z|)^3} \leq |f'(z)| \leq \frac{1+|z|}{(1-|z|)^3}, \tag{3.4.1}$$

$$\frac{|z|}{(1+|z|)^2} \leq |f(z)| \leq \frac{|z|}{(1-|z|)^2}, \tag{3.4.2}$$

$$\frac{1-|z|}{1+|z|} \leq \left|\frac{zf'(z)}{f(z)}\right| \leq \frac{1+|z|}{1-|z|}, \tag{3.4.3}$$

where the lower bounds are realized only by the functions $f_\alpha(z) = \dfrac{z}{(1+e^{-i\alpha}z)^2}$, while the upper bounds only by the functions $f_{\pi+\alpha}(z) = \dfrac{z}{(1-e^{-i\alpha}z)^2}$, where $\alpha = \arg\{z\}$.

The functions $f_\alpha(z)$, $0 \leq \alpha \leq 2\pi$, are known as the Koebe functions which map the unit disk $|z| < 1$ onto the w-plane with a slit along the ray $\arg\{w\} = \alpha$, $|w| > 1/4$. These functions are extremal in various problems in the geometric theory of univalent functions. This leads to the following result.

Theorem 3.4.1. (Koebe's 1/4-theorem) *The image domain of the unit disk $|z| < 1$ under the mapping $w = f(z) \in S$ always contains the disk $|w| < 1/4$, and the point $w = e^{i\alpha}/4$ lies on the boundary of the domain only for $f(z) = f_\alpha(z)$.*

In simple terms, this theorem states that the image of an injective analytic function $f : E \mapsto \mathbb{C}$ from the unit disk E onto a subset of the complex plane contains the disk $C(f(0), |f'(0)|/4)$.

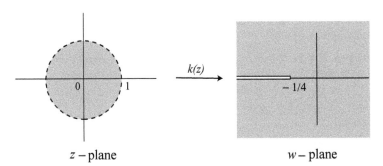

$$z-\text{plane} \qquad\qquad w-\text{plane}$$

Figure 3.4.1 Koebe's 1/4-theorem.

The inequalities (3.4.1), (3.4.2) and (3.4.3) are derived by considering the ranges for the functionals $\log f'(z)$, $\log \dfrac{f(z)}{z}$, and $\log \dfrac{zf'(z)}{f(z)}$, respectively, on $\mathcal{S}$. The inequalities (3.4.2) and (3.4.3) follow from (3.4.1) (see Jenkins [1964]).

The Koebe function is defined by

$$K(z) = \frac{z}{(1-z)^2} = \sum_{n=1}^{\infty} n z^n. \tag{3.4.4}$$

This function shows that the constant $1/4$ in the Koebe theorem cannot be improved, because the image domain does not contain the point $z = -1/4$ and therefore cannot contain any disk $C(0, \rho)$, $\rho \geq 1/4$. The rotated Koebe function is

$$K_\alpha(z) = \frac{z}{(1-\alpha z)^2} = \sum_{n=1}^{\infty} n \alpha^{n-1} z^n, \tag{3.4.5}$$

where $\alpha \in \mathbb{C}$, $|\alpha| = 1$ (α is taken as $e^{i\theta}$). The Koebe functions $K(z)$ and $K_\alpha(z)$ are univalent (*schlicht*, i.e., analytic and one-to-one) with $K(0) = 0$, $K'(0) = 1$ and $K_\alpha(0) = 0$, $K'_\alpha(0) = 1$.

PROOF OF THE $1/4$-THEOREM. Apply an affine mapping and assume that $f(0) = 0$, $f'(0) = 1$. Then $f(z) \equiv K(z) = z + a_2 z^2 + \cdots$. If $w \notin f(D)$, then $h(w) = \dfrac{wf(z)}{w - f(z)} = z + \left(a_2 + w^{-1}\right) z^2 + \cdots$ is univalent in $|z| < 1$. Applying the coefficient inequality (3.1.3) to f and h we get

$$|w|^{-1} \leq |a_2| + |a_2 + w^{-1}| \leq 4,$$

which gives $|w| \geq 1/4$. ∎

The Koebe theorem has the following geometric significance: The disk $C(0, 1/4)$ is covered by the image $f(E)$ of the unit disk E for all functions $f \in \mathcal{S}$, i.e., $C(0, 1/4) = \bigcap_{f \in \mathcal{S}} f(E)$. The disk $C(0, 1/4)$ is called the *Koebe disk*, and $1/4$ the *Koebe constant* of the class $\mathcal{S}$.

For functions regular and bounded in the unit disk, the following boundary-distortion theorem holds.

Theorem 3.4.2. (Löwner's theorem) *Let $f(z)$ be regular in the unit disk $|z| < 1$ with $f(0) = 0$, $|f(z)| < 1$ and $|f(z)| = 1$ on an arc γ of the circle $|z| = 1$. Then the length of the image of γ is not smaller than the length of γ, and these lengths are equal only for the functions $f(z) = e^{i\alpha} z$, where α is a real parameter.*

Given $f(z)$, the level curves of f are the loci $\{z : |f(z)| = c\}$, where c is a constant. The pattern of these curves tells us how large the values of $f(z)$ are, but completely ignores the direction angle associated with the polar form of $f(z)$. The bounds for families of level curves defined by $u(x, y) = $ const and $v(x, y) = $ const are available in Kulshrestha [1956].

3.5 Grunsky Inequalities

These inequalities, developed by Grunsky [1939], provide bounds for a single univalent function on the unit disk E or for a pair of univalent functions on E and its complement. In fact, they give information on the bounds of coefficients of the logarithm of a univalent function. Consider the function $f(z) \in S$ with $f(0) = 0$ and $f'(0) = 1$, and the function $g(z) = 1/f(z) \in \Sigma$ which non-vanishing univalent function on $|z| > 1$ with a simple pole at the point at infinity with residue 1. Recall that $f(z) = 1/g(z)$, and there exists a one-to-one correspondence between these two classes of functions.

Let $g \in \Sigma$. Then the function defined by (3.3.2), i.e.,

$$\log \frac{g(z) - g(\zeta)}{z - \zeta} = -\sum_{m=1}^{\infty} \sum_{n=1}^{\infty} \gamma_{mn} z^{-m} \zeta^{-n}, \quad |z| > 1, \ |\zeta| > 1, \tag{3.5.1}$$

is analytic for all $|z| > 1$ and $|\zeta| > 1$. The *Grunsky matrix* (γ_{mn}) of g is defined by (3.5.1); this matrix is symmetric and its elements are called the *Grunsky coefficient* of g. Inverting g to f, we get

$$\log \frac{g(z^{-1}) - g(\zeta^{-1})}{z^{-1} - \zeta^{-1}} = \log \frac{f(z) - f(\zeta)}{z - \zeta} - \log \frac{f(z)}{z} - \log \frac{f(\zeta)}{\zeta}. \tag{3.5.2}$$

If

$$\log \frac{f(z) - f(\zeta)}{z - \zeta} = -\sum_{m=0}^{\infty} \sum_{n=0}^{\infty} d_{mn} z^m \zeta^n, \tag{3.5.3}$$

then for $m, n > 0$ we have $d_{mn} = \gamma_{mn}$, and $d_{0n} = d_{n0}$ is given by

$$\log \frac{f(z)}{z} = \sum_{n>0} d_{0n} z^n, \quad d_{00} = 0. \tag{3.5.4}$$

For $f \in S$ and for any N complex numbers $\lambda_1, \ldots, \lambda_N$, $N \in \mathbb{Z}^+$, the following inequality, known as the *Grunsky inequality*, holds:

$$\left| \sum_{m=1}^{N} \sum_{n=1}^{N} \gamma_{mn} \lambda_m \lambda_n \right| \le \sum_{m=1}^{N} \frac{|\lambda_m|^2}{m}. \tag{3.5.5}$$

The inequality (3.5.5), which is known as the *weak Grunsky inequality*, is equivalent to the *strong Grunsky inequality*:

$$\sum_{n=1}^{\infty} n \left| \sum_{m=1}^{N} \gamma_{mn} \lambda_m \right|^2 \le \sum_{m=1}^{N} \frac{|\lambda_m|^2}{m}. \tag{3.5.6}$$

These two inequalities are also equivalent to the *generalized weak Grunsky inequality*:

$$\left|\sum_{m=1}^{N}\sum_{n=1}^{N}\gamma_{mn}\lambda_m\mu_n\right|^2 \le \sum_{m=1}^{N}\frac{|\lambda_m|^2}{m}\sum_{n=1}^{N}\frac{|\mu_n|^2}{n}, \tag{3.5.7}$$

where $\{\lambda_1,\dots,\lambda_N\}$ and $\{\mu_1,\dots,\mu_N\}$ are finite sequences of arbitrary complex numbers. The inequality (3.5.7) leads to

$$\left|\sum_{m=1}^{\infty}\sum_{n=1}^{\infty}\gamma_{mn}\sum_{i=1}^{N}\lambda_i z_i^{-n}\sum_{j=1}^{N}\mu_j\zeta_j^{-m}\right|^2$$

$$\left\{\sum_{m=1}^{\infty}\frac{1}{m}\left|\sum_{i=1}^{N}\lambda_i z_i^{-n}\right|^2\right\}\left\{\sum_{n=1}^{\infty}\frac{1}{n}\left|\sum_{j=1}^{N}\mu_j\zeta_j^{-m}\right|^2\right\}, \tag{3.5.8}$$

where λ_i and μ_j $(i,j=1,2,\dots,N)$ are $2N$ arbitrary complex numbers, and z_i,ζ_j $(i,j=1,2,\dots)$ are $2N$ points outside E. An equivalent form of inequality (3.5.8) is

$$\left|\sum_{i=1}^{N}\sum_{j=1}^{N}\lambda_i\mu_j\log\frac{g(z_i)-g(\zeta_j)}{z_i-\zeta_j}\right|^2$$

$$\le\left\{-\sum_{i=1}^{N}\sum_{j=1}^{N}\lambda_i\lambda_j\log\left(1-\frac{1}{z_i\bar{z}_j}\right)\right\}\left\{-\sum_{i=1}^{N}\sum_{j=1}^{N}\mu_i\mu_j\log\left(1-\frac{1}{\zeta_i\bar{\zeta}_j}\right)\right\}. \tag{3.5.9}$$

Grunsky inequality was used by Pederson [1968] and Ozawa [1969] to prove independently that $|a_6|\le 6$.

Consider a univalent function $f\in\mathcal{S}$ which maps the unit disk $|z|<1$ onto a starlike domain for $r\le\tanh\pi/4$. The largest value of r for which this is true is called the *radius of starlikeness* of the function f.

Theorem 3.5.1. (Grunsky's theorem) *Let $f\in\mathcal{S}$ defined on E such that $f(0)=0$. Then for all $r\le\tanh\pi/4$, the image of the disk $|z|<r$ is starlike with respect to the origin, i.e., it is invariant under multiplication by real numbers in $(0,1)$.*

3.5.1. Grunsky Bounds. The following result holds:

Theorem 3.5.2. *If $f(z)\in\mathcal{S}$ is univalent on D with $f(0)=0$, then*

$$\left|\log\frac{zf'(z)}{f(z)}\right|\le\log\frac{1+|z|}{1-|z|}. \tag{3.5.10}$$

If we take the real and imaginary parts of the inequality, we obtain

$$\left|\frac{zf'(z)}{f(z)}\right|\le\frac{1+|z|}{1-|z|},\quad\left|\arg\left\{\frac{zf'(z)}{f(z)}\right\}\right|\le\frac{1+|z|}{1-|z|}. \tag{3.5.11}$$

For fixed z, the inequalities (3.5.11) are sharp for the Koebe functions

$$K_\alpha(\zeta) = \frac{\zeta}{(1 - \bar{\alpha}\zeta)^2}, \quad |\alpha| = 1. \tag{3.5.12}$$

PROOF. This proof is based on the Goluzin inequalities [1939] for the Grunsky matrix, which uses the Löwner equation. For a function f which is normalized in D, we obtain the inequality (3.5.11) for f by taking $g(\zeta) = f\left(\zeta^{-2}\right)^{-1/2}$ where $z = \zeta^{-2}$, as follows: Since f is starlike on $|z| < 1$ if and only if $\Re\left\{\dfrac{zf'(z)}{f(z)}\right\} \geq 0$ for $|z| < r$, we have

$$\left| \arg\left\{ \frac{zf'(z)}{f(z)} \right\} \right| \leq \frac{\pi}{2}. \tag{3.5.13}$$

But from the Grunsky's inequality we have

$$\left| \arg\left\{ \frac{zf'(z)}{f(z)} \right\} \right| \leq \log\frac{1 + |z|}{1 - |z|}. \tag{3.5.14}$$

Thus the inequality (3.5.1) holds at z if $\log\dfrac{1 + |z|}{1 - |z|} \leq \dfrac{\pi}{2}$, i.e., if $|z| \leq \tanh(\pi/4)$. ∎

3.6 Polynomial Area Theorem

A direct generalization of the area theorem is known as the *polynomial area theorem*. Let the function $g(z) \in \Sigma$ map the exterior of the unit disc $|z| > 1$ onto the complement of a compact set E. The functions g is called a *full mapping* if E has measure zero. Denote the subclass of full mappings by $\tilde{\Sigma}$. The area theorem (Theorem 3.1.1) corresponds to the polynomial $p(w) = w$, where $w = h = g(z) \in \Sigma$.

Theorem 3.6.1. (Polynomial area theorem, Duren [1983]) *Let $g \in \Sigma$, let p be an arbitrary nonconstant polynomial of degree N, and let*

$$p\left(g(z)\right) = \sum_{k=-N}^{\infty} c_k z^{-k}, \quad |z| > 1. \tag{3.6.1}$$

Then

$$\sum_{k=-N}^{\infty} k|c_k|^2 \leq 0, \tag{3.6.2}$$

where equality holds if and only if $g \in \tilde{\Sigma}$. It is obvious from (3.6.2) that

$$\sum_{k=1}^{\infty} k|c_k|^2 \leq \sum_{k=1}^{N} k|c_{-k}|^2. \tag{3.6.3}$$

PROOF. we will use the following Lemma which is form of Green's theorem and is easily proved using the Cauchy-Riemann equations (1.3.4). ■

Lemma 3.6.1. *Let C be a smooth Jordan curve bounding a domain D, and $w = u + iv$. If ϕ and ψ are analytic in $\bar{D}$, then*

$$\frac{1}{2i} \int_C \overline{\phi(w)}\, \psi(w)\, dw = \iint_D \overline{\phi'(w)}\, \psi(w)\, du\, dv. \tag{3.6.4}$$

PROOF OF THEOREM 3.6.1. Let C_r denote the image under g of the circle $|z| = r > 1$, and let $E_r = \text{Int}\{C_r\}$. Then by Lemma 3.6.1,

$$0 \le \iint_{E_r} |p'(w)|^2\, du\, dv = \frac{1}{2i} \int_{C_r} \overline{p(w)}\, p'(w)\, dw$$

$$= \frac{1}{2i} \int_{|z|=r} \overline{p(g(z))}\, p'(g(z)) g'(z)\, dz = -\pi \sum_{k=-N}^{\infty} k|c_k|^2 r^{-2k}. \tag{3.6.5}$$

Let $r \to 1$; then (3.6.5) gives

$$\sum_{k=-N}^{\infty} k|c_k|^2 = -\frac{1}{\pi} \iint_E |p'(w)|^2\, du\, dv \le 0, \tag{3.6.6}$$

where equality holds if and only if E has measure zero, i.e., if and only if $g \in \tilde{\Sigma}$. ■

This proof uses the fact that p is analytic on E. A more general result, due to Duren [1983] is as follows.

Theorem 3.6.2. *Let $g \in \Sigma$, and let f be a nonconstant function analytic in some domain E_ρ, $\rho > 1$. If $f(g(z)) = \sum\limits_{k=-\infty}^{\infty} c_k z^{-k}$, $1 < |z| < \rho$, then*

$$\sum_{k=1}^{\infty} k|c_k|^2 \le \sum_{k=1}^{\infty} k|c_{-k}|^2, \tag{3.6.7}$$

where the equality holds if and only if the $g \in \tilde{\Sigma}$, provided the right-hand series is convergent.

3.7. Distortion Theorems

The bilinear transformation $z \longmapsto \dfrac{z + \zeta}{1 + \zeta z}$ maps the unit disk E onto itself. Hence, if $f(z)$, $z \in E$, is univalent, so is the function $f\left(\dfrac{z + \zeta}{1 + \zeta z}\right)$. This

function is not in the class $\mathcal{S}$. To see this, let $g_1(z) = f\left(\dfrac{z+\zeta}{1+\zeta z}\right)$, $g(0) = 0$.
Then

$$g_1'(z) = f'\left(\frac{z+\zeta}{1+\zeta z}\right)\frac{1+\zeta z - (z+\zeta)\zeta}{(1+\zeta z)^2} = f'\left(\frac{z+\zeta}{1+\zeta z}\right)\frac{1-|\zeta|^2}{(1+\zeta z)^2},$$

and $g_1'(0) = f'(\zeta)\,(1-|\zeta|^2) \neq f'(\zeta)$. Hence, $g_1 \notin \mathcal{S}$. But the function

$$g(z) = \frac{f\left(\dfrac{z+\zeta}{1+\bar{\zeta} z}\right) - f(\zeta)}{f'(\zeta)\,(1-|\zeta|^2)} \in \mathcal{S}, \tag{3.7.1}$$

since $g(0) = 0$, and $g'(0) = 1$. This function $g(z)$ is of the form $g(z) = z + \displaystyle\sum_{n=2}^{\infty} a_n\, z^n$, $z \in E$. The coefficient a_2 is called the *order* of the function in the class $\mathcal{S}$, i.e., this order is given by $a_2 = \dfrac{g''(0)}{2!}$. Now,

$$g'(z) = \frac{1}{f'(\zeta)}\frac{f'\left(\dfrac{z+\zeta}{1+\bar{\zeta} z}\right)}{(1+\bar{\zeta} z)^2},$$

$$g''(z) = \frac{1}{f'(\zeta)}\left[\frac{f''\left(\dfrac{z+\zeta}{1+\bar{\zeta} z}\right)\left(\dfrac{1-|\zeta|^2}{(1+\bar{\zeta} z)^2}\right)}{(1+\bar{\zeta} z)^2} - \frac{f'\left(\dfrac{z+\zeta}{1+\bar{\zeta} z}(2\bar{\zeta})1+\bar{\zeta} z\right)}{(1+\bar{\zeta} z)^4}\right],$$

which yields

$$g''(0) = \frac{1}{f'(\zeta)}\left[f''(\zeta)\,(1-|\zeta|^2) - 2\bar{\zeta}\,f'(\zeta)\right],$$

$$a_2 = \frac{g''(0)}{2!} = \frac{(1-|\zeta|^2)\,f''(\zeta)}{2\,f'(\zeta)} - \bar{\zeta}. \tag{3.7.2}$$

Then

$$g(z) = z + \left[\frac{(1-|\zeta|^2)\,f''(\zeta)}{2\,f'(\zeta)} - \bar{\zeta}\right]z^2 + \cdots. \tag{3.7.3}$$

Now, since $|a_2| \leq 2$, we get

$$\left|\frac{(1-|\zeta|^2)\,f''(\zeta)}{2\,f'(\zeta)} - \bar{\zeta}\right| \leq 2.$$

Since $\zeta \in E$, we replace ζ by z and obtain

$$\left|\frac{(1-|z|^2)\,f''(z)}{2\,f'(z)} - \bar{z}\right| \leq 2.$$

If we multiply both sides of this inequality by $2\,|z|$ and divide by $1 - |z|^2 > 0$, we get

$$\left| \frac{z\,f''(z)}{f'(z)} - \frac{2\,|z|^2}{1 - |z|^2} \right| \leq \frac{4\,|z|}{1 - |z|^2}.$$

Then for $|z| = r < 1$,

$$\left| \frac{z\,f''(z)}{f'(z)} - \frac{2\,r^2}{1 - r^2} \right| \leq \frac{4\,r}{1 - r^2}, \tag{3.7.4}$$

which is a circle. This means that the values of $\dfrac{z\,f''(z)}{f'(z)}$, $z \in E$, lie in a circle of radius $\dfrac{4\,r}{1 - r^2}$ and center $\dfrac{2\,r^2}{1 - r^2}$.

Recall that if w is complex, then $\Re w \leq |w|$; so

$$\Re\left\{ \frac{z\,f''(z)}{f'(z)} - \frac{2\,r^2}{1 - r^2} \right\} \leq \left| \frac{z\,f''(z)}{f'(z)} - \frac{2\,r^2}{1 - r^2} \right| \leq \frac{4\,r}{1 - r^2},$$

or

$$-\frac{4\,r}{1 - r^2} \leq \Re\left\{ \frac{z\,f''(z)}{f'(z)} \right\} - \frac{2\,r^2}{1 - r^2} \leq \frac{4\,r}{1 - r^2},$$

i.e.,

$$\frac{2\,r^2 - 4\,r}{1 - r^2} \leq \Re\left\{ \frac{z\,f''(z)}{f'(z)} \right\} \leq \frac{2\,r^2 + 4\,r}{1 - r^2}, \tag{3.7.5}$$

which, in view of the Cauchy-Riemann equations, yields

$$\frac{2\,r^2 - 4\,r}{1 - r^2} \leq \Re\left\{ z\,\frac{d}{dz}\,\log f'(z) \right\} \leq \frac{2\,r^2 + 4\,r}{1 - r^2}.$$

Then

$$\frac{r\,(2\,r - 4)}{1 - r^2} \leq r\,\frac{1}{r}\Re\left\{ \log f'(z) \right\} \leq \frac{r\,(2\,r + 4)}{1 - r^2},$$

or

$$\frac{2\,r - 4}{1 - r^2} \leq \frac{1}{r}\,\log |f'(z)| \leq \frac{2\,r + 4}{1 - r^2}. \tag{3.7.6}$$

Note that using the imaginary parts, instead of the real parts, we get

$$\frac{2\,r - 4}{1 - r^2} \leq \frac{1}{r}\,\arg\{f'(z)\} \leq \frac{2\,r + 4}{1 - r^2}. \tag{3.7.7}$$

If we integrate (3.7.6) from 0 to r and note that $f'(0) = 1$, we obtain

$$\int_0^r \frac{2\,\rho - 4}{1 - \rho^2}\,d\rho \leq \log |f'(z)| \leq \int_0^r \frac{2\,\rho + 4}{1 - \rho^2}\,d\rho,$$

or

$$\log \frac{1-r}{(1+r)^3} \le \log |f'(z)| \le \log \frac{1+r}{(1-r)^3},$$

which gives the *distortion theorem* for $|f'(z)|$:

$$\frac{1-r}{(1+r)^3} \le |f'(z)| \le \frac{1+r}{(1-r)^3}. \qquad (3.7.8)$$

Thus,

$$|f(z)| = \left| \int_0^z f'(\zeta)\, d\zeta \right| \le \int_0^r |g'(z)|\, dr \le \int_0^r \frac{1+\rho}{(1-\rho)^3}\, d\rho = \frac{r}{(1-r)^2}.$$

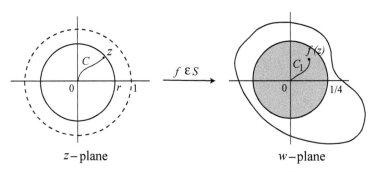

$$f \in S$$

z-plane　　　　　　　　　　w-plane

Figure 3.7.1 Mapping of Line Segment C.

Let $|f(z)| < 1/4$. By Koebe's covering theorem, the line segment C_1 joining 0 to $f(z)$ is the image of some curve C joining 0 to z (see Figure 3.7.1). Thus, $|f(z)| = \int_{C_1} |dw|$, where $C_1 = \int_C |f'(z)|\,|dz|$. Since $\int_C |dz| \geqq \int_C d|z| = r$ (length of the line segment from 0 to z), we get

$$|f(z)| = \int_C |f'(z)|\,|dz| \geqq \int_C |f'(z)\, d|z| \ge \int_0^r \frac{1-\rho}{(1+\rho)^3}\, d\rho = \frac{r}{(1+r)^2}.$$

For $|f(z)| \ge 1/4$, we have $\dfrac{r}{(1+r)^2} < \dfrac{1}{4}$ for $0 \le r < 1$. Hence, the bound $|f(z)| \ge \dfrac{r}{(1+r)^2}$ is true for $f(E)$. This yields to the *distortion theorem* for $|f(z)|$ (see Figure 3.7.2):

$$\frac{r}{(1+r)^2} \le |f(z)| \le \frac{r}{(1-r)^2}. \qquad (3.7.9)$$

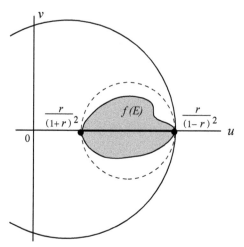

Figure 3.7.2 Distortion Theorem (3.7.9).

Theorem 3.7.1. *If* $f(z) = z + \sum_{n=2}^{\infty} a_n z^n$, *then* $|a_n| < e\,n$.

PROOF. Let $g(z) = \sqrt{f(z^2)}$. Using the area theorem,

$$\iint\limits_{|z|<r} |g'(z)|^2 \, dz \le \frac{\pi\, r^2}{(1-r^2)^2}.$$

On the other hand, if $g(z) = z + \sum_{n=2}^{\infty} b_n z^n$, $|z| < 1$, then

$$\iint\limits_{|z|<r} |g'(z)|^2 \, dx\, dy = \int_0^{2\pi} \int_0^{\rho} |g'(re^{i\theta})|^2 \, r\, dr\, d\theta$$

$$= \int_0^{\rho} r\, dr \int_0^{2\pi} |g'(re^{i\theta})|^2 \, d\theta,$$

where $z = r\,e^{i\theta}$, $0 \le \theta \le 2\pi$, and $0 \le r \le 1$. Recall that, by Parseval's theorem, if $f(z) = z + \sum_{n=2}^{\infty} a_n z^n$, $|z| < 1$, then

$$\frac{1}{2\pi} \int_0^{2\pi} |f(re^{i\theta})| \, d\theta = \sum_{n=0}^{\infty} |a_n|^2 \, r^{2n}.$$

Since in our case $g'(z) = 1 + \sum_{n=2}^{\infty} n^2 b_n^2 z^{n-1}$, we find that

$$\iint_{|z|<r} |g'(z)|^2 \, dz = \left[\int_0^{\rho} r \, dr \right] \left[2\pi \sum_{n=1}^{\infty} n^2 |b_n|^2 r^{2(n-1)} \right]$$

$$= 2\pi \sum_{n=1}^{\infty} n^2 |b_n|^2 \int_0^{\rho} r^{2n-1} \, dr = 2\pi \sum_{n=1}^{\infty} n^2 |b_n|^2 \frac{\rho^{2n}}{2n}$$

$$= \pi \sum_{n=1}^{\infty} n |b_n|^2 \rho^{2n}.$$

Now, since $\sum_{n=1}^{\infty} n |b_n|^2 r^{2n-1} \le \dfrac{r}{(1-r^2)^2}$, we get, by integrating this equality from 0 to ρ,

$$\frac{1}{2} \sum_{n=1}^{\infty} |b_n|^2 \rho^{2n} \le \int_0^{\rho} \frac{r}{(1-r^2)^2} \, dr = \frac{\rho^2}{1-\rho^2},$$

or

$$\frac{1}{2\pi} \int_0^{2\pi} \left| g\left(\rho e^{i\theta} \right) \right|^2 d\theta = \sum_{n=0}^{\infty} |b_n|^2 \rho^{2n} \le \frac{\rho}{1-\rho^2}.$$

Since $g(z) = \sqrt{f(z^2)}$, we have $g(\rho\, e^{i\theta}) = \sqrt{f\left(\rho^2 e^{2i\theta} \right)}$, and

$$\frac{1}{2\pi} \int_0^{2\pi} \left| f\left(\rho^2 e^{2i\theta} \right) \right| d\theta \le \frac{\rho^2}{1-\rho^2}.$$

Hence

$$a_n = \frac{f^{(n)}}{n!} = \frac{1}{2i\pi} \int_{|z|=r} \frac{f(z)}{z^{n+1}} \, dz = \frac{1}{2i\pi} \int_0^{2\pi} \frac{f(r\, e^{i\theta})}{r^{n+1}} e^{i(n+1)\theta} \, ir\, e^{i\theta} \, d\theta$$

$$= \frac{1}{2i\pi} \int_0^{2\pi} \frac{f(r\, e^{i\theta})}{r^n} e^{-in\theta} \, d\theta,$$

which yields

$$|a_n| = \left| \frac{1}{2\pi} \int_0^{2\pi} \frac{f(r\, e^{i\theta})}{r^n} e^{-in\theta} \, d\theta \right| \le \frac{1}{2\pi r^n} \int_0^{2\pi} \left| f(r\, e^{i\theta}) \right| d\theta$$

$$\le \frac{1}{r^n} \left(\frac{r}{1-r} \right) = \frac{1}{r^{n-1}(1-r)}. \tag{3.7.10}$$

This result holds for all r, $0 < r < 1$. To get the best bound, we minimize the right side. Let

$$f(r) = \frac{1}{r^{n-1}(1-r)} = r^{1-n}(1-r)^{-1}.$$

Then

$$f'(r) = (1-n) r^{-n}(1-r)^{-1} + r^{1-n} (1-r)^{-2} = 0,$$

or

$$\frac{1-n}{r^n (1-r)} + \frac{r^{1-n}}{(1-r^2)^2} = 0,$$

which leads to $(1-n)(1-r) + r^{1-n} r^n = 0$, or $r = 1 - 1/n$. Now, $f''(r) > 0$ for $r = 1 - 1/n$. Substituting this result into the above bound for $|a_n|$, we get

$$|a_n| \leq \frac{1}{r^{n-1}(1-r)} = \frac{1}{(1-1/n)^{n-1}(1-1+1/n)}$$

$$= \left[\frac{1}{(1-1/n)^{n-1}}\right]\left[\frac{1}{1/n}\right] < en, \tag{3.7.11}$$

since $\dfrac{1}{(1-1/n)^{n-1}} \to e$ as $n \to \infty$. ∎

3.8. Criterion for Univalency

Consider the function $g(z) \in \mathcal{S}$ defined by (3.7.1) again. Let $w(z) = \dfrac{z+\zeta}{1+\bar{\zeta} z}$, $w(0) = \zeta$. Then $w'(z) = \dfrac{1-|\zeta|^2}{(1+\bar{\zeta} z)^2}$, and $g(z)$ can be written as

$$g(z) = \frac{f(w(z)) - f(w(0))}{f'(w(0))\, w'(0)}.$$

Set $\phi(z) = f(w(z)) \equiv (f \circ w)(z)$, where

$$\phi(0) = f(w(0)),\ \phi'(z) = f'(w(z))\, w'(z),\ \phi'(0) = f' = (w(0))\, w'(0).$$

Then

$$g(z) = \frac{\phi(z) - \phi(0)}{\phi'(0)}, \quad |z| < 1. \tag{3.8.1}$$

The Maclaurin series for $\phi(z)$ is

$$\phi(z) = \phi(0) + \phi'(0)\, z + \frac{\phi''(0)}{2!} z^2 + \cdots + \frac{\phi^{(n)}(0)}{n!} z^n + \cdots, \quad |z| < 1.$$

Thus,

$$g(z) = z + \frac{\phi''(0)}{2!\, \phi'(0)} z^2 + \cdots + \frac{\phi^{(n)}}{n!\, \phi'(0)} z^n + \cdots = z + \sum_{n=2}^{\infty} b_n(\zeta)\, z^n \quad \in \mathcal{S},$$

where

$$b_n(\zeta) = \frac{\phi^{(n)}}{n!\,\phi'(0)}.$$

Now, since $g \in \mathcal{S}$, we get

$$\hat{g}(z) = \frac{1}{g(z)} = \frac{1}{z} + \sum_{n=0}^{\infty} c_n(\zeta)\, z^n \quad \in \mathcal{P}, \quad |z| < 1.$$

Note that, by the area theorem, $|c_1(\zeta)| \le 1$. Since $\hat{g}(z)\, g(z) = 1$, we obtain

$$\left(\frac{1}{z} + \sum_{n=0}^{\infty} c_n(\zeta)\, z^n \right) \left(z + \sum_{n=2}^{\infty} b_n(\zeta)\, z^n \right) = 1,$$

or, writing b_n and c_n for $b_n(\zeta)$ and $c_n(\zeta)$, respectively, we get

$$\sum_{n=2}^{\infty} b_n\, z^{n-1} + \sum_{n=0}^{\infty} c_n\, z^{n+1} + \left(\sum_{n=0}^{\infty} c_n\, z^n \right) \left(\sum_{n=2}^{\infty} b_n\, z^n \right) = 0. \qquad (3.8.2)$$

Equating the coefficients of z on both sides, we get $b_2 + c_0 = 0$ for $n = 0$, i.e.,

$$c_0(\zeta) = -b_2 = -\frac{\phi''(0)}{2\,\phi'(0)} = -\frac{1}{2}\left[(1 - |\zeta|^2)\, \frac{f''(\zeta)}{f'(\zeta)} - 2\bar{\zeta} \right]$$

$$= \bar{\zeta} - (1 - |\zeta|^2)\, \frac{f''(\zeta)}{f'(\zeta)}, \qquad (3.8.3)$$

where

$$\phi(0) = f'(w(0))\, w'(0) = (1 - |\zeta|^2)\, f'(\zeta),$$
$$\phi''(0) = f''(w(0))\, [w'(0)]^2 + f'(w(0))\, w''(0)$$
$$= f''(\zeta)\, (1 - |\zeta|^2) - 2f'(\zeta)\, \bar{\zeta}\, (1 - |\zeta|^2),$$
$$\phi'''(0) = f'''(w(0))\, [w'(0)^3] + 3f''(w(0))\, w''(0) + f'(w(0))\, w'''(0)$$
$$= f'''(\zeta)\, (1 - |\zeta|^2) - 6\bar{\zeta}\, (1 - |\zeta|^2)^2\, f''(\zeta) + 6\bar{\zeta}^2\, (1 - |\zeta|^2)\, f'(\zeta).$$

Similarly, equating the coefficients of z^2 in (3.8.3) we get $b_3 + c_1 + c_0 b_2 = 0$, which gives

$$c_1(\zeta) == c_0\, b_2 - b_3 = b_2^2 - b_3 = \left[\frac{\phi''(0)}{2\,\phi'(0)} \right]^2 - \frac{\phi'''(0)}{6\,\phi'(0)}.$$

Substituting the above computed values and simplifying we find that

$$c_1(\zeta) = \left[\bar{\zeta} - \frac{(1-|\zeta|^2)}{2} \frac{f''(\zeta)}{f'(\zeta)}\right] - \frac{(f'''(\zeta) - 6\bar{\zeta} f''(\zeta) + 6\bar{\zeta}^2 f'(\zeta))(1-|\zeta|^2)}{6(1-|\zeta|^2) f'(\zeta)}$$

$$= \frac{1}{6(1-|\zeta|^2) f'(\zeta)} \left[\frac{3}{2}(1-|\zeta|^2)^3 \frac{f''(\zeta)^2}{f'(\zeta)} - f'''(\zeta)(1-|\zeta|^2)\right]$$

$$= -\frac{1}{6}(1-|\zeta|^2)^2 \left[\frac{f'''(\zeta)}{f''(\zeta)} - \frac{3}{2}\left(\frac{f''(\zeta)}{f'(\zeta)}\right)^2\right]. \tag{3.8.4}$$

This yields

$$g(z) = \frac{1}{z} + \left\{\bar{\zeta} - \frac{1}{2}(1-|\zeta|^2) \frac{f''(\zeta)}{f'(\zeta)}\right.$$

$$\left. - \frac{1}{6}(1-|\zeta|^2)^2 \left[\frac{f'''(\zeta)}{f'(\zeta)} - \frac{3}{2}\left(\frac{f''(\zeta)}{f'(\zeta)}\right)^2\right] + \cdots\right\}. \tag{3.8.5}$$

3.9. Schwarzian Derivative

The *Schwarzian derivative* of a regular function w with respect to z is defined as

$$\{w, z\} = \left(\frac{w''}{w'}\right)' - \frac{1}{2}\left(\frac{w''}{w'}\right)^2 = \frac{w'''}{w'} - \frac{w'' w''}{(w')^2} - \frac{1}{2}\left(\frac{w''}{w'}\right)^2$$

$$= \frac{w'''}{w'} - \frac{3}{2}\left(\frac{w''}{w'}\right)^2. \tag{3.9.1}$$

Thus,

$$c_1(\zeta) = -\frac{1}{6}(1-|\zeta|^2)\{f, z\}. \tag{3.9.2}$$

Since, by the area theorem, $|c_1| \leq 1$, we get

$$\left|-\frac{1}{6}(1-|\zeta|^2)\{f, z\}\right| < 1,$$

which implies that

$$|\{f, \zeta\}| \leq \frac{6}{(1-|\zeta|^2)^2} \quad \text{with } \zeta < 1. \tag{3.9.3}$$

A necessary condition for f to be univalent in $|z| < 1$ is

$$|\{f, z\}| \leq \frac{6}{(1-|z|^2)^2}. \tag{3.9.4}$$

If f is univalent in $|z| < 1$, then the relation (3.9.4) is valid, where the equality holds for the Koebe function $k(z) = \dfrac{z}{(1-z)^2}$ on (z = real).

An alternate method for derivation for c_1 is as follows: Since

$$g(z) = \frac{f\left(\dfrac{z+\zeta}{1+\bar{\zeta}z}\right) - f(\zeta)}{f'(\zeta)\left(1 - |\zeta|^2\right)} \in \mathcal{S},$$

with $g(0) = 0$ and $g'(0) = 1$, we get

$$\frac{1}{g(z)} = \frac{f'(\zeta)(1 - |\zeta|^2)}{f\left(\dfrac{z+\zeta}{1+\zeta z}\right) - f(\zeta)} = \frac{1}{z} + c_0 + c_1 z + \cdots \quad \in \mathcal{P}.$$

Construct $f(z) = z \dfrac{1}{g(z)} = 1 + c_0 z + c_1 z^2 + \cdots$. Then

$$f(0) = \lim_{z \to 0} f(z) = \lim_{z \to 0} \frac{z}{g(z)} = \lim_{z \to 0} \frac{1}{g'(z)} = 1.$$

Thus,

$$c_0 = f'(0) = \lim_{z \to 0} \left(\frac{z}{g(z)}\right)'.$$

Comparing with the Taylor theorem we get

Theorem 3.9.1. *A sufficient condition for f to be univalent in E is given by the relation (3.9.4).*

3.10 Exercises

3.10.1. Other proofs of Theorem 3.1.1, which are variants of the above proof, are as follows:

(i) Suppose $g(z) = z + \dfrac{b_1}{z} + \dfrac{b_2}{z^2} + \cdots$ is univalent in $|z| > 1$. If $r > 1$, then the complement of the image of the disk $|z| > r$ is a bounded region $G(r)$, and

$$\text{area}\{G(r)\} = \int_{G(r)} dx\, dy = \frac{1}{2i} \int_{\partial G(r)} \bar{z}\, dz = -\frac{1}{2i} \bar{g}\, dg$$

$$= \frac{1}{2\pi r^2} - \frac{1}{2\pi} \sum_{n=1}^{\infty} n |b_n|^2 r^{2n},$$

which is positive. Letting $r \downarrow 1$, we obtain (3.1.2).

Note that equality in (3.1.2) holds if and only if the complement of the image of g has Lebesgue measure zero (i.e., zero area).

(ii) Note that the function $g(z) = \dfrac{1}{z} + b_1 z + b_2 z^2 + \cdots$ is injective on $\{0 \le |z| \le 1\}$. By direct computation for $0 < |z| = r \le 1$ we have

$$-\frac{1}{2i} \int_{|z|=r} \bar{g}(z) g'(z)\, dz = \pi r^{-2} - \pi \sum_{n=1}^{\infty} |b_n|^2 r^{2n}. \tag{3.10.1}$$

Let $G(r)$ denote the complement of $g(B(0, r)) - 0$, where $B(0, r)$ is the open disk at 0 with radius r. Then $G(r)$ is a bounded region with boundary $\partial G(r) \equiv g(\partial D_r(0))$. Then

$$\text{area}\{G(r)\} = \frac{1}{2i} \int_{\partial G(r)} \bar{z}\, dz = -\frac{1}{2i} \int_{|z|=r} \bar{g}(z) g'(z)\, dz$$

$$= \pi r^{-2} - \pi \sum_{n=1}^{\infty} n|b_n|^2 r^{2n}, \quad \text{by (3.10.1)}$$

Letting $r \to 1$, we get $\sum\limits_{n=1}^{\infty} n|b_n|^2 \le 1$. ∎

3.10.2. A *conformal equivalence* between two regions D and G in the complex plane is a one-to-one analytic function f with $f(D) = G$. Thus, $f'(z) \ne 0$ for all z in D. The Riemann mapping theorem establishes the conformal equivalence of two regions. If f is a conformal equivalence between the open sets D and G, then

$$\text{Area}\,(G) = \iint_D |f'|^2\, dz. \tag{3.10.2}$$

If f is an analytic function, then the Jacobian of f, regarded as a mapping from $\mathbb{R}^2$ into $\mathbb{R}^2$, is $|f'|^2$. Hence, if G is a simply connected region, $g : D \mapsto G$ is a Riemann map, and $g(z) = \sum\limits_n a_n z^n$ in D, then from (3.10.2)

$$\text{Area}\,(G) = \iint_D |g'|^2\, dz = \pi \sum_n n|a_n|^2. \tag{3.10.3}$$

Proof of (3.10.3): Since $g'(z) = \sum\limits_n n a_n z^{n-1}$, then since $\int_0^{2\pi} e^{i(n-m)\theta}\, d\theta = 0$ for $n \ne m$ for $r < 1$, we get, using the proof of Theorem 3.7.1,

$$\iint_D |g'|^2\, dz = \sum_n n^2 |a_n|^2 \int_0^{\pi} \int_0^1 r^{2n-2} r\, dr\, d\theta = \sum_n n^2 |a_n|^2 2\pi \int_0^1 r^{2n-1}\, dr$$

$$= 2\pi \sum_n n^2 |a_n|^2 \frac{1}{2n} = \pi \sum_n n\,|a_n|^2. ∎$$

3.10.3. A proof of the Bieberbach inequality for the general class of odd functions is as follows: Let $g(z) = \sqrt[n]{f(z^n)}$, and let $z + \sum_{k=2}^{\infty} a_k \, z^k$. Then

$$f(z^n) = z^n + \sum_{k=2}^{\infty} a_k \, z^{nk} = z^n \left[1 + \sum_{k=2}^{\infty} a_k \, z^{n(k-1)} \right],$$

$$g(z) = z \left[1 + \sum_{k=2}^{\infty} a_k \, z^{n(k-1)} \right]^{1/n} = z \, H(z),$$

where $H(z) = [h(z)]^{1/n}$, and $h(z) = 1 + \sum_{k=2}^{\infty} a_k \, z^{n(k-1)}$. Now, the Maclaurin series for $g(z)$ is

$$g(z) = g(0) + g'(0)\, z + \frac{g''(0)}{2!} z^2 + \cdots + \frac{g^n(0)}{n!} z^n + \cdots .$$

To determine the bounds on $a_n \equiv \dfrac{g^n(0)}{n!}$, note that $h(0) = 1, \ldots, h^{(n)}(0) = n!\, a_2$, $h^{2n}(0) = (2n)!\, a_3, \ldots$. Thus, after substituting in $g(z)$ where $g(z)$ has the form $g(z) = z + c_{n+1} \, z^{n+1} + c_{2n+1} \, z^{2n+1} + \cdots$, we find that $|c_{n+1}| \leq \dfrac{|a_2|}{n} \leq \dfrac{2}{n}$.

Now, we must show that $g(z) = \sqrt[n]{f(z^n)}$ is univalent in E. Assume that $g(z_1) = g(z_2)$ for $z_1 \neq z_2$, $z_1, z_2 \in E$. Then $f(z_1^n) = f(z_2^n)$. But since f is univalent, we have $z_1^n = z_2^n$, i.e., $z_1 = \mu \, z_2$, where $\mu^n = 1$. Then

$$g(z_1) = g(\mu \, z_2) = \mu \, z_2 + c_{n+1} \, \mu^{n+1} \, z_2^{n+1} + c_{2n+1} \, \mu^{2n+1} \, z_2^{2n+1} + \cdots$$

$$= \mu \left[z_2 + c_{n+1} \, z_2^{n+1} + c_{2n+1} \, z_2^{2n+1} + \cdots \right] = \mu \, g(z_2).$$

Hence, $g(z_1) = g(z_2)$ only if $\mu = 1$. But then $z_1 = z_2$, which contradicts the assumption that $z_1 \neq z_2$.

3.10.4. If $f \in S$ maps E onto the w-plane, then $f(E) \supset B(0, 1/4)$. This is the Koebe covering theorem (see Figure 3.10.1). A proof follows.

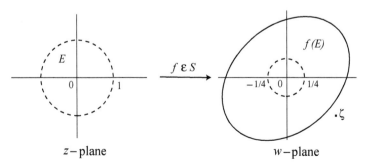

Figure 3.10.1 Koebe covering theorem.

PROOF. Let ζ be a point in the w-plane which the function $f \in \mathcal{S}$ does not take. Then $f(z) \neq \zeta$ for $z \in E$. Consider the function

$$f_1(z) = \frac{f(z)}{\zeta\left(1 - f(z)/\zeta\right)}.$$

Note that $\zeta - f(z) \neq 0$, so $1 - f(z)/\zeta \neq 0$. Since f is univalent, so is f/ζ and $1 - f/\zeta$ and $1/(1 - f/\zeta)$. Thus, $f_1(z)$ is regular and univalent in E, i.e.,

$$f_1(z) = f_1(0) + f_1'(0)\, z + \frac{f''(0)}{2!}\, z^2 + \cdots = z + \left(a_2 + \frac{1}{\zeta}\right) z^2 + \cdots \in \mathcal{S}.$$

Hence, by Gronwall's area theorem (Theorem 3.1.1), $\left|a_2 + \dfrac{1}{\zeta}\right| \leq 2$, which yields $\left|\dfrac{1}{\zeta}\right| - |a_2| \leq 2$, or $\dfrac{1}{|\zeta|} \leq 2 + |a_2| \leq 4$, or $|\zeta| \geq 1/4$, which implies that $f(E) \supset B(0, 1/4)$. Note that the Koebe function $k(z) = \dfrac{z}{(1-z)^2}$ does not take the value $-1/4$, since the solution of the equation $\dfrac{z}{(1-z)^2} = -\dfrac{1}{4}$ yields $z = -1$ which does not belong to E. ∎

3.10.5. (Goluzin's theorem) If $g(\zeta) \in \Sigma$, then for any two points ζ_1, ζ_2 on the circle $|\zeta| = \rho > 1$ the inequality

$$\left|\frac{g(\zeta_1) - g(\zeta_2)}{\zeta_1 - \zeta_2}\right| \geq 1 - \frac{1}{\rho^2} \tag{3.10.4}$$

holds, such that the equality is attained only for the functions $g(\zeta) = \zeta + C + \dfrac{e^{2i\phi}}{\zeta}$, where C is a constant and $\phi = \frac{1}{2}\left(\arg\{\zeta_1\} + \arg\{\zeta_2\}\right)$. This is one of many generalizations of the inequality (3.3.6), which are available in Goluzin [1969].

3.10.6. The Lebedev-Milin Inequality was proved by Lebedev and Milin [1965] and Milin [1977]. It was used in the proof of the Bieberbach conjecture, because the Milin conjecture implies the Robertson conjecture. This equality deals with the exponentiation of a power series, and states that if

$$\sum_{k=0}^{\infty} \beta_k z^k = \exp\left(\sum_{k=1}^{\infty} \alpha_k z^k\right), \tag{3.10.5}$$

where α_k and β_k are complex numbers, and $n \geq !$ is an integer, then

$$\sum_{k=0}^{\infty} |\beta_k|^2 \leq \exp\left(\sum_{k=1}^{\infty} k|\alpha_k|^2\right),$$

$$\sum_{k=0}^{n} |\beta_k|^2 \leq (n+1)\exp\left(\frac{1}{n+1}\sum_{m=1}^{n}\sum_{k=1}^{n}\left(k|\alpha_k|^2 - 1/k\right)\right), \tag{3.10.6}$$

$$|\beta_n|^2 \leq \exp\left(\sum_{k=1}^{n}\left(k|\alpha_k|^2 - 1/k\right)\right).$$

3.10.7. Consider a function $g(\zeta) = \zeta + b_0 + \dfrac{b_1}{\zeta} + \cdots \in \Sigma$ which is meromorphic and univalent for $|\zeta| > 1$, and for which the inequality

$$1 - \frac{1}{|\zeta_0|^2} \leq |g'(\zeta_0)| \leq \frac{|\zeta_0|^2}{|\zeta_0|^2 - 1} \tag{3.10.7}$$

holds for all $\zeta_0, |\zeta_0| > 1$. The left-side equality in (3.10.7) holds only for the functions $g_1(\zeta) = \zeta + \alpha_0 + \zeta_0\left(\zeta\bar\zeta_0\right)^{-1}$, and the right-side equality only for the functions $g_2(\zeta) = \dfrac{\zeta - \zeta_0}{1 - (\zeta\bar\zeta_0)^{-1}} + \beta_0$, where α_0, β_0 are two arbitrary fixed numbers, and the functions $w = g_1(\zeta)$ maps the domain $|\zeta| > 1$ onto the w-plane with a slit along the interval connecting the points $\alpha_0 - 2\dfrac{\zeta_0}{|\zeta_0|}$ and $\alpha_0 + 2\dfrac{\zeta_0}{|\zeta_0|}$, whereas the functions $w = g_2(\zeta)$ map the domain $|\zeta| > 1$ onto the w-plane with a slit along the circular arc $|w - \beta_0| = |\zeta_0|$ with mid-point at $\beta_0 - \zeta_0$. The inequality (3.10.7) follows from the Grunsky inequality

$$|\log g'(\zeta_0)| \leq -\log\left(1 - \frac{1}{|\zeta_0|^2}\right), \tag{3.10.8}$$

which determines the range of values of the functional $\log g'(\zeta_0)$ on the class Σ.

3.10.8. The Class Σ_0 consists of functions $g(\zeta) \in \Sigma$ with $g(\zeta) \neq 0$ for $1 < |\zeta| < \infty$. The following relation holds between functions in the class $\mathcal{S}$ and the class Σ_0: If $f(z) \in \mathcal{S}$, then $g(\zeta) = 1/f(1/\zeta) \in \Sigma_0$, and conversely, if $g(\zeta) \in \Sigma_0$, then $f(z) = 1/g(1/z) \in \mathcal{S}$. Thus, the range of functionals on Σ_0 determine the range of the corresponding functionals in $\mathcal{S}$, or conversely. For example, the range of $\log g(\zeta_0)/\zeta_0$ on Σ_0, $1 < |\zeta_0| < \infty$, is obtained from the range of $\log f(z_0)/z_0$ on $\mathcal{S}$, $0 < |z_0| < 1$.

3.10.9. Prove the strong Grunsky inequalities (3.5.6). PROOF. For a polynomial $p(w) = \sum_{n=1}^{N} \ln \Phi(w)$, where $\Phi(w) = \Phi(g(z))$ are the Faber polynomials of g, we find, in view of (9.3.2), that

$$p(g(z)) = \sum_{n=1}^{N} \lambda_n z^n + \sum_{m=1}^{\infty} \sum_{n=1}^{N} \lambda_n \gamma_{mn} z^{-m},$$

and then the result follows directly from (3.5.5).

3.10.10. Prove generalized weak Grunsky inequalities (3.5.7). PROOF. Let $\nu_n = \sum_{m=1}^{N} \beta_{mn} \lambda_m$, $n = 1, \ldots, N$. Then, in view of the strong Grunzky inequality (3.5.6), we get $\sum_{n=1}^{N} n|\nu_n|^2 \leq \sum_{m=1}^{N} m|\lambda_m|^2$. Then the Cauchy-Schwarz inequality (1.8.4) gives

$$\left| \sum_{m=1}^{N} \sum_{n=1}^{N} k\beta_{mn}\lambda_m\mu_n \right|^2 = \left| n\nu_n\mu_n \right|^2 \leq \sum_{n=1}^{N} n|\nu_n|^2 \sum_{n=1}^{N} n|\mu_n|^2$$

$$\leq \sum_{m=1}^{N} m|\lambda_m|^2 \sum_{n=1}^{N} n|\mu_n|^2.$$

3.10.11. Prove the weak Grunsky inequalities (3.5.5). This result follows from the generalized weak Grunsky inequalities (3.5.7).

3.10.12. (Minimal Area Problem) An extremal property in the conformal mapping of a simply connected region D onto a disk is connected with the minimum area problem. This problem is known as the *Bieberbach minimizing principle*. The mapping function possesses an extremal property which provides a method to compute an approximate solution for the map. Let $\mathcal{K}^1(D)$ denote the class of all functions $f \in L^2(D)$ with $f(a) = 1$, where $a \in D$. Similarly, let $\mathcal{K}^0(D)$ denote the class of all functions $f \in L^2(D)$ such that $f(a) = 0$. Note that the classes $\mathcal{K}^1(D)$ and $\mathcal{K}^0(D)$ are, respectively, a closed convex subset and a closed subspace of $L^2(D)$. Let the function

$$w = f(z) = \sum_{n=0}^{\infty} a_n (z - a)^n, \quad |z - a| < R, \quad (3.10.9)$$

which is regular in D, map D onto the disk $B(0, R)$ in the w–plane. Without loss of generality, we shall sometimes take the point a as the origin. In the class $\mathcal{K}^1$ minimize the integral

$$I = \iint_D |f'(z)|^2 \, dS_z. \quad (3.10.10)$$

SOLUTION. The Riemann mapping theorem (Theorem 2.1.1) guarantees the existence and uniqueness of the solution of this extremal problem. We will now show that the minimum area problem has a unique solution πR^2 for $f_0(z) = f'(z)$. If $f_0(a) = 0$ and $f'_0(a) = 1$, then

$$
\begin{aligned}
\text{area}(D) &= \iint_D |f'(z)|^2 \, dS_z = \int_0^R \int_0^{2\pi} |f'(r\,e^{i\theta})|^2 \, r \, dr \, d\theta \\
&= \int_0^R \sum_{n=0}^{\infty} |a_n|^2 n^2 2\pi r^{2n-1} \, dr \\
&= \pi R^2 |a_1|^2 + \pi \sum_{n=2}^{\infty} n \, |a_n|^2 \, R^{2n}.
\end{aligned}
\tag{3.10.11}
$$

Thus, in the problem of mapping by the function (3.10.9), which is regular in $B(0, R)$ and is such that $f'(a) = a_1$, the area of the mapped region D is always greater than $\pi R^2 |a_1|^2$. It is exactly equal to this value if the map $f(z)$ is linear, i.e., if $w = a_0 + a_1 z$. A particular case is when $a_1 = 1$. Then the mapping function is $w = a_0 + z$. If this linear transformation is excluded, then the mapping function can be normalized by the conditions $f(a) = 0$ and $f'(a) = 1$, by considering the function $f(z)/a_1$. In either case the minimum area of D is πR^2.

3.10.13. Show that the Schwarzian derivative for the Koebe function $k(z) = \dfrac{z}{(1-z)^2}$ satisfies $|\{k, z\}| \leq \dfrac{6}{(1-z^2)^2}$.

3.10.14. Use the above method in Exercise 3.10.13 to find the Schwarzian derivative for $f(z) = \dfrac{z}{(1-z)^3}$. Verify that $|\{f, z\}| \leq \dfrac{A}{(1-|z|^2)^4 (1-2|z|)^3}$. Note that $1 - 2|z| \neq 0$, but $1 - 2|z| > 0$, i.e., $|z| < 1/2$; the function f is univalent in $|z| < 1/2$.

3.10.15. Consider the function $g(z) \in \mathcal{S}$. Show that

$$
\frac{(1-r)}{r(1+r)} \leq \left| \frac{f'(\zeta)}{f(\zeta)} \right| \leq \frac{(1+r)}{r(1-r)}.
\tag{3.10.12}
$$

3.10.16. Let $f \in \mathcal{S}$, and $|z_1| = |z_2| = \rho < 1$. Then from (3.7.8)

$$
\frac{1-\rho}{(1+\rho)^3} \leq |f'(z_1)| \leq \frac{1+\rho}{(1+\rho)^3},
\tag{3.10.13a}
$$

and

$$
\frac{1-\rho}{(1+\rho)^3} \leq |f'(z_2)| \leq \frac{1+\rho}{(1+\rho)^3},
\tag{3.10.13b}
$$

which yields

$$
\frac{(1+\rho)^3}{1+\rho} \leq \left| \frac{1}{f'(z_2)} \right| \leq \frac{(1+\rho)^3}{1-\rho}.
\tag{3.10.13c}
$$

Combining (3.10.13a) and (3.10.13c) we obtain

$$\frac{(1-\rho)^4}{(1+\rho)^4} \leq \left|\frac{f'(z_1)}{f'(z_2)}\right| \leq \frac{(1+\rho)^4}{(1-\rho)^4}. \tag{3.10.14}$$

4

Löwner Theory

This theory deals with the slit mappings which are conformal mappings of the open disk D onto the complex plane with a curve joining 0 to ∞ removed. This leads to the Löwner differential equation which is an ordinary differential equation of first order. Any family of domains in the complex plane expanding continuously in the sense of Carathéodory to the whole plane leads to a one parameter family of conformal mappings, known as the *Löwner chain*, and a two parameter family of univalent self-mappings of the unit disk E, known as the *Löwner semigroup* which corresponds to a time-dependent analytic vector field on the disk E by a one parameter family of analytic functions with positive real part on E.

4.1 Carathéodory's Kernel Theorem

Let $\{D_n\}_{n=1}^{\infty}$ be a sequence of simply connected domains in the z-plane containing a fixed point z_0, $z_0 \neq \infty$. If there is a disk $|z - z_0| < r$, $r > 0$, contained in all D_n, then the *kernel* of the sequence $\{D_n\}$ with respect to z_0 is the largest domain D containing z_0 such that for each compact set $G \in D$ there is an N such that G belongs to D_n for all $n \geq N$. If there is no such disk, then the kernel D of the sequence $\{D_n\}$ is the point z_0 itself; in this case the sequence $\{D_n\}$ is said to have a degenerate kernel. Moreover, the sequence $\{D_n\}$ converges to D if any subsequence of $\{D_n\}$ has D as its kernel. For example, (i) if $\{D_n\}$ is an increasing sequence of connected open sets containing 0, then the kernel is the union of all these sets. and (ii) if $\{D_n\}$ is an increasing sequence of connected open sets containing 0, then, if 0 is an interior point of $D_1 \cap D_2 \cap \cdots$, the sequence converges to the component of the interior containing 0; otherwise, if 0 is not an interior point, the sequence converges to $\{0\}$.

Caratheéodory's theorem (Carathéodory [1912]) deals with the uniform convergence on compact sets of a sequence of analytic univalent functions defined on the unit disk in the z-plane. Keeping the origin $z = 0$ fixed, this uniform convergence is formulated in terms of the limiting behavior of the images of the functions. This theorem is the basis for the Löwner differential equation.

Theorem 4.1.1. (Carathéodory kernel theorem) *Let $\{f_n(z)\}$ be a sequence of analytic univalent functions on the unit disk E, normalized such that $f_n(0) = 0$ and $f'_n(0) > 0$. Then the sequence $\{f_n\}$ converges uniformly on compact sets in E to a function f if and only if the sequence $\{D_n\} = \{f_n(E)\}$ converges to its kernel which is not $\mathbb{C}$. If the kernel is the single point $\{0\}$, then $f = 0$; otherwise, the kernel is a connected open set D and f is univalent on E and $f(E) = D$.*

PROOF. Using Hurwitz's theorem (Theorem 2.2.4) and Montel's theorem (Theorem 2.2.6), check that if the sequence $\{f_n\}$ converges uniformly on compact sets to f, then each subsequence of $\{D_n\}$ has the kernel $D = f(E)$. Conversely, if the sequence $\{D_n\}$ converges to a kernel which is not $\mathbb{C}$, then by Koebe's 1/4- theorem, D_n contains the disk of radius $\frac{1}{4} f'_n(0)$ with center at 0 for each $n = 1, 2, \ldots$. Since $D \neq \mathbb{C}$, all the Koebe radii are uniformly bounded. Thus, by Koebe's distortion theorem,

$$|f_n(z)| \leq f'_n(0) \frac{|z|}{(1-|z|)^2},$$

which shows that the sequence $\{f_n\}$ is uniformly bounded on compact sets. For uniqueness, let two subsequences converge to analytic limit functions f and g. Then $f(0) = g(0)$, and $f'(0), g'(0) > 0$. Moreover, from the necessity part above, we have $f(E) = g(E)$, and thus, by Riemann mapping theorem, we have $f = g$. Hence, the sequence $\{f_n\}$ converges uniformly on compact sets. ∎

Suppose there exists a sequence of functions $\{f_n(\zeta)\}$, $f_n(\zeta_0) = z_0$, $f'_n(\zeta_0) > 0$, $n = 1, 2, \ldots$, which are analytic and univalent in the unit disk $|\zeta - \zeta_0| < 1$ and which map this disk onto the domains D_n, respectively. Then the sequences $\{f_n(\zeta)\}$ converges to a finite function $f(\zeta)$ if and only if the sequence $\{D_n\}$ converge to a kernel D which is either the point z_0 or a domain containing more than one boundary point. Moreover, this convergence is uniform on compact sets in the interior of the unit disk $|\zeta - \zeta_0| < 1$. If the limit function $f(\zeta) \not\equiv$ const, then it maps the unit disk univalently onto D, and the inverse functions $g_n(z)$, $n = 1, 2, \ldots$, are uniformly convergent on compact sets in the interior of D to the inverse function $g(z)$ of $f(\zeta)$.

The convergence of univalent functions in multiply connected domains is defined in an analogous manner.

For a sequence of unbounded domains $\{D_n\}$, $n = 1, 2, \ldots$ in the z-plane containing some fixed neighborhood of $z = \infty$, the kernel of this sequence with respect to the point at infinity is the largest domain D containing $z = \infty$ and such that any closed subdomain of it is a subset of all D_n for some $n \geq N$.

Convergence of the sequence $\{D_n\}$ to the kernel D is defined as above. The following theorem holds:

Theorem 4.1.2. (Goluzin [1969]) *Let $\{A_n\}$, $n = 1, 2, \ldots$ be a sequence of domains in the z-plane containing $z = \infty$ and converging to a kernel A, and suppose that the functions $\zeta = f_n(z)$, $n = 1, 2, \ldots$, map them univalently onto corresponding domains B_n containing $z = \infty$, and $f_n(\infty) = \infty$, $f'_n(\infty) = 1$ for $n = 1, 2, \ldots$. Then the sequence $\{f_n(z)\}$, $n = 1, 2, \ldots$, converges uniformly on compact sets in the interior of A to a univalent function $f(z)$ if and only if the sequence $\{B_n\}$, $n = 1, 2, \ldots$, has a kernel B and converges to this kernel. Then in this case $\zeta = f(z)$ maps A univalently onto B.*

Note that there are other theorems on the convergence of sequences of univalent functions, for which see Goluzin [1969], and Duren [1983: ch. 03].

4.1.1. Löwner Chains. Let f and g be analytic univalent function on E with $f(0) = 0 = g(0)$. Then f is said to be *subordinate* to g if and only if there is a univalent self-mapping (automorphism) ϕ that fixes 0 such that $f(z) = (g \circ \phi)(z) = g(\phi(z))$ for $|z| < 1$. A necessary and sufficient condition for existence of such a mapping ϕ is that $f(E) \subseteq g(E)$. To prove, note that necessity is obvious. To prove sufficiency, since ϕ is defined by $\phi(z) = g^{-1}(f(z))$, and by definition ϕ is analytic univalent self-mapping of E with $\phi(0) = 0$, the map $\phi(z)$ satisfies $0 < |\phi'(0)| \leq 1$ and maps every disk $E_r : |z| < r$, $0 < r < 1$, into itself, we have $|f'(0)| \leq |g'(0)|$ and $f(E_r) \subseteq g(E_r)$.

The *Löwner chain* is defined as follows: For $t \geq 0$ let $G(t)$ be a family of domains in $\mathbb{C}$ containing 0, such that $G(s) \subset G(t)$ if $s < t$, $G(t) = \cup_{s<t}G(s)$, and $G(\infty) = \mathbb{C}$. Then $G(s_n) \to G(t)$ if $s_n \uparrow t$, in the sense of the Carathéodory kernel theorem (Theorem 4.1.1). In particular, if $D = E$, then there are unique univalent maps $f(z, t)$, where $f(z, t) = G(t)$, $f(0, t) = 0$, $\partial_z f(0, t) = 1$ by the Riemann mapping theorem, which are uniformly continuous on compact subsets of $[0, \infty) \times E$. Moreover, the function $f'(0, t) \equiv a(t)$ is positive, strictly increasing and continuous. By reparametrization we can assume that $a(t) = e^t$. Then $f(z, t) = e^t z + a_2(t)z^2 + \cdots$ are univalent mappings that form a *Löwner chain*.

If $f(D, s)$ is a Löwner chain, then $f(D, s) \subset f(D, t)$ for $s < t$. Then there is a unique univalent mapping $\phi(z, s, t) \equiv \phi_{s,t}(z)$ of the disk that fixes 0 and is such that $f(z, s) = f(\phi_{s,t}(z, t))$. Since the mappings $\phi_{s,t}(z)$ are unique, we get $\phi_{s,t} \circ \phi_{t,r} = \phi_{s,r}$ for $s \leq t \leq r$, which is a semigroup property. The mappings $\phi_{s,t}(z)$ constitute a *Löwner semigroup*. The self-mappings depend continuously on s and t, and satisfy $\phi_{t,t}(z) = z$.

Note that the Löwner chain can be recovered from the Löwner semigroup by taking the limit: $f(z, s) = \lim\limits_{t \to \infty} e^t \phi_{s,t}(z)$.

Given any univalent self-mapping $\phi(z)$ of D with fixed 0, we can construct a Löwner semigroup $\phi_{s,t}(z)$ such that $\phi_{0,1}(z) = \phi(z)$. Similarly, given a univalent function g on D, $g(0) = 0$, such that $g(D)$ contains the closed unit disk $E \cup \partial E$, there is a Löwner chain $f(z,t)$ such that $f(z,0) = f(z), f(z,1) = g(z)$. This result follows if ϕ or g extend continuously to ∂D, in general, if the mapping $f(z)$ is replaced by $f(rz)/r$ and then compactness of the domain $D \cup \partial D$ is used.

4.1.2. Löwner's Differential Equation. Löwner's differential equation can be derived by two methods:

(i) By the Löwner chain: Since $f(z,t) = f(\phi_{s,t}(z),t)$, the function $f(z,t)$ satisfies the differential equation $\partial_t f(z,t) = a\, p(z,t)\partial_z f(z,t)$ with initial condition $f(z,t)|_{t=0} = f(z)$. The Picard-Lindelöf existence theorem (Coddington and Levinson [1955]) guarantees the existence of solutions of these initial value problems, and also that these solutions are analytic in z.

(ii) By the Löwner semigroup: Let $w_s(z) = \partial_t \phi_{s,t}(z)|_{t=s}$. Then $w_s(z) = -zp(z,s)$, $\Re\{p(z,s)\} > 0$ for $|z| < 1$, and $w(t) = \phi_{s,t}(z)$ satisfies the first-order ordinary differential equation:

$$\frac{dw}{dt} = -w\,p(w,t), \quad \text{with the initial value } w(s) = z. \tag{4.1.1}$$

Given any univalent self-mapping $\phi(z)$ of D with fixed 0, we can construct a Löwner semigroup $\phi_{s,t}(z)$ such that $\phi_{0,1}(z) = \phi(z)$. Similarly, given a univalent function g on D, $g(0) = 0$, such that $g(D)$ contains the closed unit disk ∂E, there is a Löwner chain $f(z,t)$ such that $f(z,0) = z, f(z,1) = g(z)$. This result follows if ϕ or g extend continuously to ∂D, in general by replacing the mapping $f(z)$ by $f(rz)/r$ and then using compactness of the domain ∂D.

This equation plays an important rôle in de Branges proof of the Bieberbach conjecture.

4.1.3. Löwner's Parametric Method. This method introduced by Löwner [1923] was the first attempt to solve the Bieberbach conjecture on the bounds for the coefficients of analytic univalent functions of the class $\mathcal{S}$. It turned out that the this conjecture was eventually solved by de Branges [1984] using this very method, as explained in §10.2. The modern form of the Löwner's method is due to Kufarev [1943] and Pommerenke [1965; also 1975, ch.6], which is as follows. Let $f(z,0) = z + \sum_{n=1}^{\infty} a_n z^n \in \mathcal{S}$ be defined on the unit disk E. This function can be regarded as member of the family $\{f(z,t)\}_{t \geq 0}$ defined on E, where $t > 0$ is a real parameter such that (i) $f(z,t) = e^t z + a(t)z^2 + \cdots$ for any $t > 0$, and (ii) $f(D,s) \subset f(D,t)$ for $0 \leq s \leq t$. This family is the classical *Löwner chain*. It has the property that every function in this family is differentiable in the parameter t, and independently of z, almost everywhere on $[0,+\infty)$, and satisfies the first-order partial differential equation, known as the *Löwner equation,*

$$\frac{\partial f(z,t)}{\partial t} = z\,p(z,t)\frac{\partial f(z,t)}{\partial z}, \qquad (4.1.2)$$

where the function $p(z,t)$ is measurable with respect to $t \in [0,+\infty)$ for all $z \in E$, with $p(0,t) = 1$ and $\Re\{p(z,t)\} > 0$ almost everywhere for $t \in [0,=\infty)$, and is analytic in $z \in E$. Eq (4.1.1) is known as the *characteristic equation* for (4.1.2), with initial condition $w(s) = z$, thus providing a well-posed initial value problem. The equation (4.1.2) is called the *Löwner equation,* and its right-hand side as the *associated vector field* (see §4.3 for more on the Löwner-Kufarev equation). The family $\{\phi_{s,t}\}$ is continuous in $t \in [s,+\infty)$ for each $s \geq 0$ such that

$$\phi_{s,s}(z) = z,\ s \geq 0, \quad \text{and} \quad \phi_{s,t} = \phi_{u,t} \cdot \phi_{s,u},\ 0 \leq s \leq u \leq t < +\infty. \quad (4.1.3)$$

Geometrically, Eq (4.1.2) represents an outward (potential) flow in the plane, where the vector z provides the direction of the outward normal to the circle $C_r : |z| = r$, and $z\dfrac{\partial f}{\partial z}$ gives the direction of the outward normal to the curve $f(C_r)$ at the point $f(z,t)$, whereas the velocity vector $\dfrac{\partial f}{\partial t}$ makes an acute angle (less than $\pi/2$) with the normal at that point. (see Figure 4.1.1).

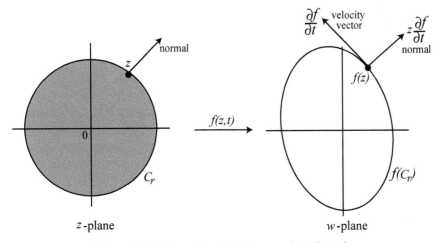

Figure 4.1.1 Geometric Significance of Eq (4.1.2).

The derivation of Eq (4.1.2) is as follows: For $0 \leq s < t$, define $\phi(z,s,t) = f^{-1}(z,t) \circ f(z,s) = e^{t-s}z + \cdots$. The function ϕ maps $|z| < 1$ into $|w| < 1$, but not onto $|w| < 1$, such that $z = 0$ goes into $w - 0$. Thus, by Schwarz lemma, we have $|\phi(z,s,t)| < |z| = |\phi(z,s,s)|$ for all $z \neq 0$, which physically implies an inward flow on the unit disk. Assume that $\dfrac{\partial \phi}{\partial t}$ exists and is analytic in z. Then the angle between the vector $\dfrac{\partial \phi}{\partial t}$ for $t = s$ and the vector $-z$ must be bounded by $\pi/2$. Hence,

$$\left.\frac{\partial \phi}{\partial t}\right|_{t=s} = -z\,p(z,s) \quad \Re\{p(z,s)\} > 0, \text{ and } p(0,s) = 1, \tag{4.1.4}$$

where $p(z,s)$ is analytic in z.

The definition of ϕ gives $f(z,t) \circ \phi(z,s,t) = f(z,s)$, which after differentiating with respect to t and setting $t = s$ gives

$$\frac{\partial f(z,t)}{\partial t} + \frac{\partial f(z,t)}{\partial z}\frac{\partial \phi}{\partial t} = 0. \tag{4.1.5}$$

Then combining (4.1.5) and (4.1.4) gives the Löwner equation (4.1.2).

The Löwner differential equation (4.1.2) holds for arbitrary Löwner chains. Note that the function $f(z,t)$ satisfies Lipschitz condition in t,[1] and Eq (4.1.2) holds for almost all t. Conversely, every partial differential equation (4.1.2) determines a Löwner chain of conformal maps.

4.2 Löwner's Theorem

The importance of Löwner's theory lies in its application to prove the Robertson and the Milin conjecture which led to the final proof of the Bieberbach conjecture by de Branges [1984]. To prove Löwner's theorem, we need the following two lemmas.

Lemma 4.2.1. *Let f be a holomorphic function in E with $f(0) = 0$, $|f(z)| < 1$ for $z \in E$, and $|f(z)| = 1$ on an arc γ of length l on the boundary $\partial E = \{|z| = 1\}$. Then as z moves along γ, the mapping $w = f(z)$ traces an arc Γ on $|w| = 1$ of length $L \geq l$.*

PROOF. Suppose z describes the arc γ in the positive (i.e., counter-clockwise) direction. Then $f(z)$ moves in a non-negative direction such that the increase in $\arg\{f(z)\} \geq \arg\{z\}$, where equality holds for $f(z) = az$, $|a| = 1$. ∎

Lemma 4.2.2. *Suppose $w = f(z)$ is holomorphic in E with $f(0) = 0$ and $|f(z)| < 1$ for $z \in E$. Let D be the domain contained in E, which is bounded by the arc γ and the arc of the circle orthogonal to the unit circle $|z| = 1$ and passing through the endpoints of γ. Then the map of D under $w = f(z)$ lies in the part G of $|w| < 1$, which is bounded by the arc Γ and the arc of the circle orthogonal to the circle $|w| = 1$ and passing through the endpoints of Γ.*

---

[1] i.e., there exists a constant $M > 0$ such that $|f(z,t) - f(z,s)| < M|t - s|$.

PROOF. Let z_0 be any point in D and let $w_0 = f(z_0)$. Map the unit disks $|z| < 1$ and $|w| < 1$ onto the unit disks $|z'| < 1$ and $|w'| < 1$, respectively, such that z_0 and w_0 are mapped into the origins. Let γ' and Γ' be the arc which are the images of γ and Γ, respectively, and D' and G' be the respective domains under this mapping. Then the mapping $w = f(z)$ is transformed into $w' = f(z')$, satisfying the conditions of Lemma 4.2.1, such that L' (length of Γ') $\geq l'$ (length of γ'). Since D' contains $z' = 0$, so must G' contain $w' = 0$, i.e., w_0 lies in G. ■

Theorem 4.2.1. (Löwner's theorem) *Let the function* $f(z) = \beta(z + a_2 z^2 + \cdots) \in \mathcal{S}$ *be defined on the slit domain* (s), *which is the bounded domain denoted by* B. *If we annex to* B *the points belonging to the inner end of the cut, we obtain a family of bounded domains* B_t, *which depends continuously on a real parameter* t, $0 \leq t \leq t_0$, *such that* $B_0 = B$, $B_{t_0} = \{|w| < 1\}$, *and* $B_{t'} \subset B_{t''}$ *for* $t' < t''$. *Let* $w = g(z,t)$ *map* E *onto* B_t *such that* $g(0,t) = 0, g'(0,t) > 0$. *Then* $g(z,0) = f(z)$ *and* $g(z,t_0) = z$. *Also, the function*

$$f(z,t) = g^{-1}[g(z,0),t] \tag{4.2.1}$$

maps E *onto a domain* B *for each* t, $0 \leq t \leq t_0$, *and satisfies the conditions* $f(0,t) = 0, f'(0,t) > 0$, *with* $f(z,0) = z, f(z,t_0) = g(z,0) = f(z)$. *Write* $f(z,t) = \beta(t)[z + a_2(t)z^2 + \cdots]$, *where* $\beta(t) > 0$ *is a continuous function of* t, $0 \leq t \leq t_0$, *and* $\beta(0) = 1$; *thus*, $\beta(t) \downarrow 0$ *as* $t \uparrow [0, t_0]$. *Let* $\beta(t) = e^{-t}$, $0 \leq t \leq t_0 = \ln(1/\beta)$. *Then* $f(z,t)$ *satisfies the differential equation*

$$\frac{\partial f(z,t)}{\partial t} = -f(z,t)\frac{1 + k(t)f(z,t)}{1 - k(t)f(z,t)}, \tag{4.2.2}$$

where $k(t)$ *is a continuous function of* t, $|k(t)| = 1$. *Eq (4.2.2) is known as the Löwner's differential equation.*

PROOF. This proof for Eq (4.2.2) is based on Goluzin [1939]. Let

$$h(z, t', t'') = g^{-1}[g(z,t'), t''] = f[f^{-1}(z,t'), t'']$$
$$= e^{t' - t''} z + \cdots, \quad 0 \leq t' \leq t'' \leq t_0. \tag{4.2.3}$$

If $g(z,t)$ is continuous in E, and $g'(z,t)$ continuous in B_t, then $h(z, t't'')$ is continuous in $E \cup \partial E$ for any t', t'' chosen in $[0, t_0]$.

Let $\lambda(t)$ be the point on the circle $|z| = 1$, which is mapped by $w = g(z,t)$ into the tip of the slit inside $|w| < 1$. Also, let $S_{t't''}$ denote a slit in the unit disk $|w| < 1$ obtained from the mapping of $|z| = 1$ by the function $w = h(z, t', t'')$, and let the corresponding arc in E be $\gamma_{t't''}$. Thus, the mapping $h(z, t', t'')$ carries the point $\lambda(t')$ into the tip of the slit $S_{t't''}$ lying in $|w| < 1$, and the endpoint of the arc $\gamma_{t't''}$ into the point $\lambda(t'')$ lying on the circle $|w| = 1$ (see Figure 4.2.1). Thus, we find that (i) for fixed $t' = t$ and $t'' \to t$, the arc $\gamma_{t't''}$

reduces to the point $\lambda(t)$; and (ii) for fixed $t'' = t$ and $t' \to t$, the slit $S_{t't''}$ reduces to the point $\lambda(t)$.

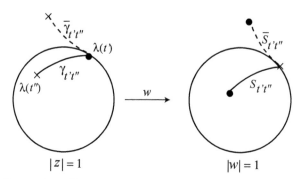

Figure 4.2.1 Slit Mappings in and outside the Unit Disk.

Now we will show that for $t' = t$ and $t'' \to t$, or for $t'' = t$ and $t' \to t$, the function $h(z, t', t'')$ approaches z uniformly in the unit disk E. If we continue $h(z, t', t'')$ analytically into $|z| > 1$ across the arc $\bar{\gamma}_{t't''}$ which is complementary to the arc $\gamma_{t't''}$ with respect to $|z| = 1$, then the function $h(z, t', t'')$ so continued maps the z-plane cut along the arc $\gamma_{t't''}$ onto the w-plane cut along the slit $S_{t't''}$ and along the arc $\bar{S}_{t't''}$ which is obtained from $S_{t't''}$ by reflection in $|w| = 1$. In the limiting case, either $\gamma_{t't''} + \bar{\gamma}_{t't''}$ or $S_{t't''} + \bar{S}_{t't''}$ reduces to the point $\lambda(t)$, as follows:

CASE 1: When $\gamma_{t't''}$ reduces to the point $\lambda(t)$, there exists a sequence $\{t''_m\}$, $m = 1, 2, \ldots$, $t'' \to t' = t$, such that the function $h(z, t', t'')$ converges uniformly to a univalent function in any bounded domain in the z-plane not containing the point $\lambda(t)$. Then as $t'' \to t$ continuously, the function $h(z, t', t'')$ converges to z uniformly in any closed bounded part of the plane excluding the point $\lambda(t)$, in particular, in any closed part of E excluding $\lambda(t)$. In fact, if $w = h(z, t', t'')$ converges continuously, the limit function is not a constant since its derivative at $z = 0$ tends to 1. Hence, the limit function is univalent in the entire z-plane excluding $\lambda(t)$ and preserves 0 and ∞ and therefore, equal to z. If $w = h(z, t', t'')$ does not converge, sequences can be found corresponding to different univalent functions, which is a contradiction since there cannot be two univalent functions defined in the entire z-plane minus one point, which preserve 0 and ∞ and have derivatives at $z = 0$ equal to 1. To show that the convergence is uniform, consider a circle $C_\varepsilon = \{z : |z - \lambda(t)| < \varepsilon\}$ of radius $\varepsilon > 0$ with center at the point $\lambda(t)$. On this circle the convergence is uniform, since, for sufficiently small $|t'' - t'|$, $|h(z, t', t'') - z| < \varepsilon$, and so, on the circle C_ε,

$$|h(z, t', t'') - \lambda(t)| \le |h(z, t', t'') - z| + |z - \lambda(t)| < 2\varepsilon. \qquad (4.2.4)$$

Since $h(z, t', t'')$ is univalent, the inequality (4.2.4) holds in the circle C_ε, which implies that $|h(z,t',t'') - z| \leq |h(z,t',t'') - \lambda(t)| + |z - \lambda(t)| < 3\varepsilon$ in C_ε, i.e., the convergence is uniform in E.

CASE 2: The same argument shows that as $t' \to t'' = t$ continuously, the function $z = h^{-1}(w,t',t'')$ converges uniformly to w inside and on the boundary of the domain obtained by removing $S_{t't''}$ from the unit disk $|w| < 1$. But then the map of $S_{t't''}$, which is $\gamma_{t't''}$, also reduces to the point $\lambda(t)$, and thus, the argument of case 1 can be repeated. Hence, in both cases $h(z,t',t'')$ converges to z uniformly in $|z| \leq 1$, and in particular, $\lambda(t)$ is a continuous function of t.

Let

$$\phi_1(z,t',t'') = \frac{\phi(z,t',t'') - \phi_0}{1 - \phi_0\phi_1(z,t',t'')}, \quad \phi_2(z,t',t'') = \frac{\phi_1(z,t',t'')}{z}, \qquad (4.2.5)$$

where

$$\phi(z,t',t'') = \frac{h(z,t',t'')}{z}, \quad \phi_0 = \phi(0,t',t'').$$

The function $w = \phi_1(z,t',t'')$ transforms the arc $\bar{\gamma}_{t't''}$ (complementary to the arc $\gamma_{t't''}$) into an arc $\Gamma^{(1)}_{t't''}$ on $|w| = 1$ such that, by Lemma 4.2.1, the length $l\left(\Gamma^{(1)}_{t't''}\right) \geq l\left(\bar{\gamma}_{t't''}\right)$. Since $h(z,t',t'')$ converges to z uniformly in $E \cup \partial E$, we see that for sufficiently small $|t' - t''|$, $\phi(z,t',t'') \neq -1$, i.e., $\phi_1(z,t',t'') \neq -1$ in $E \cup \partial E$. Thus, the arc $\Gamma^{(1)}_{t't''}$ must cover nearly the entire unit circle except for a small arc containing the point -1.

The function $w = \phi_2(z,t',t'')$ maps $\bar{\gamma}_{t't''}$ onto an arc $\Gamma^{(2)}_{t't''}$ of the circle $|w| = 1$, which shrinks to the point $-1/\lambda(t)$, since at an endpoint of the arc $\bar{\gamma}_{t't''}$, z is nearly equal to $\lambda(t)$ and ϕ is nearly equal to -1. Thus, by Lemma 4.2.2, the mapping by ϕ_2 of the part $\Delta_{t't''}$ of the unit disk E bounded by $\bar{\gamma}_{t't''}$ and the arc of the orthogonal circle through the endpoints of the arc $\bar{\gamma}_{t't''}$ lies in the part $G_{t't''}$ of the unit disk $|w| < 1$ bounded by $\Gamma^{(2)}_{t't''}$ and the arc of the orthogonal circle through its endpoints. Since $G_{t't''}$ shrinks into the point $-1/\lambda(t)$, we see that $\phi_2(z,t',t'') \to -1/\lambda(t)$, $\phi_1(z,t',t'') \to -z/\lambda(t)$ uniformly in any closed part of E. Thus, since $\phi(z,t',t'') = h(z,t',t'')/z$, $\phi_0 = e^{t'-t''}$, we get

$$\frac{\phi_1 - 1}{\phi_1 + 1} = \frac{(\phi - 1)(1 + \phi_0)}{(\phi + 1)(1 - \phi_0)} \to -\frac{1 + \dfrac{z}{\lambda(t)}}{1 - \dfrac{z}{\lambda(t)}} \quad \text{as } t' - t'' \to 0,$$

which, by setting $k(t) = 1/\lambda(t)$, $|k(t)| = 1$, yields

$$\frac{h(z,t',t'') - z}{t'' - t'} \to -z\frac{1 + k(t)z}{1 - k(t)z}. \qquad (4.2.6)$$

Since $h(z, t', t'') = h\left[f(\zeta, t').t', t''\right] = f(\zeta, t'')$, Eq (4.2.6) becomes

$$\frac{f(\zeta, t'') - f(\zeta, t')}{t'' - t'} = -f(\zeta, t)\frac{1 + k(t)f(\zeta, t)}{1 - k(t)f(\zeta, t)}, \qquad (4.2.7)$$

where $t = t'$ or t'' depending on the variable held fixed. Hence, f is differentiable from both sides, and thus, replacing ζ by z in (4.2.7) we obtain (4.2.2). ∎

The parametric differential equation (4.2.2) can be written as the initial value problem

$$\frac{\partial w}{\partial t} = -w\frac{1 + k(t)w}{1 - k(t)w}, \qquad w = f(z, t),$$
$$f(z, 0) = z, \qquad (4.2.8)$$

with a unique solution $w = f(z)$ for $t = t_0$. Conversely, given a continuous function $k(t)$, $|k(t)| = 1$, $0 \leq t' \leq t_0$, the solution $w = f(z, t)$, $f(0, t) = z$, of the corresponding initial value problem is analytic and univalent and bounded by 1 in E. In fact, in view of (4.2.8), we have

$$\frac{\partial |w|}{\partial t} = -|w|\,\Re\left\{\frac{1 + k(t)w}{1 - k(t)w}\right\} = -|w|\frac{1 - |w|^2}{|1 - k(t)w|^2|}, \qquad (4.2.9)$$

i.e, for fixed $z \in E$, $|w| = |f(z, t)|$ is a decreasing function of t. Conversely, $w = f(z, t)$ stays, for increasing t, inside the domain of univalency of the right-hand side of (4.2.2) and so is analytic, univalent and bounded by 1 in the unit disk E.

To show that the solution of the initial value problem (4.2.8) is unique, let there be two points z_1 and z_2 for a certain t, such that $f(z_1, t) = f(z_2, t) = w_0$. Then, in view of the uniqueness of the solution of the initial value problem (4.2.8), we must have $f(z_1, 0) = f(z_2, 0)$, i.e., $z_1 = z_2$. Hence, the function $f(z, t)$ is univalent for all t, $0 \leq t \leq t_0$.

An alternative method to derive the Löwner's differential equation (4.2.8) is to consider the function

$$\psi(z) = \psi(z, t', t'') = \log\frac{h(z, t', t'')}{z}, \qquad (4.2.10)$$

where the branch of the log is chosen such that $\psi(0) = t' - t''$. The function $\psi(z)$ is analytic on a domain D. Using the definition (4.2.3) of $h(z, t', t'')$, we find that $\Re\{\psi(z)\} = 0$ on $\gamma_{t't''}$ and $\Re\{\psi(z)\} < 0$ on $\bar{\gamma}_{t't''}$. Then by Schwarz integral formula (Theorem 2.7.2)

$$\psi(z) = \frac{1}{2\pi}\int_{\lambda(t')}^{\lambda(t'')}\Re\{\psi(e^{i\theta})\}\frac{e^{i\theta} + z}{e^{i\theta} - z}\,d\theta, \qquad (4.2.11)$$

such that in particular,

$$t' - t'' = \psi(0) = \frac{1}{2\pi} \int_{\lambda(t')}^{\lambda(t'')} \Re\{\psi(e^{i\theta})\} \, d\theta. \qquad (4.2.12)$$

Replacing z in (4.2.11) by $f(z,t')$ and recalling that $h(z,t'),t',t'') = f(z,t)$, we get

$$\log \frac{f(z,t)}{f(z,t')} = \frac{1}{2\pi} \int_{\lambda(t')}^{\lambda(t'')} \Re\{\psi(e^{i\theta})\} \frac{e^{i\theta} + f(z,t')}{e^{i\theta} - f(z,t')} d\theta. \qquad (4.2.13)$$

Then, by the mean value theorem applied to the real and imaginary parts of (4.2.13), we obtain

$$\begin{aligned}
\log \frac{f(z,t)}{f(z,t')} = \frac{1}{2\pi} \Big[&\Re\Big\{ \frac{e^{i\tau'} + f(z,t')}{e^{i\tau'} - f(z,t')} \Big\} \\
&+ i \Im\Big\{ \frac{e^{i\tau''} + f(z,t')}{e^{i\tau''} - f(z,t')} \Big\} \Big] \int_{\lambda(t')}^{\lambda(t'')} \Re\{\psi(e^{i\theta})\} \, d\theta,
\end{aligned} \qquad (4.2.14)$$

where $\lambda(t') = e^{i\tau'}$ and $\lambda(t'') = e^{i\tau''}$ are the two endpoints of $\gamma_{t't''}$. Let $t'' \to t'$; the arc $\gamma_{t't''}$ shrinks to the point $\lambda(t')$. Then using (4.2.12), we get

$$\frac{\partial}{\partial t'}\{\log f(z,t')\} = -\frac{\lambda(t') + f(z,t')}{\lambda(t') - f(z,t')}, \qquad (4.2.15)$$

where the derivative is taken from the right. Similarly, if $t' \to t''$, then the derivative in (4.2.15) is from the left. Since $k(t) = 1/\lambda(t)$, the function $k(t)$ has the same properties as $\lambda(t)$, that is, $k(t)$ is continuous and $|k(t)| = 1$. Thus, Eq (4.2.15) becomes the Löwner differential equation (4.2.2). ∎

Eq (4.2.8) is known as the *radial Löwner equation* defined on the unit disk E and can be written as

$$\frac{\partial}{\partial t} f(z,t) = -f(z,t)\frac{f(z,t) + \lambda(t)}{f(z,t) - \lambda(t)}, \quad f(z,0) = z, \qquad (4.8.16)$$

where $\lambda : [0,T] \mapsto \mathbb{R}$ is a continuous function of z and t, $z \in E$; in fact, $\lambda(t) = 1/k(t)$, $|\lambda(t)| = 1$.

4.2.1 Chordal Löwner Equation. This equation is defined on the upper half-pane $\mathbb{U}^+$ in two ways: (a) by running time forwards, yielding the *forward Löwner equation*, and (b) by running time backwards, yielding the *backward Löwner equation*.

(a) **Forward Löwner Equation.** The equation, defined on $\mathbb{U}^+$, has the form

$$\frac{\partial}{\partial t} f(z,t) = \frac{2}{f(z,t) - \lambda(t)}, \quad f(z,0) = z, \qquad (4.8.17)$$

where λ is the same function defined in Eq (4.8.16). In view of the existence and uniqueness theorem for differential equations (see below), every $z \in \mathbb{U}^+$ corresponds to some time interval $[0, t_0)$ such a unique solution of the initial value problem (4.8.17) exists.

(b) **Backward Löwner Equation.** Just as the forward Löwner equation takes the curve γ and moving it along the real axis as time moves forward, so in the case of backward Löwner equation we start at the ending time T and move backwards to time 0, thus resulting in some curve γ from the previously 'empty' (at time T) upper half-plane under the mapping $g(z,t)$. Such functions are generated by the *backward Löwner equation*

$$\frac{\partial g(z,t)}{\partial t} = \frac{-2}{f(z,t) - \xi(t)}, \quad g(z,0) = z, \qquad (4.8.18)$$

where ξ is real and continuous. Notice the difference between Eqs (4.8.17) and (4.8.18): Besides the obvious negative sign on the right hand side, these two equations are related such that if T is the largest possible value for t, then letting $\lambda(T - t) = \xi(t)$ we get one equation from the other. Although it is quite natural to think that the function $g(z,t)$ generated by Eq (4.8.18) will be inverse of $g(z,t)$ generated by Eq (4.8.18), yet this is not generally true. Thus, the curve γ generated by (4.8.18) is not necessarily the same curve γ generated by (4.8.17). However, it is true that $g(z,T) = f^{-1}(z,T)$, where T is the final time.

Now, going back to the forward equation (4.8.17), let us set $\lambda = \frac{3}{2} - \frac{3}{2}\sqrt{1 - 8t}$. Then the line of singularities begins to track out a semicircle of radius $1/2$ centered at $z = 1/2$. Note that at time $t = \frac{1}{8}$ the generated curve touches the real axis.

4.3 Löwner-Kufarev Equation

The Löwner-Kufarev theorem is a slight modification of Löwner's theorem. The subtle differences will become clear, and it will be found that Löwner-Kufarev theorem leads to additional results and conformal inequalities.

Theorem 4.3.1. *If $f \in S^\star$ and $p \in P$ for $z \in E$, then for any real $m > 0$ the principal branch of the function*

$$w(z) = \left\{ \int_0^z p(\zeta)\, \zeta^{-1} \left[g(\zeta) \right]^m d\zeta \right\}^{1/m} \qquad (4.3.1)$$

belongs to the class S and is a solution of the Löwner-Kufarev equation

$$\frac{\partial f(z,t)}{\partial t} = -f(z,t)\,p(z,t), \qquad (4.3.2)$$

under the initial condition $f(z,0) = z$, where $f(z,t) = e^{-t}z + c_3\,z^3 + \cdots$ is the solution of the integral equation (4.3.2) and, for a fixed value of t, represents a function in the class S,

$$p(z,t) = 1 + \alpha_1(t)\,z + \cdots \in \mathcal{P}, \quad \text{for } z \in E,$$

and is piecewise continuous with respect to the parameter t, $0 \leq t, \infty$.

Theorem 4.3.2. *Let $p(z,t) = \alpha_1(t)\,z + \cdots \in \mathcal{P}$ for $z \in E$ and is a piecewise continuous with respect to t, $0 \leq t < \infty$. Then the solution of the equation (4.3.2) which satisfies the initial condition $f(z,0) = z$ is a function $w = f(z,t)$ such that $f(0,t) = 0$ and $f'(0,t) = e^{-it}$, and for each $t \in [0,\infty)$ is a regular univalent function of z in E. Moreover, the limit*

$$f(z) = \lim_{t \to \infty} e^t f(z,t), \quad f(0) = 0, f'(0) = 1$$

exists and is also a function with these properties:

$$p(z,t) = \frac{1 + k(t)}{1 - k(t)},$$

where $k(t)$ is an arbitrary piecewise continuous function , and $|k(t)| = 1$ for $0 \leq t < \infty$.

4.4 Applications

An important application of Löwner's theorem is an elaborate proof for the bound on $|a_3|$ for univalent functions $f \in S$, as proved in the following theorem. Another application of the Löwner method is the Fekete-Szegö theorem which disproved the Littlewood-Paley conjecture on modulus estimates of coefficients of odd univalent functions.

Theorem 4.4.1. *Let $f(z) = z + a_2 z^2 + a_3 z^3 + \cdots \in S$. Then $|a_3| \leq 3$, and equality holds if and only if f is the Koebe function $K(z)$ or one of its rotations.*

PROOF. The proof is in two parts. First, will prove that $\Re\{a_3\} \leq 3$, which will imply that $|a_3| \leq 3$, since $f \in S$ is invariant under rotations. Recall that $f(z) = \lim_{t \to \infty} f(z,t)$, where $f(z,t)$ is the solution of the Löwner initial value problem (4.2.8). Since $f(z,t) = e^{-t}\left[z + a_2(t)z^2 + a_3(t)z^3 + \cdots\right]$, and $a_n(0) = 0$, we have $\lim_{n \to \infty} a_n(t) = a_n$ for $n = 2, 3, \ldots$. Expanding both sides of the Löwner differential equation in (4.2.2) and equating the coefficients, we get

$$a_2'(t) = -2e^{-t} k(t), \tag{4.4.1}$$

$$a_3'(t) = -2e^{-2t} k^2(t) - 4e^{-t} k(t)a_2(t)$$
$$= -2e^{-2t} k^2(t) + 2a_2(t)a_2'(t), \text{ using (4.4.1).} \tag{4.4.2}$$

Integrating (4.4.1) we get

$$a_2 = \int_0^\infty a_2'(t)\, dt = -2 \int_0^\infty e^{-t} k(t)\, dt, \tag{4.4.3}$$

which, using the fact that $|k(t)| = 1$, yields $|a_2| \leq 2 \int_0^\infty e^{-t}\, dt = 2$, where equality holds if and only if $k(t) = \text{const}$, which is the result of $f(z)$ being a rotation of the Koebe function. Next, integrating Eq (4.4.2) we get

$$a_3 = -2 \int_0^\infty e^{-2t} k^2(t)\, dt + 4 \left[\int_0^\infty e^{-t} k(t)\, dt \right]^2. \tag{4.4.4}$$

Set $k(t) = e^{i\theta(t)}$. Then

$$\Re\{a_3\} \leq 2 \int_0^\infty e^{-2t} \left[1 - 2\cos^2 \theta(t) \right]\, dt + 4 \left[\int_0^\infty e^{-t} \cos\theta(t)\, dt \right]^2.$$

Then, by Cauchy-Schwarz inequality, we have

$$\begin{aligned}
\Re\{a_3\} &\leq 1 - 4 \int_0^\infty e^{-2t} \cos^2 \theta(t)\, dt \\
&\quad + 4 \left[\int_0^\infty e^{-t}\, dt \right] \left[\int_0^\infty e^{-t} \cos^2 \theta(t)\, dt \right] \\
&= 1 + 4 \int_0^\infty \left(e^{-t} - e^{-2t} \right) \cos^2 \theta(t)\, dt \bigg] \\
&\leq 1 + 4 \int_0^\infty \left(e^{-t} - e^{-2t} \right)\, dt = 3,
\end{aligned}$$

where equality holds if and only if $\cos^2 \theta(t) = 1$. Since $k(t)$ is continuous, and $|k(t)| = 1$, it will happen when either $k(t) = 1$ or $k(t) = -1$, which corresponds to the Koebe function $K(z) = \dfrac{z}{(1+z)^2}$ or $K(z) = \dfrac{z}{(1-z)^2}$.

Next, we prove that $|a_3| \leq 3$ for $f \in \mathcal{S}$. As shown above, without loss of generality we can assume that $a_3 = 3$ for some function $f \in \mathcal{S}$. But for such a function f, there exists a sequence of single slit mappings $\{f_n\}$ such that f_n

converges locally to f. Since each f_n is a slit mapping, the coefficients $a_2(f_n)$ and $a_3(f_n)$ of f_n can, in view of (4.4.3) and (4.4.4), be written as

$$a_2(f_n) = -2 \int_0^\infty e^{-t} k_n(t)\, dt,$$

$$a_3(f_n) = -2 \int_0^\infty e^{-2t} k_n^2(t)\, dt + 4\left\{ \int_0^\infty e^{-t} k_n(t)\, dt \right\}^2, \tag{4.4.5}$$

such that $a_2(f_n) \to a_2$ and $a_3(f_n) \to a_3$ as $n \to \infty$. Let $k(t) = e^{i\theta_n(t)}$. Then

$$\Re\{a_3(f_n)\}$$

$$= -2 \int_0^\infty e^{-2t} \cos 2\theta_n\, dt + 4\left\{ \int_0^\infty e^{-t} \cos\theta_n\, dt \right\}^2 - 4\left\{ \int_0^\infty e^{-t} \sin\theta_n\, dt \right\}^2$$

$$= 1 - 4 \int_0^\infty e^{-2t} \cos^2\theta_n\, dt + 4\left\{ \int_0^\infty e^{-t} \cos\theta_n\, dt \right\}^2 - 4\left\{ \int_0^\infty e^{-t} \sin\theta_n\, dt \right\}^2$$

$$= 3 - 4\,(I_1 + I_2 + I_3), \tag{4.4.6}$$

where

$$I_1 = \int_0^\infty \left(e^{-t} - e^{-2t} \right)\left(1 - \cos^2\theta_n \right) dt \geq 0,$$

$$I_2 = \int_0^\infty e^{-t}\, dt \int_0^\infty e^{-t} \cos^2\theta_n\, dt - \left\{ \int_0^\infty e^{-t} \cos\theta_n\, dt \right\}^2 \geq 0, \tag{4.4.7}$$

$$I_3 = \left\{ \int_0^\infty e^{-t} \sin\theta_n\, dt \right\}^2 \geq 0.$$

Since $a_3(f_n) \to 3$ implies $\Re\{a_3(f_n)\} \to 3$, we find that $I_1 \to 0, I_2 \to 0$, and $I_3 \to 0$ as $n \to \infty$.

Now, since $I_3 \to 0$, we have

$$\lim_{n\to\infty} \int_0^\infty e^{-t} \sin\theta_n\, dt = 0.$$

Since $I_1 \to 0$, we have

$$\lim_{n\to\infty} \int_0^\infty \left(e^{-t} - e^{-2t} \right)\left(1 - \cos^2\theta_n \right) dt = 0.$$

The integrand in this integral being non-negative, we have

$$\lim_{n\to\infty} \int_c^\infty \left(e^{-t} - e^{-2t} \right)\left(1 - \cos^2\theta_n \right) dt = 0.$$

Since $\dfrac{e^{-t} - e^{-2t}}{1 - e^{-c}} \geq e^{-t}$ for $t > c$, we get $\displaystyle\lim_{n \to \infty} \int_c^\infty e^{-t} \left(1 - \cos^2 \theta_n\right) dt = 0$,

as $\displaystyle\int_0^c e^{-t} \left(1 - \cos^2 \theta_n\right) dt \leq c$. Take $c = \varepsilon/2$, where $\varepsilon > 0$ is a small number.

Then for a sufficiently large $N > 0$ such that $\displaystyle\int_0^c e^{-t} \left(1 - \cos^2 \theta_n\right) dt < \varepsilon/2$

for $n > N$. Thus, $\displaystyle\int_0^\infty e^{-t} \left(1 - \cos^2 \theta_n\right) dt < \varepsilon$ when $n > N$, which implies
that

$$\lim_{n \to \infty} \int_0^\infty e^{-t} \cos^2 \theta_n \, dt = 1.$$

Since $I_2 \to 0$, we have

$$\lim_{n \to \infty} \left\{ \int_0^\infty e^{-t} \cos \theta_n \, dt \right\}^2 = \lim_{n \to \infty} \int_0^\infty e^{-t} \cos^2 \theta_n \, dt = 1.$$

Thus, $\displaystyle\lim_{n \to \infty} \left| \int_0^\infty e^{-t} \cos \theta_n \, dt \right| = 1$. Combining these results, we find from
(4.4.5) that

$$|a_2(f_n)|^2 = \left| -2 \int_0^\infty e^{-t} k_n(t) \, dt \right|^2$$

$$= \left\{ -2 \int_0^\infty e^{-t} \cos \theta_n \, dt \right\}^2 + \left\{ 2 \int_0^\infty e^{-t} \sin \theta_n \, dt \right\}^2,$$

which yields $\displaystyle\lim_{n \to \infty} |a_2(f_n)| = 2$, i.e., $\displaystyle\lim_{n \to \infty} |a_2(f_n)| = |a_2| = 2$, and hence from
(4.2.2), $|a_3| = 3$ if and only if f is the Koebe function or one of its rotations. ∎

It is not easy to use this method further to prove that $|a_4| \leq 4$. However, Nehari [1974] proved this estimate by using Löwner's method exclusively after 50 years of Löwner's original work.

Theorem 4.4.2. (Fekete-Szegö theorem) *For every* $f(z) = z + a_2 z^2 + a_3 z^3 + \cdots \in \mathcal{S}$, *and* $0 < \alpha < 1$, *the following estimate holds:*

$$\left| a_3 - \alpha a_2^2 \right| \leq 1 + 2e^{-2\alpha/(1-\alpha)}. \tag{4.4.8}$$

This estimate is sharp for each fixed α.

PROOF. Since $\arg\{a_3\}$ and $\arg\{a_2^2\}$ both change by the same amount under a rotation of $f(z)$, it would suffice to consider $\Re\{a_3 - a_2^2\}$. Then from (4.4.3) and (4.4.4), with $k(t) = e^{i\theta(t)}$, we get

$$\Re\{a_3 - a_2^2\} = 4(1 - \alpha)\left\{ \left[\int_0^\infty e^{-t} \cos \theta(t) \, dt \right]^2 - \left[\int_0^\infty e^{-t} \sin \theta(t) \, dt \right]^2 \right\}$$

$$- 4 \int_0^\infty e^{-2t} \cos^2 \theta(t) \, dt + 1$$

$$= 4(1 = \alpha)\left\{ \int_0^\infty \phi(t) \, dt \right\}^2 - 4 \int_0^\infty |\phi(t)|^2 \, dt + 1,$$

where $\phi(t) = e^{-t} \cos \theta(t)$. If $\int_0^\infty |\phi(t)|^2 \, dt = \left(\lambda + \frac{1}{2}\right) e^{-2\lambda}$, then by the Valiron-Landau lemma (Exercise 4.6.2), we get

$$\Re\{a_3 - a_2^2\} \leq 4e^{-2\lambda}\left[(1-\alpha)(\lambda+1)^2 - \left(\lambda + \tfrac{1}{2}\right)\right] + 1. \tag{4.4.9}$$

The right-hand side of inequality (4.4.9) attains its maximum value of $1 + 2\exp\left\{\dfrac{-2\lambda}{1-\lambda}\right\}$ when $\lambda = \dfrac{\alpha}{1-\alpha}$, and the inequality (4.4.8) is proved. To show that the estimate (4.4.8) is sharp, let $\lambda = \dfrac{\alpha}{1-\alpha}$, and choose $\theta(t)$, $-\pi/2 > \theta(t) < \pi/2$, such that $e^{-t} \cos \theta(t) = \psi(t)$. We choose

$$\cos \theta(t) = \begin{cases} e^{t-\lambda} & \text{if } 0 \leq t \leq \lambda, \\ 1 & \text{if } \lambda < t < \infty. \end{cases}$$

and

$$\int_0^\infty e^{-t} \sin \theta(t) \, dt = 0. \tag{4.4.10}$$

Let τ, $0 < \tau < \lambda$, be such that $-\pi/2 < \theta(t) \leq 0$ for $\tau \leq t \leq \lambda$ and $0 < \theta(t) < \pi/2$ for $0 \leq t \leq \tau$. Then

$$\sin \theta(t) = \begin{cases} \left[1 - e^{2(t-\lambda)}\right]^{1/2} & \text{if } 0 \leq t \leq \tau, \\ -\left[1 - e^{2(t-\lambda)}\right]^{1/2} & \text{if } \tau \leq t \leq \lambda, \\ 0 & \text{if } \lambda < t < \infty. \end{cases}$$

Substituting these values of $\sin \theta(t)$ into (4.4.10), we can find a suitable value of τ so that (4.4.10) holds. ∎

Corollary 4.4.1. We have $c_5 = \frac{1}{2}\left(a_3 - \frac{1}{4}a_2^2\right)$, i.e., $|c_5| \leq \frac{1}{2} + e^{-2/3} \approx 1.013$. This result disproved the Littlewood-Paley conjecture on the estimate of modulus of coefficients c_{2n-1} of odd univalent functions $h(z)$, defined by

$$h(z) = \sqrt{f(z^2)} = \sum_{n=1}^\infty c_{2n-1} z^{2n-1}. \tag{4.4.11}$$

Theorem 4.4.3. *If $h(z)$ is an odd univalent function, defined by (4.4.11), then*

$$|c_1|^2 + |c_3|^2 + |c_5|^2 \leq 3. \tag{4.4.12}$$

PROOF. Assume that $c_5 \geq 0$, and since $c = 1$, we will show that $|c_3|^2 + \left[\Re\{c_5\}\right]^2 \leq 2$. Note that

$$c_3 = \frac{1}{2}a_2 = -\int_0^\infty e^{-t} k(t) \, dt,$$

$$c_5 = \frac{1}{2}\left(a_3 - \frac{1}{4}a_2^2\right) = \frac{3}{2}\left\{\int_0^\infty e^{-t} k(t) \, dt\right\}^2 - \int_0^\infty e^{-2t} k^2(t) \, dt.$$

Set $k(t) = e^{i\theta(t)}$, and $\int_0^\infty e^{-t} k(t)\,dt = u + iv$. Then

$$|c_3|^2 = u^2 + v^2,$$

$$c_5 = \Re\{c_5\} = \frac{3}{2}\left(u^2 - v^2\right) - 2\int_0^\infty e^{-2t} \cos^2 \theta(t)\,dt + \frac{1}{2}.$$

Set $\int_0^\infty e^{-2t} \cos^2 \theta(t)\,dt = \left(\lambda + \frac{1}{2}\right) e^{-2t}$, $0 \le \lambda < \infty$. Then, the Valiron-Landau lemma (Exercise 4.6.2), we get $|u| \le (\lambda+1)\,e^{-\lambda}$. But since $u^2 + v^2 \le 1$, we obtain $|c_3|^2 + |c_5|^2 \le m(\lambda)$, where

$$m(\lambda) = \min\left\{1 - v^2, (\lambda+1)^2\,e^{-2\lambda}|\right\} + v^2 + \frac{1}{4}\left[\beta(\lambda) + 1 = 3v^2\right]^2,$$

$$\beta(\lambda) = \left(3\lambda^2 + 2\lambda + 1\right)e^{-2\lambda}.$$

Note that $0 \le \beta(\lambda) \le 2e^{-2\lambda} \approx 1.026$ for $0 \le \lambda < \infty$. Now, we consider the following two cases:

CASE 1. If $0 \le 1 - v^2 \le (\lambda+1)^2\,e^{-2\lambda}$, then $m(\lambda) = 1 + \frac{1}{4}\left[\beta(\lambda) + 1 = 3v^2\right]^2$, which is a quadratic polynomial in v^2 for each fixed λ, and this polynomial attains its maximum at the endpoints of the interval. Thus, at the end $v^2 = 1 - (\lambda+1)^2\,e^{-2\lambda}$, we get $m(\lambda) = 1 + \left[\left(3\lambda^2 + 4\lambda + 2\right)e^{-2\lambda} = 1\right]^2 \le 2$, while at the end $v^2 = 1$ we get $m(\lambda) = 1 + \frac{1}{4}\left[\beta(\lambda) - 2\right]^2 \le 2$.

CASE 2. If $(\lambda+1)^2\,e^{-2\lambda} \le 1 - v^2 \le 1$, then $m(\lambda) = (\lambda+1)^2\,e^{-2\lambda} + v^2 + \frac{1}{4}\left[\beta(\lambda) + 1 - 3v^2\right]^2$, which is again a quadratic polynomial in v^2 that attains its maximum at the endpoints of the interval. Thus, at end $v^2 = 1 - (\lambda+1)^2\,e^{-2\lambda}$, we get $m(\lambda) \le 2$, as shown in Case 1, while at the end $v^2 = 0$, we get $m(\lambda) = (\lambda+1)^2\,e^{-2\lambda} + \frac{1}{4}\left[\beta(\lambda) + 1\right]^2$.

Now, since $m'(\lambda) = \lambda(1 - 3\lambda)\left(3\lambda^2 + 2\lambda + 1\right)e^{-4\lambda} = \lambda(5\lambda + 1)\,e^{-2\lambda}$, we find that $m'(\lambda), 0$ for $1/3 \le \lambda < \infty$, whereas for $0 < \lambda < 1/3$ we obtain $m; (\lambda) < -3\lambda^2\left(3\lambda^2 + \lambda + 2\right)e^{-2\lambda} < 0$. Hence, $m(\lambda) \le m(0) = 2$ for all $o \le \lambda < \infty$. ∎

The next application of the Löwner method to obtain the coefficients estimates is as follows.

Theorem 4.4.5. (Goluzin theorem) *Let* $f(z) = z + a_2 z^2 + a_3 z^3 + \cdots \in \mathcal{S}$. *Then*

$$-1 \le |a_3| - |a_2| \le \frac{3}{4} + e^{-\lambda_0}\left(2e^{-\lambda_0} - 1\right) \approx 1.029. \qquad (4.4.13)$$

PROOF. Using the inequality (3.1.2) (Gronwall area principle) for any function $g \in \Sigma$, we have $|b_1| = |a_2|^2 - |a_3| \le 1$. First we prove the left-hand side of (4.4.13): If $|a_2| < 1$, then $|a_2| - |a_3| \le 1$. If $|a_2| \ge 1$, then, using

$|a_2^2 - a_3| \le 1$, we get $|a_2| - |a_3| - |a_2|^2 - |a_3| + |a_2| (1 - |a_2|) \le 1$, which shows that the left-hand side inequality holds. Next, we use the Löwner method to prove the right-hand side of (4.4.13): Rotate the function such that $a + 3 \ge 0$, write $a_2 = -2(u + iv)$, and let $k(t) = e^{i\theta(t)}$. Then, using (4.4.3) and (4.4.4), we get

$$
\begin{aligned}
|a_3| - |a_2| &= \Re\{a_3\} - |a_2| \\
&= -4 \int_0^\infty e^{-2t} \cos^2 \theta(t)\, dt + 1 + 4 \left(u^2 - v^2\right) - 2 \left[u^2 + v^2\right]^{1/2} \\
&\le 1 + 4u^2 - 2|u| - 4 \int_0^\infty e^{-2t} \cos^2 \theta(t)\, dt.
\end{aligned}
$$

Since the integral is taken as $\int_0^\infty e^{-2t} \cos^2 \theta(t)\, dt = (\lambda + \frac{1}{2}) e^{-2\lambda}$, $0 \le \lambda < \infty$, so by the Valiron-Landau lemma (Exercise 4.6.2), we get $|u| \le (\lambda + 1) e^{-\lambda}$. Now, we consider two cases.

CASE 1. If $(\lambda + 1) e^{-\lambda} \le \frac{1}{2}$, then $4u^2 - 2|u| \le 0$, giving $|a_3| = |a_2| \le 1$.

CASE 2. If $(\lambda + 1) e^{-\lambda} > \frac{1}{2}$, then $4u^2 - 2|u|$ is positive and increasing in $|u|$ over the interval $\frac{1}{2} < |u| < \infty$. Thus,

$$
\begin{aligned}
|a_3| - |a_2| &\le 1 + 4(\lambda + 1)^2 e^{-2\lambda} - 2(\,l + 1) e^{-\lambda} - 4 \left(\lambda + \frac{1}{2}\right) e^{-2\lambda} \\
&= 1 + 2 \left(2\lambda^2 + 2\lambda + 1\right) e^{-2\lambda} - 2(\lambda + 1) e^{-\lambda}.
\end{aligned}
$$

Now, we find the maximum of the function $\phi(\lambda) = 2 \left(2\lambda^2 + 2\lambda + 1\right) e^{-2\lambda} - 2(\lambda + 1) e^{-\lambda}$ over the interval $0 \le \lambda < \infty$. Since $\phi'(\lambda) = \lambda e^{-\lambda} \left(1 - 4\lambda e^{-\lambda}\right)$, which shows that the function $\phi(\lambda)$ attains its maximum at the points $\lambda = 0, \lambda_0, \lambda_1$, where $0 < \lambda_0 < 1$ and $\lambda_1 > 2$ are the roots of the equation $4\lambda e^{-\lambda} = 1$. However, since $(\lambda_1 + 1) e^{-\,l_1}, \frac{1}{2}$, we drop λ_1. Taking the second derivative of $\phi(\lambda)$, i.e.,

$$
\phi''(\lambda) =!1 - \lambda) e^{-\lambda} \left(1 - 4la\, e^{-\lambda}\right)^2 + 4\lambda(\lambda - 1) e^{-2\lambda},
$$

we find that $\phi''(\lambda_0) = (\lambda_0 - 1) e^{-\lambda_0} < 0$, and λ_0 is a local maximum of $\phi(\lambda)$, which the numerical value $\lambda_0 \approx 0.3574$. Since $\lambda_0 e^{-\lambda_0} = \frac{1}{4}$, we get the right-had side (4.4.13). The uniqueness can be proved the same way as in the Valiron-Landau lemma (Exercise 4.6.2). ∎

4.5 Slit Mappings

Let $f \in \mathcal{S}$, and so is the function $\dfrac{1}{\rho} f(\rho z)$, $0 < \rho < 1$, where $\rho \to 1$. Let the function $f(z) = \beta(z +_2 z^2 + \cdots$, $\beta > 0$, map the unit disk E onto a region bounded by a Jordan curve, which together with its boundary lies entirely within the unit disk $|w| < 1$ such that the directions at the origin $z = 0$ are preserved under this mapping. This region is known as the bounded region and will be denoted by B.

A domain in the w-plane obtained by removing from the unit disk $|w| < 1$ a Jordan arc Γ starting from a point on the circle $|w| = 1$ and extending into the interior of this disk is known as a *slit domain*. If the arc Γ does not pass through $w = 0$, then such domains are called the *domain* (s). It is easy to see that any bounded domain B can be approximated by a domain (s) in the sense that the limit of functions mapping E onto the domain (s) and preserving $z - 0$ and the directions there, will map E onto the bounded domain B and satisfy the same conditions at $w = 0$. In fact, to see this, pass a cut from any point on $|w| = 1$ to any point P on the boundary ∂B, and then cut from P around ∂B back to a point Q which is near P. As $Q \to P$, the corresponding domain (s) has the required property in view of the theorem on convergence.

The Carathéodory kernel theorem (Theorem 4.1.1) shows that the slit mappings are dense in $\mathcal{S}$. The Löwner chains of slit mappings of the unit disk E is defined as a conformal mapping of the complex plane $\mathbb{C}$ minus a simple Jordan curve going to infinity. Then a family is obtained by gradually erasing this curve, and Eq (4.1.2) becomes

$$\frac{dw}{dt} = -w\frac{k(t) + w}{k(t) - w}, \qquad (4.5.1)$$

where $k : [0, +\infty) \mapsto \mathbb{R}$ is a continuous function. The associated family $\{\phi_{s,t}\} = f^{-1}(z,t) \circ f(z,s)$ maps E onto itself with a slit (the Jordan curve) starting from the boundary. The mappings $\phi_{s,t}$ are normalized by $\phi_{s,t}(0) = 0, \phi'_{s,t}(0) > 0$.

The Löwner's differential equation (4.2.2) describes the evolution of a normalized Löwner chain. An important case is that of *slit domains* $D_t = D \backslash \gamma(t, t_1)$, where $D \subseteq \mathbb{C}$ is simply connected and γ parametrizes a simple arc that is contained in D except for one endpoint $\gamma(t_1) \in \partial D$. In view of Eq (4.5.1), we have

$$\partial_t f = z\frac{k(t) + z}{k(t) - z}\,\partial_z f, \qquad (4.5.2)$$

where $k(t) = f^{-1}(\gamma(t), t)$ is known as the *driving term* of the Löwner's equation. Since $k(t)$ is continuous, it generates a sequence of increasing domains. In fact, Eq (4.2.2) generates conformal maps onto increasing domains assuming measurability of f in t. Also, Kufarev [1947] has shown that the domains D_t obtained from Eq (4.5.2) are not necessarily slit domains if k is only assumed to be continuous. The condition on k that guarantees slit domains is that $k \in \mathrm{Lip}\left(\frac{1}{2}\right)^1$ (see Marshall and Rhode [2005]).

---

[1] The Lipschitz condition $\mathrm{Lip}(\alpha)$ is defined in the Notation at the beginning.

Using $w = f(z,t)$ in Eq (4.1.1) we obtain an initial-value problem

$$\frac{d}{dt}f(z,t) = \frac{2}{f(z,t) - k(z,t)}, \quad f(z,0) = z. \qquad (4.5.3)$$

The solution of this problem yields a family of conformal maps $w = f(z,t)$ which map a subdomain of the upper half z-plane into the upper half w-plane, such that the function $f(z,t)$ satisfies the asymptotic condition at $t = 0$:

$$f(z) = z + \frac{a}{z} + O(z^{-2}) \quad \text{as } z \to \infty, \quad a = 2t. \qquad (4.5.4)$$

This function maps a region D in the upper half-plane $\mathbb{U}^+ : \{z : \Im\{z\} > 0\}$ onto the entire upper half-plane. At $t = 0$, the function $f(z)$ is the identity map which maps the upper half z-plane onto the upper half w-plane. For each value of the parameter $t > 0$, the function $f(z,t)$ defines a new mapping domain D_t which maps onto the upper half w-plane, such that ∂D_t is mapped onto the real axis of the w-plane. The chain property of conformal mapping (§2.3.1) which is precisely the composition property also applies to the function $f(z,t)$. For example, consider two conformal maps $f^I(z,t)$ and $f^{II}(z,t)$ generated by the forcing functions $k_1(t)$ and $k_2(t)$, respectively, which exist in the intervals $[0,T_1]$ and $[0,t_2]$, respectively. Let the maps generated for each forcing functions be denoted by $f_I = f_{t_1}^I(z,t)$ and $f_{II} = f_{t_2}^{II}(z,t)$, respectively. Then the composite forcing function becomes

$$k(t) = \begin{cases} k_1(t) & \text{for } 0 < t < t_1, \\ k_2(t - t_1) & \text{for } t_1 < t < t_1 + t_2. \end{cases} \qquad (4.5.5)$$

Then , by the composition rule, the composite forcing function generates a map $f(z,t)$ such that at $t = t_1 + t_2$ it is

$$f_{t_1+t_2}(z,t) = f_{II}\left(f_I(z,t)\right). \qquad (4.5.6)$$

In fact, first we fix z and calculate $f(z,t)$ for $0 < t \le t_1$ from (4.5.3) using the initial condition $f(z,0) = z$, and then repeat the calculation for $t_1 < t \le t_2$ with the initial condition $f(z,t_1) = f_I(z,t)$. As a result we find Eq (4.5.5) at $t = t_2$.

An immediate consequence of this composition property is that the mapping set $\{D_t\}$ gets 'smaller' as t gets larger. In other words, $D_s \subset D_t$ if $s < t$. In fact, the mapping domain get smaller when some point $z_c(t)$ passes out of the domain of analyticity of $f(z,t)$, which happens as the denominator in Eq (4.5.3) passes through zero or the points at which $f(z_c,t) = k(t)$. Since $k(t)$ is real, the point $z_c(t)$ is always located at the edge of the domain D_t that is mapped to the upper half w-plane $\mathbb{U}^+$ by $f(z,t)$. If $k(t)$ is continuous, as t increases the singularities trace out a continuous curve in $\mathbb{U}^+$. We will call this curve simply as *trace*. Thus, the composition property implies that if a point is in the trace now, it will remain in it for all subsequent values of t. The trace is, therefore, a permanent path which shows where the singularities arise and what points have been removed (i.e., slit out) from the mapping domain. Hence, the properties of the bounding curve that is composed of the trace and the real axis are as follows: (i) if $k(t)$ is smooth enough such that its derivatives exist everywhere, the bounding curve never intersects itself; (ii) if $k(t)$ is periodic, the bounding curve is a self-similar curve (see Figure 4.5.1 for an example in which (a) $k(t) = \sin 2\pi t$, and (b) $k(t) = t \sin 2\pi t$); and (iii) the bounding curve can intersect itself finitely many times only if $k(t)$ is sufficiently singular, i.e., it satisfies the Löwner's condition (Löwner [1923]) that for some t,

$$\lim_{s \to 0^+} \left| \frac{k(t-s) - k(t)}{s^{1/2}} \right| > 4. \tag{4.5.7}$$

An example for $k(t) = c$ (const) is given in Exercise 4.6.4. Modern developments and applications of slit mappings are discussed in Chapter 11.

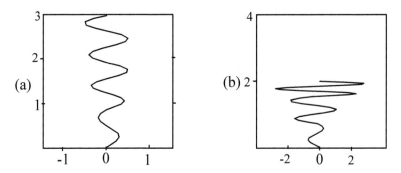

Figure 4.5.1 Slit mappings when $k(t)$ is periodic.

More results on slits mappings, mostly dealing with recent developments, are given in chapter 11.

4.6 Exercises

4.6.1. Another proof of Theorem 4.1.1 is as follows: Let $\{D_n\}$, $n = 1, 2, \ldots$, be a sequence of simply connected domains in the z-plane containing a fixed point $z_0 \neq \infty$. If there exists a disk $|z - z_0| < \rho$, $\rho > 0$, belonging to all D_n, then the kernel of the sequence $\{D_n\}$ with respect to z_0 is the largest domain D containing z_0 and such that for each compact set G belonging to D there is an N such that G belongs to D_n for all $n \geq N$. A largest domain is the one that contains any other domain having the same property. If there is no such disk, then the kernel D of the sequence $\{D_n\}$ will be the point

z_0, in which case the sequence $\{D_n\}$ has a degenerate kernel. A sequence of domains $\{D_n\}$, $n = 1, 2, \ldots$ converges to the kernel D if any subsequence of $\{D_n\}$ has the kernel D.

4.6.2. Prove the Valiron-Landau lemma, which states that if $\phi(t)$ is a real continuous function on $t > 0$, such that $|\phi(t)| \leq e^{-t}$, and $\int_0^\infty |\phi(t)|^2 \, dt = \left(\lambda + \frac{1}{2}\right) e^{-2\lambda}$ for $0 \leq \lambda < \infty$, then $\left| \int_0^\infty |\phi(t)|^2 \, dt \right| \leq (\lambda+1) e^{-\lambda}$. The equality holds if and only if $\phi(t) = \pm\psi(t)$, where $\psi(t) = \begin{cases} e^{-\lambda}, & \text{if } 0 \leq t < \lambda, \\ e^{-t}, & \text{if } \lambda < t < \infty. \end{cases}$

PROOF. For $|\phi(t)| \leq e^{-t}$, we have $0 < \int_0^\infty \phi^2(t) \, dt < \frac{1}{2}$. Note that the function $\left(t + \frac{1}{2}\right) e^{-2t}$ decreases as t increases from 0 to ∞. Thus, there exists a unique value of λ for which (4.3.11) holds. Since $\psi(t)$ is a continuous function such that $|\psi(t)| \leq e^{-t}$, and since

$$\int_0^\infty |\psi(t)|^2 \, dt = \left(\lambda + \frac{1}{2}\right) e^{-2\lambda}, \quad \int_0^\infty \psi(t) \, dt = (\lambda + 1) e^{-\lambda},$$

the function $F(t) = \big[\psi(t) - |\phi(t)|\big]\big[2e^{-\lambda} - \psi(t) - |\phi(t)|\big]$ is non-negative for all $t \geq 0$. Thus,

$$0 \leq \int_0^\infty F(t) \, dt = 2e^{-\lambda}\left\{ \int_0^\infty \psi(t) \, dt - \int_0^\infty |\phi(t)| \, dt \right\}$$
$$- \int_0^\infty \psi^2(t) \, dt + \int_0^\infty |\phi(t)|^2 \, dt - 2e^{-\lambda}\left\{ (\lambda + 1) e^{-\lambda} - \int_0^\infty |\phi(t)| \, dt \right\},$$

and $\left| \int_0^\infty \phi(t) \, dt \right| \leq \int_0^\infty |\phi(t)| \, dt \leq (\lambda + 1) e^{-\lambda}$, where the equality holds if and only if $\phi(t) = \pm\psi(t)$. ∎

4.6.3. The Robertson conjecture states that for any odd function $h(z) = \sqrt{f(z^2)} = z + c_3 z^3 + \cdots \in S$ the inequalities $1 + |c_3 2|^2 + \cdots + |c_{2n-1}|^2 \leq n$ for $n = 2, 3, \ldots$. Note that the Robertson conjecture is itself an estimate of the coefficients of square roots of the Riemann mapping functions which are power series with distinct values in the unit disk E. Use the Löwner method to prove the Robertson conjecture for $n = 3$. PROOF. The proof follows from Theorem 4.3.3.

4.6.4. Let $k(t)$ be a constant, say $k(t) = c$. Then Eq (4.4.3) has the solution

$$f(z, t) = c + \left[(z - c)^2 + 4t\right]^{1/2}. \tag{4.6.1}$$

From Figure 4.6.1, it is clear that the singularity at time t is at $z_c(t) = c + 2it^{1/2}$. Thus, the trace is a straight line which extends from $z = c$ to

$z = c + 2it^{1/2}.$

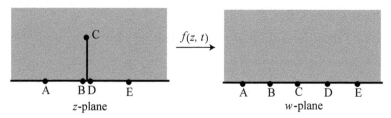

Figure 4.6.1 Slit Mapping for $k(t) = c$.

4.6.5. A Löwner family is a family of Riemann mapping functions $F(z,t)$, indexed by a positive parameter t, such that each power series $F(z,t)$ has the constant coefficient zero and coefficient of z equal to t and such that $f(z,a)$ is subordinate to $F(z,b)$ when $a < b$. PROOF. Use the fact that if $f(z)$ and $g(z)$ are normalized univalent functions, then $f(z)$ is subordinate to $g(z)$ if and only if the region onto which $f(z)$ maps the unit disk E is contained in the region onto which $g(z)$ maps E.

5

Higher-Order Coefficients

We will discuss the early efforts to solve the coefficients problem that became an inspiration after Löwner's proof for the third coefficient as presented in the previous chapter. Löwner's method became difficult to solve the coefficient problem for $n \geq 4$. Schiffer developed a regular variation method in 1938 for solving initial value problems in the complex plane, which aimed at getting the solution of the varied (iterated) problem obtain the solution of the original problem as the variation parameter approaches a specific limit. A varied problem may arise either due to a small variation in the governing differential equation or due to a slight variation of the boundary of the domain under consideration. This method provides approximate solutions for initial and/or boundary value problems by using a small parameter, say δ, where the solution for $\delta = 0$ is known. We will present Schiffer's method to obtain the bound $|a_4| \leq 4$ as presented in Garabedian and Schiffer [1955].

The other important development is based on the unitary Grunsky matrix and related application by Pederson [1967] as well as by Charzyński and Schiffer [1960] to obtain the bound for a_4. Later Pederson [1968] and Ozawa [1969], both independently, used the Grunsky inequalities and showed independently that $|a_6| \leq 6$. Later Pederson and Schiffer [1972] proved that $|a_5| \leq 5$.

Although the development of these methods to solve the coefficients problem are significant, they lacked the insight to approach a proof of the Bieberbach conjecture in the right direction. However, these methods are not only historically relevant, they provide us with an admirable mathematical development per se.

5.1 Variation Method

Schiffer [1938a] developed a variation method that could solve extremal problems within $\mathcal{S}$. While in variational calculus one of the major problems is the question about the existence of a solution, the existence is not an issue since $\mathcal{S}$ is compact. The difficulty lies in the fact that suitable comparison functions are difficult to find since $\mathcal{S}$, in spite of its useful properties, is after all merely a linear space. The main idea in Schiffer's boundary variation is to compose a univalent function f with a family of almost identical mappings in $F(D)$. As a result Schiffer obtained a differential equation that the extremal function characterizes. To clarify, consider a continuous linear functional L, and let $\Re\{L(f_0)\} = \max\limits_{f \in \mathcal{S}} \Re\{L(f)\}$. Then Schiffer's method established that $f(D) = \mathbb{C}\backslash\Gamma$, where Γ is an analytic arc that satisfies the differential inequality

$$\frac{1}{\Gamma^2(t)} L\left(\frac{f^2}{f - \Gamma(t)}\right) d\Gamma^2(t) > 0, \tag{5.1.1}$$

which implies that Γ is the trajectory of a quadratic differential. Using the Schwarz reflection principle (§2.5) for the coefficient functional $L = \Re\{a_n(f)\}$, we find the mapping function f that appears in the *Schiffer's differential equation*:

$$\left(\frac{zf'(z)}{f(z)}\right) P_n\left(\frac{1}{f(z)}\right) = R_n(z), \tag{5.1.2}$$

where P_n is the polynomial of degree $n - 1$ with the generating function

$$\frac{\varsigma\,(f(z))^2}{1 - \varsigma f(z)} = \sum_{k=2}^{\infty} P_k(\varsigma) z^k,$$

and R_n is the rational function

$$R_n(z) = (n - 1)a_n + \sum_{k=1}^{n-1}\left(k a_k z^{n-k} + k \bar{a}_k z^{n-k}\right).$$

Thus, every solution f of the Bieberbach coefficient problem must satisfy the Schiffer differential equation (5.1.2). In particular, $f(D) = \mathbb{C}\backslash\Gamma$. For every $n \geq 2$, the Koebe function is a solution of Eq (5.1.2), but this solution is not unique.

The function $f(z, t)$, with t as a parameter, $0 \leq t \leq t_0$, satisfies the Löwner differential equation (4.2.2) which is

$$\frac{\partial f(z, t)}{\partial t} = -f(z, t)\frac{1 + k(t)f(z, t)}{1 - k(t)f(z, t)}, \tag{5.1.3}$$

where $f(z, 0) = z$, and $k(t)$ is the forcing function ($k(t) = e^{i\theta}$), and $\theta = \theta(t)$ is any real continuous function in $0 \leq t \leq t_0$. We will vary $\theta(t)$, i.e., we will replace $\theta(t)$ by $\theta(t) + \delta\theta(t)$, where $\delta\theta(t)$ is continuous in $0 \leq t \leq t_0$. In view of the theory of differential equations, this yields a function $f + \delta f$ of the same class such that for a fixed z the magnitude $|\delta f|$ does not exceed in order the magnitude of $\delta\theta$ for a small variation of $\theta(t)$. Let

$$\frac{\partial \log f}{\partial t} = \Phi(kf), \tag{5.1.4}$$

Neglecting the higher order terms in $\delta\theta$, the variation $f + \delta f$ is given for a small $\delta\theta$ by

$$\frac{\partial \log(f + \delta f)}{\partial t} = \Phi(kf) + \Phi'(kf)\,(k\delta f + ikf\delta\theta)\,. \qquad (5.1.5)$$

Thus, subtracting Eq (5.1.4) from (5.1.5), we get

$$\frac{\partial(\delta f/f)}{\partial t} = \Phi'(kf)\,(k\delta f + ikf\delta\theta)\,. \qquad (5.1.6)$$

However, since

$$\frac{\partial \log(f'/f)}{\partial t} = -\frac{2kf}{(1 - kf)^2} = kf\Phi'(kf), \qquad (5.1.7)$$

by eliminating $\Phi'(kf)$ between Eqs (5.1.6) and (5.1.7), we obtain

$$\frac{\partial \log(f'/f)}{\partial t} = -\frac{f'}{f}\frac{\partial(f'/f)}{\partial t}\Big(\frac{\delta f}{f} + i\delta\theta\Big),$$

or

$$\frac{f}{f'}\frac{\partial \log(f'/f)}{\partial t} + \frac{\delta f}{f}\frac{\partial(f'/f)}{\partial t} = -\frac{\partial(f/f')}{\partial t}\,i\delta\theta,$$

or

$$\frac{\partial(\delta f/f')}{\partial t} = -\frac{\partial(f/f')}{\partial t}\,i\delta\theta. \qquad (5.1.8)$$

Integrating Eq (5.1.8) with respect to t from 0 to t_0, we obtain *Schiffer's variation formula* for $f(z, t_0)$:

$$\delta f = -f(z, t_0)\int_0^{t_0}\frac{\partial(f/f')}{\partial t}\,i\delta\theta(t)\,dt. \qquad (5.1.9)$$

Let $f(z, t_0) = \sum_{n=1}^{\infty} a_n(t_0)\,z^n$. To find the variation of the real part of $a_n(t_0)$, note that

$$\delta\Re\{a_n(t_0)\} = \Re\Big\{\frac{1}{2\pi i}\int_{|z|=r|}\frac{\delta f(z, t_0)}{z^{n+1}}\,dz\Big\}, \qquad (5.1.10)$$

where $|z| = r$, $0 < r < 1$, is some small circle in E. Then

$$\delta\Re\{a_n(t_0)\} = -\int_0^{t_0}\Re\Big\{\frac{1}{2\pi}\int_{|z|=r}\frac{f'(z, t_0)}{z^{n+1}}\frac{\partial\frac{f(z, t)}{f'(z, t)}}{\partial t}\,dz\Big\}\,\delta\theta(t)\,dt. \qquad (5.1.11)$$

Now, suppose that among all functions $f(z, t_0)$ of the class S there exists one function, say $f_0(z, t_0)$, of the class which has the nth coefficient $a_n(t_0)$ with the largest real part. Then for such a coefficient we have $\Re\{a_n\} = 0$. Since $\delta\theta(t)$ is arbitrary, we obtain from (5.1.11)

$$\Re\left\{\frac{1}{2\pi}\int_{|z|=r}\frac{f_0'(z, t_0)}{z^{n+1}}\frac{\partial \dfrac{f_0(z, t)}{f_0'(z, t)}}{\partial t}\, dz\right\} = 0,$$

which, when integrated with respect to t from 0 to t ($t \le t_0$), yields

$$\Re\left\{\frac{1}{2\pi}\int_{|z|=r}\frac{f_0'(z, t_0)}{z^{n+1}}\left[\frac{f_0(z, t)}{f_0'(z, t)} - z\right] dz\right\} = 0, \quad 0 \le t \le t_0. \qquad (5.1.12)$$

Thus, we have a necessary condition that the function $f_0'(z, t_0)\left[\dfrac{f_0(z, t)}{f_0'(z, t)} - z\right]$ has a real nth coefficient for all t, $0 \le t \le t_0$. It is obvious that this condition is satisfied for any $n \ge 1$ by the function $f(z, t)$ for a constant $k(t)$ equal to ± 1.

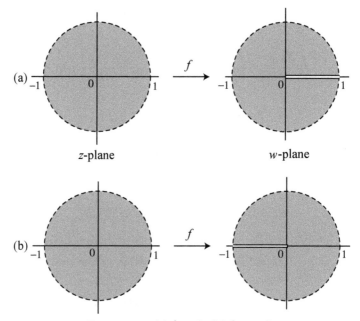

Figure 5.1.1 (a) $k = 1$; (b) $k = -1$.

The two such cases are as follows:

CASE 1. $k = 1$: the function $f(z, t)$ maps $|z| < 1$ onto $|w| < 1$ with a slit from $w = 1$ extending along the real axis toward the origin (see Figure 5.1.1(a)).

CASE 2. $k = -1$: the function $f(z, t)$ maps $|z| < 1$ onto $|w| < 1$ with a slit from $w = 0$ extending along the real axis toward $w = -1$ (see Figure 5.1.1(b)).

5.2 Fourth Coefficient

Garabedian and Schiffer [1955] used Löwner's theorem and Schiffer's variation method to prove $|a_4| \leq 4$ for $f \in \mathcal{S}$.

Theorem 5.2.1. (Garabedian and Schiffer [1955]) *For $f \in \mathcal{S}$, we have* $|a_4| \leq 4$.

PROOF. It is known that from the theory of normal families of univalent functions that there exists an extremal function $f(z)$ for which $|a_4|$ is maximum. Without loss of generality, assume that the forth coefficient of this function is positive; if not, then replacing $f(z)$ by $e^{-i\theta} f\left(re^{i\theta}\right)$ and choosing a suitable value of θ, we will have a function with $a_4 > 0$. Since the original proof is very long, we will provide a detailed outline of the proof in two parts: the first part is based on the Löwner's theorem followed by an enormous amount of numerical computation, while the second part uses Schiffer's variation technique to solve a system of nonlinear boundary value problem. Let $f(z)$ denote the extremal function throughout this proof which is based on the following steps.

Part I. This part consists of the following four steps and yields the bound on the fourth coefficient as $|a_4| \leq 4.0142$, given by inequality (5.2.35). Although this part does not use the variation method, it develops some useful results to be used in Part II which uses the variation method to solve a nonlinear boundary value problem in $\mathbb{C}$.

STEP 1. We need the following eight lemmas.

Lemma 5.2.1. *An extremal function $f(z)$, with $|a_4| > 0$ maximum, satisfies the ordinary differential equation*

$$\frac{z^2 f'(z)^2}{f(z)^2} \left[\frac{1}{f(z)^3} + \frac{3a_2}{f(z)^2} + \frac{2a_3 + a_2^2}{f(z)} \right] = \left(\frac{1}{z^3} + \frac{2a_2}{z^2} + \frac{3a_3}{z} + 3a_4 + 3\bar{a}_3 z + 2\bar{a}_2 z^2 + z^3 \right).$$
(5.2.1)

The disk E is mapped by the extremal function onto the exterior of a set of analytic arcs in the w-plane which satisfy the differential equation

$$\frac{d^2 w}{dz^2} \left[\frac{1}{w^3} + \frac{3a_2}{w^2} + \frac{2a_3 + a_2^2}{w} \right] \leq 0.$$
(5.2.2)

These two differential equations can be derived from Schiffer's interior variation technique.

Lemma 5.2.2. *Let the extremal function $f(z)$ maps E onto the exterior of a single analytic slit. Then there exists a real-valued analytic function*

$\phi(t)$, $0 < t < 1$, such that the coefficients a_2, a_3, a_4 of the extremal function $f(z)$ have the Löwner representation

$$a_2 = 2 \int_0^1 e^{i\phi(t)} \, dt, \tag{5.2.3}$$

$$a_3 = -2 \int_0^1 t \, e^{i\phi(t)} \, dt + a_2^2, \tag{5.2.4}$$

$$a_4 = 2 \int_0^1 t^2 \, e^{i\phi(t)} \, dt + 4 \int_0^1 \int_{t_1=t_2}^1 t_1 \, e^{2i\phi(t_1)} \, e^{i\phi(t_2)} \, dt_1 \, dt_2 + 3a_2 a_3 - 2a_2^3. \tag{5.2.5}$$

For proof, see §4.4, Schaeffer et al. [1949] and Schaeffer and Spencer [1950].

Lemma 5.2.3. *The coefficients a_2 and a_3 of the extremal function $f(z)$ satisfy two equations defined by the nonlinear boundary value problem*

$$\dot{a} = -\tfrac{4}{3}k, \quad \dot{b} = -\tfrac{2}{3}tk^2, \tag{5.2.6}$$

$$\Im\{t^2 k^3 - atk^2 + bk\} = 0, \tag{5.2.7}$$

$$a(0) = 2a_2, \; b(0) = \tfrac{1}{3}\left(2a_3 + a_2^2\right), \; a(1) = \tfrac{4}{3}a_2, \; b(1) = a_3. \tag{5.2.8}$$

Here $a(t)$ and $b(t)$ are unknown functions defined on the interval $[0, 1]$, and $k = k(t) = e^{i\phi(t)}$, and dot denotes the time derivative. The proof is based on Löwner's parametric representation (§4.1.3) with details in Garabedian and Schiffer [1955: 430-431].

Lemma 5.2.4. *There is an angle ψ such that*

$$a_3 = \frac{a_4}{2} e^{i\psi} + a_2 e^{-i\psi} + \frac{1}{3} e^{4i\psi} - \frac{2}{3} e^{-2i\psi} - \frac{\bar{a}_2}{3} e^3 i\psi. \tag{5.2.9}$$

Lemma 5.2.5. *The function*

$$f\left(z^{-2}\right)^{-1/2} = z + \frac{b_1}{z} + \frac{b_3}{z^3} + \frac{b_5}{z^5} + \cdots, \quad 0 < |z| < 1, \tag{5.2.10}$$

is univalent and in view of the classical area theorem (§3.1),

$$\sum_{n=1}^{\infty} (2n - 1)|b_{2n-1}|^2 \leq 1. \tag{5.2.11}$$

Lemma 5.2.6. *The coefficients a_5, a_6 and a_7 of $f(z)$ can be expressed in terms of the coefficients a_2, a_3 and a_4 by*

$$a_5 = \tfrac{2}{5}a_2 a_4 + \tfrac{3}{5}\bar{a}_3, \tag{5.2.12}$$

$$a_6 = \tfrac{6}{7}a_3 a_4 - \tfrac{16}{35}a_2^2 a_4 - \tfrac{3}{7}a_2 a_3^2 + \tfrac{4}{7}a_2^2 a_3 - \tfrac{1}{7}a_2^5 + \tfrac{6}{35}a_2 \bar{a}_3 + \tfrac{2}{7}\bar{a}_2, \tag{5.2.13}$$

$$a_7 = a_4^2 - \tfrac{54}{35}a_2 a_3 a_4 + \tfrac{64}{105}a_2^3 a_4 - \tfrac{4}{9}a_3^3 + \tfrac{40}{21}a_2^2 a_3^2 - \tfrac{10}{7}a_2^4 a_3$$
$$+ \tfrac{19}{63}a_2^6 - \tfrac{8}{35}a_2^2 \bar{a}_3 + \tfrac{2}{5}a_3 \bar{a}_3 + \tfrac{4}{63}a_2 \bar{a}_2 + \tfrac{1}{9}. \tag{5.2.14}$$

Lemma 5.2.7. *The coefficients a_2, a_3 and a_4 of $f(z)$ satisfy the inequality*

$$|\tfrac{1}{2}a_2|^2 + 3|\tfrac{1}{2}a_3 - \tfrac{3}{8}a_2^2|^2 + 5|\tfrac{1}{2}a_4 - \tfrac{3}{2}a_2a_4 + \tfrac{5}{16}a_2^2|^2$$

$$+ 7|\tfrac{11}{20}a_2a_4 - \tfrac{3}{10}\bar{a}_3 + \tfrac{3}{8}a_3^2 - \tfrac{15}{16}a_2^2a_3 + \tfrac{35}{128}a_2^4|^2$$

$$+ 9|\tfrac{9}{28}a_3a_4 - \tfrac{229}{560}a_2^2a_4 - \tfrac{1}{7}\bar{a}_2 + \tfrac{51}{140}a_2\bar{a}_3 - \tfrac{81}{112}a_2a_3^2 + \tfrac{181}{224}a_2^3a_4$$

$$- \tfrac{313}{1792}a_2^5|^2 + 11|\tfrac{1}{8}a_4^2 - \tfrac{9}{56}a_2a_3a_4 - \tfrac{239}{3360}a_2^3a_4 - \tfrac{23}{126}a_2\bar{a}_2 - \tfrac{1}{4}a_3\bar{a}_3$$

$$+ \tfrac{179}{560}a_2^2\bar{a}_3 + \tfrac{13}{144}a_3^3 - \tfrac{493}{1344}a_2^2a_3^2 + \tfrac{157}{1792}a_2^4a_3 + \tfrac{2087}{64512}a_2^6 + \tfrac{1}{18}|^2 \le 1.$$

$$(5.2.15)$$

Lemma 5.2.8. *The coefficients a_2, a_3 and a_4 satisfies the inequalities*

$$|a_2| \le 2, \ |a_3| \le 3, \ |a_2^2 - a_3| \le 1, \ |a_4 - 2a_2a_3 + \tfrac{13}{12}a_2^3| \le \tfrac{2}{3}. \qquad (5.2.16)$$

The first three inequalities are given in the previous two chapters and the last inequality can be found in Nehari [1953].

STEP 2. Establish bounds on $|a_2 - 2|$ and $|a_3 - 3|$. We need the following six Lemmas.

Lemma 5.2.9. *Let $\lambda = a_3 - \tfrac{3}{4}a_2^2$. Then*

$$\tfrac{1}{4}|a_2|^2 + \tfrac{3}{4}|\lambda|^2 + \tfrac{5}{4}|a_4 - \tfrac{3}{2}a_2\lambda - \tfrac{1}{2}a_2^3|^2$$

$$+ 7|\tfrac{11}{20}a_2a_4 - \tfrac{7}{32}a_2^4 - \tfrac{9}{40}\bar{a}_2^2 - \tfrac{3}{10}\bar{\lambda} + \tfrac{3}{8}\lambda^2 - \tfrac{3}{8}a_2^2\lambda|^2$$

$$+ 9|\tfrac{47}{280}a_2^2a_4 + \tfrac{1}{7}\bar{a}_2 - \tfrac{153}{560}a_2\bar{a}_2^2 - \tfrac{11}{448}a_2^5 + \left(\tfrac{31}{112}a_2^3 - \tfrac{9}{28}a_4\right)\lambda$$

$$\tfrac{51}{140}a_2\bar{\lambda} + \tfrac{81}{112}a_2\lambda^2|^2 + 11|\tfrac{1}{8}a_4^2 + \tfrac{83}{1680}a_2^3a_4 + \tfrac{111}{1120}a_2^2\bar{a}_2^2$$

$$- \tfrac{23}{126}a_2\bar{a}_2 - \tfrac{283}{4032}a_2^6 + \tfrac{13}{144}\lambda^3 - \tfrac{1}{4}\lambda\bar{\lambda}$$

$$+ \left(\tfrac{9}{56}a_2a_4 - \tfrac{3}{16}\bar{a}_2^2 - \tfrac{139}{448}a_2^4\right)\lambda + \tfrac{37}{280}a_2^2\bar{\lambda} - \tfrac{55}{336}a_2^2\lambda^2 + \tfrac{1}{18}|^2 \le 1.$$

$$(5.2.17)$$

This is a direct consequence of (5.2.15) when we replace a_3 by $a_3 = \lambda + \tfrac{3}{4}a_2^2$.

Lemma 5.2.10. *The terms $|b_1|^2 + 3|b_3|^2 + 5|b_5|^2 \le 1$ alone in (5.2.17) lead to the estimates*

$$|a_3| \ge 1.67, \quad |\lambda| \le 0.64. \qquad (5.2.18)$$

Lemma 5.2.11. *Let $a_2 = Ae^{i\alpha}$ and normalize $f(z)$ so that $0 \le \alpha \le \pi/3$. Then*

$$0 \le \alpha \le 0.22. \qquad (5.2.19)$$

The normalization is obtained by the substitution $\overline{f(\bar{z})}$ or $e^{-i\theta}f\left(ze^{i\theta}\right)$, $\theta = \pm\pi/3$.

Lemma 5.2.12. *The terms $|b_1|^2 + 3|b_3|^2 + 5|b_5|^2 + 7|b_7|^2 \le 1$ imply the estimates*

$$|a_2| \ge 1.92, \quad |\lambda| \le 0.324. \qquad (5.2.20)$$

Lemma 5.2.13. *We have* $0 \le \alpha \le 0.104$, $0 \le \Im\{a_2\} \le 0.208$.

Lemma 5.2.14. *The terms*

$$|b_1|^2 + 3|b_3|^2 + 5|b_5|^2 + 7|b_7|^2 + 9|b_9|^2 \le 1, \tag{5.2.21}$$

alone in (5.2.17) imply the estimate $|a_2| \ge 1.95$. *For proof, see Garabedian and Schiffer* [1955: 437–439].

STEP 3. Estimate the imaginary parts, by expanding the main inequality (5.2.15) in terms of the differences between the coefficients a_2, a_3 and a_4 and the conjectured values 2, 3, and 4. We need the following five lemmas.

Lemma 5.2.15. *If* $a_2 = 2 - \delta$, $a_3 = 3 - \eta$, $a_4 = 4 + \epsilon$, *then*

$$
\begin{aligned}
&|1 - \tfrac{1}{2}\delta|^2 + 3|\tfrac{3}{2}\delta - \tfrac{3}{2}\eta - \tfrac{3}{8}\delta^2|^2 + 5|\tfrac{3}{2}\delta - \tfrac{3}{2}\eta - \tfrac{1}{2}\epsilon - \tfrac{15}{8}\delta^2 + \tfrac{3}{4}\eta\delta + \tfrac{5}{16}\delta^2|^2 \\
&+ 7|\tfrac{3}{10}\delta + \tfrac{3}{2}\eta + \tfrac{3}{10}\bar{\eta} + \tfrac{11}{10}\epsilon + \tfrac{15}{4}\delta^2 - \tfrac{15}{4}\delta\eta + \tfrac{3}{8}\eta^2 \\
&\qquad\qquad - \tfrac{11}{20}\delta\eta + \tfrac{15}{16}\delta^2\eta - \tfrac{35}{16}\delta^2 + \tfrac{35}{128}\delta^4|^2 \\
&+ 9|\tfrac{221}{70}\delta - \tfrac{1}{7}\bar{\delta} - \tfrac{13}{14}\eta + \tfrac{51}{70}\bar{\eta} + \tfrac{47}{70}\epsilon + \tfrac{149}{140}\delta^2 - \tfrac{75}{14}\delta\eta - \tfrac{51}{140}\delta\bar{\eta} \\
&\qquad + \tfrac{81}{56}\eta^2 + \tfrac{9}{28}\eta\epsilon - \tfrac{229}{140}\delta\epsilon - \tfrac{511}{112}\delta^3 + \tfrac{543}{112}\delta^2\eta - \tfrac{81}{112}\delta\eta^2 \\
&\qquad\qquad + \tfrac{229}{560}\delta^2\epsilon - \tfrac{181}{224}\delta^3 + \tfrac{1565}{896}\delta^4 - \tfrac{313}{1792}\delta^5|^2 \\
11&|\tfrac{2143}{630}\delta - \tfrac{23}{63}\bar{\delta} - \tfrac{31}{7}\eta + \tfrac{37}{70}\bar{\eta} - \tfrac{293}{210}\epsilon + \tfrac{23}{126}\delta\bar{\delta} - \tfrac{8419}{840}\delta^2 - \tfrac{179}{140}\delta\eta \\
&+ \tfrac{75}{14}\delta\eta + \tfrac{1}{4}\eta\bar{\eta} + \tfrac{55}{84}\eta^2 - \tfrac{1}{8}\epsilon^2 - \tfrac{13}{35}\delta\epsilon + \tfrac{9}{28}\eta\epsilon - \tfrac{1481}{672}\delta^4 \\
&- \tfrac{11}{112}\delta^2\eta + \tfrac{179}{560}\delta^2\eta - \tfrac{493}{336}\delta\eta^2 + \tfrac{13}{144}\eta^3 + \tfrac{239}{560}\delta^2\epsilon - \tfrac{9}{56}\delta\eta\epsilon - \tfrac{1481}{672}\delta^4 \\
&- \tfrac{157}{224}\delta^3\eta + \tfrac{493}{1344}\delta^2\eta^2 - \tfrac{239}{3360}\delta^3\epsilon + \tfrac{157}{1792}\delta^4\eta + \tfrac{2087}{5376}\delta^5 - \tfrac{2087}{64512}\delta^6|^2 \le 1. \\
&\tag{5.2.22}
\end{aligned}
$$

This inequality is obtained by direct substitution of δ, η and ϵ into ((5.2.15).

Lemma 5.2.16. *Using the notation of the previous lemma, set* $\delta = p_i P$, *and* $\eta = q - iQ$. *Then*

$$0 \le P \le 0.078, \quad -0.067 \le Q \le 0.078. \tag{5.2.23}$$

The proof follows from the previous lemma, using only the imaginary parts of b_3, b_5 and b_7; for details, see Garabedian and Schiffer [1955: 440–441].

Lemma 5.2.17. *The increments* η *and* ϵ *of the coefficients* a_3 *and* a_4 *satisfy*

$$0 \le \eta \le 0.116, \quad 0 \le \epsilon \le 0.45. \tag{5.2.24}$$

These inequalities follow from (5.2.23), which yields, in view of Lemmas 5.2.14 and 5.2.15, the bound $\epsilon + 3.8788\eta \le 0.4494$.

Lemma 5.2.18. *We have*

$$-0.059 \le 0.75Q - P \le 0.0402. \tag{5.2.25}$$

Lemma 5.2.19. *Set* $S = 0.75Q - P$. *Then*

$$P^2 \leq 0.123p, \quad Q^2 \leq 0.123p, \quad S^2 \leq 0.07p, \qquad (5.2.26)$$

where $p \geq 0$ *is defined, in view of (5.2.16), by the inequality*

$$\epsilon + 4q \leq 7p + 2pq - 2PQ + 7.5P^2 - 6.5p^2 + 1.08333p^2 - 3.25pP^2. \quad (5.2.27)$$

STEP 4. Determine the discriminant condition, based on the following four lemmas whose proofs can be found in Garabedian and Schiffer [1955: 442–444].

Lemma 5.2.20. *The angle* ψ *satisfies*

$$1.5\epsilon + 3q \cos \psi - 2p \cos 2\psi = -(1 - \cos \psi) \left[(2 \cos \psi - 1)^2 + 1 \right]$$
$$+ (3Q - 4P) \sin \psi + 4P(1 - \cos \psi) \sin \psi, \qquad (5.2.28)$$
$$h(\psi) = -(\tfrac{1}{2}\epsilon + \tfrac{2}{3}p + \tfrac{4}{3}p \cos^2 \psi) \sin \psi + Q - \tfrac{4}{3}P + \tfrac{4}{3}P(1 - \cos^3 \psi), \qquad (5.2.29)$$

where $h(\psi) \left(\tfrac{2}{3} - \tfrac{8}{3} \cos^2 \psi + \tfrac{8}{3} \cos^3 \psi \right) \sin \psi$. *Note that (52.28) is obtained from (5.4.8) and (5.2.29) from the imaginary part of (5.2,9) after replacing* a_2, a_3 *and* a_4 *by the new variables* p, q, P, Q *and* ϵ.

Lemma 5.2.21. *The angle* ψ *lies in the interval* $-0.122 \leq \psi \leq 0.087$.

Lemma 5.2.22. *The quantities* p, q, P, Q *and* ϵ *satisfy the inequality*

$$\tfrac{3}{2}\epsilon + 3q - 2p \leq \tfrac{1}{2}(3Q - 4P)^2.$$

Lemma 5.2.23. *We have* $q \leq 0.05$ *and* $\epsilon \leq 0.1$.

STEP 5. Linearize and refine the estimates for p, q, P, Q and ϵ. We have the following four lemmas which we provide without proofs as they can be obtained from Garabedian and Schiffer [1955: 444–449].

Lemma 5.2.24. *The quantities* p, q, P *and* Q *satisfy*

$$p \geq 100.653p^2 - 30.291pq + 35.04q^2 - 156.2p^3 - 493.6p^4 + 93.056P^2$$
$$- 101.81PQ + 41.681Q^2 + \min\{0, 5.777PQ\}. \qquad (5.2.30)$$

Lemma 5.2.25. *In view of (5.2.21), the quantities* p, q, ϵ, P, Q *satisfy*

$$p \leq 0.0109, \quad q \leq 0.0073, \quad \epsilon \leq 0.0146, \quad P \leq 0.0094, \quad -0.0081 \leq Q \leq 0.0141. \qquad (5.2.31)$$

Lemma 5.2.26. *The following estimate holds:*

$$p \geq 94.2p^2 + 255P^2 - 501PQ + 303Q^2. \qquad (5.2.32)$$

Lemma 5.2.27. *We have the estimates:*

$$p \leq 0.0107, \quad q \leq 0.0071, \quad \epsilon \leq 0.0142, \quad P \leq 0.0075,$$

$$-0.003 \leq Q \leq 0.0069, \quad |S| \leq 0.0036. \tag{5.2.33}$$

Lemma 5.2.28. *The combination of a_2 and a_3 appearing in the boundary conditions (5.2.8) satisfy*

$$0 \leq \Re\{4 - 2a_2\} \leq 0.0214, \quad 0 \leq \Im\{2a_2\} \leq 0.015,$$
$$0 \leq \Re\{\tfrac{10}{3} - \tfrac{1}{3}(2a_2 + a_2^2)\} \leq 0.019, \quad -0.002 \leq \Im\{\tfrac{1}{3}(2a_3 + a_2^2)\} \leq 0.0146,$$
$$0 \leq \Re\{\tfrac{8}{3} - \tfrac{4}{3}a_2\} \leq 0.0143, \quad 0 \leq \Im\{\tfrac{4}{3}\} \leq 0.01,$$
$$0 \leq \Re\{3 - a_3\} \leq 0.0071, \quad -0.003 \leq \Im\{a_3\} \leq 0.0069.$$
$$\tag{5.2.34}$$

Lemma 5.2.29. *From the discriminant condition (5.2.9) and the inequality (5.2.15) based on the area theorem (5.2.11) and the recursion formulas (5.2.12), (5.2.13) and (5.2.14), we get the bound*

$$|a_4| \leq 4.0142. \tag{5.2.35}$$

Part II. This part has the following four steps, uses the variation method, yielding the bound $|a_4| \leq 4$ below in Lemma 5.2.40.

STEP 1. Solve the nonlinear boundary value problem of Lemma 5.2.3 by perturbation method. So far we have a set of equations for the coefficients a_2 and a_3 of the extremal function $f(z)$, and these conditions are satisfied in the case of the Koebe function $K(z) = \sum_{n=1}^{\infty} nz^n$, for which $a_2 = 2$, $a_3 = 3$ and

$$K(t) \equiv 1, \quad a(t) = 4 - \tfrac{4}{3}t, \quad b(t) = \tfrac{10}{3} - \tfrac{1}{3}t^2.$$

Thus, we will apply the perturbation method in the neighborhood of the known solution $K = 1$, and determine whether there exists other solutions of the same set of nonlinear equations. Let us set

$$K = e^{i\phi}, \quad a(t) = u(t) + iv(t), \quad b(t) = x(t) + iy(t). \tag{5.2.36}$$

Then the boundary value problem (5.2.3) reduces to the real form

$$u'(t) = -\tfrac{4}{3}\cos\phi, \quad x'(t) = -\tfrac{2}{3}t\cos 2\phi,$$
$$v'(t) = -\tfrac{4}{3}\sin\phi, \quad y'(t) = -\tfrac{2}{3}t\sin 2\phi, \tag{5.2.37}$$
$$t^2\sin 3\phi - t(u\sin 2\phi + v\cos 2\phi) + x\sin\phi + y\cos\phi = 0,$$

with the boundary conditions

$$u(1) = \tfrac{2}{3}u(0), \quad x(1) = \tfrac{3}{2}x(0) - \tfrac{1}{8}\left[u(0)^2 - v(0)^2\right],$$
$$v(1) = \tfrac{2}{3}v(0), \quad y(1) = \tfrac{3}{2}y(0) - \tfrac{1}{4}u(0)v(0). \tag{5.2.38}$$

Let us introduce the functions

$$p(t) = \tfrac{16}{3}t^2 - 9t + \tfrac{10}{3}, \quad U(t) = u(t) - 4 + \tfrac{4}{3}t, \quad X(t) = x(t) - \tfrac{10}{3} + \tfrac{1}{3}t^2. \tag{5.2.39}$$

The polynomial $p(t)$ is positive-definite and attains its minimum $\tfrac{1}{3}$ at $t = \tfrac{3}{4}$. The functions $U(t)$ and $X(t)$ are the deviation of the solutions $u(t)$ and $x(t)$ from the corresponding known solutions (5.2.36). In view of the above notation, we can rewrite (5.2.37) in the form

$$\sin\phi \left[p(t) + X(t) - 2tU(t) - 16\left(\tfrac{4}{3}t^2 - t\right)\sin^2(\phi/2) + 16t^2\sin^4(\phi/2)\right.$$
$$\left. 4tU(t)\sin^2(\phi/2) + 2yv(t)\sin\phi - y(t)\tan(\phi/2)\right] = tv(t) - y(t). \tag{5.2.40}$$

Moreover, we have the differential equations

$$U'(t) = \tfrac{8}{3}\sin^2(\phi/2), \quad X'(t) = \tfrac{4}{3}t\sin^2\phi. \tag{5.2.41}$$

The uniqueness of the solution can be made by an estimation of the form

$$|\sin(\phi/2)| \le \tfrac{1}{2}\rho,$$

where ρ is sufficiently small.

STEP 2. Estimates by means of ρ are provided by (5.2.37) and Lemma 5.2.28 as

$$|\tan\phi(0)| = \frac{|y(0)|}{|x(0)|} \le 0.0045,$$

and hence, either $|\phi(0)| < 0.0045$ or $|\phi(0) - \pi| < 0.0045$. In both cases we get from (5.2.40) and Lemma 5.2.28 that

$$2.4|\sin\phi(t)| \le 0.04 \quad \text{in the interval } 0 \le t \le 0.1.$$

Thus, if $|\phi(0) - \pi| < 0.0045$, we would have $\cos\phi(t) < 0$ in the above interval, since $\phi(t)$ is continuous. By (5.2.3) this would then imply that $\Re\{a_2\} < 1.8$, which contradicts (5.2.34). Thus, $|\phi(t))| < 0.0045$. If we take ρ in the interval $0.1 \le \rho < 1$, then $|\sin\tfrac{1}{2}\phi(0)| < \tfrac{1}{2}\rho$. Let next $[0, T_1]$ be the largest interval in $0 \le t \le \tfrac{1}{2}$ for which the inequality (5.2.41) is still satisfied. This leads to the following result.

Lemma 5.2.30. *We have in the interval* $0 \le t \le T_1$

$$0 \le U(t) - U(0) \le \tfrac{2}{3}t\rho^2, \quad 0 \le X(t) - X(0) \le \tfrac{2}{3}t^2\rho^2, \tag{5.2.42}$$

where the left-hand inequalities hold in the entire interval $0 \le t \le 1$.

From the differential equations (5.2.37) for $v(t)$ and $y(t)$ we infer

Lemma 5.2.31. *We have in the interval* $0 \le t \le T_1$

$$|v(t) - v(0)| \le \tfrac{4}{3}\rho t, \quad |y(t) - y(0)| \le \tfrac{2}{3}\rho t^2. \tag{5.2.43}$$

Note that (5.2.38) implies the boundary condition $U(1) = \tfrac{2}{3}U(0)$. By (5.2.34), $U(0) \le 0$ and, since $U(t)$ increases monotonically, we have the estimate $U(0) \le U(t) \le \tfrac{2}{3}U(0)$ for $0 \le t \le 1$. We now rewrite (5.2.40) in the form

$$F(t)\sin\phi = tv - y, \tag{5.2.44}$$

where

$$\begin{aligned}
F(t) = {} & p(t) + X(0) + [X(t) - X(0)] - 2tU(t) + 16\left(t - \tfrac{4}{3}t^2\right)\sin^2(\phi/2) \\
& + 16^2\sin^4(\phi/2) + 4tU(t)\sin^2(\phi/2) + 2tv(0)\sin\phi - y(0)\tan(\phi/2) \\
& + 2t\left[v(t) - v(0)\right]\sin\phi - [y(t) - y(0)]\tan(\phi/2).
\end{aligned} \tag{5.2.45}$$

In view of Lemmas 5.2.30 and 5.2.31 , and Eq (5.2.45), we have for t in the interval $0 \le t \le T_1$,

$$F(t) \ge p(t) - |X(0)| - 2t|v(0)|\rho - \frac{|y(0)|\rho}{(4 - \rho^2)^{1/2}} - \tfrac{8}{3}\rho^2 t^2 - \tfrac{2}{3}\frac{\rho^2 t^2}{(4 - \rho^2)^{1/2}}. \tag{5.2.46}$$

Since $p(t)$ attains its minimum at $t = \tfrac{3}{4}$, we have the estimate for $0 \le t \le T_1$ as

$$F(t) \ge \tfrac{2}{3} - |X(0)| - |v(0)|\rho - \frac{|y(0)|\rho}{(4 - \rho^2)^{1/2}} - \tfrac{2}{3}\rho^2 - \tfrac{1}{6}\frac{\rho^2}{(4 - \rho^2)^{1/2}},$$

which, in view of Lemma 5.2.28, and taking $\rho = \tfrac{1}{7}$, gives

Lemma 5.2.32. *In the entire interval* $0 \le t \le T_1$ *in which* $|\sin(\phi/2)| \le \tfrac{1}{14}$ *holds, we have*

$$F(t) \ge 0.629. \tag{5.2.47}$$

Now we will make estimates that are valid in an interval extending to the left from $t = 1$. From Lemma 5.2.21 and the identity $\phi(1) = \psi$, we know that $|\phi(1)| \le 0.122$. However, in view of (5.2.29), (5.2.40) and Lemma 5.2.27, this bound can be improved to $|\phi(1)| < 0.008$. Since $0.01 < \rho < 1$, we get

$|\sin\frac{1}{2}\phi(1)| < \frac{1}{2}\rho$. So let $[T_2, 1]$ be the largest interval in $\frac{1}{2} \le t \le 1$ for which the inequality (5.2.41) is still valid. Then from (5.2.37) we obtain

Lemma 5.2.33. *We have in the interval* $T_2 \le t \le 1$

$$|v(t) - v(1)| \le \tfrac{4}{3}\rho(1-t), \quad |y(t) - y(1)| \le \tfrac{2}{3}\rho(1-t^2). \qquad (5.2.48)$$

Eq (5.2.45) can now be replaced by the equivalent form in which $v(0)$ and $y(0)$ are replace by $v(1)$ and $y(1)$. This leads to the estimate in the interval $T_2 \le t \le 1$ as

$$F(t) \ge \tfrac{16}{3}t^2 - 8t + \tfrac{10}{3} - |X(0)| - 4\rho^2 \max\left\{0, t - \tfrac{4}{3}t^2\right\}$$
$$- 2|v(1)|\rho t - \frac{|y(1)|\rho}{(4-\rho^2)^{1/2}} - \tfrac{8}{3}\rho^2 t(1-t) - \tfrac{2}{3}(1-t^2)\frac{\rho^2}{(4-\rho^2)^{1/2}}.$$

Again, set $\rho = \frac{1}{7}$ and by Lemma 5.2.28 we get

$$F(t) \ge 3.307 - 8.0573t + 5.3945t^2 - \max\left\{0, \frac{(12-16t)t}{147}\right\} \ge 0.298,$$

since the minimum of the quadratic on the right occurs at $t \approx 0.746807$. This gives

Lemma 5.2.34. *In the entire interval* $T_2 \le t \le 1$ *in which* $|\sin(\phi/2)| \le \frac{1}{14}$, *we have*

$$F(t) \ge 0.298. \qquad (5.2.49)$$

STEP 3. Integrate the differential equation system. Define a new function $w(t)$ by

$$w(t) = tv(t) - y(t). \qquad (5.2.50)$$

Then Eq (5.2.44) can be written as $\in \phi = w(t)/F(t)$, and Eqs (5.2.37) lead to the system of differential equations

$$\frac{dv}{dt} = -\frac{4}{3F(t)}w(t), \quad \frac{dw}{dt} = v(t) - w(t)\frac{8t\sin^2(\phi/2)}{3F(t)}. \qquad (5.2.51)$$

We introduce the functions

$$r_1(t) = \exp\left[\frac{8}{3}\int_0^t \frac{t\sin(\phi/2)}{F(t)}\,dt\right], \quad r_2(t) = \exp\left[-\frac{8}{3}\int_t^1 \frac{t\sin(\phi/2)}{F(t)}\,dt\right],$$
$$(5.2.52)$$

and define

$$w(t) = r_j(t)w(t), \quad j = 1, 2. \qquad (5.2.53)$$

Then for $j = 1, 2$ the system (5.2.51) reduces to

$$\frac{dv}{dt} = -R_j(t)w_j, \quad \frac{dw_j}{dt} = r_j(t)v, \qquad (5.2.54)$$

where $R_j(t) = \dfrac{4}{3F(t)r_j(t)}$. Now using the above assumption that $\rho = \dfrac{1}{7}$, we find that

$$1 \le r_1(t) \le 1.003, \quad 0 < R_1(t) \le 2.12 \quad \text{in the interval } [0, T_1]$$
$$0.983 \le r_2(t) \le 1, \quad 0 < R_2(t) \le 4.552 \quad \text{in the interval } [T_2, 1] \ (5.2.55)$$

To simplify calculation, the following change of scale is made by introducing the new independent variables

$$s_1 = \int_0^t r_1(t)\, dt, \quad s_2 = 1 - \int_t^1 r_2(t)\, dt.$$

Then the system of differential equations simplifies to

$$\frac{dv}{ds_j} = -\frac{R_j}{r_j} w_j, \quad \frac{dw_j}{ds_j} = v,$$

or

$$\frac{d^2 w_j}{ds_j^2} + \frac{R_j}{r_j} w_j = 0. \qquad (5.2.56)$$

Then, using the Sturm-Liouville theory, we obtain

Lemma 5.2.35. *The solution $w_1(s_1)$ has at most one zero in the interval $[0, s_1(T_1)]$ and the solution $w_2(s_2)$ has at most one zero in the interval $[s_2(T_2), 1]$.*

Lemma 5.2.36. *In the intervals $0 \le s_1 \le s(T_1)$ and $s_2(T_2) \le s_2 \le 1$ we have the estimates, respectively, as*

$$|w_1(s_1)| \le s_1|v(0)| + |y(0)|, \quad |w_2(s_2)| \le |w_2(1)| + (1 - s_2)|w_2'(1)|. \quad (5.2.57)$$

The estimate $|\sin(\phi/2)| \le \frac{1}{2}(0.035)$ holds in the entire interval $) \le t \le \frac{1}{2}$, and $|\sin(\phi/2)| \le \frac{1}{2}(0.0336)$ holds in the entire interval $\frac{1}{2} \le t \le 1$. Moreover, $0 < R(t) \le 2.07$ in the interval $0 \le t \le \frac{1}{2}$ and $0 < R(t) \le 4.48$ in the entire interval $0 \le t \le 1$ The proofs of these estimates are available in Garabedian and Schiffer [1955: 455-459].

STEP 4. To prove the uniqueness of the nonlinear boundary value problem defined by

$$\frac{dv}{dt} = -R(t)w, \quad \frac{dw}{dt} = r(t)v, \qquad (5.2.58)$$

with the boundary conditions $v(1) = \frac{2}{3}v(0)$, $w(1) = r(1)[\frac{2}{3}v(0) + \frac{3}{2}w(0)$ $+\frac{1}{4}u(0)v(0)]$, obtained from (5.2.38), (5.2.50), and (5.2.51). After integration by parts we obtain the identity

$$\int_{t_1}^{t_2} \frac{1}{R(t)} \left(\frac{dv}{dt}\right)^2 dt = -\left[v(t)w(t)\right]_{t_1}^{t_2} + \int_{t_1}^{t_2} r(t)v(t)^2\, dt, \qquad (5.2.59)$$

or, with $t_1 = 0$ and $t_2 = 1$,

$$\int_{t_1}^{t_2} \frac{1}{R(t)} \left(\frac{dv}{dt}\right)^2 dt + \frac{4}{9}r(1)v(0)^2 + [r(1) - 1]\, v(0)w(0) + \frac{1}{6}r(1)u(0)v(0)^2$$

$$= \int_0^1 r(t)v(t)^2\, dt. \qquad (5.2.60)$$

We have proved

Lemma 5.2.37. *A non-trivial solution $v(t)$ of the system (5.2.58) can vanish at most once in the interval $0 \le t \le 1$.*

PROOF. Suppose $v(t_1) = v(t_2) = 0$ for $0 \le t_1 \le t_2 \le 1$. Then using identity (5.2.59) we get

$$\int_{t_1}^{t_2} \frac{1}{R(t)} \left(\frac{dv}{dt}\right)^2 dt = \int_{t_1}^{t_2} rv^2\, dt,$$

and using the bounds on $r(t)$ and $R(t)$ in Lemma 5.2.36, we obtain the inequality

$$\frac{1}{4.48} \int_{t_1}^{t_2} \left(\frac{dv}{dt}\right)^2 dt \le 1.002 \int_{t_1}^{t_2} v^2\, dt. \qquad (5.2.61)$$

Again, using the inequality

$$\int_{t_1}^{t_2} \left(\frac{dv}{dt}\right)^2 dt \ge \frac{\pi^2}{(t_2 - t_1)^2} \int_{t_1}^{t_2} v^2\, dt,$$

which holds for all continuously differentiable functions $v(t)$ vanishing at t_1 and t_2, we obtain from (5.2.61) the absurd relation $\dfrac{\pi^2}{4.48} \le 1.002\,(t_2 - t_1)^2 \le$ 1.002, which proves the lemma. ∎

We have assumed that $v(0) \ge 0$. Now, if $v(0) = 0$, then in the boundary conditions we would also have $v(1) = 0$ and therefore, $v(t) \equiv 0$ by Lemma 5.2.37. Thus, the extremal function would be the Koebe function $K(z) = \sum_{n=1}^{\infty} nz^n$, and we would have nothing to prove. Thus, we may restrict ourselves to the case $v(0) > 0$, which would imply $v(1) > 0$ in the boundary conditions, and hence, $v(t)$ cannot vanish in the interval $0 \le t \le 1$ by

Lemma 5.2.37. Moreover, since $v(t)$ decreases from $v(0)$ to $\frac{2}{3}v(0)$, we conclude from the first equation (5.2.58) that $w(t)$ must be positive somewhere in this interval. From the second equation (5.2.58) we find that $w(t)$ increases monotonically, and hence, $w(1) \geq 0$. This implies by the boundary conditions that $w(0) \geq -\left[\frac{4}{9} + \frac{1}{6}u(0)\right] v(0)$. Therefore, we replace the identity (5.2.60) by the inequality

$$\int_0^1 \left(\frac{dv}{dt}\right)^2 dt + v(0)^2 \left[\frac{4}{9} + \frac{1}{6}u(0)\right] \leq \int_0^1 r(t)v(t)^2 \, dt.$$

Using the estimates on $r(t)$ and $R(t)$ in Lemma 5.2.36, and putting $l = \frac{4}{9} + \frac{1}{6}u(0) = \frac{10}{3} - \frac{1}{3}p$, we find that

$$\int_0^1 \frac{1}{R(t)} \left(\frac{dv}{dt}\right)^2 dt + l\, v(0)^2 \leq 1.002 \int_0^1 v(t)^2 \, dt.$$

By Lemma 5.2.22, we have $l \geq 1.107$, which proves the following result.

Lemma 5.2.38. *If $v(t) \not\equiv 0$, then*

$$\frac{\int_0^1 \frac{1}{R(t)} \left(\frac{dv}{dt}\right)^2 dt + 1.107\, v(0)^2}{\int_0^1 v(t)^2 \, dt} \leq 1.002. \qquad (5.2.62)$$

On the other hand, the following lemma can be easily proved.

Lemma 5.2.39. *For all functions $V(t)$ which are piecewise continuously differentiable and not identically zero in the interval $0 \leq t \leq 1$ and which satisfy the boundary condition*

$$V(1) = \tfrac{2}{3}V(0), \qquad (5.2.63)$$

we have

$$\frac{\int_0^1 \frac{1}{R(t)} \left(\frac{dv}{dt}\right)^2 dt + 1.107\, V(0)^2}{\int_0^1 V(t)^2 \, dt} > 1.08.$$

PROOF. By Lemma 5.2.36, we have

$$\frac{1}{R(t)} \geq \begin{cases} (0.695) & \text{for } 0 \leq t \leq \frac{1}{2}, \\ (0.472)^2 & \text{for } \frac{1}{2} \leq t \leq 1, \end{cases}$$

and so it suffices to calculate

$$k^2 = \min \frac{\int_0^{1/2}(0.695)^2 V'(t)^2\, dt + \int_{1/2}^1 (0.472)^2 V'(t)^2\, dt + 1.107V(0)^2}{\int_0^1 V(t)^2\, dt},$$

$$(5.2.64)$$

and to show that $k^2 > 1.08$. Using the calculus of variations, we obtain the following characterization for the extremal function $V(t)$ which yields the minimum in (5.2.64). Since $V(t)$ satisfies the differential equations

$$V'' + \frac{k^2}{(0.695)^2} V = 0, \quad \text{for } 0 \le t \le \tfrac{1}{2}, \tag{5.2.65}$$

$$V'' + \frac{k^2}{(0.472)^2} V = 0, \quad \text{for } \tfrac{1}{2} \le t \le 1, \tag{5.2.66}$$

with the saltus conditions

$$(0.695)^2 V'(\tfrac{1}{2} - 0) = (0.472)^2 V'(\tfrac{1}{2} + 0), \quad V(\tfrac{1}{2} - 0) = V(\tfrac{1}{2} + 0), \tag{5.2.67}$$

and the natural boundary condition

$$\tfrac{2}{3}(0.472)^2 V'(1) = (0.695)^2 V'(0) - 1.107 V(0), \tag{5.2.68}$$

together with the side condition (5.2.63) (which is always assumed). The solutions of the Eqs (5.2.65)–(5.2.66) are

$$V(t) = \begin{cases} C_1 \cos \dfrac{k}{0.695}(t - \tfrac{1}{2}) + \dfrac{1}{0.695} D_1 \sin \dfrac{k}{0.695}(t - \tfrac{1}{2}), & \text{for } 0 \le t \le \tfrac{1}{2}, \\[2mm] C_2 \cos \dfrac{k}{0.472}(t - \tfrac{1}{2}) + \dfrac{1}{0.472} D_2 \sin \dfrac{k}{0.472}(t - \tfrac{1}{2}), & \text{for } \tfrac{1}{2} \le t \le 1. \end{cases}$$

The saltus condition (5.2.67) gives $C + 1 = C - 2, D_1 = D - 2$. On the other hand, the boundary conditions (5.2.63) and (5.2.68) yield the system of linear equations

$$\begin{aligned} & C_1 \cos \frac{k}{0.944} + \frac{1}{0.472} D_1 \sin \frac{k}{0.944} = \frac{2}{3}\Big[C_1 \cos \frac{k}{1.39} - \frac{1}{0.695} \sin \frac{k}{1.39} \\ & - C_1 \frac{2}{3}(0.472)\sin \frac{k}{0.944} + \frac{2}{3} D_1 k \cos \frac{k}{0.944} \Big] \\ & = C_1\Big[0.695 k \sin \frac{k}{1.39} - 1.107\cos \frac{k}{1.39} + D_1\Big[k \cos \frac{k}{1.39} - \frac{1.107}{0.695}\sin \frac{k}{1.39}\Big]. \end{aligned}$$

This homogeneous system of linear equations has non-trivial solutions only if its determinant

$$\begin{aligned} \Delta = & -\tfrac{4}{3}k + \Big[\tfrac{13}{18} + \tfrac{0.695}{0.994} + \tfrac{2}{9}\tfrac{0.472}{0.695}\Big] k \cos \tfrac{1}{2}k\Big(\tfrac{1}{0.695} + \tfrac{1}{0.472}\Big) \\ & + \Big[\tfrac{13}{18} - \tfrac{0.695}{09.44} - \tfrac{2}{9}\Big] k \cos \tfrac{1}{2}k\Big(\tfrac{1}{0.695} + \tfrac{1}{0.472}\Big) \\ & + 1.017\Big[\tfrac{1}{2}\Big(\tfrac{1}{0.695} + \tfrac{1}{0.472}\Big)\Big] \sin \tfrac{1}{2}k\Big(\tfrac{1}{0.695} + \tfrac{1}{0.472}\Big) \\ & + 1.017\Big[\tfrac{1}{2}\Big(\tfrac{1}{0.695} - \tfrac{1}{0.472}\Big)\Big] \sin \tfrac{1}{2}k\Big(\tfrac{1}{0.695} - \tfrac{1}{0.472}\Big) = 0. \end{aligned} \tag{5.2.69}$$

Thus, for k we get the transcendental relation

$$1.20703 \cos(1.77875k) - 0.123694 \cos(0.339897)$$
$$+ \frac{1.4768}{k} \sin(1.7785k) + \frac{0.2822}{k} \sin(0.339897k) = 1.$$
$$(5.2.70)$$

This equation has no roots in the interval $0 \leq k \leq 1.04$, whence $k > 1.04$, and the lemma is proved. ∎

Since the inequalities (5.2.61) and (5.2.63) are contradictory, we therefore finally obtain the following result.

Lemma 5.2.40. *There does not exist a solution of the boundary value problem (5.2.37) whose initial values satisfy the estimates of Lemmas 5.2.27 and 5.2.28, except for the solution with the exact initial values*

$$u(0) = 4, \quad v(0) = 0, \quad x(0) = \tfrac{10}{3}, \quad y(0) = 0. \qquad (5.2.71)$$

This completes the proof of Theorem 5.2.1. This method of proof can be used in different kinds of difficult problems, if one can go through all the calculations which were carried out by Garabedian and Schiffer in pre-electronic computer era. However, in modern times such calculations should be very easy, and remember these authors' final remark: "However, we must resist the temptation to spend a lifetime working out a wealth of examples, each with its own special twist!"

5.3 Grunsky Matrix

A complex square matrix U is said to be *unitary* if $UU^* = U^*U = I$, where U^* is the complex conjugate transpose of U and I is the identity matrix. The matrices U and U^* are each unitary. Some of the properties of U are: (i) $\langle Ux, Uy \rangle = \langle x, y \rangle$; (ii) $|\det U| = 1$; an the eigenvalues of U are orthogonal. The columns of U form an orthogonal basis of $\mathbb{C}^n$ with respect to the usual inner product. U is an isometry with respect to the usual norm, and U is a normal matrix, i.e., $UU^* = U^*U$, with eigenvalues lying on the unit circle. The matrix U is invertible, i.e., $U^{-1} = U^*$. The matrix U can also be written as $U = e^{iH}$, where H is a Hermitian matrix. An example of a 2×2 unitary matrix is $U = e^{i\varphi} \begin{bmatrix} a & b \\ -b^\star & a^\star \end{bmatrix}$, with $\det U = e^{2i\varphi}$, and $|a|^2 + |b|^2 = 1$.

The *unitary Grunsky inequalities* are defined as follows: The Faber polynomials Φ_n associated with a univalent function f is defined by

$$\Phi_n\left(\frac{1}{f(z)}\right) = \frac{1}{z^n} - \sum_{m=1}^{\infty} \gamma_{mn} z^m, \qquad (5.3.1)$$

where the coefficients γ_{mn} are defined in the series expansion (3.3.2). Let $C_{mn} = \sqrt{mn}\, \gamma_{mn}$. Grunsky [1939] showed that the series (3.3.2) converges for $|z| < 1, \zeta < 1$, if and only if the symmetric infinite matrix $\mathbf{C} = (C_{mn})$, known as the *Grunsky unitary matrix*, satisfies

$$\left| \sum_{m,n=1}^{\infty} C_{mn} x_m x_n \right| \leq \sum_{n=1}^{\infty} |x_n|^2, \tag{5.3.2}$$

for every complex vector $\mathbf{x} = (x_1, \dots, x_n, \dots)$. Later, Schur [1945] showed that every quadratic form satisfying (5.3.2) has a matrix of the form

$$\mathbf{C} = \mathbf{U}^* \mathbf{e} \mathbf{U}, \tag{5.3.3}$$

where $\mathbf{U}$ is unitary, $\mathbf{U}^*$ its complex conjugate transpose, and $\mathbf{e}$ is a real diagonal matrix with elements $0 \leq e_i \leq 1$. This result shows that there is a relationship between Grunsky matrices and unitary matrices.

Let $f(z)$ be analytic and univalent. Then f is said to define a *slit mapping* if the complement of the range of f has measure zero (with respect to the ordinary Lebesgue measure in the complex plane).

Theorem 5.3.1. *Suppose that $f(z) \in \mathcal{S}$ in the unit disk E. Then f is univalent if and only if*

$$\sum_{n=1}^{\infty} \left| \sum_{m=1}^{\infty} x_m C_{mn} \right|^2 \leq \sum_{n=1}^{\infty} |x_n|^2 \tag{5.3.4}$$

for every complex vector $\mathbf{x} = (x_1, \dots, x_n, \dots)$. Equality holds for all sequences if and only if f defines a slit mapping.

PROOF. In view of the Cauchy-Schwarz inequality (1.8.4), the inequality (5.3.2) is implied by the apparently stronger inequality (5.3.4). The left hand side of (5.3.4) is equal to $\|\mathbf{C}\mathbf{x}\|^2$. Then, using (5.3.3) we get

$$\|\mathbf{C}\mathbf{x}\|^2 = \langle \mathbf{U}^* \mathbf{e} \mathbf{U} \mathbf{x}, \mathbf{U}^* \mathbf{e} \mathbf{U} \mathbf{x} \rangle = \langle \mathbf{e} \mathbf{U}^* \star \mathbf{U}^* \mathbf{e} \mathbf{U} \mathbf{x}, \mathbf{U} \mathbf{x} \rangle$$
$$= \langle \mathbf{e}^2 \mathbf{U} \mathbf{x}, \mathbf{U} \mathbf{x} \rangle \leq \|\mathbf{e}\|^2 \|\mathbf{U}\mathbf{x}\|^2 \leq \|\mathbf{x}\|^2,$$

since $\|\mathbf{e}^2\| \leq 1$ and $\mathbf{U}$ is unitary. Further, equality for slit mappings follows from the proof of Grunsky inequality (§3.5). ∎

This result leads to the following theorem.

Theorem 5.3.2. (Pederson [1967]) *If f is univalent in the unit disk E, then f defines a univalent slit mapping if and only if the matrix $\mathbf{C}$ is unitary.*

PROOF. (*Necessity:*) Suppose f is univalent. Let $x_n = 1$ and $x_m = 0$ if $m \neq n$. Since (5.3.4) is an equality for finite sequences, we get

$$\sum_{n=1}^{\infty} |C_{mn}|^2 = 1, \tag{5.3.5}$$

which means that the rows of the matrix $\mathbf{C}$ have norm 1. Let $x_n = 1$ and $x_m = 0$ for $m \neq n$ or k ($k \neq n$). Then $\sum_{n=1}^{\infty} |C_{mn} + x_k C_{km}|^2 = 1 + |x_k|^2$. If we expand the left hand side of the above inequality and use (5.3.5), we get

$$\Re\left\{ \sum_{n=1}^{\infty} C_{mn} \bar{C}_{kn} \bar{x}_k \right\} = 0,$$

or, since x_k is an arbitrary complex number, $\sum_{n=1}^{\infty} C_{mn} \bar{C}_{kn} = 0$, i.e., the rows of $\mathbf{C}$ are positive orthogonal. Hence, $\mathbf{C}$ is unitary.

(*Sufficiency:*) This follows from the fact that if $\mathbf{C}$ is unitary, then (5.3.4) is an equality. ∎

The question as to for which univalent functions a truncated Grunsky matrix is unitary finds an answer in the following result.

Theorem 5.3.3. (Pederson [1967]) *Let f be an analytic function in E with Grunsky matrix $\mathbf{C} = (C_{jk})$. If there exists a finite set of integers $1 = \alpha_1 < \alpha_2 < \cdots < \alpha_n$ such that the matrix $(C_{\alpha_j \alpha_k})$ is unitary, then $\mathbf{C}$ is diagonal. Moreover,*

$$f(z) = \frac{z}{e^{i\theta} z^2 + az + 1}$$

where θ is a real constant, and a is constant.

PROOF. Denote $\tilde{\mathbf{C}} = (C_{\alpha_k \alpha_k}), j, k = 1, 2, \ldots, n$. Since $\tilde{\mathbf{C}}$ and $\mathbf{C}$ are both unitary, we have $C_{\alpha_j \beta} = 0$ for $\beta \neq \alpha_k$ for some k. In particular, since the first Faber polynomial is given by $\Phi_1(w) = w - b_0$, b_0 constant (§9.3), we get

$$\frac{1}{f(z)} = a + \frac{1}{z} + \sum_{j=1}^{m} b_{1\alpha_j} z^{\alpha_j}, \quad b_{1\alpha_m} \neq 0, \ m \leq n.$$

The α_nth Faber polynomial has the form

$$\Phi_{\alpha_n}(w) = w^{\alpha_n} + \gamma_1 w^{\alpha_n - 1} + \cdots + \gamma_{\alpha_n - 1} w + \gamma_{\alpha_n}.$$

Thus, $b_{\alpha_n, \alpha_m \alpha_n} = b_{1\alpha_m}^{\alpha_n}$. But since $b_{\alpha_n, k} = 0$ for $k > \alpha_n$, we get $\alpha_m = 1$. Then using the unitary property of $\mathbf{C}$, we get $b_{11} = e^{i]theta}$. Hence, $\frac{1}{f(z)} = a + \frac{1}{z} + e^{i\theta} z$. It is easy to see from (3.3.2) that if $1/f(z)$ differs from $1/g(z)$ by a constant, then f and g have the same Grunsky matrix. In particular, if $a = 0$, then $\mathbf{C}$ is diagonal. ∎

Let C_m denote the mth row vector and $\Delta \mathbf{C} = \mathbf{C} - \mathbf{I}$, where $\mathbf{I}$ is the identity matrix, then we have an alternate form of Theorem 5.3.2 which is more useful.

Theorem 5.3.4. (Pederson [1967]) *If* $\mathbf{C} = (C_{mn})$ *is a symmetric unitary matrix and* $C_{mn} = r_{mn} + i\,s_{mn}$, *where* r_{mn} *and* s_{mn} *are real, then*

$$r_{mn} - \delta_{mn} = -\frac{1}{2}\left(\Delta C_m, \Delta C_n\right), \tag{5.3.6}$$

where δ_{mn} *denotes the Kronecker delta.*

PROOF. Since $\mathbf{C}$ and $\mathbf{I}$ are unitary, we have

$$\begin{aligned}
\delta_{mn} &= (C_m, C_n) = (\Delta C_m + I_m, \Delta C_n + I_n) \\
&= (\Delta C_m, \Delta C_n) + (\Delta C_m, I_n) + (I_m, \Delta C n) + (I_m, I_n) \\
&= (\Delta C_m, \Delta C_n) + r_{mn} + r_{nm} - \delta_{mn} - i\left(s_{nm} - s_{mn}\right).
\end{aligned}$$

The result follows from the symmetry of $\mathbf{C}$. ∎

5.3.1 Fourth Coefficient Problem. We have the following result.

Theorem 5.3.5. (Pederson [1967]) *If* $f(z) \in S$ *in the unit disk* E, *then*

$$\left|a_4 + \alpha a_2\left(a_3 - \tfrac{3}{2}a_2^2\right)\right| \le 4 \tag{5.3.7}$$

for all real α *such that* $|\alpha + 2| \le \sqrt{75/17}$. *Equality holds only for the Koebe function.*

PROOF. The polynomial $p(a) = a_4 + \alpha a_2\left(a_3 - \tfrac{3}{2}a_2^2\right)$ is homogeneous of degree 3 in the sense that replacing a_k by $e^{i(k-1)\theta}a_k$ yields a factor of $e^{3i\theta}$ in front of $p_\alpha(a)$. Thus, we can assume that if f is the extremal function, then

$$p_\alpha(a) \ge 0, \quad 0 \le \Re\{a_2\} \le 2. \tag{5.3.8}$$

It is obvious that f satisfies a Schiffer differential equation of the type (5.1.2) and therefore, f defines a slit mapping as does $\sqrt{f(z^2)}$. By direct computation we have

$$C_{11} = \frac{a_2}{2},$$

$$C_{13} = \frac{\sqrt{3}}{2}\left(a_3 - \frac{3}{4}a_2^2\right); \quad \text{giving} \quad a_4 = \frac{2}{3}C_{33} = \frac{10}{3}C_{11}^3 + \frac{8}{\sqrt{3}}C_{11}C_{13},$$

where (C_{jk}) is the Grunsky matrix of $\sqrt{f(z^2)}$. Set $C_{jk} = r_{jk} + is_{jk}$, $t = r_{11}$, $\Delta p_\alpha = p_\alpha(a_2, a_3, a_4) - p_\alpha(2, 3, 4)$, and using (5.3.8), we get

$$\Delta p_\alpha = \frac{2}{3}\left(r_{33} - 1\right) + \frac{10}{3}\left(t^3 - 1\right) - 10ts_{11}^2 + \frac{4\lambda t}{\sqrt{3}}r_{13} - \frac{4\lambda}{\sqrt{3}}s_{11}s_{13},$$

where $\lambda = \alpha + 2$ and $0 \leq t \leq 1$. Then, in view of Theorem 5.3.4, we find that

$$\frac{2}{3}(r_{33} - 1) + \frac{4\lambda t}{\sqrt{3}} r_{13} = -\frac{1}{3} \|\Delta C_3\|^2 - \frac{2\lambda t}{\sqrt{3}} (\Delta C_3, \Delta C_1)$$

$$= -\left\| \frac{1}{\sqrt{3}} \Delta C_3 + \lambda t \Delta C_2 \right\|^2 + \lambda^2 t^2 \|\Delta C_1\|^2 \leq 2\lambda^2 t^2 (1 - t).$$
$$(5.3.9)$$

Hence,

$$-10ts_{11}^2 - \frac{4\lambda}{\sqrt{3}} s_{11} s_{13} = -t\left(10s_{11}^2 + \frac{4\lambda}{\sqrt{3}} s_{11} s_{13}\right)$$

$$-\frac{4\lambda(1-t)}{\sqrt{3}} s_{11} s_{13} \leq \frac{2}{15} t\lambda^2 s_{13}^2 + \frac{2}{\sqrt{3}} |\lambda|(1-t)\left(s_{11}^2 + s_{13}^2\right).$$
$$(5.3.10)$$

Now, by Theorem 5.3.1, with $x_1 = 1, x_n = 0$ otherwise, we have $r_{11}^2 + s_{11}^2 + s_{13}^2 \leq 1$. After substituting this inequality into (5.3.10) we get

$$-10ts_{11}^2 - \frac{4\lambda}{\sqrt{3}} s_{11} s_{13} \leq \left(\frac{2}{15} t\lambda^2 + \frac{2}{3} |\lambda|(1-t)\right)(1-t^2).$$
$$(5.3.11)$$

Substituting (5.3.10) and (5.3.11) into (5.3.9) we obtain

$$\Delta p_\alpha \leq \frac{10}{3}(t^3 - 1) + 2\lambda^2 t^2 (1-t) + \left(\frac{2}{15} t\lambda^2 + \frac{2}{\sqrt{3}} |\lambda|(1-t)\right)(1-t^2). \quad (5.3.12)$$

It is obvious that the above estimate is monotone in λ for $0 \leq t \leq 1$. Choosing $|\lambda| = \sqrt{75/17}$, which is the largest value for which the right hand side of (5.3.12) is negative near $t = 1$, and using the crude estimate $\sqrt{25/17} < 3/2$, we obtain

$$\Delta p_\alpha \leq -(1-t)^2\left(\frac{1}{3} + \frac{157}{51} t\right) \leq 0,$$

with equality only if $t = 0$. ∎

Bombieri 1963] proved that

$$\liminf_{t \to 1^-} \frac{4 - \Re\{a_4\}}{1-t} > 1.6.$$

But from the estimate (5.3.12) we get

$$\liminf_{t \to 1^-} \frac{4 - \Re\{a_4\}}{1-t} > \frac{14}{15},$$

which can be slightly improved by considering the contribution of the imaginary parts of the first two components of $\Delta C_3/\sqrt{3} + 2t\Delta C_1$. Jenkins and

Ozawa [1967] used Theorem 5.3.1 to derive he local results for the sixth and eighth coefficients, by choosing special values of the parameters rather than using the unitary property.

Besides these works, Charsyński and Schiffer [1960] used Grunsky inequalities to prove $|a_4| \leq 4$. We will, however, not give the details here, and refer the reader to their work.

5.4 Higher-Order Coefficients

We remark that Garabedian, Ross and Schiffer [1965], and Pederson [1967] obtained partial results to the solution of the sixth coefficient problem. If the coefficients are real, the bound $|a_6| \leq 6$ readily follows. If a_2 and a_3 are real, the Pederson method gives the same estimates as for real coefficients. Schiffer [1967] also gave a similar result. For a_2 real and positive, the Pederson method can be used to prove $\Re\{\Delta a_6\} \leq 0$; this estimate differs from the estimate for real coefficients by $\lambda(t)t\left(1 - t^2\right)$, where $\lambda(t)$ is the largest eigenvalue of a 2×2 matrix. Ozawa [1965] proved the result for a_2 real by using the classical form of Grunsky inequality (§3.5.1). In the general case, if f is normalized such that $|\arg\{a_2\}| \leq \pi/5$, then $\Delta\Re\{a + 6\} \leq Q(t)$, $t = \frac{1}{2}\Re\{a + 2\}$, where $Q(t)$ depends on t and the largest values of 3×4 matrices.

Pederson [1968] and Ozawa [1969], independently, used Grunsky inequalities and proved $|a_6| \leq 6$. The fifth coefficient took a little longer; finally, Pederson and Schiffer [1972] proved $|a_5 \leq 5$. These works, although significant in own right, took the research only so far, and failed to resolve the Bieberbach conjecture.

5.5 Exercises

5.5.1. Proof of Lemma 5.2.4.: PROOF. The derivative $f'(z)$ of the extremal function has a zero on the unit circle at the point that corresponds to the finite end of the analytic slit in the w-plane bounding the extremal region. Let this zero be denoted by $e^{-i\psi}$. Then in view of Lemma 5.2.1, this zero is a double zero of the right-hand side of the differential equation (5.2.1), which yields

$$- e^{-3i\psi} + 2e^{-2i\psi} - 3a_3e^{-i\psi} + 3a_4 - 3\bar{a}_3e^{i\psi} + 2\bar{a}_2e^{2i\psi} - e^{3i\psi} = 0,$$

$$\tag{5.5.1}$$

$$3e^{-3i\psi} - 4a_2e^{-2i\psi} + 3a_3e^{-i\psi} - 3\bar{a}_3e^{i\psi} + 4\bar{a}_2e^{2i\psi} - 3e^{3i\psi} = 0. \tag{5.5.2}$$

After subtracting (5.5.1) from (5.5.2) and dividing the difference by $6e^{-i\psi}$, we obtain (5.2.9). ∎

5.5.2. Proof of Lemma 5.2.6: Substitute the power series $f(z) = z + \sum_{n=2}^{\infty} a_n z^n$ into the differential equation (5.2.1) and equate the corresponding powers of z. The calculations are reduced by using the expansion

$$-\frac{1}{4}\frac{d}{dz}\left[\frac{1}{f(z)^4}+\frac{4a_2}{f(z)^3}+\frac{4a_3+2a_2^2}{f(z)^2}\right]$$

$$=\frac{1}{z^5}-\frac{a_4}{z^2}+\left(a_6-2a_2a_5-3a_3a_4+aa_2^2a_4+3a_2a_3^2-4a_2^3a_3+a_2^5\right)$$

$$+\left(2a_7-4a_2a_6-6a_3a_5+8a_2^2a_5-5a_4^2+24a_2a_3a_4-16a_2^3a_4+4a_3^3\right.$$

$$\left.-24a_2^2a_3^2+22a_2^4a_3-5a_2^6\right)z+\cdots,\tag{5.5.3}$$

which is based on the theory of Faber polynomials (§9.3). By substituting the expansion (5.5.3) into the left hand side of (5.2.1) and multiplying out the resulting power series in z, we obtain

$$\frac{1}{z^3}+\frac{2a_2}{z^2}+\frac{3a_3}{z}+3a_4+(5a_5-2a_2a_4)z+(7a_6-2a_2a_5-6a_3a_4+4a_2^2a_4$$

$$+3a_2a_3^2-4a_2^3a_3+a_2^5)z^2+(9a_7-2a_2a_6-6a_3a_5+4a_2^2a_5-9a_4^2$$

$$+18a_2a_3a_4-8a_2^3a_4+4a_3^3a_4+4a_3^3-18a_2^2a_3^2+14a_2^4a_3-3a_2^6)z^3+\cdots.$$

Equating the coefficients of z, z^2 and z^3 on both sides of the above equality, we get the results. ∎

5.5.3. Proof of Lemma 5.2.7: From (5.2.10) we get $b_1=-\frac{1}{2}a_2$, $b_3=-\frac{1}{2}a_3+\frac{3}{8}a_2^2$, $b_5=-\frac{1}{2}a_43+\frac{3}{4}a_2a_3-\frac{5}{16}a_2^3$, $b_7=-\frac{1}{2}a_5+\frac{3}{4}a_2a_4+\frac{3}{8}a_3^2-\frac{15}{16}a_2^2a_3+\frac{35}{128}a_2^4$, $b_9=-\frac{1}{2}a_6+\frac{3}{4}a_2a_5+\frac{3}{4}a_3a_4-\frac{15}{16}a_2a_3^2+\frac{35}{32}a_2^3a_3-\frac{63}{256}a_2$, $b_{11}=-\frac{1}{2}a_7+\frac{3}{4}a_2a_6+\frac{3}{4}a_3a_5+\frac{3}{8}a_4^2-\frac{15}{16}a_2^2a_5-\frac{15}{8}a_2a_3a_4-\frac{5}{16}a_3^3+\frac{35}{32}a_2^3a_4+\frac{105}{64}a_2^2a_3^2-\frac{315}{256}a_2^4a_3+\frac{231}{1024}a_2^6$. Substituting the expressions (5.2.12)–(5.2.14) into the above values and obtain formulas for the coefficients $b_1,b_3,\dots,b_{11}$ in terms of only a_2,a_3 and a_4. These formulas together with the first six terms in the inequality (5.2.11) gives the result. ∎

5.5.4. Proof of Lemma 5.2.8: Consider the extremal problem $\max\Re\{a_4-2a_2a_3+\frac{13}{12}a_2^3\}$, with the extremal region bounded by analytic arcs which satisfy the the differential equation

$$\frac{d^2w}{w^2}\left[\frac{1}{w^2}+\frac{a_2}{w^2}+\frac{a_2^2}{4w}\right]=\left(\frac{dw}{w}\left[\frac{1}{w^{3/2}}+\frac{a_2}{2w^{1/2}}\right]\right)^2\le0,\tag{5.5.4}$$

This equation is similar to Eq (5.2.2). Since the left-hand side of Eq (5.5.4) is a perfect square, it can be easily integrated, which shows that $\Re\{2w^{-3/2}+3a_2w^{-1/2}\}=0$ along the extremal arcs. This result, together with the Schwarz reflection principle (§2.5), we find that the only extremal functions are the Koebe function $K(z)=z/\left(1-e^{i\theta}z\right)^2$, and the mapping is $w=\left(z^{-3}-2+z^3\right)^{-1/2}$, for which equality holds in the last inequality. ∎

5.5.5. Proof of Lemma 5.2.10: Set $A = |a_2|, L = |\lambda|$. Then by (5.2.11) we have

$$\left(4 - A^2 - 3L^2\right)^{1/2} - 5^{1/2}\left(4 - 1.5AL - 0.5A^3\right) \geq 0. \tag{5.5.5}$$

By maximizing the left-hand side of (5.2.19) with respect to L, we get $2L\left(4 - A^2 - 3L^2\right)^{-1/2} = 5^{1/2}A$, and hence,

$$L = 5^{1/2}A\left(\frac{4 - a^2}{4 + 15A^2}\right)^{1/2}, \quad 4 - A^2 - 3L^2 = 4\frac{4 - A^2}{4 + 15A^2}.$$

Thus, A satisfies the inequality $\left(\frac{4}{5} + 3A^2\right)^{1/2}\left(4 - A^2\right)^{1/2} + A^3 \geq 8$. Since this relation is not satisfied in the interval $0 \leq A \leq 1.67$, which establishes the bound on a_2. The bound on $|\lambda|$ is obtained from the estimate $3|la|^2 \leq 4 - A^2 \leq 4 - (1 - 67)^2 = 1.2111$. ∎

5.5.6. Proof of Lemma 5.2.11: By Lemma 5.2.10, $|a_4 - 1.5a_2\lambda - 0.5a_2^3| \leq (0.24222)^{1/2} < 0.493$, which gives $|a_4 - 0.5a_2^2| \leq 2.413$ and $|\sin 3\alpha| \leq 0.604$, which yields (5.2.19). ∎

5.5.7. Proof of Lemma 5.2.12: Using the notation in Lemma 5.2.10, setting $\mu = a_4 - \frac{3}{2}a_2\lambda - \frac{1}{2}a_2^3$, and $M = |\mu|$, and applying the triangle inequality, we have

$$\left(4 - A^2 - 3L^2 - 5M^2\right)^{1/2}$$
$$\geq (28)^{1/4}|\tfrac{3}{10}a_2a_4 - \tfrac{3}{32}a_2^4 - \tfrac{9}{40}\bar{a}_2^2 + \tfrac{1}{4}a_2\mu - \tfrac{3}{8}\bar{\lambda} + \tfrac{3}{8}\lambda^2|$$
$$\geq (28)^{1/4}[\tfrac{6}{5}A\cos\alpha - \tfrac{3}{32}A^4\cos 4\alpha - \tfrac{9}{40}A^2\cos 2\alpha - \tfrac{1}{4}AM$$
$$- \tfrac{3}{8}L\cos\theta + \tfrac{3}{8}L^2\cos 2\theta], \tag{5.5.6}$$

where $\lambda = Le^{i\theta}$. Set $\alpha = 0$ in (5.5.6). Since the right-hand side is an increasing function of α, $0 \leq \alpha \leq 0.22$, and since $A \geq 1.67$, we get $-\frac{6}{5}A\sin\alpha + \frac{3}{8}A^4\sin 4\alpha + \frac{9}{20}A^2\sin 2\alpha \geq 0$. Also, $\frac{3}{8}L^2\cos 2\theta - \frac{3}{10}L\cos\theta = \frac{3}{4}\left(L\cos\theta - \frac{1}{5}\right)^2 - \frac{3}{8}L^2 - \frac{3}{100} \geq -\frac{3}{8}L^2 - \frac{3}{100}$. Hence,

$$\left(4 - A^2 - 3L^2 - 5M^2\right)^{1/2} - (28)^{1/2}[1.2A - 0.09375A^4$$
$$- 0.225A^2 - 0.3 - 0.25AM - 0.375L^2] \geq 0. \tag{5.5.7}$$

We maximize the left hand side of this inequality and find for M the worst value

$$M = \left(\frac{28A^2 - 7A^4 - 21A^2L^2}{100 + 35vA^2}\right),$$

and this (5.5.7) yields

$$\left(4 - A^2 - 3L^2\right)^{1/2}\left(100 + 35A^2\right)^{1/2}$$
$$- 7^{1/2}\left[24A - 1.875A^4 - 0.6 - 4.5A^2 - 7.5L^2\right] \geq 0. \tag{5.5.8}$$

Maximizing with respect to L^2 gives $L^2 = \frac{8}{7} - \frac{2}{5}A^2$. There are two cases according as $A^2 < \frac{20}{7}$, or $A^2 > \frac{20}{7}$. In the first case, (5.5.8) reduces to $\frac{80}{7} - 24A + 1.875A^4 + 2.5A^2 + 0.6 \geq 0$, which is not satisfied in the interval $1.67 \leq A \leq (20/7)^{1/2}$, and thus, by Lemma 5.2.10 the other case must prevail. But then $L = 0$ is the least favorable possibility and (5.5.8) yields

$$[(4 - A^2)(100 + 35A^2)/7]^{1/2} - 24A + 1.875A^4 + 4.5A^2 + 0.6 \geq 0. \quad (5.5.9)$$

But the above inequality is not satisfied in the interval $(20/7)^{1/2} \leq A \leq 1.02$, which establishes (5.2.20); the upper bound (5.2.20) on L follows from the lower estimate on A and the inequality $3L^2 \leq 4 - A^2$. ∎

5.5.8. Proof of Lemma 5.2.13: Using Lemma 5.2.10, we get $|a_4 - 0.5a_2^2| \leq 3L + (0.8 - 0.2A^2)^{1/2} \leq 1.2225$, which gives $\sin 3\alpha \leq 0.30563$, and the lemma follows. ∎

5.5.9. Proof of Lemma 5.2.18: Set $S = 0.75Q - P$. Then $S \geq 0$ and $0.0494 \geq 5(1.9394S)^2 + 7(1.3S)^2 \geq 30.635S^2$, which establishes the upper bound on S. If $S \leq 0$ and $-S \geq 0.343P$, then

$$0.0494 \geq 3(0.6666S - 0.78788P)^2 + 5(1.9394S - 0.66512P)^2$$
$$= 20.13968S^2 + 9.7478PS + 4.07417P^2 \geq 14.309S^2,$$

and $S \geq -0.059$. However, when $0 \geq S \geq -0.343P$, then $S \geq -0.027$ by lemma 5.2.16. ∎

5.5.10. Proof of Lemma 5.2.20: Formula (5.2.28) is obtained from (5.2.6) and formula (5.2.29) is obtained from the imaginary part of (5.2.9) by replacing a_2, a_3 and a_4 by the new variables p, q, P, Q and ϵ. ∎

5.5.11. Prove Lemma 5.2.23. PROOF. These bounds follow from Lemma 5.2.22 and the inequalities (5.2.23) and (5.2.24). ∎

5.5.12. Proof of Lemma 5.2.35: Suppose there are two points s_j' and s_j'' at which w_j vanishes in one of these two intervals. Then we would find on integrating by parts that

$$\int_{s_j'}^{s_j''} w_j'(s_j)^2 \, ds_j = \int_{s_j'}^{s_j''} \frac{R_j}{r_j} w_j(s_j)^2 \, ds_j,$$

and, thus, by (5.2.54),

$$\int_{s_j'}^{s_j''} \frac{(w_j')^2 \, ds_j}{\int_{s_j'}^{s_j''} w_j^2 \, ds_j} \leq 4.7, \quad w_j(s_j') = w_j(s_j'') = 0.$$

But it is known from calculus of variations that the left-hand ratio is at least $\pi^2 (s_j'' - s_j')^{-2}$. Hence, $\frac{\pi^2}{4.7} \leq (s_j'' - s_j')^2 < 1.1$, which is absurd. Thus, $w_j(s_j)$ vanishes at lease once in the given interval. In fact, the lemma implies that $\omega(t)$ vanishes at most once in the given interval. ∎

6

Subclasses of Univalent Functions

We will present certain important subclasses of univalent functions, especially for the coefficient problem as an effort to highlight and possibly solve the Bieberbach conjecture. Although a lot of analytic development and methods evolved out of this line of enquiry, the Bieberbach conjecture sill remained unresolved.

6.1 Basic Classes

In view of the Riemann mapping theorem, a simply connected domain D can be mapped conformally onto the open unit disk $E = \{z : |z| < 1\}$. The map is unique if $f(0) = 0$ and $f'(0) = 1$. A function f is univalent in E if and only if $f(z_1) \neq f(z_2)$ for $z_1 \neq z_2$, $z_1, z_2 \in E$. If f has at most one simple zero at $z = 0$, then f is univalent in E. The Taylor's series of f at $z = 0$ is

$$f(z) = f(0) + f'(0)\, z + \frac{f''(0)}{2!}\, z^2 + \cdots + \frac{f^{(n)}(0)}{n!}\, z^n + \cdots = \sum_{n=0}^{\infty}, \quad |z| < 1.$$

If f is univalent, so is $\dfrac{f(z) - a_0}{a_1}$, $a_1 \neq 0$. If f has a simple zero at $z = 0$, then

$$f(z) = a_1\, z + a_2\, z^2 + \cdots = z\,(a_1 + a_2\, z + \cdots),$$

and
$$\frac{f(z)}{a_1} = z + a_1'\, z^2 + \cdots, \quad a_1 \neq 0.$$

If f is univalent in E, so is $\dfrac{f(z)}{a_1}$. We shall normalize f by $f(0) = a_0 = 0$ and $f'(0) = a_1 = 1$. Thus, a normalized univalent function in E has the series representation

$$f(z) = z + a_2\, z^2 + a_3\, z^3 \cdots = z + \sum_{n=2}^{\infty} a_n\, z^n, \quad z \in E. \qquad (6.1.1)$$

The class of all such functions shall be denoted by $\mathcal{S}$.

Consider a meromorphic function f in $0 < |z| < 1$ with a simple pole at

$z = 0$ with residue 1. Such a function is represented as

$$f(z) = \frac{1}{z} + b_0 + b_1 z + \cdots, \qquad |z| < 1. \tag{6.1.2}$$

Under the mapping $z \longmapsto \dfrac{1}{z}$, we obtain

$$g(z) \equiv f\left(\frac{1}{z}\right) = z + b_0 + \frac{b_1}{z} + \frac{b_2}{z^2} + \cdots, \qquad |z| > 1. \tag{6.1.3}$$

We shall denote the class of all such functions by Σ.

Let $f \in \mathcal{S}$, i.e., $f(z) = z + a_2 z^2 + a_3 z^3 \cdots$, which gives

$$f\left(\frac{1}{z}\right) = \frac{1}{z} + \frac{a_2}{z^2} + \frac{a_3}{z^3} + \cdots.$$

Then

$$\frac{1}{g(z)} = \frac{1}{\dfrac{1}{z} + \dfrac{a_2}{z^2} + \dfrac{a_3}{z^3} + \cdots} = \frac{1}{\dfrac{1}{z}\left(1 + \dfrac{a_2}{z} + \dfrac{a_3}{z^2} + \cdots\right)}$$

$$= z\left(1 + \frac{a_2}{z} + \frac{a_3}{z^2} + \cdots\right)^{-1} \quad \text{if } |1/z| < 1$$

$$= z\left\{1 - a_2 - \frac{a_3}{z} + \frac{a_4}{z^2} + \cdots\right\} = z - a_2 - \frac{a_3}{z} + \frac{a_4}{z^2} + \cdots, \quad |z| > 1. \tag{6.1.4}$$

We have introduced the following basic classes of univalent functions:

Class	Representation		Defined by
$\mathcal{S}:$	$f(z) = z + \displaystyle\sum_{n=2}^{\infty} a_n z^n,$	$|z| < 1$	(6.1.1)
$\Sigma:$	$g(z) = z + \displaystyle\sum_{n=0}^{\infty} \dfrac{b_n}{z^n},$	$|z| > 1$	(6.1.3)

The following relation holds between functions in the class $\mathcal{S}$ and the class Σ: If $f(z) \in \mathcal{S}$, then $g(z) = 1/f(1/z) \in \Sigma$, and conversely, if $g(z) \in \Sigma$, then $f(z) = 1/g(1/z) \in \mathcal{S}$. Thus, the range of functionals on Σ determine the range of the corresponding functionals on $\mathcal{S}$, or conversely. For example, the range of $\log g(z)/z$ on Σ, $1 < |z| < \infty$, is obtained from the range of $\log f(z)/z$ on $\mathcal{S}$, $0 < |z| < 1$.

Theorem 6.1.1. *Any conformal mapping of $E = \{z : |z| < 1\}$ onto itself is a bilinear transformation of the form*

$$T(z) = e^{i\theta} \frac{z - z_0}{1 - \bar{z}_0 z}, \qquad T(z_0) = 0, \tag{6.1.5}$$

for some fixed $z_0 \in E$, $\theta \in [0, 2\pi]$, and only T is a conformal map of E onto itself.

PROOF. First, show that under the map T, $|z| = 1$ implies $|T(z)| = 1$: Since

$$|T(z)| = \left| \frac{z - z_0}{1 - \bar{z}_0 z} \right| = \frac{|z - z_0|}{|z| \, |z^{-1} - \bar{z}_0|},$$

and since $|z| = 1$ and $z^{-1} = \bar{z}$, we get $|T(z)| = \dfrac{|z - z_0|}{|z - \bar{z}_0|} = 1$. Note that the only singularity of $T(z)$ is at $z = \bar{z}_0^{-1}$ which lies outside E. Thus, by the maximum modulus principle (§1.5), T maps E onto itself. However, the inverse mapping

$$T^{-1}(w) = e^{-i\theta} \left\{ \frac{w - \left(-e^{i\theta} z_0\right)}{1 - \left(-e^{i\theta} \bar{z}_0\right) w} \right\}$$

is also a map from E onto itself, which shows that T is a conformal map of E onto itself. Now, to prove uniqueness, let $S : E \mapsto E$ be any conformal map, and let $z_0 = S^{-1}(0)$, and $\theta = \arg\{S'(z_0)\}$. Since $T(z_0) = 0$ and $\theta = \arg\{T(z_0)\}$ also, we see that $T'(z) = e^{i\theta} \left\{ \dfrac{1 - |z_0|^2}{(1 - \bar{z}_0 z)^2} \right\}$, which at $\bar{z} = z_0$ is equal to $e^{i\theta} \left\{ \dfrac{1}{1 - |z_0|^2} \right\}$, i.e., a real constant times $e^{i\theta}$. Thus, by the uniqueness theorem of conformal mapping $S = T$. ∎

The conformal mapping $T(z) : E \mapsto E : T(z_0) = 0$ will denote the *analytic automorphism*, as shown in Figure 6.1.1.[1] The automorphic transformation (6.1.5) is also denoted by $\mathrm{Aut}(E)$, indicating that this mapping is an automorphism of E onto itself.

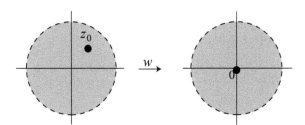

Figure 6.1.1 Conformal Map of E onto E.

6.1.1 Bergman Metric and Projections. The Bergman kernel function of E is

$$K(z, \bar{\zeta}) = \frac{1}{\pi \left(1 - z\bar{\zeta}\right)^2}, \quad z, \zeta \in E. \tag{6.1.6}$$

---

[1] For definition and set-theoretic concept of automorphism, see Bourbaki [1968: ch. 4].

This kernel function has the property that

$$\frac{\partial}{\partial z} K(z, \bar{z}) = \overline{\frac{\partial}{\partial z} K(z, \bar{z})}. \tag{6.1.7}$$

The *Bergman metric* (or the *Poincaré-Bergman metric*) of E is

$$ds^2 = \frac{\partial^2}{\partial z \partial \bar{z}} \log K(z, \bar{z}) = \frac{2|dz|^2}{(1 - |z|^2)}, \tag{6.1.8}$$

where $K(z, \bar{z})$ is the Bergman reproducing kernel. The Bergman metric is invariant under the automorphism $T(z)$ defined by (6.1.5). The *Bergman projection* $\mathfrak{P}\{f(z)\}$ and its associated projection $\overline{\mathfrak{P}}\{f(z)\}$ for the function $f \in \mathcal{S}$ are defined by

$$\mathfrak{P}\{f(z)\} = \int_E \frac{f(\zeta)}{(1 - z\bar{\zeta})^2} \, dA(\zeta), \quad \text{for } z, \zeta \in E, \tag{6.1.9}$$

$$\overline{\mathfrak{P}}\{f(z)\} = \int_E \frac{f(\zeta)}{(1 - \bar{z}\zeta)^2} \, dA(\zeta), \quad \text{for } z, \zeta \in E, \tag{6.1.10}$$

where $dA(\zeta) = \dfrac{dx \, dy}{\pi}$. These definitions will be useful in §11.4.

6.2 Functions with Positive Real Part

Consider the function

$$H(z) = \frac{1 + z}{1 - z} = \frac{2}{1 - z} - 1 = 1 + 2z + 2z^2 + 2z^3 + \cdots. \tag{6.2.1}$$

This function is a Möbius transformation that map E onto the right half-plane $\mathbb{H}^+ = \{w \in \mathbb{C} : |Re\{w\} > 0\}$. In fact,

$$H\left(e^{i\theta}\right) = \frac{1 + e^{i\theta}}{1 - e^{i\theta}} = \frac{e^{-i\theta/2} + e^{i\theta/2}}{e^{-i\theta/2} - e^{i\theta/2}} = i \cot \frac{\theta}{2},$$

which yields

$$\Re\left\{\frac{1 + re^{i\theta}}{1 - re^{i\theta}}\right\} = \Re\left\{\frac{\left(1 + re^{i\theta}\right)\left(1 - re^{i\theta}\right)}{\left(1 - re^{i\theta}\right)\left(1 - re^{i\theta}\right)}\right\} = \frac{1 - r^2}{|1 - re^{i\theta}|^2}$$

$$= \frac{1 - r^2}{1 - 2r \cos \theta + r^2} = p(r, \theta).$$

Thus, obviously, $\Re\{H\left(re^{i\theta}\right)\} > 0$. The function $p(r, \theta)$ represents the Poisson kernel which appears in Poisson's integral formula (2.7.1).

Let p be an analytic function, and let $\mathcal{P} = \{p : p(0) = 1, \Re\{p(z)\} > 0\}$ be the set of functions with positive real part. The function $H(z) = \dfrac{1+z}{1-z}$, defined by (6.2.1), belongs to the set $\mathcal{P}$. Moreover,

$$\mathcal{P} = \{p : p(0) = 1, p(E) \subset \mathbb{H}^+\} = \left\{p : p \prec \frac{1+z}{1-z}\right\}. \qquad (6.2.2)$$

The set $\mathcal{P}$ is a compact and convex subset of the set of analytic functions, such that

$$\frac{1-r}{1+r} \leq |p(re^{i\theta})| \leq \frac{1+r}{1-r}, \qquad 0 < r < 1. \qquad (6.2.3)$$

Theorem 6.2.1. *If $p(z) = 1 + p_1 z + p_2 z^2 + \cdots \in \mathcal{P}$, then $|p_n| \leq 2$ for all $n \in \mathbb{N}$. Moreover, $\max_{p \in \mathcal{P}} |a_n(p)| = 2$, where $a_n(p) = p^{(n)}/(n!)$. For $n = 1$, equality holds for $p(z) = \dfrac{1 + e^{i\theta} z}{1 - e^{i\theta} z} = H\left(e^{i\theta} z\right)$ for a $\theta \in \mathbb{R}$.*

PROOF. The case $n = 1$ follows from Theorem E.1(i). The rest of the statement follows from Littlewood theorem E.3. For $n = 2$, let $p \in \mathcal{P}$. Then the even function

$$q(z) = \frac{1}{2}\left[p(z) + p(-z)\right] = \sum_{m=0}^{\infty} p + 2m z^{2m} \in \mathcal{P}.$$

Also, we have

$$s(z) = \sum_{k=0}^{\infty} s_k z^k = q\left(\sqrt{z}\right) = \sum_{m=0}^{\infty} p_{2m} z^m \in \mathcal{P}, \qquad \Re\{r\} > 0.$$

Thus, $s_1 = p_2$, which yields $|p_1| = |s_1| \leq 2$. For any n let $\omega_n = e^{2\pi i i/n}$ be the n roots of unity. Then for $p \in \mathcal{P}$,

$$q(z) = \frac{1}{n} \sum_{j=1}^{n} p\left(\omega_n^j z\right) = \sum_{k=0}^{\infty} \frac{1}{n} \sum_{j=1}^{n} \omega_n^{jk} p_k z^k = \sum_{m=0}^{\infty} p_{nm} z^{nm} \in \mathcal{P}.$$

Hence, $s(z) = q\left(z^{1/n}\right) \in \mathcal{P}$, which gives $s_1 = p_n$, and the proof is complete. ∎

Theorem 6.2.2. (Riesz-Herglotz Representation) *An analytic function $p(z) = 1 + p_1 z + p_2 z^2 + \cdots$, $z \in E$ is in the class $\mathcal{P}$ if there is an increasing function $\mu(t)$ in $[0, 2\pi]$ with $\mu(2\pi) - \mu(0) = 1$, such that*

$$p(z) = \int_0^{2\pi} \frac{1 + e^{-it} z}{1 - e^{-it} z}\, d\mu(t). \qquad (6.2.4)$$

PROOF. We must show that $p \in \mathcal{P}$ implies the representation (6.2.4). Let $p \in \mathcal{P}$. In view of Cauchy integral formula, we get

$$p(z) = \frac{1}{2\pi} \int_0^{2\pi} \frac{re^{it} + z}{re^{it} - z} \Re\{p(re^{it})\} \, dt = \int_0^{2\pi} \frac{re^{it} + z}{re^{it} - z} \, d\mu(t, r),$$

where

$$\mu(t, r) = \frac{1}{2\pi} \int_0^t \Re\{p(re^{i\tau})\} \, d\tau$$

is an increasing function in $[0, 2\pi]$ with $\mu(0, r) = 0$ and $\mu(2\pi, r) = 1$. But this follows from the property $p(0) = 1$. The rest of the proof follows from the limiting process as $r \to 1$. ∎

An alternate proof is given in Exercise 6.8.2. This theorem leads to the following formula.

Theorem 6.2.3. (Herglotz formula) *If $p \in \mathcal{P}$, then there exists an increasing function $\mu(t)$, $0 < \mu < 2\pi$, of bounded variation 1, i.e., $\int_{-\pi}^{\pi} \mu(t) = 1$ such that*

$$p(z) = \frac{1}{2\pi} \int_{-\pi}^{\pi} \frac{1 + ze^{-it}}{1 - ze^{-it}} \, d\mu(t) + ip(0). \tag{6.2.5}$$

Theorem 6.2.4. *Let $p(z) = 1 + p_1 z + p_2 z^2 + \cdots \in \mathcal{P}$, and let $\mu(t) \in [0, 2\pi]$ be an increasing function with $\mu(2\pi) - \mu(0) = 1$. Then the coefficients*

$$p_n = 2 \int_0^{2\pi} e^{-int} \, d\mu(t). \tag{6.2.6}$$

The proof follows from the (6.2.1).

Theorem 6.2.5. *Let $p(z) = 1 + p_1 z + p_2 z^2 + \cdots \in \mathcal{P}$. Then $|p_n| \leq 2$, where equality holds for the function*

$$p(z) = \sum_{k=1}^{n} \mu_k \frac{e^{i\alpha + 2\pi ik/n} + z}{e^{i\alpha + 2\pi ik/n} - z}, \quad \alpha \in \mathbb{R}, \ \mu_k > 0, \ \sum_{k=1}^{n} \mu_k = 1.$$

PROOF. From Theorem 6.2.3, we get

$$|p_n| = \left| 2 \int_0^{2\pi} e^{-int} d\mu(t) \right| \leq 2 \int_0^{2\pi} d\mu(t) = 2.$$

Equality holds only when $\mu(t)$ is constant. ∎

6.3 Functions in Class $\mathcal{S}_0$

Let $\mathcal{S}_0$ be a subfamily of $\mathcal{S}$. For any $f \in \mathcal{S}_0$, let $g(z) = f(\phi(z))$, where $\phi(z)$ is of the form (6.1.5). Then $w(z) = \dfrac{g(z) - g(0)}{g'(0)} \in \mathcal{S}$. If $w \in \mathcal{S}_0$ for any mapping $\phi(z)$ of the form (6.1.5), then, according to Pommerenke [1964], the class $\mathcal{S}_0$ is known as a *linear invariant family*.

Theorem 6.3.1. (Pommerenke [1975]) *If $f(z) \in \mathcal{S}_0$ with the series expansion (6.1.9), then*

$$\left| \log \frac{f'(z)}{\sqrt{K(z,\bar{z})/K(0,0)}} \right| \leq C(\mathcal{S}_0) \log \frac{1+|z|}{1-|z|}, \tag{6.3.1}$$

where $C(\mathcal{S}_0) = \sup\{|a_2| : f \in \mathcal{S}_0\}$. For $z \in E$, $z \neq 0$, the equality in (6.3.1) holds only for the function for which $|a_2| = C(\mathcal{S}_0)$.

PROOF. Let ϕ_ζ be a mapping of the form (6.1.5), and $\phi_\zeta(0) = \zeta \in E$. Define $g(w) = f(\phi_\zeta(w))$. Since the Taylor series expansion of $g(w)$ at $w = 0$ is $g(w) = \sum\limits_{n=0}^{\infty} \dfrac{g^{(n)}(0)}{n!} w^n$, we have $g'(0) = f'(\zeta)\phi_\zeta(0)$, and $g''(0) = f''(\zeta)\left(\phi_\zeta(0)\right)^2 + f'(\zeta)\phi_\zeta''(0)$. Define the normalized function $g(w)$ as

$$G(w) = \frac{g(w - g(0))}{g'(0)} = w + \frac{1}{2}\left[\frac{f''(\zeta)}{f'(\zeta)} \phi_\zeta'(0) + \frac{\phi_\zeta''(0)}{\phi_\zeta'(0)} \right] w^2 + \cdots \in \mathcal{S}_0.$$

Set

$$c_2 = \frac{1}{2}\left[\frac{f''(\zeta)}{f'(\zeta)} \phi_\zeta'(0) + \frac{\phi_\zeta''(0)}{\phi_\zeta'(0)} \right], \tag{6.3.2}$$

which upon integration with respect to ζ, gives

$$\frac{d}{d\zeta} \log f'(\zeta) = \frac{2c_2}{\phi_\zeta(0)} - \frac{\phi_\zeta''(0)}{\phi_\zeta'(0)}. \tag{6.3.3}$$

The property of the Bergman kernel function

$$K(w, \bar{w}) = |\phi_\zeta;(w)|^2 K\left(\phi_\zeta(w), \overline{\phi_\zeta(w)} \right),$$

when differentiated with respect to w and then letting $w \to 0$, yields

$$0 = \frac{\partial}{\partial w} K(w, \bar{w}) \Big|_{w=0}$$

$$= K\left(\zeta, \bar{\zeta}\right) \overline{\phi_\zeta'(0)}\, \phi_\zeta''(0) + |\phi_\zeta'(0)|^2 \frac{\partial}{\partial w} K\left(\phi_\zeta(w), \overline{\phi_\zeta(w)} \right) \Big|_{w=0}$$

$$= K(\zeta, \bar{\zeta}) \overline{\phi_\zeta'(0)}\, \phi_\zeta''(0) + |\phi_\zeta'(0)|^2 \frac{\partial}{\partial \zeta} K(\zeta, \bar{\zeta})\, \phi_\zeta'(0),$$

whence we get

$$\frac{\phi_\zeta''(0)}{\phi_\zeta'(0)} = \frac{\frac{\partial}{\partial\zeta}K(\zeta,\bar\zeta)}{K(\zeta,\bar\zeta)}\,\phi_\zeta'(0). \tag{6.3.4}$$

Substituting (6.3.4) into (6.3.3), we obtain

$$\frac{d}{d\zeta}\log f'(\zeta) = \frac{2c_2}{\phi_\zeta(0)} + \frac{\frac{\partial}{\partial\zeta}K(\zeta,\bar\zeta)}{K(\zeta,\bar\zeta)}. \tag{6.3.5}$$

For a fixed $z \in E$, let $\zeta = tz$, $0 < t < 1$. Then, by the property (6.1.7), we get from (6.3.5)

$$\frac{d}{dt}K(\zeta,\bar\zeta) = \frac{\partial}{\partial\zeta}K(\zeta,\bar\zeta)\frac{d\zeta}{dt} + \frac{\partial}{\partial\bar\zeta}K(\zeta,\bar\zeta)\frac{d\bar\zeta}{dt} = 2\Re\Big\{\frac{\partial}{\partial\zeta}K(\zeta,\bar\zeta)\,z\Big\},$$

where $\Im\Big\{\dfrac{\partial}{\partial\zeta}K(\zeta,\bar\zeta)\,z\Big\} = 0$. Thus, $\dfrac{\partial}{\partial\zeta}K(\zeta,\bar\zeta) = \dfrac{1}{2z}\dfrac{d}{dt}K(\zeta,\bar\zeta)$, and (6.3.5) becomes

$$\frac{d}{d\zeta}\log f'(\zeta) = \frac{2c_2}{\phi_\zeta(0)} + \frac{1}{2}\frac{d}{d\zeta}\log K(\zeta,\bar\zeta),$$

or

$$\frac{d}{d\zeta}\log\frac{f'(\zeta)}{\sqrt{K(\zeta,\bar\zeta)}} = \frac{2c_2}{\phi_\zeta'(0)}. \tag{6.3.6}$$

Integrating both sides of (6.3.6) with respect to t from 0 to 1, we find that

$$\frac{1}{2}\log\frac{f'(z)/f'(0)}{\sqrt{K(z,\bar z)/K(0,0)}} = \int_0^1 \frac{c_2}{\phi_\zeta'(0)}\,dt,$$

which yields

$$\Big|\log\frac{f'(z)}{\sqrt{K(z,\bar z)/K(0,0)}}\Big| = 2\Big|\int_0^1 \frac{c_2 z}{\phi_\zeta'(0)}\,dt\Big| \le 2C(\mathcal{S}_0)\int_0^1 \frac{|z|}{|\phi_\zeta'(0)|}\,dt. \tag{6.3.7}$$

Let $w = 0$ in (6.3.7); then $K(0,0) = |\phi_\zeta'(0)|^2 K(\zeta,\bar\zeta)$. Since, by (6.1.6), $|\phi)\zeta'(0)| = 1 - |\zeta|^2$, substituting it into (6.3.7) yields (6.3.1). Equality holds in (6.3.1) only if $|c_2| = C(\mathcal{S}_0)$ on the radius from 0 to z, where c_2 is defined by (6.3.2), and hence in particular at 0, i.e., $|a_2| = C(\mathcal{S}_0)$. ∎

Corollary 6.3.1. *For $f \in \mathcal{S}_0$, we have*

$$\sqrt{\frac{K(z,z)}{K(0,0)}}\Big[\frac{1-|z|}{1+|z|}\Big]^{C(\mathcal{S}_0)} \le |f'(z)| \le \sqrt{\frac{K(z,z)}{K(0,0)}}\Big[\frac{1+|z|}{1-|z|}\Big]^{C(\mathcal{S}_0)}, \tag{6.3.8}$$

where $C(\mathcal{S}_0) = \sup\{|a+2| : f \in \mathcal{S}\}$. For $z \in E, z \neq 0$, equality holds only for the function f for which $|a_2| = C(\mathcal{S}_0)$. In particular, using (6.1.6), the inequality (6.3.8) is equivalent to

$$\frac{(1-|z|)^{C(\mathcal{S}_)-1}}{(1+|z|)^{C(\mathcal{S}_0)-1}} \leq |f'(z)| \leq \frac{(1+|z|)^{C(\mathcal{S}_)-1}}{(1-|z|)^{C(\mathcal{S}_0)-1}}. \tag{6.3.9}$$

Corollary 6.3.2. For $f \in \mathcal{S}_0$, the inequality

$$\frac{1}{2C(\mathcal{S}_0)}\left[1 - \left(\frac{1-|z|}{1+|z|}\right)^{C(\mathcal{S}_0)}\right] \leq |f(z)| \leq \frac{1}{2C(\mathcal{S}_0)}\left[\left(\frac{1+|z|}{1-|z|}\right)^{C(\mathcal{S}_0)} - 1\right] \tag{6.3.10}$$

holds, where $C(\mathcal{S}_0) = \sup\{|a_2| : f \in \mathcal{S}_0\}$. For $z \in E, z \neq 0$, equality holds for the function for which $|a_2| = C(\mathcal{S}_0)$.

PROOF. Let $f \in \mathcal{S}_0$, $z = r e^{i\theta}$, $0 < r < 1$. Since $f(0) = 0$, we have $f(z) = \int_0^r f'\left(\rho e^{i\theta}\right) e^{i\theta} d\rho$. Using (6.3.9), we get

$$|f(z)| \leq \int_0^r |f'\left(\rho e^{i\theta}\right)| \, d\rho \leq \int_0^r \frac{(1+\rho)^{C(\mathcal{S}_0)-1}}{(1-\rho)^{C(\mathcal{S}_0)-1}} \, d\rho$$
$$= \frac{1}{2C(\mathcal{S}_0)}\left[\left(\frac{1_r}{1-r}\right)^{C(\mathcal{S}_0)} - 1\right],$$

which is the right-hand side of (6.3.10). To get the left-hand side of (6.3.10), let $m(r) = \min\limits_{|z|=r} |f(z)|$. Note the f is a mapping from the z-plane to the w-plane, under which the image of $|z| < r$ in the w-plane contains the disk $|w| < m(r)$. Thus, there exists a curve γ from 0 to the circle $|z| = r$ such that $m(r) = \int_\gamma |f'(z)| \, |dz|$. Since γ intersects all circles $|z| = \rho < r$, the lower bound for $|f'(z)|$ is given by

$$m(r) \geq \int_0^r \frac{(1+\rho)^{C(\mathcal{S}_0)-1}}{(1-\rho)^{C(\mathcal{S}_0)-1}} \, d\rho = \frac{1}{2C(\mathcal{S}_0)}\left[1 - \left(\frac{1-|z|}{1+|z|}\right)^{C(\mathcal{S}_0)}\right]. \blacksquare$$

Corollary 6.3.3. The image of the unit disk E under the mapping $f \in \mathcal{S}_0$ contains the disk $C\left(0, 1/(2C(\mathcal{S}_0))\right)$, such that $|a_2| = C(\mathcal{S}_0)$ under this mapping.

Let Σ denote the normalized family of meromorphic univalent functions defined on the domain $\Delta = \{z \in \mathbb{C} : |z| > 1\}$ with the series expansion

$$g(z) = z + b_0 + \frac{b_1}{z} + \frac{b_2}{z^2} + \cdots, \tag{6.3.11}$$

such that $g(z) \neq 0$ for all $z \in \Delta$. The function $g \in \Sigma$ is meromorphic outside E except for a pole with residue 1 at infinity. A function $g \in \Sigma$ has the property that it maps $|z| > 1$ onto $\mathbb{C}$ minus a set of (two-dimensional) Lebesgue measure zero. Using an appropriate translation the family Σ can be transformed into another normalized family Σ_0, with series expansion (6.3.11).

Note that if $f \in \mathcal{S}$, then

$$g(z) = \left\{ f\left(\frac{1}{z}\right) \right\}^{-1} = z - a_2 + \left(a_2^2 - a_3\right) z^{-1} + \cdots \in \Sigma_0. \qquad (6.3.12)$$

Conversely, any function $g \in \Sigma_0$ can be transformed into a function $f \in \mathcal{S}$.

6.4 Typically Real Functions

Let $\mathcal{T} \subset \mathcal{S}$ denote the class of typically real functions in E. A function f is said to belong to the class $\mathcal{T}$ if and only if f has real coefficients in its Taylor series expansion about a point $z_0 \in E$. Recall that if $f(z) = a_0 + a_1 z + a_2 z^2 + \cdots$, a_i complex, $|z| < R$, then $f(\bar{z}) = a_0 + a_1 \bar{z} + a_2 \bar{z}^2 + \cdots$, and $\overline{f(\bar{z})} = \bar{a}_0 + \bar{a}_1 z + \bar{a}_2 z^2 + \cdots$. Then $f(z) = \overline{f(\bar{z})}$ only if all a_i are real. Thus, $f \in \mathcal{T}$ implies that $f(z) = \overline{f(\bar{z})}$, or $\overline{f(z)} = f(\bar{z})$. The class $\mathcal{T}$ was introduced by Rogosinski [1939].

Theorem 6.4.1. (Dieudonné [1931]) *If* $f(z) = z + \sum_{n=2}^{\infty} a_n z^n \in \mathcal{T}$ *for* $|z| < 1$, *then* $|a_n| \le n$ *for* $n = 2, 3, \dots$.

PROOF. Let $z = re^{i\theta}$, $r < 1$, $0 \le \theta < 2\pi$, and $f(z) = u(z) + i v(z)$. Then

$$v(z) = \Im\{f(z)\} = \Im\left\{ r e^{i\theta} + \sum_{n=2}^{\infty} a_n r^n e^{in\theta} \right\} = r \sin\theta + \sum_{n=2}^{\infty} a_n r^n \sin n\theta,$$

thus, $v\left(r e^{i\theta}\right) = \sum_{n=1}^{\infty} a_n r^n \sin n\theta$, where $a_1 = 1$.. Multiplying its both sides by $\frac{2}{\pi} \sin m\theta$ and integrating from 0 to π, we get

$$\frac{2}{\pi} \int_0^\pi v\left(r e^{i\theta}\right) \sin m\theta \, d\theta = \frac{2}{\pi} \int_0^\pi \sum_{n=1}^{\infty} a_n r^n \sin n\theta \sin m\theta$$

$$= \frac{2}{\pi} \int_0^\pi a_n r^n \sin^2 n\theta \, d\theta = \frac{2}{\pi} a_n r^n \frac{\pi}{2} = a_n r^n,$$

where we have used the orthogonality relation

$$\int_0^\pi \sin n\theta \sin m\theta \, d\theta = \begin{cases} 0 & \text{if } n \ne m, \\ \pi/2 & \text{if } n = m. \end{cases}$$

Also note that $|\sin(n+1)\theta| \le |\sin n\theta| |\cos\theta| + |\cos n\theta| \sin\theta|$. Since $|\cos\theta| \le 1$, $|\sin\theta| \le 1$, we find that $|\sin(n + 1)\theta| \le |\sin n\theta| + |\sin\theta|$, i.e., $|\sin n\theta| \le$

$|\sin(n-1)\theta| + |\sin\theta|$, or, by induction, $|\sin n\theta| \leq n|\sin\theta|$, where equality holds only for $n = 1$. Thus,

$$\left|a_n\, r^n\right| = \left|\frac{2}{\pi}\int_0^{\pi} v\left(r\,e^{i\theta}\right)\sin n\theta\, d\theta\right| \leq \frac{2}{\pi}\int_0^{\pi}\left|v\left(r\,e^{i\theta}\right)\right| n|\sin\theta|\, d\theta$$
$$\leq \frac{2n}{\pi}\int_0^{\pi}\left|v\left(r\,e^{i\theta}\right)\right|\sin\theta\, d\theta,$$

where $\sin\theta \geq 0$ in $[0,\pi]$. Now, since $v\left(r\,e^{i\theta}\right) = \Im\{f(z)\} = \dfrac{f(re^{i\theta}) - f(re^{-i\theta})}{2i} \neq 0$ for $0 < r < 1$ and for $0 \leq \theta < 2\pi$, we find that $v\left(r\,e^{i\theta}\right)$ is a continuous function of θ. Also, note that

$$r = |a_1|r = \frac{2}{\pi}\int_0^{\pi}\left|v\left(r\,e^{i\theta}\right)\right|\sin\theta, \tag{6.4.1}$$

where $a_1 = 1$, and the equality holds since $n = 1$. Thus,

$$\left|a_n\, r^n\right| = n\left[\frac{2}{\pi}\int_0^{\pi}\left|v\left(r\,e^{i\theta}\right)\right|\sin\theta\right] = n\,r,$$

by using (6.4.1). Let $r \to 1$. Then the above inequality yields $|a_n| \leq n$. ∎

The functions $f \in \mathcal{T}$ f are characterized by domains that are symmetric with respect to the real axis, Dieudonné [1931] and Rogosinski [1932] proved independently, by using different methods, that $|a_n| \leq n$ for $f \in \mathcal{T}$. Details can be found in their respective research papers.

Consider the function $\dfrac{1+z}{1-z} \in \mathcal{P}$, for which $\Re\left\{\dfrac{1+z}{1-z}\right\} > 0$ for all $z \in E$ (see §6.2). Also, recall the Koebe function $K(z) = \dfrac{z}{(1-z)^2} \in \mathcal{S}$. The Koebe function $K(z)$ also belongs to the class $\mathcal{T}$; in fact,

$$K(z) = \frac{z}{(1-z)^2}\cdot\frac{1+z}{1-z}\cdot\frac{1-z}{1+z} = \left(\frac{z}{1-z^2}\right)\left(\frac{1+z}{1-z}\right) \in \mathcal{T}.$$

This suggests that if $f \in \mathcal{T}$, then $f(z) = \dfrac{z}{1-z^2}\,g(z)$, $z \in E$, where $g \in \mathcal{P}$. Then $g(z) = \dfrac{1-z^2}{z}\,f(z)$, $z \in E$, and since $\Re\{g(z)\} > 0$ for $z \in E$, then

$$\Re\left\{\frac{1-z^2}{z}\,f(z)\right\} > 0, \quad z \in E, \quad \text{or} \quad \Im\left\{i\frac{1-z^2}{z}\,f(z)\right\} > 0, \quad z \in E.$$

Note that $\Im\{z\}\Im\{w\} = \Im\left\{\dfrac{z_2 - |z|^2}{2iz}\,w\right\}$. This result can be verified by taking $z = x+iy$ and $w = u+iv$. Then $\Im\left\{\dfrac{1-z^2}{iz}\,f(z)\right\} > 0$, or

$$\Im\left\{\frac{1-z^2}{2iz}\,f(z)\right\} > 0, \quad \text{for } |z| \leq r < 1.$$

Letting $r \to 1$ and noting that then $|z| = 1$, we get $\Im\left\{\dfrac{z^1 - |z|^2}{2iz} f(z)\right\} > 0$, which implies that

$$\Im\{z\}\,\Im\{f(z)\} > 0. \qquad (6.4.2)$$

Thus we have proved:

Theorem 6.4.2. *If $f \in \mathcal{T}$, then the condition (6.4.2) holds for all $z \in E$.*

Note that $\Im\{z\}\,\Im\{f(z)\} > 0$ only if either $\Im\{z\} > 0$ and $\Im\{f(z)\} > 0$, or $\Im\{z\} < 0$ and $\Im\{f(z)\} < 0$. This means that $\Im\{f(z)\}$ changes sign at $\theta = 0, \pi$, where $z = re^{i\theta}$; that is,

Theorem 6.4.3. *If $f \in \mathcal{T}$, then $\overline{f(z)} = f(\bar{z})$.*

This theorem implies that the series expansion of f has real coefficients.

Theorem 6.4.4 (Herglotz Formula) *If $g \in \mathcal{P}$, then there exists an increasing function $\alpha(t)$, $-\infty < \alpha < \infty$, of bounded variation 1, i.e., $\displaystyle\int_{-\pi}^{\pi} \alpha(t) = 1$ such that*

$$g(z) = \frac{1}{2\pi}\int_{-\pi}^{\pi} \frac{1 + ze^{-it}}{1 - ze^{-it}}\, d\alpha(t) + ig(0). \qquad (6.4.3)$$

Now, we shall present Robertson's integral representation for $f \in \mathcal{T}$. Since

$$g(z) = \frac{1}{2\pi}\int_{-\pi}^{\pi} \frac{1 + ze^{-it}}{1 - ze^{-it}} \cdot \frac{1 - ze^{it}}{1 - ze^{it}}\, d\alpha(t) + ig(0)$$

$$= \frac{1}{2\pi}\int_{-\pi}^{\pi} \frac{1 - 2i\sin t - z^2}{1 - 2z\cos t + z^2}\, d\alpha(t) + ig(0)$$

$$= \frac{1}{2\pi}\int_{-\pi}^{\pi} \frac{1 - z^2}{1 - 2z\cos t + z^2}\, d\alpha(t) - \frac{i}{\pi}\int_{-\pi}^{\pi} \frac{z\sin t}{1 - 2z\cos t + z^2}\, d\alpha(t) + ig(0)$$

$$= \frac{1}{\pi}\int_{0}^{\pi} \frac{1 - z^2}{1 - 2z\cos t + z^2}\, d\alpha(t) + ig(0),$$

where the imaginary part of the integral above is zero as it contains an odd function $(\sin t)$ of t while $\alpha'(t)$ is an even function of t. Hence, if $f \in \mathcal{T}$ then

$$f(z) = \frac{z}{1 - z^2}\, g(z) = \frac{1}{\pi}\int_{0}^{\pi} \frac{1 - z^2}{1 - 2z\cos t + z^2}\, d\alpha(t). \qquad (6.4.4)$$

6.5 Starlike Functions

A domain D is said to be *starlike* with respect to a point $z_0 \in D$ if the line segment joining the point z_0 to any point $z \in D$ lies entirely inside D. An analytic property of a starlike domain is: If D is starlike with respect to a point $z_0 \in D$ and if $z \in D$, then $tz \in D$ for $0 < t < 1$. The class $\mathcal{S}^* \subset \mathcal{S}$ is called the class of starlike functions, which defined as follows: $f \in \mathcal{S}^*$ if the domain $f(E)$ is starlike with respect to the origin, i.e., if f maps E onto a starlike domain with respect to the origin. It is a geometrical property of this mapping. The concept of starlike functions is always with respect to a certain point in E. Some starlike domains are represented in Figure 6.5.1.

Lemma 6.5.1. *Let $f \in \mathcal{S}$. Then $f \in \mathcal{S}^*$ if and only if $f(E)$ is starlike with respect to the origin.*

PROOF. Let $D \equiv f(E)$ and $D_r \equiv f(E_r)$, where $E = \{|z| < 1\}$ and $E_r = \{|z| < r < 1\}$. Suppose $f \in \mathcal{S}^*$. If $w \in D$, then $tw \in D$ for $0 < t < 1$, because D is starlike. Then $f^{-1}(tw) = f^{-1}(tf(z)) = g(z)$ is analytic in E and $|g(z)| < 1$. Also, $g(0) = f^{-1}(tf(0)) = 0$. Thus, by the Schwarz lemma, $|g(z)| \le |z|$ for $z \in E$. Choose a point w_1 in D_r (Figure 6.5.2). Then $f(z_1) = w_1$ for some $z_1 \in E_r$ ($|z_1| < r$). For an arbitrary t, $0 < t < 1$,

$$\left| f^{-1}(tw_1) \right| = \left| f^{-1}(tf(z_1)) \right| = g(z_1) \le |z_1| < r,$$

whence $tw_1 \in D_r$. Since this is true for all $w_1 \in D_r$ and all t, $0 < t < 1$, we conclude that $D_r = f(E_r)$ is starlike with respect to $w = 0$.

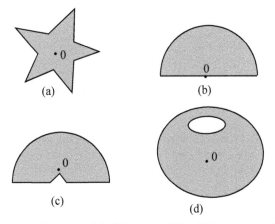

Figure 6.5.1 Domains (a)–(c) are starlike; (d) is not starlike.

To prove sufficiency, we shall show that if $f(E)$ is starlike then $f \in \mathcal{S}^*$. Suppose on the contrary that $f \notin \mathcal{S}^*$. Let $w_0 \in D \equiv f(E)$ (Fig 6.5.2 (b)). Then there is a t_0, $0 < t_0 < 1$, such that $t_0 w_0 \notin D$. Now choose a disk E_r such that $w_0 \in D_r$. Obviously, $D_0 \subset D$ and $t_0 w_0 \notin D_r$, which implies that

D_r is not starlike, i.e., f does not map E_r onto a starlike domain. ∎

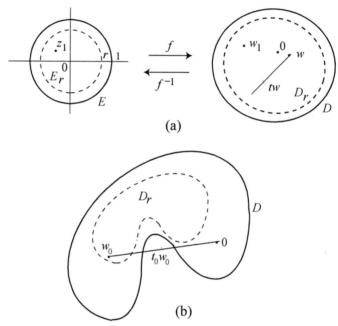

(a)

(b)

Figure 6.5.2 Domain D_r.

Thus, we have shown that if $f \in \mathcal{S}^\star$, then f maps $\{|z| < 1\}$ onto a starlike domain with respect to the origin.

6.5.1 Analytical Formulation of Class $\mathcal{S}^\star$. The following theorem holds:

Theorem 6.5.1. *Let* $f \in \mathcal{S}$. *Then* $f \in \mathcal{S}^\star$ *if and only if* $\Re\left\{ \dfrac{z\, f'(z)}{f(z)} \right\} > 0$ *for* $z \in E$.

PROOF. Let $f \in \mathcal{S}^\star$. Then by the above lemma, $f(E)$ is starlike with respect to 0. Let $z = r\, e^{i\theta}$, $r < 1$, $0 \le \theta < 2\pi$. This means that in $0 \le \theta < 2\pi$ the radius vector from $w = 0$ to $w = f(r\, e^{i\theta})$ lies entirely in $f(E_r)$. Thus, $\arg\{f(r\, e^{i\theta})\}$ is a strictly increasing function of θ, for otherwise the radius vector will intersect $\partial f(E_r)$ in at least two points (see Figure 6.5.3). This means that $\dfrac{\partial}{\partial \theta}\{\arg\{f(r\, e^{i\theta})\}\} > 0$, i.e., $\dfrac{\partial}{\partial \theta}\{\Im \log f(r\, e^{i\theta})\} > 0$. Since

$\log f(r\,e^{i\theta}) = \log\left|f(r\,e^{i\theta})\right| + i\,\arg\left\{f(r\,e^{i\theta})\right\}$, $0 \le \theta < 2\pi$, we find that

$$\Im\left\{\frac{\partial}{\partial\theta}\log\{f(r\,e^{i\theta})\}\right\} = \Im\left\{\frac{f'(r\,e^{i\theta})}{f(r\,e^{i\theta})}\,i\,r\,e^{i\theta}\right\} = \Re\left\{\frac{r\,e^{i\theta}\,f'(r\,e^{i\theta})}{f(r\,e^{i\theta})}\right\}$$

$$= \Re\left\{\frac{z\,f'(z)}{f(z)}\right\} > 0.$$

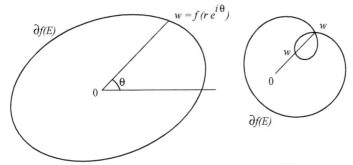

Figure 6.5.3 Radius Vector from 0 to $f(r\,e^{i\theta})$.

We will assume that $f'(0) = \alpha \ne 0$. Then f has a simple zero (and no poles) at the origin. If f has a simple zero (at 0) and no poles in E_r, then by the argument principle $\Delta_{|z|=r}f(z) = 2\pi$, i.e., the winding number is 1, which means that ∂D_r winds around the origin only once. Also, the condition $\Re\left\{\frac{z\,f'(z)}{f(z)}\right\} > 0$ means that $\arg\{f(z)\}$ is an increasing function of $\arg\{z\}$, i.e., $\arg\{f(z)\}$ increases as $\arg\{z\}$ increases. Thus, ∂D_r is a simple closed curve (contour), i.e., D_r is starlike with respect to the origin. Hence, $f \in S^\star$. $\blacksquare$

Note that the example $f(z) = z^2$ which is not conformal (and thus not univalent), although for this function $\Re\left\{\frac{z\,f'(z)}{f(z)}\right\} = 2 > 0$ and $f(0) = 0$. This means that only the necessity part of this theorem which tells that $\Re\left\{\frac{z\,f'(z)}{f(z)}\right\} = 2 > 0$ implies $f(E)$ is starlike but it does not guarantee that f is univalent. Also, a non-univalent function that maps E onto a starlike domain has a zero derivative at the origin. For example, if $f(z) = z + \sum_{n=2}^{\infty} a_n\,z^n$, $|z| < 1$, then

$$\frac{z\,f'(z)}{f(z)} = \frac{1 + \sum_{n=2}^{\infty} n\,a_n\,z^{n-1}}{1 + \sum_{n=2}^{\infty} a_n\,z^{n-1}} = 1 + \alpha_1\,z + \alpha_2\,z^2 + \cdots . \blacksquare \qquad (6.5.1)$$

Theorem 6.5.2. (Löwner [1917]) *Let $f \in \mathcal{S}$ with the series expansion (6.1.1). Then*

$$\frac{1}{(1+r)^2} \le |f'(z)| \le \frac{1}{(1-r)^2}, \qquad (6.5.2)$$

$$\frac{r}{1+r} \le |f(z)| \le \frac{r}{1-r}. \qquad (6.5.3)$$

Moreover, $f(E) \supset C(0, 1/2)$, where $C(0, 1/2)$ denotes the disc with center at the origin and radius $1/2$. For $z \in E$, $z \ne 0$, the equality holds if and only if $f(z) = z/(1-z)$ or one of its rotations.

PROOF. These inequalities follow from Corollaries 6.3.1, 6.3.2 and 6.3.3. ∎

Theorem 6.5.3. *If $f \in \mathcal{S}^\star$, i.e., $f(z) = z + \sum\limits_{n=2}^{\infty} a_n z^n$, then $|a_n| \le 2$.*

The extremal function is $p_0(z) = 1 + 2\varepsilon \sum\limits_{n=1}^{\infty} z^n = 1 + 2\varepsilon \left(\dfrac{z}{1-z} \right) = \dfrac{1 + (2\varepsilon - 1)z}{1-z}$. Let $\varepsilon = 1$. Then $p(z) = \dfrac{1+z}{1-z} \in \mathcal{P}$. To verify this, we note that $p(1) = 1$,

$$\Re\{p(z)\} = \Re\left\{\frac{1+z}{1-z}\right\} = \Re\left\{\frac{(1+z)(1-\bar{z})}{(1-z)(1-\bar{z})}\right\} = \Re\left\{\frac{1 - z\bar{z} + z - \bar{z}}{|1-z|^2}\right\}$$

$$= \Re\left\{\frac{1 - |z|^2 + 2i\Im\{z\}}{|1-z|^2}\right\} = \frac{1 - |z|^2}{|1-z|^2}.$$

Since $|z| < 1$, we have $1 - |z|^2 > 0$, $|1-z|^1 > 0$, and thus, $\Re\{p(z)\} = \dfrac{1 - |z|^2}{|1-z|^2} > 0$.

Theorem 6.5.4. (Nevanlinna [1920-21]) *If $f(z) = z + \sum\limits_{n+2}^{\infty} a_n z^n \in \mathcal{S}^\star$, $z \in E$, then $|a_n| \le n$ for $n = 1, 2, \ldots$.*

PROOF. If $f \in \mathcal{S}^\star$, then $\Re\left\{\dfrac{z f'(z)}{f(z)}\right\} > 0$, $z \in E$. Then $p(z) = \dfrac{z f'(z)}{f(z)} \in \mathcal{P}$,

$$p(z) = \frac{z f'(z)}{f(z)} = \frac{z\left[1 + \sum_{n=2}^{\infty} n a_n z^{n-1}\right]}{z + \sum_{n=2}^{\infty} a_n z^n}$$

$$= \frac{1 + \sum_{n=2}^{\infty} n a_n z^{n-1}}{1 + \sum_{n=2}^{\infty} a_n z^{n-1}} = 1 + \sum_{n=2}^{\infty} \alpha_n z^{n-1}.$$

Since $p(z) = 1 + \sum\limits_{n=1}^{\infty} \alpha_n z^n$, we have

$$\left(1 + \sum_{n=1}^{\infty} \alpha_n z^n\right)\left(1 + \sum_{n=2}^{\infty} a_n z^{n-1}\right) = 1 + \sum_{n=2}^{\infty} n a_n z^{n-1}.$$

After equating the coefficients of powers $k - 1$ on each side, we get

$$k\, a_k = a_k + a_{k-1}\,\alpha_1 + a_{k-2}\,\alpha_2 + \cdots + a_2\,\alpha_{k-2} + \alpha_{k-1},$$

or

$$(k - 1)a_k = a_{k-1}\,\alpha_1 + a_{k-2}\,\alpha_2 + \cdots + a_2\,\alpha_{k-2} + \alpha_{k-1};$$

thus,

$$(k - 1)\,|a_k| \le |a_{k-1}|\,|\alpha_1| + |a_{k-2}|\,|\alpha_2| + \cdots + |a_2|\,|\alpha_{k-2}| + |\alpha_{k-1}|.$$

Since $|\alpha_k| \le 2$ from the previous theorem, we have

$$(k - 1)|a_k| \le 2\left[|a_{k-1}| + |a_{k-2}| + \cdots + |a_2| + 1\right].$$

We already know that $|a_2| \le 2$ from a previous theorem. Now, assume that $a_k| \le k$ for $k = 2, 3, \dots, n - 1$. For $k = n$,

$$(n - 1)|a_n| \le 2\left[|a_{n-1}| + |a_{n-2}| + \cdots + |a_2| + 1\right]$$
$$\le 2\left[(n - 1) + (n - 2) + \cdots + 2 + 1\right]$$
$$\le 2\left[\frac{(n - 1)\,n}{2}\right] = (n - 1)\,n.$$

which yields $|a_n| \le n$. Thus, by induction, $|a_n| \le n$ is valid for $n = 2, 3, \dots$. ∎

This theorem proves that the Bieberbach conjecture holds for starlike functions which is a rather large subset of $\mathcal{S}$. Nevanlinna's result (Theorem 6.5.4) gave some hope that the Bieberbach's conjecture may be true. However, at this time it was not known whether this conjecture was true for any $n > 2$. Löwner [1923] proved that $|a_3| \le 3$ (Theorem 6.5.1), and the search for a proof of this conjecture continued.

Note that the *Koebe function* defined by $K(z) = \dfrac{z}{(1 - z)^2}$ is the extremal function for the classes $\mathcal{S}^\star$ and $\mathcal{T}$. Since $K(z) = \dfrac{z}{(1 - z)^2} = \sum\limits_{n=1}^{\infty} n z^n$ yields

$$K'(z) = \frac{1}{(1 - z)^2} + \frac{2z}{(1 - z)^3} = \frac{1 + z}{(1 - z)^3}, \text{ we have}$$

$$\Re\left\{\frac{z\,K'(z)}{K(z)}\right\} = \Re\left\{\frac{1 + z}{1 - z}\right\} > 0,$$

where $\dfrac{1+z}{1-z} \in \mathcal{P}$. The distortion theorem in this case (class $\mathcal{S}^\star$) is the same as for the class $\mathcal{S}$.

The conformal mapping under a function $w = f(z)$ is represented in Figure 6.4.2, where $z_1 - z_0 = r\,e^{i\theta_1}$, $z_2 - z_0 = r\,e^{i\theta_2}$, and $\theta_1 \to \alpha_1$, $\theta_2 \to \alpha_2$ as $r \to 0$. Let $w_1 - w_0 = \rho_1\,e^{i\phi_1}$, $w_2 - w_0 = \rho_2\,e^{i\phi_2}$. Then

$$f'(z_0) = \lim_{z_1 \to z_0} \frac{w_1 - w_0}{z_1 - z_0} = \lim \frac{\rho_1\,e^{i\phi_1}}{r\,e^{i\theta_1}}$$

$$= \lim \frac{\rho_1}{r}\,e^{i(\phi_1 - \theta_1)} = \left(\lim_{z_2 \to z} \frac{\rho_2}{r}\,e^{i(\phi_2 - \theta_2)} \right) = R\,e^{i\delta} \neq 0.$$

Thus, we have a rotation by an angle $\delta = \arg\{f'(z_0)\}$ and a magnification by $R = |f'(z_0)|$. The mapping of a convex region E which is starlike with respect to every point into a region $f(E)$ under the conformal map $z \longmapsto f(z)$ is shown in Figure 6.5.4.

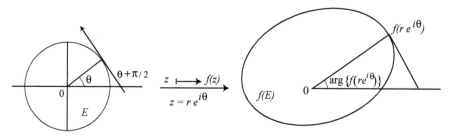

$z \longmapsto f(z)$
$z = r\,e^{i\theta}$

Figure 6.5.4 Mapping $z \longmapsto f(z)$.

Example 6.5.1. A function $f \in \mathcal{S}^\star$ if and only if $\Re\left\{ \dfrac{z f'(z)}{f(z)} \right\} > 0$ for $z \in E$. Then there exists a unique $p \in \mathcal{P}$ such that $\dfrac{z f'(z)}{f(z)} = p(z)$, $z \in E$. If we integrate from 0 to z, we get

$$\int_0^z \frac{z f'(z)}{f(z)}\, dz = \int_0^z \frac{p(z)}{z}\, dz.$$

Now, since $\dfrac{d}{dz} \log \dfrac{f(z)}{z} = \dfrac{z}{f(z)} \left\{ \dfrac{f'(z)}{z} - \dfrac{f(z)}{z^2} \right\} = \dfrac{f'(z)}{f(z)} - \dfrac{1}{z}$, so

$$\log \frac{f(z)}{z} = \int_0^z \frac{p(z)}{z}\, dz - \frac{1}{z} = \int_0^z \frac{1}{z} \int_{-\pi}^{\pi} \frac{1 + z e^{-it}}{1 - z e^{-it}}\, dm(t) - \int_{-\pi}^{\pi} \frac{1}{z}\, dm(t)$$

$$= \int_0^z \int_{-\pi}^{\pi} \frac{1}{z} \left(\frac{1 + z e^{-it}}{1 - z e^{-it}} - 1 \right) dm(t)\, dz$$

$$= \int_{-\pi}^{\pi} \int_0^z \frac{1}{z} \left(\frac{1 + z e^{-it}}{1 - z e^{-it}} - 1 \right) dm(t)\, dz$$

$$= \int_{-\pi}^{\pi} \int_0^z \frac{1}{z} \left(\frac{1 + ze^{-it} - 1 + ze^{-it}}{1 - ze^{-it}} \right) dz\, dm(t)$$

$$= \int_{-\pi}^{\pi} \int_0^z \frac{1}{z} \left(\frac{2ze^{-it}}{1 - ze^{-it}} \right) dz\, dm(t)$$

$$= \int_{-\pi}^{\pi} \int_0^z \frac{2e^{-it}}{1 - ze^{-it}}\, dz\, dm(t) = \int_{-\pi}^{\pi} \log \frac{1}{(1 - ze^{-it})^2}\, dm(t)$$

$$= -2 \int_{-\pi}^{\pi} \log \left(1 - ze^{-it} \right) dm(t) = \log \left(1 - \varepsilon z \right)^{-2}, \quad \text{where } \varepsilon = e^{-it}.$$

This means that $f(z) = \dfrac{z}{(1 - \varepsilon z)^2}$, which can be compared with the Koebe function.

6.5.2 Radius of Starlikeness. This radius, denoted by $r_\star(f)$, is the upper bound of the radii r of the circles $|z| \le r$ which are mapped by a function $w = f(z) \in \mathcal{S}$ onto starlike domains with respect to $w = 0$. Thus, the radius of starlikeness is defined for $f \in \mathcal{S}^\star$ by

$$r_\star(f) = \sup \left\{ r : \Re \left\{ \frac{z f'(z)}{f(z)} \right\} > 0,\ |z| \le r \right\}. \tag{6.5.4}$$

If $r_\star(f) \ge 1$, we say that f is starlike on the unit disk E.

Note that the equality $\Re \left\{ \dfrac{z f'(z)}{f(z)} \right\} = 0$ in (6.5.4) cannot hold for $z \in E$, because otherwise the function f will reduce to a constant. The condition $\Re \left\{ \dfrac{z f'(z)}{f(z)} \right\} > 0$ does not guarantee that the function f will be univalent in E and $f(E)$ will be a starlike domain with respect to the origin (i.e., the segment $[0, w] \in f(E)$ for all $w \in f(E)$), unless we impose the additional condition that $f'(0) \ne 0$. Using the subordination, the class $\mathcal{S}^\star$ is defined as follows: if $f \in \mathcal{S}$, then $f \in \mathcal{S}^\star$ if and only if $\dfrac{z f'(z)}{f(z)} \prec \dfrac{1 + z}{1 - z}$, $z \in E$.

As mentioned in the previous section, any circle $|z| \le r$ is mapped onto a starlike domain if on $|z| = r$, $z = re^{i\theta}$,

$$\frac{\partial \arg\{f(z)\}}{\partial \theta} = \Re \left\{ \frac{z f'(z)}{f(z)} \right\} \ge 0, \tag{6.5.5}$$

or

$$\left| \arg \left\{ \frac{z f'(z)}{f(z)} \right\} \right| \le \frac{\pi}{2}. \tag{6.5.6}$$

Using the inequality (6.5.4), we find that $r_\star$ is the solution of the equation

$$\log \frac{1+r}{1-r} = \frac{\pi}{2}, \text{ i.e.,}$$

$$r_\star = \frac{1 - e^{-\pi/2}}{1 + e^{-\pi/2}} = \tanh \frac{\pi}{4} = 0.65\ldots. \tag{6.5.7}$$

There are two subclasses of starlike functions: Let μ be an increasing function on the interval $[a, b]$. Define $M[a, b] = \{\mu : [a, b] \mapsto \mathbb{R}, \int_a^b d\mu(t) = \mu(b) - \mu(a) = 1\}$. The function $f \in \mathcal{S}$, $z \in E$, belongs to the class $\mathcal{S}^\star$ if and only if there exists a function $\mu \in M[a, b]$ such that

$$f(z) = z \exp\left\{-2 \int_0^{2\pi} \log\left(1 - ze^{-\mu}\right) d\mu(t)\right\}, \quad z \in E.$$

Then the two subclasses of the class $\mathcal{S}^\star$ are: (i) the subclass $\mathcal{S}^\star(\alpha)$ of starlike functions of order α, $0 \le \alpha < 1$, and (ii) the subclass $\mathcal{S}^\star[\alpha]$ of strongly starlike functions of order α, $0 < \alpha \le 1$.

A function $f \in \mathcal{S}^\star(\alpha)$, $0 \le \alpha < 1$, is a starlike function of order α if

$$\Re\left\{\frac{zf'(z)}{f(z)}\right\} > \alpha, \quad z \in E. \tag{6.5.8}$$

A function $f \in \mathcal{S}^\star[\alpha]$, $0 < \alpha \le 1$, is a strongly starlike function of order α if

$$\left|\arg\left\{\frac{zf'(z)}{f(z)}\right\}\right| < \frac{\pi}{2}\alpha, \quad z \in E. \tag{6.5.9}$$

Obviously, $\mathcal{S}^\star(0) = \mathcal{S}^\star = \mathcal{S}^\star[1]$.

6.6 Functions with Real Coefficients

The following results hold:

Lemma 6.6.1. (Carathéodory) *If $f = 1 + c_1 z + c_2 z^2 + \cdots + c_n z^n + \cdots$ is analytic in E and has positive real part, then $|c_n| \le 2$ for $n = 1, 2, \ldots$. This inequality is sharp for each n.*

PROOF. Note that $\Re\{f\} \ge 0$ if and only if $|1 + f(z)| \ge |1 - f(z)|$. Consider $g(z) = \dfrac{f(z) - 1}{f(z) + 1} = \dfrac{c_1}{2} z + \cdots$. Then $|g(z)| \le 1$ and $g(0) = 0$. By Schwarz lemma, $|g(z)| = \left|\frac{c_1}{2} z + \cdots\right| \le |z|$, which yields

$$|c_1| \le 2. \tag{6.6.1}$$

Let ω_j, $j = 1, \ldots, n$ be the distinct n roots of unity. Then

$$\Re\left(\frac{1}{n} \sum_{j=1}^n f\left(\omega_j z^{1/n}\right)\right) = \Re\{1 + c_n z + \cdots\} \ge 0.$$

Then , using (6.6.1), we get $|c_n| \le 2$ for $n = 2, 3, \ldots$. The inequality is sharp for the function

$$\phi(z) = \frac{1+z}{1-z} = 1 + 2\sum_{n=1}^{\infty} z^n. \quad \blacksquare$$

Theorem 6.6.1. *If $f \in \mathcal{S}$ and all the coefficients a_n in the series expansion are real, then $|a_n| \le n$ for all $n = 2, 3, \ldots$.*

PROOF. Let $z_1 = r\, e^{i\theta}$, $z_2 = r\, e^{-i\theta}$, $\theta \ne 0$ for $0 < r < 1$. Then

$$\frac{f(z_1) - f(z_2)}{z_1 - z_2} = 1 + \sum_{n=2}^{\infty} a_n \frac{z_1^n - z_2^n}{z_1 - z_2}$$

$$= 1 + \sum_{n=2}^{\infty} a_n\, r^{n-1} \frac{\sin n\theta}{\sin \theta} \ne 0.$$

Here the expression $\vartheta(r, \theta) \equiv 1 + \sum_{n=2}^{\infty} a_n\, r^{n-1} \dfrac{\sin n\theta}{\sin \theta}$ is real and non-zero when $\theta \ne 0$ because all a_n are real, and $\vartheta(0, \theta) = 1$. Hence, $\vartheta(r, \theta) \ge 0$ when $0 \le r < 1$ and $\theta \ne 0$. Also, we have

$$2\sin^2 \theta \vartheta(r, \theta) = 1 + a_2 r \cos \theta + \left(a_3 r^2 - 1\right) \cos 2\theta$$
$$+ \left(a_4 r^2 - a_2\right) r \cos 3\theta + \cdots + \left(a_n r^2 - a_{n-2}\right) r^{n-3} \cos(n-1)\theta + \cdots \ge 0$$

For a fixed r, $0 < r < 1$, the function

$$F(z) = 1 + a_2 r z + \left(a_3 r^2 - 1\right) z^2 + \left(a_4 r^2 - a_2\right) r z^3 + \cdots$$
$$+ \left(a_n r^2 - a_{n-2}\right) r^{n-3} z^{n-1} + \cdots$$

is such that $\Re\{F(z)\} \ge 0$. Then by Lemma 6.6.1, we get

$$|a_2| \le 2;$$
$$|a_3 r^2 - 1| \le 2;$$
$$|a_4 r^2 - a_2|r \le 2;$$
$$\vdots$$
$$|a_n r^2 - a_{n-2}|r^{n-3} \le 2; \cdots.$$

Let $r \to 1$, and the proof is complete by induction. $\blacksquare$

6.7 Functions in Class $\mathcal{S}_\alpha$

Let $\mathcal{S}_\alpha$ denote the class of α-spiral functions in E. Then $f \in \mathcal{S}_\alpha$ if and only if

$$\Re\left\{e^{i\alpha} \frac{z f'(z)}{f(z)}\right\} > 0, \quad \text{for } z \in E, \text{ and } |\alpha| < \frac{\pi}{2}. \tag{6.7.1}$$

Note that $\mathcal{S}_0 \equiv \mathcal{S}^*$. A simple question arises as to why the condition in the above definition does not read as

$$\Re\left\{e^{i\alpha}\, \frac{zf'(z)}{f(z)}\right\} \geq 0.$$

Note that equality is then obtained for $|\alpha| = \pi/2$, or for $f(z) = z$ and $|\alpha| < \pi/2$. We, therefore, exclude these cases from further consideration and adopt (6.7.1) as the condition for the class $\mathcal{S}_\alpha$.

Recall that if $f \in \mathcal{S}$, then f is univalent, $f(0) = 0$, $f'(0) = 1$, and

$$\lim_{z \to 0} \frac{zf'(z)}{f(z)} = \lim_{z \to 0} \frac{zf''(z) + f'(z)}{f'(z)} = \lim_{z \to z} \left(\frac{zf''(z)}{f'(z)} + 1\right) = 1.$$

Now, if $p \in \mathcal{P}$ with $p(0) = 1$, then $\Re\{p(z)\} > 0$ for $z \in E$. Consider

$$\sec\alpha \left[e^{i\alpha}\, \frac{zf'(z)}{f(z)} - i\sin\alpha\right]_{z=0} = 1 \equiv p(0).$$

Therefore, there is a unique $p \in \mathcal{P}$ such that

$$\sec\alpha \left[e^{i\alpha}\, \frac{zf'(z)}{f(z)} - i\sin\alpha\right] = p(z), \quad z \in E,$$

or

$$\frac{zf'(z)}{f(z)} = e^{-i\alpha}\left[p(z)\cos\alpha + i\sin\alpha\right], \tag{6.7.2}$$

which can be written as

$$\frac{zf'(z)}{f(z)} = \frac{p(z)\cos\alpha + i\sin\alpha}{\cos\alpha + i\sin\alpha} = \frac{p(z) + i\tan\alpha}{1 + i\tan\alpha}.$$

Set $h = i\tan\alpha$. Then

$$\frac{zf'(z)}{f(z)} = \frac{p(z) + h}{1 + h}. \tag{6.7.3}$$

Before we study the coefficient problem and the distortion and rotation for functions in the class f_α, we will introduce Clunie's method that is useful in such problem. This method has also been used in §7.3.

6.7.1 Clunie's Method

Let $w(z)$ be a regular function in E such that $|w(z)| < 1$, $w(0) = 0$; it satisfies the Schwarz lemma: $|w(z)| \leq |z|$, $z \in E$. Then

$$\left|\frac{1 + w(z)}{1 - w(z)}\right| \leq \left|\frac{1 + z}{1 - z}\right|,$$

where $\dfrac{1 + w(z)}{1 - w(z)} \in \mathcal{P}$ for $z \in E$. Then from (6.7.2) we have

$$e^{i\alpha} \frac{z f'(z)}{f(z)} - i \sin \alpha = \cos \alpha \frac{1 + w(z)}{1 - w(z)},$$

or

$$z f'(z) - f(z) = \left[z f'(z) + e^{-2i\alpha} f(z) \right] w(z). \tag{6.7.4}$$

Since $f(z) = z + \sum\limits_{n=2}^{\infty} a_n z^n$, $f'(z) = 1 + \sum\limits_{n=2}^{\infty} n a_n z^{n-1}$, and $z f'(z) = z + \sum\limits_{n=2}^{\infty} n a_n z^n$, so by substituting these expressions into (6.7.4) we get

$$z + \sum_{k=2}^{\infty} k a_k z^k - \left(z + \sum_{k=2}^{\infty} a_k z^k \right) = \left[z + \sum_{k=2}^{\infty} k a_k z^k + e^{-2i\alpha} \left(z + \sum_{k=2}^{\infty} a_k z^k \right) \right] w(z),$$

or

$$\sum_{k=2}^{\infty} (k-1) a_k z^k = \left[\left(1 + e^{-2i\alpha} \right) z + \sum_{k=2}^{\infty} \left(k + e^{-2i\alpha} \right) a_k z^k \right] w(z).$$

Put $c = e^{-2i\alpha} = 2 \cos \alpha\, e^{-i\alpha} - 1$. Then the above equality becomes

$$\sum_{k=2}^{\infty} (k-1) a_k z^k = \left[(1+c) z + \sum_{k=2}^{\infty} (k+c) a_k z^k \right] w(z),$$

or

$$\sum_{k=2}^{n} (k-1) a_k z^k + \sum_{k=n+1}^{\infty} (k-1) a_k z^k = \Big[(1+c) z +$$
$$\sum_{k=2}^{n-1} (k+c) a_k z^k + \sum_{k=n}^{\infty} (k+c) a_k z^k \Big] w(z),$$

or

$$\sum_{k=2}^{n} (k-1) a_k z^k + \sum_{k=n+1}^{\infty} b_k z^k = \left[(1+c) z + \sum_{k=2}^{n-1} (k+c) a_k z^k \right] w(z),$$

where the series $\sum\limits_{k=n+1}^{\infty} b_k z^k$ is absolutely and uniformly convergent in the compact set E.

6.7.2 Coefficient Problem for $\mathcal{S}_\alpha$. We shall use Clunie's method. Let

$$f(z) = z + \sum_{n=2}^{\infty} a_n z^n \in \mathcal{S}_\alpha.$$

Then Eq (6.7.4) holds for a regular function $w(z)$. Recall Parseval's identity (Exercise 1.10.7) which states $\dfrac{1}{2\pi} \displaystyle\int_0^{2\pi} \left|f(re^{i\theta})\right|^2 d\theta = \sum_{n=1}^{\infty} |a_n|^2 r^{2n}$, and Schwarz lemma (Theorem 1.5.8) $|w(z)| \leq |z| < 1$ for $z \in E$. Put $z = re^{i\theta}$, $r < 1$, $0 \leq \theta < 2\pi$. Then from (6.7.4) we have

$$\sum_{k=2}^{n}(k-1)^2\left|a_k\right|^2 r^{2k} \leq \sum_{k=2}^{n}(k-1)^2\left|a_k\right|^2 r^{2k} + \sum_{k=n+1}^{\infty}\left|b_k\right|^2 r^{2k}$$

$$\leq \frac{1}{2\pi}\int_0^{2\pi}\left|(1+c)\,re^{i\theta} + \sum_{k=2}^{n-1}(k+c)a_k r^k e^{ik\theta}\right|^2 \left|w\left(re^{i\theta}\right)\right|^2 d\theta$$

$$\leq \frac{1}{2\pi}\int_0^{2\pi}\left|(1+c)\,re^{i\theta} + \sum_{k=2}^{n-1}(k+c)a_k r^k e^{ik\theta}\right|^2 d\theta$$

$$= \left|1+c\right|^2 r^2 + \sum_{k=2}^{n-1}\left|k+c\right|^2\left|a_k\right|^2 r^{2k},$$

which as $r \to 1$ yields, for all $n = 2, 3, \ldots$,

$$\sum_{k=2}^{n}(k-1)^2\left|a_k\right|^2 \leq \left|1+c\right|^2 + \sum_{k=2}^{n-1}\left|k+c\right|^2\left|a_k\right|^2. \tag{6.7.5}$$

Taking $n = 2, 3, \ldots$ in (6.7.5) successively; we obtain,
for $n = 2$: $|a_2| \leq |1+c|$;
for $n = 3$: $|a_3| \leq \dfrac{1}{2}|1+c|\,|2+c|$;
for $n = 4$: $|a_4| \leq \dfrac{1}{3!}|1+c|\,|2+c|\,|3+c|$;
and hence, by induction, for $n = 2, 3, \ldots$,

$$|a_n| \leq \frac{|1+c|\,|2+c|\,|3+c|\cdots|n-1+c|}{(n-1)!}$$

$$= \frac{1}{(n-1)!}\prod_{k=1}^{n-1}|k+c| = \prod_{k=1}^{n-1}\frac{|k+c|}{k}, \tag{6.7.6}$$

where equality is attained by $f(z) = z(1-z)^{c+1} = z(1-z)^{-2\cos\alpha\,e^{-i\alpha}}$. Thus,

Theorem 6.7.1. *If* $f(z) = z + \sum_{n=2}^{\infty} a_n z^n \in \mathcal{S}_\alpha$, *then for all* $n = 2, 3, \ldots$

$$|a_n| \leq \prod_{k=0}^{n-2} \frac{|k + 2 \cos \alpha \, e^{-i\alpha}|}{k}. \tag{6.7.7}$$

The inequality is sharp for $f(z) = z(1 - z)^{-2 \cos \alpha \, e^{-i\alpha}}$.

If $f \in \mathcal{S}_\alpha$, then the β-*spiral radius* of f is defined is the largest number r such that for some $p \in \mathcal{P}$ and $|z| < r < 1$

$$\Re \left\{ e^{i(\beta - \alpha)} \left[\cos \alpha \, p(z) + i \, \sin \alpha \right] \right\} > 0. \tag{6.7.8}$$

To determine the β-spiral radius of $f \in \mathcal{S}_\alpha$, let $B(z) = e^{i(\beta - \alpha)} \left[\cos \alpha \, p(z) + i \, \sin \alpha \right]$, and let $p(z) = \dfrac{1+z}{1-z} \in \mathcal{P}$. Recall the following result: $w = \dfrac{az + b}{cz + d}$ maps the circle $|z - z_0| = r$ univalently onto $|w - w_0| = \rho$, where

$$w_0 = \frac{(az_0)(\overline{cz_0} + \overline{d}) - acr^2}{|cz_0 + d|^2 - |c|^2 r^2}, \qquad \rho = \frac{r \, |ad - bc|}{\left| |cz_0 + d|^2 - |c|^2 r^2 \right|}.$$

Then

$$\begin{aligned}
B(z) &= e^{i(\beta - \alpha)} \left[\cos \alpha \, \frac{1+z}{1-z} + i \, \sin \alpha \right] \\
&= e^{i(\beta - \alpha)} \, \frac{\cos \alpha \, (1+z) + i \, \sin \alpha \, (1 - z)}{1 - z} \\
&= e^{i(\beta - \alpha)} \, \frac{(\cos \alpha - i \, \sin \alpha) z + \cos \alpha + i \, \sin \alpha}{1 - z} \\
&= e^{i(\beta - \alpha)} \, \frac{e^{-i\alpha} z + e^{i\alpha}}{1 - z}.
\end{aligned}$$

Here $a = e^{i(\beta - 2\alpha)}$, $b = e^{i\beta}$, $c = -1$, $d = 1$, and $z_0 = 0$. Then $B(z)$ maps $|z| \leq r$ univalently onto a closed disk $|w - w_0| \leq \rho$, with center at

$$w_0 = \frac{e^{i\beta} + e^{i(\beta - 2\alpha)} r^2}{1 - r^2} = \frac{e^{i\beta} \left(1 + e^{-2i\alpha} r^2 \right)}{1 - r^2},$$

and radius

$$\rho = \frac{\left| e^{i(\beta - \alpha)} + e^{i\beta} \right| r}{1 - r^2} = \frac{\left| e^{i\beta} \right| \left| 1 + e^{-2i\alpha} \right| r}{1 - r^2} = \frac{\left| 1 + e^{-2i\alpha} \right| r}{1 - r^2}.$$

Hence, $\Re \left\{ B(z) \right\} > 0$ if and only if

$$\Re \left\{ \frac{e^{i\beta} \left(1 + e^{-2i\alpha} r^2 \right)}{1 - r^2} \right\} > \frac{\left| 1 + e^{-2i\alpha} \right| r}{1 - r^2},$$

i.e., if and only if

$$\Re\left\{e^{i\beta}\left(1 + e^{-2i\alpha}\,r^2\right)\right\} > \left|1 + e^{-2i\alpha}\right|r.$$

Note that the left side is equal to

$$\Re\left\{(\cos\beta + i\,\sin\beta)\left[1 + (\cos 2\alpha - i\,\sin 2\alpha)\,r^2\right]\right\}$$
$$\Re\left\{(\cos\beta + i\,\sin\beta)\left[(1 + \cos 2\alpha\,r^2) - i\,\sin 2\alpha\,r^2\right]\right\}$$
$$= \cos\beta\,(1 + \cos 2\alpha\,r^2) + \sin\beta\,\sin 2\alpha\,r^2,$$

and the right side is equal to

$$\left|1 + \cos 2\alpha - i\,\sin 2\alpha\right|r = r\,\sqrt{(1 + \cos 2\alpha)^2 + \sin^2 2\alpha}.$$

An alternate method is as follows:

$$w_0 = \frac{e^{i\beta} + e^{i(\beta - 2\alpha)}\,r^2}{1 - r^2} = \frac{e^{i(\beta - \alpha)} + \left[e^{i\alpha} + e^{-i\alpha}\,r^2\right]}{1 - r^2}$$
$$= \frac{e^{i\beta}}{1 - r^2}\left\{\cos\alpha + i\,\sin\alpha + \cos\alpha\,r^2 - i\,\sin\alpha\,r^2\right\}$$
$$= e^{i(\beta - \alpha)}\left\{\cos\alpha\,\frac{1 + r^2}{1 - r^2} + i\,\sin\alpha\right\},$$

and

$$\rho = \frac{\left|e^{i(\beta - \alpha)} + e^{i\beta}\right|r}{1 - r^2} = \frac{\left|e^{i(\beta - \alpha)}\left(e^{-i\alpha} + e^{i\alpha}\right)\right|r}{1 - r^2}$$
$$= \left|e^{i(\beta - \alpha)}\,2\cos\alpha\right|\frac{r}{1 - r^2} = 2\left|e^{i(\beta - \alpha)}\,\cos\alpha\right|\frac{r}{1 - r^2}.$$

Hence, $\Re\{B(z)\} > 0$ if and only if

$$\Re\left\{e^{i(\beta - \alpha)}\left[\cos\alpha\,\frac{1 + r^2}{1 - r^2} + i\,\sin\alpha\right]\right\} > \left|2\cos\alpha\,e^{i(\beta - \alpha)}\right|\frac{r}{1 - r^2},$$

where $|\alpha| < \pi/2$, i.e.,

$$\cos(\beta - \alpha)\cos\alpha\,\frac{1 + r^2}{1 - r^2} - \sin(\beta - \alpha)\sin\alpha > \frac{2\cos\alpha\,r}{1 - r^2},$$

which yields

$$\cos(\beta - \alpha)\cos\alpha\,(1 + r^2) - \sin\alpha\,(1 - r^2) > 2\cos\alpha\,r,$$

or

$$\cos(\beta - \alpha)\, r^2 - 2\cos\alpha\, r + \cos\beta > 0.$$

Theorem 6.7.2. *The β-spiral radius of $f \in \mathcal{S}_\alpha$ is the smallest positive root of the equation*

$$\cos(\beta - \alpha)\, r^2 - 2\cos\alpha\, r + \cos\beta = 0. \tag{6.7.9}$$

Corollary 6.7.1. *The 0-spiral radius of $f \in \mathcal{S}_\alpha$ is the smallest positive root of the equation*

$$\cos(\beta - \alpha)\, r^2 - 2\cos\alpha\, r + 1 = 0. \tag{6.7.10}$$

Then, if r_0 denotes the 0-spiral radius of $f \in \mathcal{S}_\alpha$, then setting $\beta = 0$ and solving (6.7.9), we get

$$r_0 = \frac{\cos\alpha - \sqrt{\sin^2\alpha}}{\cos 2\alpha} = \frac{1}{\cos\alpha + |\sin\alpha|}.$$

As an example, consider $\alpha = \pm\pi/4$. Then $\cos\alpha = \sin\alpha = \pm 1/\sqrt{2}$, and $r_0 = 1/\sqrt{2}$. For some other examples, see Exercises 6.8.3– 6.8.5.

6.7.3 Univalence Preservation in $\mathcal{S}_\alpha$ A univalence preserving transformation is

$$g(z) = \frac{f\left(\dfrac{z+a}{1+\bar{a}\, z}\right) - f(a)}{f'(a)\, (1 - |a|^2)}, \quad z \in E\ a \in E, \tag{6.7.11}$$

where $f \in \mathcal{S}$, that is, if $f \in \mathcal{S}$, then $g \in \mathcal{S}$. Note that f is univalent, with $f(0) = 0$ and $f'(0) = 1$. Also, $f \in \mathcal{S}_\alpha$ if and only if

$$\Re\left\{e^{i\alpha}\, \frac{z f'(z)}{f(z)}\right\} > 0, \quad z \in E.$$

PROBLEM. Find a transformation that preserves the membership in the class $\mathcal{S}_\alpha$.

Theorem 6.7.3. *If $f \in \mathcal{S}_\alpha$ and $a \in E$, then*

$$g(z) = \frac{a z\, f\left(\dfrac{z+a}{1+\bar{a}\, z}\right)}{f(a)(z+a)\, (1+\bar{a}\, z)^c}, \quad z \in E, \tag{6.7.12}$$

where $c = e^{-2i\alpha}$, is also in $\mathcal{S}_\alpha$.

PROOF. We must show that $\Re\left\{e^{i\alpha}\dfrac{zg'(z)}{g(z)}\right\} > 0$. Recall that $z \longmapsto \dfrac{z+a}{1+\bar{a}\,z}$ maps E onto E. Let ρ, $0 < \rho < 1$, be real. Consider

$$g_\rho(z) = \frac{az\, f\left(\rho\,\dfrac{z+a}{1+\bar{a}\,z}\right)}{f(\rho a)(z+a)\,(1+\bar{a}\,z)^c}, \quad z \in E.$$

Then

$$g_\rho'(z) = \frac{a}{f(\rho a)}\,\frac{1}{(z+a)^2\,(1+\bar{a}\,z)}\left\{\left[\rho z\,\frac{1-|a|^2}{(1+az)^2}\,f'\left(\rho\frac{z+a}{1+\bar{a}z}\right)\right.\right.$$
$$\left.\left. +\, f\left(\rho\frac{z+a}{1+\bar{a}z}\right)\right] - z\, f\left(\rho\frac{z+a}{1+\bar{a}z}\right)\left[c\bar{a}(z+a) + (1+\bar{a}z)\right]\right\},$$

and

$$\frac{zg_\rho'(z)}{g_\rho(z)} = \rho z\,\frac{1-|a|^2}{(1+\bar{a}z)^2}\,\frac{f'\left(\rho\dfrac{z+a}{1+\bar{a}z}\right)}{f\left(\rho\dfrac{z+a}{1+\bar{a}z}\right)} + 1 - \frac{z\left[c\bar{a}\,(z+a) + (1+\bar{a}z)\right]}{(z+a)(1+\bar{a}z)}.$$

Thus,

$$e^{i\alpha}\frac{zg_\rho'(z)}{g_\rho(z)} = e^{i\alpha}\,\rho z\,\frac{1-|a|^2}{(1+\bar{a}z)^2}\,\frac{f'\left(\rho\dfrac{z+a}{1+\bar{a}z}\right)}{f\left(\rho\dfrac{z+a}{1+\bar{a}z}\right)} + e^{-i\alpha} - e^{i\alpha}\,\frac{z\left[c\bar{a}\,(z+a) + (1+\bar{a}z)\right]}{(z+a)(1+\bar{a}z)}.$$

Let $\zeta = \rho\,\dfrac{z+a}{1+\bar{a}z}$. Then $\zeta \in E$, and

$$\frac{e^{i\alpha}g_\rho'(z)}{g_\rho(z)} = e^{i\alpha}\,\frac{\zeta f'(\zeta)}{f(\zeta)}\,\frac{z\left(1-|a|^2\right)}{(z+a)(1+\bar{a}z)} + e^{i\alpha} - e^{i\alpha}\,\frac{z\left[c\bar{a}\,(z+a) + (1+\bar{a}z)\right]}{(z+a)(1+\bar{a}z)}.$$

Let $z = e^{i\theta}$ (z is on the boundary). Then, since $\dfrac{1}{1+\bar{a}z}\,\dfrac{1+a\bar{z}}{1+a\bar{z}} = \dfrac{1+a\bar{z}}{|1+az|^2}$, we have

$$e^{i\alpha}\frac{zg_\rho'(z)}{g_\rho(z)} = e^{i\alpha}\,\frac{\zeta f'(\zeta)}{f(\zeta)}\,\left(1-|a|^2\right)\,\frac{1+ae^{-i\theta}}{|1+\bar{a}e^{i\theta}|^2}$$
$$+\, e^{i\alpha}\,\frac{(z+a)(1+\bar{a}z) - zc\bar{a}(z+a) - z(1+\bar{a}z)}{(z+a)(1+\bar{a}z)}$$
$$= e^{i\alpha}\,\frac{\zeta f'(\zeta)}{f(\zeta)}\,\left(1-|a|^2\right)\,\left(\frac{1}{|e^{i\theta}+a|^2}\right)$$
$$+\, e^{i\alpha}\,\frac{(e^{-i\theta}+a)\left[a + |a|^2 e^{i\theta} - 2e^{-2i\alpha}\bar{a}e^{i\theta} - 2e^{-2i\alpha}|a|^2\right]}{|1+ae^{-i\theta}|^2}$$
$$= A + B.$$

Now, note that B can be written as

$$B = e^{i\alpha} \left\{ 1 - \frac{z\left[c\bar{a}(z+a) + (1+\bar{a}z)\right]}{(z+a)(1+\bar{a}z)} \right\}$$

$$= e^{i\alpha} \left\{ 1 - \frac{1 + \bar{a}e^{i\theta} + \bar{a}e^{i\theta}e^{-2i\alpha} + e^{-2i\alpha}|a|^2}{\left|1 + ae^{-i\theta}\right|^2} \right\}$$

$$= \frac{1}{\left|1 + ae^{-i\theta}\right|^2} \left\{ ae^{-i\theta+i\alpha} + \bar{a}e^{i\theta+i\alpha} + |a|^2 e^{i\alpha} - ae^{i\theta+i\alpha} - \bar{a}e^{i\theta-i\alpha} - e^{-i\alpha}|a|^2 \right\}$$

$$= \frac{1}{\left|1 + ae^{-i\theta}\right|^2} \left\{ a\,e^{-i\theta+i\alpha} - \bar{a}\,e^{i\theta-i\alpha} + |a|^2 e^{i\alpha} - |a|^2 e^{-i\alpha} \right\}$$

$$= \frac{1}{\left|1 + ae^{-i\theta}\right|^2} \left\{ 2i\Im\left\{ e^{-i\theta+i\alpha} \right\} + |a|^2\Im\left\{ e^{-i\alpha} \right\}.$$

Hence,

$$\Re\left\{ e^{i\alpha}\, \frac{zg'_\rho(z)}{g_\rho(z)} \right\} > 0,$$

which implies that $g_\rho(z) \in \mathcal{S}_\alpha$. Let $\rho \to 1$. Then $g_\rho(z) \equiv g(z) \in \mathcal{S}_\alpha$. ∎

Theorem 6.7.4. (Distortion Theorem) *If* $g(z)z + \sum\limits_{n=2}^{\infty} a_n\, z^n \in \mathcal{S}_\alpha$, *then*

$$|a_2| \leq |1+c| = \left|1 + e^{-2i\alpha}\right| = \left|2\cos\alpha\, e^{-i\alpha}\right| = 2\cos\alpha,$$

and

$$a_2 = \frac{g''(0)}{2!} = \left(1 - |a|^2\right)\frac{f'(a)}{f(a)} - \frac{1 + c|a|^2}{a}.$$

Replace a by z and use the bound $|a_2| \leq 2\cos\alpha$ and $|z| = r < 1$. Then

$$\left|(1 - r^2)\frac{g'(z)}{g(z)} - \frac{1 + cr^2}{z}\right| \leq 2\cos\alpha,$$

or

$$\left|\frac{zg'(z)}{g(z)} - \frac{1 + cr^2}{z}\right| \leq \frac{2r\cos\alpha}{1 - r^2}.$$

Using $-|w| \leq \Re\{w\} \leq |w|$, we get

$$-\frac{2r\cos\alpha}{1 - r^2} \leq \Re\left\{ \frac{zg'(z)}{g(z)} - \frac{1 + cr^2}{1 - r^2} \right\} \leq \left|\frac{zg'(z)}{g(z)} - \frac{1 + cr^2}{1 - r^2}\right| \leq \frac{2r\cos\alpha}{1 - r^2},$$

or

$$\frac{\Re\{1 + cr^2\} - 2r\cos\alpha}{1 - r^2} \leq \Re\left\{ \frac{zg'(z)}{g(z)} \right\} \leq \frac{\Re\{1 + cr^2\} - 2r\cos\alpha}{1 - r^2}.$$

Now $\Re\{1 + cr^2\} = 1 + r^2 \cos 2\alpha$, and we obtain

$$\frac{1 + r^2 \cos 2\alpha - 2r \cos \alpha}{1 - r^2} \leq \Re\left\{\frac{zg'(z)}{g(z)}\right\} \leq \frac{1 + r^2 \cos 2\alpha + 2r \cos \alpha}{1 - r^2}$$

$$= r \frac{\partial}{\partial r} \Re\{\log g'(z)\}.$$

Then

$$\int_0^z \frac{1 + \rho^2 \cos 2\alpha - 2\rho \cos \alpha}{\rho(1 - \rho^2)} \, d\rho \leq \log |g'(z)| \leq \int_0^z \frac{1 + \rho^2 \cos 2\alpha + 2\rho \cos \alpha}{\rho(1 - \rho^2)} \, d\rho,$$

which will give the bounds on $|g'(z)|$.

6.8 Exercises

6.8.1. If $g(z) = z + b_0 + \dfrac{b_1}{z} + \dfrac{b_2}{z^2} + \cdots \in \Sigma$, then prove the area theorem:

$$\sum_{n=1}^{\infty} n \, |b_n|^2 \leq 1.$$

PROOF. Let C be the image of $|z| = r > 1$ under g, and let R be the region bounded by the closed curve C (Figure 6.8.1). Then the area $A(r)$ of the region R is given by $A(r) = \iint_R du \, dv > 0$, where $g(z) = u + iv$ and $z = x + iy$. In view of Green's theorem in the plane we have

$$\int_C P \, du + Q \, dv = \iint_R \left(\frac{\partial Q}{\partial u} - \frac{\partial P}{\partial v}\right) dx \, dy.$$

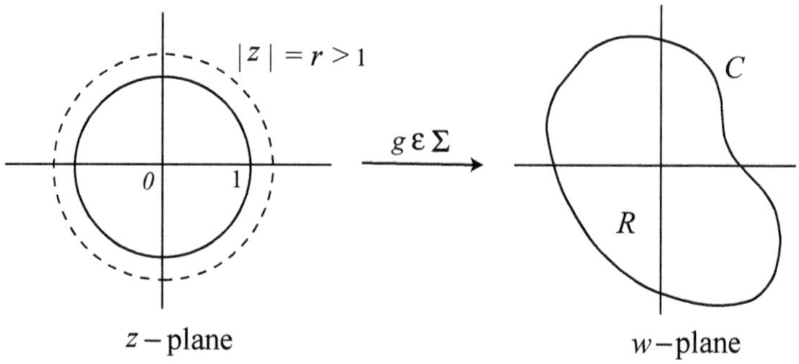

Figure 6.8.1 Mapping $g \in \Sigma$.

Since we want the integrand on the right side to be equal to 1, we choose $P = -v/2$, $Q = u/2$. Then

$$A(r) = \iint_R du \, dv = \int_C \left(-\frac{v}{2} \, du + \frac{u}{2} \, dv\right)$$

$$= \frac{1}{2} \int_C (u \, dv - v \, du) = \frac{1}{2} \int_0^{2\pi} \left(u \frac{\partial v}{\partial \theta} - v \frac{\partial u}{\partial \theta}\right) d\theta > 0,$$

where $z = r\,e^{i\theta}$, $0 \le \theta \le 2\pi$. Using the polar form (1.3.5) of the Cauchy-Riemann equations we have $g'(z) = \dfrac{e^{-i\theta}}{ir}\dfrac{\partial g}{\partial \theta} = \dfrac{1}{iz}\dfrac{\partial g}{\partial \theta}$, and

$$\frac{1}{2}\int_{|z|=r}\overline{g(z)}\,g'(z)\,dz = \frac{1}{2}\int_0^{2\pi}(u-iv)\left(\frac{\partial u}{\partial \theta}+i\frac{\partial v}{\partial \theta}\right)d\theta$$

$$= \frac{1}{2}\int_0^{2\pi}\left\{\left(u\frac{\partial u}{\partial \theta}+v\frac{\partial v}{\partial \theta}\right)+i\left(u\frac{\partial v}{\partial \theta}-v\frac{\partial u}{\partial \theta}\right)\right\}d\theta.$$

Thus, since $g(z) = z + \displaystyle\sum_{n=0}^{\infty} b_n\,z^{-n}$, we get

$$A(r) = \Im\left\{\frac{1}{2}\int_{|z|=r}\overline{g(z)}\,g'(z)\,dz\right\} = \Im\left\{\frac{1}{2}\int_{|z|=r}\left(\bar z + \sum_{m=0}^{\infty}\bar b_m\,\bar z^{-m}\right)g'(z)\,dz\right\}$$

$$= \Im\left\{\frac{1}{2}\int_{|z|=r}\left(\bar z + \sum_{m=0}^{\infty}\bar b_m\,\bar z^{-m}\right)\left(1 - \sum_{n=1}^{\infty}\frac{nb_n}{z^{n+1}}\right)dz\right\}$$

$$= \Im\left\{\frac{1}{2}\int_{|z|=r}\bar z\,dz - \frac{1}{2}\int_{|z|=r}\left(\frac{\displaystyle\sum_{n=1}^{\infty}n|b_n|^2\,r^{-2n}}{z}\right)dz\right\}$$

$$= \Im\left\{\frac{1}{2}\left(2\pi ir^2 - 2\pi i\sum_{n=1}^{\infty}n|b_n|^2 r^{-2n}\right)\right\} = \pi\left(r^2 - \sum_{n=1}^{\infty}n|b_n|^2 r^{-2n}\right) \ge 0,$$

Which gives $\displaystyle\sum_{n=1}^{\infty}n|b_n|^2 \le 1$, where $\displaystyle\int_{|z|=r}\bar z^{-m}\,z^{-n-1}\,dz = \begin{cases}2\pi i\,r^{-2n} & \text{if } n=m \\ 0 & \text{if } n\ne m,\end{cases}$

and $1 - \displaystyle\sum_{n=1}^{\infty}n|b_n|^2 \ge 0$ as $r \to 1^+$. ■

6.8.2. Consider $h(z) = \sqrt{f(z^2)} = \exp\left\{\dfrac{1}{2}\log f(z^2)\right\}$, where $f \in \mathcal{S}$. The function $h(z)$ so defined is sometimes called the *square root transform* of $f \in \mathcal{S}$. Since f has a zero at $z = 0$, we write

$$h(z) = z\sqrt{\frac{f(z^2)}{z^2}} = z\sqrt{\frac{1}{z^2}\left(z^2 + \sum_{n=2}^{\infty}a_n\,z^{2n}\right)}$$

$$= z\sqrt{1 + \sum_{n=2}^{\infty}a_n\,z^{2n-2}} = z + \frac{a_2}{2}z^3 + \cdots = z + \sum_{n=2}^{\infty}c_{2n-1}\,z^{2n-1}. \tag{6.8.1}$$

Thus, $h(z)$ is an *odd* function, i.e., $h(-z) = -h(z)$. To show that $h(z)$ is univalent, we must show that $z_1 \neq z_2$ implies $h(z_1) \neq h(z_2)$ for distinct $z_1, z_2 \in \Sigma$. In fact, if $h(z_1) = h(z_2)$, then $f(z_1^2) = f(z_2^2)$. Since $f \in S$, we have $z_1^2 = z_2^2$, i.e., $z_1 = \pm z_2$. Now, $z_1 = -z_2$, since $z_1 \neq z_2$. Since h is an odd function, $h(z_1) = h(-z_2) = -h(z_2)$. Hence $h(z_1) = 0 = h(z_2)$, which contradicts the assumption that f has only one zero because this implies that $f(z_1^2) = 0 = f(z_2^2)$. Thus, $h(z) \in S$. If $f(z) = z + \sum\limits_{n=2}^{\infty} a_n z^n$, then

$$\frac{1}{f(z)} = \frac{1}{z} - a_2 + (a_2^2 - a_3) z + \cdots = \frac{1}{z} + \sum_{n=2}^{\infty} b_n z^n \in \mathcal{P}.$$

and by the area theorem, $\sum\limits_{n=1}^{\infty} n |b_n|^2 \leq 1$.

6.8.3. We know that $p \in \mathcal{P}$ if and only if $\Re\{p(z)\} > 0$ for $z \in E$ and $p(0) = 1$. The function $p(z)$ has the Herglotz representation

$$p(z) = \frac{1}{2\pi} \int_{-\pi}^{\pi} \frac{1 + ze^{-it}}{1 - ze^{-it}} \, dm(t),$$

where $m(t)$ is nondecreasing and $\int_{-\pi}^{\pi} dm(t) = 2\pi$, or $\int_{-\pi}^{\pi} dm(t) = 1$. Note that $m(t)$ is a positive measure, i.e., $m(t) \geq 0$ for $t \in [-\pi, \pi]$. Choose $m(t)$ as the degenerate positive measure

$$m(t) = \begin{cases} 0, & \text{if } -\pi \leq t < \tau, \\ 1, & \text{if } \tau \leq t \leq \pi. \end{cases} \tag{6.8.2}$$

Then by the first mean value theorem we have

$$p(z) = \frac{1}{2\pi} \int_{-\pi}^{\pi} \frac{1 + ze^{-it}}{1 - ze^{-it}} \, dm(t) = \frac{1 + \varepsilon z}{1 - \varepsilon z},$$

where $\varepsilon = e^{-i\tau}$, $|\varepsilon| = 1$. If $\tau = 0$, then $p(z) = \dfrac{1 + z}{1 - z}$, which is known as the *dominant function*. Thus, $p(z)$ is dominant in the class $\mathcal{P}$ in the sense that this function guarantees the sharpness of the coefficient bounds. It is not necessarily true that a dominant function is extremal — this must be proved.

6.8.4. $f \in S_\alpha$ if and only if $\Re\left\{e^{i\alpha} \dfrac{zf'(z)}{f(z)}\right\} > 0$, for $z \in E$, and $|\alpha| < \dfrac{\pi}{2}$. Then there is a unique $p \in \mathcal{P}$ such $\dfrac{zf'(z)}{f(z)}$ is of the form $\dfrac{zf'(z)}{f(z)} = \cos\alpha \, e^{-i\alpha} p(z) + i\sin\alpha$, or,

$$\frac{zf'(z)}{f(z)} = \frac{p(z) + h}{1 + h}, \tag{6.8.3}$$

where $h = 2 \tan \alpha$. Note that

$$\frac{1}{1+h} = \frac{1}{1+i\tan\alpha} \frac{1-i\tan\alpha}{1-i\tan\alpha} = \frac{\cos\alpha - i\sin\alpha}{\sec^2\alpha} = \cos\alpha\, e^{-i\alpha}.$$

Then integrating (6.8.3) from 0 to z we get

$$\begin{aligned}
\log \frac{f(z)}{z} &= \int_0^z \frac{p(z) + h - (1+h)}{(1+h)\,z}\, dz \\
&= \frac{1}{1+h} \int_0^z \frac{1}{z} \left\{ \int_{-\pi}^{\pi} \left(\frac{1 + z\,e^{-it}}{1 - z\,e^{-it}} \right) dm(t) \right\} dz \\
&= \cos\alpha\, e^{i-t} \int_{-\pi}^{\pi} \left\{ \int_0^z \frac{2e^{-it}}{1 - z\,e^{-it}}\, dz \right\} dm(t) \\
&= -2\cos\alpha\, e^{i-t} \int_{-\pi}^{\pi} \log\left(1 - z\,e^{-it} \right) dm(t).
\end{aligned}$$

If we choose $m(t)$ as in (6.8.2), then

$$\log \frac{f(z)}{z} = -2\cos\alpha\, e^{-it} \log\left(1 - \varepsilon\, z \right), \quad \text{where } \varepsilon = e^{-it}, \tag{6.8.4}$$

thus, $f(z) = z\,(1 - \varepsilon\, z)$.

6.8.5. To determine the 'dominant' function for the class $\mathcal{T}$, let $f \in \mathcal{T}$. Then there is a unique $p \in \mathcal{P}$ such that

$$f(z) = \frac{z}{1-z^2}\, p(z) = \frac{z}{1-z^2} \int_{-\pi}^{\pi} \frac{1 + ze^{-it}}{1 - ze^{-it}}\, dm(t), \quad z \in E.$$

Choose $m(t)$ as in (6.8.2). Then

$$f(z) = \frac{z}{1-z^2} \frac{1 + \varepsilon z}{1 - \varepsilon z}, \quad \varepsilon = e^{-it}.$$

Note that for $\varepsilon = 0$, we have $f(z) = \dfrac{z}{(1-z)^2}$ which is the Koebe function. If

$$f(z) = z + \sum_{n=2}^{\infty} a_n\, z^n \in \mathcal{T}, \quad \text{then } |a_n| \le n \text{ for all } n, \text{ and}$$

$$\frac{r}{(1+r)^2} \le |f(z)| \le \frac{r}{(1-r)^2}, \quad |z| = r < 1;$$

and $f'(z) = \dfrac{1+z}{(1-z)^3}$ which gives

$$\frac{1-r}{(1+r)^3} \le |f'(z)| \le \frac{1+r}{(1-r)^3}, \quad |z| = r < 1.$$

6.8.6. Consider the class $\mathcal{P}$ with $p_2 = p_3 = \ldots = 0$. Then $|p_1|^2 \leq 1$, i.e., $|p_1| \leq 1$. However, since $f(z) = \dfrac{1}{z} + p_0 + p_1 z$, we can translate this function so that $p_0 = 0$. Then $f(z) = \dfrac{1}{z} + p_1 z$, where $|p_1| \leq 1$. What is the geometry under the mapping? SOLUTION. Without loss of generality, take $p_1 = e^{i\alpha}$. Then, the equality in $|p_1| \leq 1$ holds only if $p_1 = e^{i\alpha}$ for some real α. The simplest case is when $p_1 = 1$. Then $f(z) = \dfrac{1}{z} + z$, or

$$u + i v = r\, e^{i\theta} + \frac{1}{r} e^{-i\theta} = \left(r + \frac{1}{r}\right) \cos\theta + \left(r - \frac{1}{r}\right) \sin\theta,$$

which leads to the ellipse $u = \left(r + \dfrac{1}{r}\right) \cos\theta, \quad v = \left(r - \dfrac{1}{r}\right) \sin\theta$, or

$$\frac{u}{\left(r + \dfrac{1}{r}\right)^2} + \frac{v}{\left(r - \dfrac{1}{r}\right)^2} = 1.$$

This ellipse reduces to the line segment $[-2, 2]$ as $r \to 1$ (see Figure 6.8.2(a).).

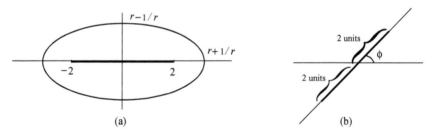

(a)   (b)

Figure 6.8.2 Ellipse into a Line Segment.

If $\phi = \arg\{w\}$, then we have a rotation (Figure 6.8.2(b)). The inequality $|p_n| \leq 1$ is *sharp* for the function

$$f(z) = \frac{1}{z} + e^{i\alpha} z \quad \in \mathcal{P}, \tag{6.8.5}$$

which is an extremal function for this class.

Note that in the class Σ, if we take $p_2 = p_3 = \ldots = 0$, we get the same result as above, since $g(z) = z + p_0 + \dfrac{p_1}{z} \in \Sigma$, where $|p_1| \leq 1$. Thus,

$$\Sigma: \quad g(z) = \frac{1}{f(1/z)}, \text{where } f \in \mathcal{P}. \tag{6.8.6}$$

7

Generalized Convexity

The concept of generalized convexity was introduced by P. Moçanu in 1969. We will investigate certain subclasses of this type and establish their mapping properties, as related to extremum, distortion, and rotation of mappings under these subclasses.

7.1 Convex Functions

Let f be analytic in E with $f'(z) \neq 0$ for all z in the domain $0 < |z| < 1$, and let $f(C_r)$ denote the image of the circle $|z| = r$, $0 < r < 1$. Then the curve $f(C_r)$ is called *convex* if the angle $\psi(r, \theta) = \dfrac{\pi}{2} + \arg\{z f'(z)\}$, $z = r e^{i\theta}$, between the tangents at the point $f(z)$ to the curve $f(C_r)$ and the positive real axis is an increasing function of θ, $0 \leq \theta < 2\pi$. Obviously, the function f is convex on the circle $|z| = r$ if $f(C_r)$ is a convex curve.

Let $\mathcal{K}$ denote the class of regular univalent functions $f(z)$, $z \in E$, such that $f(E)$ is convex. The $\mathcal{K}$ is called the class of convex univalent functions in E, where $\mathcal{K} \subset \mathcal{S}^\star$. In fact, if $f \in \mathcal{K}$, then, since every convex domain is starlike, $f \in \mathcal{S}^\star$. But if $f \in \mathcal{S}^\star$, then, since a starlike domain need not be convex, f need not belong to $\mathcal{K}$. Hence, $\mathcal{K} \subset \mathcal{S}^\star \subset \mathcal{S}$.

A domain D is said to be *convex*, if the line segment joining any two points in the domain D is contained in D. A convex domain D is starlike at every point in D. For example, consider the Koebe function $K(z) = \dfrac{z}{(1-z)^2} \in \mathcal{S}^\star$. This function maps E onto a starlike domain which is not convex, since it does not take the point $-\frac{1}{4}$; but it takes the points $-\frac{1}{4} \pm i$ because the line segment joining $-\frac{1}{4} + i$ and $-\frac{1}{4} - i$ does not lie in $f(E)$.

It is known that if $f(z) = z + \displaystyle\sum_{n=2}^{\infty} a_n z^n$, $z \in E$, is in $\mathcal{S}^\star$, then $|a_n| \leq n$ for $n = 2, 3, \ldots$ (Theorem 6.4.4). We will sharpen this result for $\mathcal{K}$ and show that for this class $|a_n| \leq 1$ for all n. The following notation is used:

$E = \{z : |z| < 1\}; D = f(E); E_r = \{z : |z| < r < 1\};$ and $D_r = f(E_r)$.

Lemma 7.1.1. *Let $f \in \mathcal{S}$. Then $f \in \mathcal{K}$ if and only if D_n is convex.*

PROOF. To prove necessity ($\Rightarrow$), choose $w_1, w_2 \in D_r$ and show that the line segment $t\, w_1 + (1-t)\, w_2$, $0 < t < 1$, is also in D_r. There exist two points z_1, z_2 in E_r such that $w_1 = f(z_1)$, $w_2 = f(z_2)$ (see Figure 7.1.1(a)). Without loss of generality, we assume that $|z_1| \le |z_2| < r$. Then $\left|\dfrac{z_1}{z_2}\right| < 1 \in E_r$. Let

$$g(z) = t\, f\left(\frac{z_1}{z_2}\, z\right) + (1-t)\, f(z), \; 0 < t < 1. \text{ Then } g(E) \subset D.$$

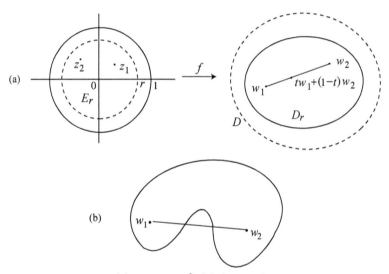

(a)

(b)

Figure 7.1.1 (a) Mapping f; (b) A Non-Convex Domain.

The function $h(z) = f^{-1}\big(g(z)\big)$ is regular in E,

$$|h(z)| = \left|f^{-1}\big(g(z)\big)\right| < 1,$$
$$h(0) = f^{-1}(g0)) = f^{-1}\big[t\, f(0) + (1-t)\, f(0)\big] = f^{-1}\big(f(0)\big) = 0.$$

By Schwarz lemma, $|h(z)| \le |z|$, $z \in E$. In particular,

$$h(z_2) = f^{-1}\big(g(z_2 = \left|f^{-1}\big(t\, f(z_1) + (1-t)\, f(z_2)\big)\right| = \left|f^{-1}\big(t\, w_1 + (1-t)\, w_2\big)\right|,$$

i.e., $\big|h(z_2)\big| \le |z_2| < r$. Also, $D_r \subset D$ implies that there exists a $z_0 \in E_r$ such that $t\, w_1 + (1-t)\, w_2 = f(z_0)$ for given t, $0 < t < 1$. Then $f^{-1}\big(f(z_0)\big) = z_0 \in E_r$. Since this is true of every value of t, $0 < t < 1$, we have $t\, w_1 + (1-t)\, w_2 \in D_r$.

For sufficiency ($\Leftarrow$), we shall prove by negation. We shall assume that $f \in \mathcal{K}$ and show that $f(E)$ is not convex. If $f \in \mathcal{K}$, there exist two points in

D such that the line segment joining them is not contained in D (see Figure 7.1.1(b)). Now choose a disk E_r such that D_r contains these two points. Then $D_r \subset D$ implies that the line segment joining these points will not be contained in D, i.e., $D \equiv f(E)$ is not convex. ∎

The following result provides an analytical characterization of the class $\mathcal{K}$:

Theorem 7.1.1. *If* $f \in \mathcal{K}$, *then*

$$\Re\left\{1 + \frac{z\,f''(z)}{f'(z)}\right\} > 0, \quad z \in E. \tag{7.1.1}$$

PROOF. By lemma, $D \equiv f(E)$ is convex. Geometrically this means (see Figure 7.1.2), that the function $f(z)$, $z = r\,e^{i\theta}$, $0 \leq \theta < 2\pi$, maps ∂E_r onto a simple closed curve such that the tangent at $f\left(r\,e^{i\theta}\right)$ moves in a counterclockwise direction along $\partial f(E_r) \equiv \partial D_r$ as θ increases. The angle which the tangent vector makes with the real axis in the w-plane is $\frac{\pi}{2} + \theta + \arg\{f'(r\,e^{i\theta})\}$ which means that

$$\frac{\partial}{\partial \theta}\left\{\frac{\pi}{2} + \theta + \arg\left\{f'\left(r\,e^{i\theta}\right)\right\}\right\} > 0, \tag{7.1.2}$$

and

$$1 + \Im\left\{\frac{\partial}{\partial \theta}\log f'\left(r\,e^{i\theta}\right)\right\} > 0$$

$$1 + \Im\left\{\frac{ir e^{i\theta}\,f''\left(r\,e^{i\theta}\right)}{f'\left(r\,e^{i\theta}\right)}\right\} > 0$$

$$1 + \Re\left\{\frac{r e^{i\theta}\,f''\left(r\,e^{i\theta}\right)}{f'\left(r\,e^{i\theta}\right)}\right\} > 0$$

$$\Re\left\{1 + \frac{z\,f''(z)}{f'(z)}\right\} > 0. \;\blacksquare$$

Note that $\frac{\pi}{2} + \theta + \arg\left\{f'\left(r\,e^{i\theta}\right)\right\} = \arg\{iz f'(z)\}$. Then the condition (7.1.2) can be written as $\frac{\partial}{\partial \theta}\arg\{iz f'(z)\} > 0$, or $\frac{\partial}{\partial \theta}\arg\{z f'(z)\} > 0$. ∎

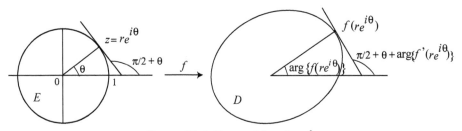

Figure 7.1.2 Convex Mapping f.

Compare the conclusion of this theorem with the result

$$f \in \mathcal{S}^* \iff \frac{\partial}{\partial \theta} \arg\{f(z)\} > 0. \tag{7.1.3}$$

Hence, $zf'(z) \in \mathcal{S}^*$, i.e., $zf'(z)$ maps E onto a starlike domain with respect to the origin.

The *radius of convexity* $r_c(f)$ for the function f is defined by

$$r_c(f) = \sup\left\{r : \Re\left\{1 + \frac{zf''(z)}{f'(z)}\right\} > 0, |z| \leq r\right\}. \tag{7.1.4}$$

If $r_c(f) \geq 1$, we say that the function f is *convex in E* (or simply *convex*), and f will satisfy the condition (7.1.1) which implies that $f'(z) \neq 0$ for all $0 < |z| < 1$. However, this condition does not guarantee that the function f is univalent in E; for example, $f(z) = z^2$ satisfies the condition (7.1.1) but is not univalent in E. For more details, see §7.2.

Corollary 7.1.1. $f \in \mathcal{K} \iff zf'(z) \in \mathcal{S}^*, z \in E$.

Theorem 7.1.2. *If $f(z) = z + \sum_{n=2}^{\infty} a_n z^n \in \mathcal{K}, z \in E$, then $|a_n| \leq 1$ for all n.*

PROOF. By Corollary 7.1.1,

$$zf'(z) = z + \sum_{n=2}^{\infty} na_n z^n \in \mathcal{S}^*.$$

Hence, $|na_n| \leq n$ which implies the result. ∎

This inequality is sharp for

$$f(z) = z + \sum_{n=2}^{\infty} z^n = \frac{z}{1-z},$$

which is called the extremal function. Now, since $f'(z) = \dfrac{1}{(1-z)^2}$ and $zf'(z) = \dfrac{z}{(1-z)^2}$ which is the Koebe function, this extremal function is in the class $\mathcal{S}^*$ and maps $|z| < 1$ onto a convex domain.

Note that the 1/4-covering theorem holds for both $f \in \mathcal{S}$ and $f \in \mathcal{S}^*$, but for $f \in \mathcal{K}$ this becomes the 1/2-covering theorem, i.e., $f \in \mathcal{K} \implies f(E) \subset D(0, 1/2)$.

Theorem 7.1.3. (Marx [1932-33]) *Let $f \in \mathcal{K}$ and let $f(z) \neq \zeta$ for $z \in E$. Then $|\zeta| \geq 1/2$.*

PROOF. Let $g(z) = [f(z) - \varsigma]^2$, $z \in E$. Then g is univalent in E. Let $z_1, z_2 \in E$, $z_1 \neq z_2$. Then

$$g(z_1) - g(z_2) = [f(z_1) - \varsigma]^2 - [f(z_2) - \varsigma]^2$$
$$= \left[(f(z_1) - \varsigma) - (f(z_2) - \varsigma) \right] \left[(f(z_1) - \varsigma) + (f(z_2) - \varsigma) \right]$$
$$= \left[f(z_1) - f(z_2) \right] \left[f(z_1) + f(z_2) - 2\varsigma \right].$$

Since f is univalent in E, so $f(z_1) \neq f(z_2)$ for $z_1 \neq z_2$, and since $f \in \mathcal{K}$, we have $\dfrac{f(z_1) + f(z_2)}{2} \neq \varsigma$, i.e., $f(z_1) + f(z_2) \neq 2\varsigma$. Thus, $g(z_1) - g(z_2) \neq 0$, or $g(z_1) \neq g(z_2)$. Hence, g is univalent in E. Next,

$$g(z) = [\varsigma - f(z)]^2 = \varsigma^2 \left[1 - \frac{f(z)}{\varsigma} \right]^2$$
$$= \varsigma^2 - 2\varsigma z + \cdots, \quad \text{by Taylor's series expansion.}$$

Then the normalized function

$$h(z) = \frac{g(z) - \varsigma^2}{-2\varsigma} = z + \cdots \in \mathcal{S},$$

where h is univalent in E, $h(0) = 0$ and $h'(0) = 1$. Note that $g(z) \neq 0$ for $f(z) \neq \varsigma$. Also, $g(z) \neq 0$ for $z \in E$. This means that $h(z) \neq \dfrac{\varsigma}{2}$. Then, by the 1/4-covering theorem (since $h \in \mathcal{S}$), we get $\left| \dfrac{\varsigma}{2} \right| \geq \dfrac{1}{4}$, or $|\varsigma| \geq 1/2$. ∎

Theorem 7.1.4. (Distortion theorem) *If $f \in \mathcal{K}$, then*

$$\frac{r}{1+r} \leq |f(z)| \leq \frac{r}{1-r}, \tag{7.1.5}$$

$$\frac{1}{(1+r)^2} \leq |f'(z)| \leq \frac{1}{(1-r)^2}. \tag{7.1.6}$$

Note that if $f \in \mathcal{K}$, then $|a_n| \leq 1$ for all n. In particular, $|a_2| \leq 1$, where ord $(f) = |a_2|$. and equality holds for the function $f(z) = \dfrac{z}{1 + e^{i\alpha} z}$, $\alpha \in \mathbb{R}$, $z \in E$. This theorem establishes that the class $\mathcal{K}$ is a compact set. Let $r \to 1$ in the left inequality in (d.6). Then the *Koebe constant* for the class $\mathcal{K}$ is $\frac{1}{2}$.

A function $f \in \mathcal{S}$ is said to be *convex of order* α, $0 \leq \alpha < 1$, if

$$\Re\left\{ 1 + \frac{z f''(z)}{f'(z)} \right\} > \alpha, \quad z \in E. \tag{7.1.7}$$

The class of all such functions is denoted by $\mathcal{K}(\alpha)$.

7.1.1 Radius of Convexity. Recall that $\mathcal{K} \subset \mathcal{S}^\star \subset \mathcal{S}$. Since not all functions in the class $\mathcal{S}$ are convex in E ($|z| < 1$), the radius of convexity is a 'measure' of the degree of convexity of each function in $\mathcal{S}$.

To each function $f \in \mathcal{S}$ we associate a positive real number $R = R(f)$ which is the largest radius R for which the disk $D(0, R)$ ($|z| < R \leq 1$) is mapped by f onto a convex domain. Note that if $E_R = \{z : |z| < R \leq 1\}$, then $f(E_R)$ is convex. Obviously, $f \in \mathcal{K}$ if and only if $R(f) = 1$, i.e., if $f \in \mathcal{K}$, then $f(E)$ is convex; therefore, $R = 1$. ($R = 1$ means $E = D(0, 1)$ which is the open unit disk.)

Theorem 7.1.5. *If $f \in \mathcal{S}$, then f maps the disk $|z| < 2 - \sqrt{3}$ onto a convex domain.*

This theorem means that the radius of convexity for the set of all functions in the class $\mathcal{S}$ is $2 - \sqrt{3}$.

PROOF. $f \in \mathcal{K}$ if and only if $\Re\left\{1 + \dfrac{z\,f''(z)}{f'(z)}\right\} > 0$ for $0 < z < 1$. We also know from the distortion theorem for the class $\mathcal{S}$ that

$$\frac{2r^2 - 4r}{1 - r^2} \leq \Re\left\{\frac{z\,f''(z)}{f'(z)}\right\} \leq \frac{2r^2 + 4r}{1 + r^2}. \tag{7.1.8}$$

Hence,

$$\Re\left\{1 + \frac{z\,f''(z)}{f'(z)}\right\} = 1 + \Re\left\{\frac{z\,f''(z)}{f'(z)}\right\}$$

$$\geq 1 + \frac{2r^2 - 4r}{1 - r^2} = \frac{1 - 4r + r^2}{1 - r^2}. \tag{7.1.9}$$

The smallest positive zero of $1 - 4r + r^2$ is the required $R(f)$. Thus, the above inequality implies that for $R < 2 - \sqrt{3}$,

$$\Re\left\{1 + \frac{z\,f''(z)}{f'(z)}\right\} > 0. \tag{7.1.10}$$

Hence, if $f \in \mathcal{S}$, then $f \in \mathcal{K}$ for the disk $|z| < R = 2 - \sqrt{3}$. ∎

The extremal function is the Koebe function $K(z)$ since $\Re\left\{1 + \dfrac{z\,k''(z)}{k'(z)}\right\} > 0$ for $|z| < 2 - \sqrt{3}$ and is equal to zero at $z = 2 - \sqrt{3}$.

In an analogous manner, we can define the radius of starlikeness as follows: If $f \in \mathcal{S}^\star$, then $\Re\left\{\dfrac{z\,f'(z)}{f(z)}\right\} > 0$, $|z| < r < 1$.

Recall $\mathcal{K} \subset \mathcal{S}^\star \subset \mathcal{S}$. Look at the mapping under f shown in Figure 7.2.1.

What can we say about the rest of the domain in the w-plane?

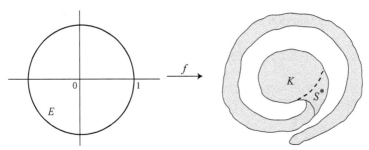

Figure 7.2.1 Mapping $f \in \mathcal{K}$.

7.2 Close-to-Convex Functions

Let $\mathcal{C}$ denote the class of *close-to-convex* univalent functions f in E. Note that $\mathcal{S}^* \subset \mathcal{C} \subset \mathcal{S}$. We shall later study the classes of alpha-convex functions $(0 \le \alpha \le 1)$ and the Bazilevich functions, which lie between $\mathcal{C}$ and $\mathcal{S}$, in that order. The class of the Bazilevich functions is the largest known class of univalent functions.

A univalent function f belongs to the class $\mathcal{C}$ if there exists a function $g \in \mathcal{K}$ and an $\alpha \in \mathbb{R}$ such that $\Re\left\{e^{i\alpha}\dfrac{f'(z)}{g'(z)}\right\} > 0$ for $z \in E$. By specific choices of the function $g \in \mathcal{K}$ it can be shown that the class $\mathcal{C}$ contains some of the subclasses of $\mathcal{S}$; viz., (i) Take $g(z) = f(z) \in \mathcal{K}$, and let $\alpha = 0$. Then $\Re\left\{\dfrac{f'(z)}{g'(z)}\right\} = 1 > 0$. Thus, $f \in \mathcal{K} \implies f \in \mathcal{C}$, i.e., every convex function in E is close-to-convex; (ii) take $g(z) = z \in \mathcal{K}$. Then $\Re\left\{\dfrac{f'(z)}{g'(z)}\right\} = \Re\{f'(z)\} > 0$, i.e., if the derivative of f has a positive real part, then f is close-to-convex in E. Also, if $f' \in \mathcal{P}$, then $f \in \mathcal{C}$; and (iii) take $g(z) = \displaystyle\int_0^z \dfrac{f(\zeta)}{\zeta}\, d\zeta$, $f \in \mathcal{S}^*$. Then $g \in \mathcal{K}$. In fact, in this case we have

$$g'(z) = \frac{f(z)}{z}, \quad f(0) = 0, \quad g''(z) = \frac{f'(z)}{z} - \frac{f(z)}{z^2} = \frac{zf'(z) - f(z)}{z^2}.$$

Then

$$\Re\left\{1 + \frac{g''(z)}{g'(z)}\right\} = \Re\left\{1 + \frac{zf'(z) - f(z)}{z^2}\right\} = \Re\left\{1 + \frac{zf'(z)}{f(z)} - 1\right\}$$

$$= \Re\left\{\frac{zf'(z)}{f(z)}\right\} > 0, \quad \text{since } f \in \mathcal{S}^*.$$

Also,

$$\Re\left\{\frac{f'(z)}{g'(z)}\right\} = \Re\left\{\frac{zf'(z)}{f(z)}\right\} > 0.$$

Hence, $f \in \mathcal{S}^* \implies f \in \mathcal{C}$, i.e., a starlike function is close-to-convex.

To prove the converse, we will show by an example that there exists a close-to-convex function which is not starlike. Such a function is

$$f(z) = z + \frac{1}{2} z^2 + \frac{1}{3} z^3.$$

According to the Bieberbach conjecture, if $f(z) = z + \sum\limits_{n=2}^{\infty} a_n z^n \in \mathcal{C}$, then $|a_n| \le n$ for all n. Associate with f a function g given by $g(z) = -\log(1-z)$, which has a principal branch. Since $\log(1-z) = 0$ at $z = 0$, we will show that $g(z) = \log(1-z)$ is in the class $\mathcal{K}$. Since $zg'(z) = z/(1-z)$ is starlike, the logarithmic derivative gives that $g \in \mathcal{K}$. Also f is univalent in $\mathcal{K}$:

$$\Re\left\{\frac{f''(z)}{g'(z)}\right\} = \Re\left\{(1 + z + z^2)(1 - z)\right\} = \Re\left\{1 - z^2\right\} > 0.$$

Hence, $f \in \mathcal{C}$.

Now, to show that $f \notin \mathcal{S}^*$: If f were in $\mathcal{S}^*$, then $\Re\left\{\dfrac{zf'(z)}{f(z)}\right\}$ should be positive. But

$$\Re\left\{\frac{zf'(z)}{f(z)}\right\} = \Re\left\{\frac{z(1 + z + z^2)}{1 + \frac{z}{2} + \frac{z^2}{3}}\right\} = \Re\left\{\frac{(1 + z + z^2)\left(1 + \frac{\bar{z}}{2} + \frac{\bar{z}^2}{3}\right)}{\left|1 + \frac{z}{2} + \frac{z^2}{3}\right|^2}\right\}.$$

Let $z = r e^{i\theta}$. Then

$$\Re\left\{(1 + z + z^2)\left(1 + \frac{\bar{z}}{2} + \frac{\bar{z}^2}{3}\right)\right\}$$

$$= \Re\left\{[1 + r(\cos\theta + i\sin\theta) + r^2(\cos 2\theta + i\sin 2\theta)]\cdot\right.$$

$$\left.[1 + \frac{r}{2}(\cos\theta - i\sin\theta) + \frac{r^2}{3}(\cos 2\theta - i\sin 2\theta)]\right.$$

$$= 1 + \frac{3}{2}\cos\theta\, r + \left(\frac{1}{2} + \frac{4}{3}\cos 2\theta\right) r^2 + \frac{5}{6}\cos\theta\, r^3 + \frac{1}{3} r^4$$

$$= A(r, \theta), \quad \text{say}.$$

Thus,

$$\Re\left\{\frac{zf'(z)}{f(z)}\right\} = \frac{A(r, \theta)}{\left|1 + \frac{\bar{z}}{2} + \frac{\bar{z}^2}{3}\right|^2}.$$

We need only show that $A(r, \theta) < 0$ at some point in E. To do this, note that $A(r, \theta)$ is continuous in $E \cup \partial E$. Thus, it would suffice to show that $A(1, \theta_0) < 0$ for some θ_0. This will then imply that $A(r, \theta_0) < 0$ for all r in $1 - \delta < r < 1$. Now

$$A(1, \theta) = 1 + \frac{3}{2} \cos \theta + \left(\frac{1}{2} + \frac{4}{3} \cos 2\theta \right) + \frac{5}{6} \cos \theta + \frac{1}{3}$$

$$= \frac{11}{6} + \frac{7}{3} \cos \theta + \frac{4}{3} \cos 2\theta,$$

whence $\dfrac{d}{d\theta} A(1, \theta) = -\dfrac{7}{3} \sin \theta - \dfrac{8}{3} \sin 2\theta = 0$ yields $\sin \left(\dfrac{7}{3} + \dfrac{11}{3} \cos \theta \right) = 0$,

which has the roots $\theta = 0, \pi, \arccos \left(-\dfrac{7}{16} \right)$. Moreover, $\dfrac{d^2}{d\theta^2} A(1, \theta) < 0$ for

$\cos \theta = -\dfrac{7}{16}$. This means that $A(1, \theta)$ attains its minimum at $\cos \theta_0 = -\dfrac{7}{16}$.

At this point θ_0, we have $\cos 2\theta_0 = 2 \cos^2 \theta_0 - 1 = -\dfrac{79}{128}$. Thus,

$$A(1, \theta_0) = \frac{11}{6} - \frac{7}{3} \frac{7}{16} - \frac{4}{3} \frac{79}{128} = -\frac{1}{96}.$$

Hence, we have shown that $\Re \left\{ \dfrac{z f'(z)}{f(z)} \right\} < 0$ at some point $z = r e^{i\theta_0}$, where $1 - \delta < r < 1$, and $\cos \theta_0 = -\dfrac{7}{16}$, which shows that $f \notin \mathcal{S}^*$. ∎

Theorem 7.2.1. (Reade [1955]) If $f(z) = z + \displaystyle\sum_{n=2}^{\infty} a_n z^n \in \mathcal{C}$, $z \in E$, then $|a_n| \leq n$ for all n.

Reade's result [1955] on close-to-convex functions generalizes Nevanlinna's result [1920].

PROOF. Suppose $g(z) = z + \displaystyle\sum_{n=2}^{\infty} b_n z^n \in \mathcal{K}$, $z \in E$. Then we know that $|b_n| \leq 1$ for all n. Let

$$\frac{f'(z)}{g'(z)} = 1 + \sum_{n=1}^{\infty} \alpha_n z^n. \tag{7.2.1}$$

Note that $\Re \left\{ \dfrac{f'(z)}{g'(z)} \right\} > 0$ for $z \in E$, and the function $\dfrac{f'(z)}{g'(z)} \in \mathcal{P}$, so that $|\alpha_n| \leq 2$ for all n. From (7.2.1) we have

$$\frac{f'(z)}{g'(z)} = \frac{1 + \displaystyle\sum_{n=2}^{\infty} n a_n z^{n-1}}{1 + \displaystyle\sum_{n=2}^{\infty} n b_n z^{n-1}} = 1 + \sum_{n=2}^{\infty} \alpha_n z^n,$$

or

$$1 + \sum_{n=2}^{\infty} n\,a_n\,z^{n-1} = \left(1 + \sum_{n=2}^{\infty} n\,b_n\,z^{n-1}\right)\left(1 + \sum_{n=2}^{\infty} \alpha_n\,z^n\right).$$

Then by equating coefficients of z^{n-1} on both sides, we get

$$n\,a_n = n\,b_n + (n-1)\,b_{n-1}\,\alpha_1 + (n-2)\,b_{n-2}\,\alpha_2 + \cdots + 2b_2\,\alpha_{n-2} + \alpha_{n-1},$$

which yields

$$\left|n\,a_n\right| = n\left|b_n\right| + (n-1)\left|b_{n-1}\right|\left|\alpha_1\right| + (n-2)\left|b_{n-2}\right|\left|\alpha_2\right| + \cdots + 2\left|b_2\right|\left|\alpha_{n-2}\right| + \left|\alpha_{n-1}\right|.$$

Now, each $\left|b_n\right| \leq 1$ and each $\left|\alpha_n\right| \leq 2$, so that

$$\begin{aligned}
\left|n\,a_n\right| &= n + (n-1)\left|\alpha_1\right| + \cdots + +2\left|\alpha_{n-2}\right| + \left|\alpha_{n-1}\right| \\
&= n + 2\big[(n-1) + (n-2) + \cdots + 2 + 1\big] \\
&= n + 2 \cdot \frac{(n-1)n}{2} = n^2,
\end{aligned}$$

which implies that $\left|a_n\right| \leq n$ for all $n = 2, 3, \ldots$. ∎

Since $\mathcal{S}^* \subset \mathcal{C}$, then, in view of this theorem, this result proved earlier for the class $\mathcal{S}^*$ is a consequence of this theorem.

We summarize all results in the form of Table 7.2.1.

Table 7.2.1.

Class	Condition	Tangency
$f \in \mathcal{S}^* \iff$	$\Re\left\{\dfrac{z f'(z)}{f(z)}\right\} > 0$	$\dfrac{\partial}{\partial\theta}\left(\arg\left\{f'(r\,e^{i\theta})\right\}\right) > 0$
$f \in \mathcal{K} \iff$	$\Re\left\{1 + \dfrac{z f''(z)}{f'(z)}\right\} > 0$	$\dfrac{\partial}{\partial\theta}\left(\theta + \arg\left\{f'(r\,e^{i\theta})\right\}\right) > 0$
$f \in \mathcal{C} \iff$	$\Re\left\{\dfrac{z f'(z)}{g'(z)}\right\} > 0,\ g \in \mathcal{K}$	see Theorem 7.2.3

Lemma 7.2.1. *If $\phi(z)$ is analytic in a convex domain E and $0 < \Re\left\{\phi'(z)\right\}$ for all $z \in E$, then ϕ is univalent in E.*

This lemma shows that the class of analytic functions whose derivative has a positive part in E is univalent in E.

PROOF OF LEMMA. Choose two points $z_0, z_1 \in E$, $z_0 \neq z_1$. Since E is convex the line segment $z = z_0 + t(z_1 - z_0)$, $0 \leq t \leq 1$, must lie in E.

Integrating along this line segment from z_0 to z_1 we have

$$\phi(z_0) - \phi(z_1) = \int_{z_0}^{z_1} \phi'(z)\,dz = \int_{z_0}^{z_1} \phi'\left(z_0 + t(z_1 - z_0)\right)\,dz$$

$$= \int_0^1 \phi'[z + t(z_1 - z_0)]\,(z_1 - z_0)\,dt.$$

Dividing both sides by $(z_1 - z_0)$ and taking the real part, we get

$$\Re\left\{\frac{\phi(z_1) - \phi(z_0)}{z_1 - z_0}\right\} = \Re\left\{\int_0^1 \phi'[z + t(z_1 - z_0)]\,(z_1 - z_0)\,dt\right\}$$

$$= \int_0^1 \Re\left\{\phi'\left(z_0 + t(z - 1 - z_0)\right)\right\}\,dt > 0.$$

Then $\phi(z_1) \neq \phi(z_0)$ for $z_1 \neq z_0$. ∎

Theorem 7.2.2. *If $f \in \mathcal{C}$, then f is univalent in E.*

PROOF. We will use Clunie's method as described in §6.8.1. Recall the equality (6.8.4) which leads to

$$\sum_{k=2}^{n}(k - 1)a_k\,z^k + \sum_{k=n+1}^{\infty} b_k\,z^k = \left[2z + \sum_{k=2}^{\infty}(k + 1)a_k\,z^k\right]w(z), \qquad (7.2.2)$$

where the infinite series $\displaystyle\sum_{k=n+1}^{\infty} b_k\,z^k$ is absolutely and uniformly convergent in E (which is compact). Again, in view of Parseval's identity (Exercise 1.10.7) which states that if $f(z) = \displaystyle\sum_{n=0}^{\infty} a_n\,z^n$ converges for $|z| < R$, then for $0 < r < R$ (with $z = r\,e^{i\theta}$, $0 \le \theta < 2\pi$)

$$\frac{1}{2\pi}\int_0^{2\pi}\left|f(r\,e^{i\theta})\right|^2\,d\theta = \sum_{n=0}^{\infty}\left|a_n\right|^2 r^{2n},$$

and in view of Cauchy-Schwarz inequality (1.8.4) which is

$$|z_1 w_1 + \cdots + z_n w_n| \le \sqrt{\left|z_1\right|^2 + \cdots + \left|z_n\right|^2} \cdot \sqrt{\left|w_1\right|^2 + \cdots + \left|w_n\right|^2},$$

we have

$$\left[2z + \sum_{k=2}^{n-1}(k + 1)a_k\,z^k\right]w(z) \le \left|2z + \sum_{k=2}^{n-1}(k + 1)a_k\,z^k\right||z|$$

$$= \left|2z^2 + \sum_{k=2}^{n-1}(k + 1)a_k\,z^{k+1}\right|. \qquad (7.2.3)$$

Let $z = r e^{i\theta}$, $0 < r < 1$, $0 \leq \theta < 2\pi$. Then integrating (7.2.3) from 0 to 2π we get

$$\frac{1}{2\pi} \int_0^{2\pi} \left| \sum_{k=2}^n (k-1) a_k r^k e^{ik\theta} + \sum_{k=n+1}^\infty b_k r^k e^{ik\theta} \right|^2 d\theta$$

$$\leq \frac{1}{2\pi} \int_0^{2\pi} \left| 2r^2 e^{2i\theta} + \sum_{k=2}^{n-1} (k+1) a_k r^k e^{i(k+1)\theta} \right|^2 d\theta.$$

Note that

$$\frac{1}{2\pi} \int_0^{2\pi} \left| f(re^{i\theta}) \right|^2 d\theta = \sum_{k=0}^\infty |a_k|^2 r^{2k},$$

where

$$f\left(re^{i\theta}\right) = \sum_{k=0}^\infty a_k r^k e^{ik\theta},$$

which is equivalent to

$$\sum_{k=2}^n (k-1)^2 |a_k|^2 r^{2k} + \sum_{k=n+1}^\infty |b_k|^2 r^{2k} \leq 4r^2 + \sum_{k=2}^{n-1} (k+1)^2 |a_k|^2 r^{2(k+1)}.$$

Since the right side of the above inequality is positive, let $r \to 1^-$. Then

$$\sum_{k=2}^n (k+1)^2 |a_k|^2 \leq 4r^2 + \sum_{k=2}^{n-1} (k+1)^2 |a_k|^2. \qquad (7.2.4)$$

Returning to a previous step, we see that

$$\frac{1}{2\pi} \int_0^{2\pi} \left| 2re^{i\theta} + \sum_{k=2}^{n-1} (k+1) a_k r^k e^{ik\theta} \right|^2 d\theta$$

$$\geq \frac{1}{2\pi} \int_0^{2\pi} \left| 2re^{i\theta} + \sum_{k=2}^{n-1} (k+1) a_k r^k e^{ik\theta} \right|^2 |w(re^{i\theta})\})|^2 d\theta, \text{ since } |w(z)| < |z| <$$

$$\geq \frac{1}{2\pi} \int_0^{2\pi} \left| \sum_{k=2}^n (k-1) a_k r^k e^{ik\theta} + \sum_{k=n+1}^\infty b_k r^k e^{ik\theta} \right|^2 d\theta$$

$$\geq \sum_{k=2}^n (k-1)^2 |a_k|^2 r^{2k} + \sum_{k=n+1}^\infty |b_k|^2 r^{2k}$$

$$\geq \sum_{k=2}^n (k-1)^2 |a_k|^2 r^{2k}.$$

Then using the method of induction, we get

$$\sum_{k=2}^{n}(k-1)^2|a_k|^2 \leq 4 + \sum_{k=2}^{n-1}(k+1)|a_k|^2. \qquad (7.2.5)$$

From this result we find that for $n = 2$ we get $|a_2| \leq 2$, and for $n = 3$ we get $|a_3| \leq 3$. Therefore, we assume that $|a_k| \leq k$ for $k = 2, 3, \ldots, n-1$. Then

$$1^2|a_2|^2 + 2^2|a_3|^2 + \cdots + (n-1)^2|a_n|^2 \leq 4 + 3^2|a_2|^2 + 4^2|a_3|^2 + \cdots + n^2|a_{n-1}|^2,$$

or

$$(n-1)^2|a_n|^2 \leq 4 + (3^2-1^2)|a_2|^2 + (4^2-2^2)|a_3|^2 + \cdots + [n^2-(n-1)^2]|a_{n-1}|^2$$
$$\leq 4 + (3^2-1^2)2^2 + (4^2-2^2)3^2 + \cdots + [n^2-(n-1)^2](n-1)^2$$

since we assumed that $|a_k| \leq k$ for $k = 2, 3, \ldots,$

$$= 2^2 + 3^2 2^2 - 1^2 2^2 + 4^2 3^2 - 2^2 3^2 + \cdots + n^2(n-1)^2 - (n-1)^2(n-1)^2$$
$$= n^2(n-1)^2,$$

which yields $|a_n|^2 \leq n^2$, or $|a_n| \leq n$.

Let $\phi = f\left(g^{-1}(z)\right)$. Then $f = \phi \circ g$ and $f(z) = \phi(g(z))$. Since ϕ is analytic in a convex domain E and g is univalent in E since $g \in K$, then f is analytic in E. Also, $f'(z) = \phi'(g(z)) g'(z)$; so $\dfrac{f'(z)}{g'(z)} = \phi'(g(z))$, and hence, $\Re\left\{\dfrac{f'(z)}{g'(z)}\right\} = \Re\{\phi'(g(z))\} > 0$ for $z \in E$. By the Lemma 7.2.1, ϕ is univalent. Therefore, $f = \phi \circ g$ is also univalent in E. ∎

Theorem 7.2.3. $f \in C$ if and only if

$$\int_{\theta_1}^{\theta_2} \Re\left\{1 + \frac{re^{i\theta} f''\left(re^{i\theta}\right)}{f'\left(re^{i\theta}\right)}\right\} d\theta > -\pi \qquad (7.2.6)$$

for $0 \leq r < 1$, $\theta_1 < \theta_2$; or, equivalently,

$$\arg\left\{\frac{z_2 f'(z_2)}{z_1 f'(z_1)}\right\} > -\pi \quad \text{for } z_1, z_2 \in E. \qquad (7.2.7)$$

PROOF. Since $f \in C$ if and only if there exists a $g \in K$ such that $\Re\left\{\dfrac{f'(z)}{g'(z)}\right\} > 0$, $z \in E$.

Let $p(z) = \arg\{f'(z)\}$ and $q(z) = \arg\{g'(z)\}$, for $z \in E$, be chosen such that they are continuous in θ. Then

$$|p(z) - q(z)| < \frac{\pi}{2}. \tag{7.2.8}$$

Note that $f'(z)$ and $g'(z)$ do not have zeros in E. Now, let for a fixed r and all real θ

$$P(r, \theta) = \theta + p\left(re^{i\theta}\right),$$
$$Q(r, \theta) = \theta + q\left(re^{i\theta}\right).$$

Then, by (7.2.8), we have

$$\left|P(r, \theta) - Q(r, \theta)\right| < \frac{\pi}{2}. \tag{7.2.9}$$

Since $g \in \mathcal{K}$, we have $\dfrac{\partial}{\partial \theta} \arg\left\{ire^{i\theta}\, f'\left(re^{i\theta}\right)\right\} > 0$, i.e., $\dfrac{\partial Q}{\partial \theta} > 0$, which implies that $Q(r, \theta)$ is monotone increasing in θ for a fixed r, i.e., $Q(r, \theta_1) < Q(r, \theta_2)$ for $\theta_1 < \theta_2$; thus,

$$\begin{aligned}
P(r, \theta_1) - P(r, \theta_2) &= [P(r, \theta_1) - Q(r, \theta_1)] - [P(r, \theta_2) - Q(r, \theta_2)] \\
&\quad - [Q(r, \theta_1) - Q(r, \theta_2)] \\
&< [P(r, \theta_1) - Q(r, \theta_1)] - [P(r, \theta_2) - Q(r, \theta_2)],
\end{aligned}$$

since $Q(r, \theta_1) - Q(r, \theta_2) < 0$. So

$$\begin{aligned}
\left|P(r, \theta_1) - P(r, \theta_2)\right| &< \left|P(r, \theta_1) - Q(r, \theta_1) - (P(r, \theta_2) - Q(r, \theta_2))\right| \\
&< \frac{\pi}{2} + \frac{\pi}{2} = \pi, \quad \text{by (7.2.9)}.
\end{aligned}$$

Hence, $-\pi < P(r, \theta_1) - P(r, \theta_2) < \pi$. Now,

$$\begin{aligned}
\frac{\partial}{\partial \theta}\left\{\arg\{f'(re^{i\theta})\}\right\} &= \frac{\partial}{\partial \theta}\Im\left\{\log f'(re^{i\theta})\right\} = \Im\left\{\frac{\partial}{\partial \theta}\log f'(re^{i\theta})\right\} \\
&= \Im\left\{ire^{i\theta}\frac{f''(re^{i\theta})}{f'(re^{i\theta})}\right\} = \Re\left\{re^{i\theta}\frac{f''(re^{i\theta})}{f'(re^{i\theta})}\right\},
\end{aligned}$$

which gives

$$\begin{aligned}
\int_{\theta_1}^{\theta_2}\left\{1 + \frac{\partial}{\partial \theta}\arg\left\{f'(re^{i\theta})\right\}\right\} d\theta &= \theta_2 - \theta_1 + \arg\{f'(re^{i\theta_2})\} - \arg\{f'(re^{i\theta_1})\} \\
&= (\theta_2 - \theta_1) - \left[\arg\{f'(re^{i\theta_1})\} - \arg\{f'(re^{i\theta_2})\}\right]. \tag{7.2.10}
\end{aligned}$$

Since $-\pi < P(r,\theta_1) - P(r,\theta_2) < \pi$, we have

$$-\pi < \theta_1 + p(r,\theta_1) - \theta_2 - p(r,\theta_2) < \pi$$
$$-\pi < \theta_1 + \arg\{f'(re^{i\theta_1})\} - \theta_2 - \arg\{f'(re^{i\theta_2})\} < \pi.$$

Thus,

$$(\theta_1 - \theta_2) + \arg\{f'(re^{i\theta_1})\} - \arg\{f'(re^{i\theta_2})\} < \pi,$$

and

$$(\theta_2 - \theta_1) - \left[\arg\{f'(re^{i\theta_1})\} - \arg\{f'(re^{i\theta_2})\}\right] > -\pi,$$

which means that

$$\int_{\theta_1}^{\theta_2} \left\{1 + \frac{\partial}{\partial\theta} \arg\{f'(re^{i\theta})\}\right\} d\theta > -\pi.$$

Also,

$$\int_{\theta_1}^{\theta_2} \left\{1 + \frac{\partial}{\partial\theta} \arg\{f'(re^{i\theta})\}\right\} d\theta$$
$$= (\theta_2 - \theta_1) + \arg\{f'(re^{i\theta_2})\} - \arg\{f'(re^{i\theta_1})\}$$
$$= \arg\{re^{i\theta_2} f'(re^{i\theta_2})\} - \arg\{re^{i\theta_1} f'(re^{i\theta_1})\}$$
$$= \arg\{z_2 f'(z_2)\} - \arg\{z_1 f'(z_1)\}, \quad \text{taking } z_{1,2} = re^{i\theta_{1,2}},$$

which implies that

$$\arg\left\{\frac{z_2 f'(z_2)}{z_1 f'(z_1)}\right\} > -\pi. \blacksquare$$

7.3 γ-Spiral Functions

Let $f(z) \in S$ be holomorphic and univalent in the open unit disk E with $f(0) = 0$ and $f'(0) = 1$. Then f belongs to the subclass $S(\beta,\gamma) \subset S$ if and only if there exist real numbers β and γ, $0 \le \beta < 1$, $|\gamma| < \pi/2$, such that

$$\Re\left\{e^{i\gamma} \frac{zf'(z)}{f(z)}\right\} > \beta\cos\gamma, \quad z \in E. \tag{7.3.1}$$

This definition is a generalization of Špaček's condition on spirallike functions in E (see Špaček [1933]). For more details, see Kulshrestha [1973b].

If $f \in S(\beta,\gamma)$, we have, with appropriate normalizing factors,

$$\sec\gamma\left[e^{i\gamma} \frac{zf'(z)}{f(z)} - \beta\cos\gamma - \sin\gamma\right]_{z=0} = 1 - \beta, \tag{7.3.2}$$

which enables us to express members of the subclass $\mathcal{S}(\beta, \gamma)$ in terms of functions in the class P. Thus, $f \in \mathcal{S}(\beta, \gamma)$ if and only if there exists a function p with positive real part, such that

$$\frac{zf'(z)}{f(z)} = \frac{(1-\beta)\cos\gamma p(z) + \beta\cos\gamma + i\sin\gamma}{\cos\gamma + i\sin\gamma}, \quad z \in E. \tag{7.3.3}$$

By introducing a complex number $h = (\beta + i\tan\gamma)/(1 - \beta)$, the condition (7.3.3) can be written as

$$\frac{zf'(z)}{f(z)} = \frac{p(z) + h}{1 + h}, \quad z \in E, \tag{7.3.4}$$

which after differentiation gives

$$1 + \frac{zf''(z)}{f'(z)} = \frac{zp'(z)}{p(z) + h} + \frac{p(z) + h}{1 + h}, \quad z \in E. \tag{7.3.5}$$

7.4 Generalized Convexity

The concept of generalized convexity was introduced by Moçanu [1969] for a particular subclass of functions in $\mathcal{S}$. The class of *generalized α-convex* functions in E is defined as follows: Let f be in the class $\mathcal{S}$, $z \in E$. Then f is a generalized α-convex function in E, i.e., $f \in \mathcal{S}(\alpha, \beta, \gamma)$, if and only if f satisfies (7.3.1) and there exist real numbers $\alpha \geq 0$ such that for some $p \in P$,

$$(1-\alpha)\frac{zf'(z)}{f(z)} + \alpha\left(1 + \frac{zf''(z)}{f'(z)}\right) = \frac{p(z) + h}{1 + h} + \alpha\frac{zp'(z)}{p(z_+h)}, \tag{7.4.1}$$

where h is a complex number defined in (7.3.4). This defining property follows from (7.3.3) and (7.3.4). These functions are called *α-convex* if they are holomophic in E for $0 \leq \alpha \leq 1$ and satisfy there the conditions that $f(z)f'(z)/z \neq 0$ and

$$\Re\left\{(1-\alpha)\frac{zf'(z)}{f(z)} + \alpha\left[1 + \frac{zf''(z)}{f'(z)}\right]\right\} \geq 0, \quad z \in E. \tag{7.4.2}$$

In this case we say that the functions f belong to the subclass $\mathcal{S}(\alpha, \beta, \gamma)$ for $z \in E$. As shown in Moçanu [1969], these functions are starlike in E for $\alpha \geq 0$ and convex for $\alpha \geq 1$.

Another generalization of the Moçanu condition (7.4.2) is obtained by replacing it by

$$\Re\left\{e^{i\gamma}\left[(1-\alpha)\frac{zf'(z)}{f(z)} + \alpha\left\{1 + \frac{zf''(z)}{f'(z)}\right\}\right]\right\} > \beta\cos\gamma, \quad z \in E, \ \alpha \geq 0. \tag{7.4.3}$$

The property (7.4.3) then defines a class of generalized convex functions which leads to the Bazilevich functions (see Bazilevich [1955]), although the Bazilevich class of functions differs in general from the class $\mathcal{S}(\alpha, \beta, \gamma)$ defined above.

We will discuss mapping properties of functions of the class $\mathcal{S}(\alpha, \beta, \gamma)$, especially those related to their extremum, distortion, and rotation theorems, which reduce to well known results for particular values of the parameters α, β and γ. The following seven theorems hold (all in Kulshrestha [1973b]).

Theorem 7.4.1. (Kulshrestha [1973b]) *Let $f \in \mathcal{S}(\alpha, \beta, \gamma)$. Then the set of all possible values of $\log(f(z)/z)$ for a fixed z, $|z| \leq r < 1$, lies in the convex image of $|z| = r$ under the mapping*

$$w = \log\left\{(1+h)^\alpha (1-z)^{-2/(1+h)}\right\}. \tag{7.4.4}$$

PROOF. The condition (7.4.2) when integrated from 0 to z gives

$$\log\left[\left(\frac{f(z)}{z}\right)^{1-\alpha} (f'(z))^\alpha\right] = \frac{1}{1+h} \int_0^z \frac{p(x)-1}{x}\, dx + \alpha \log\left[p(z) + h\right],$$

which, in view of the relation $p(z) + h = (1+h)\dfrac{zf'(z)}{f(z)}$, becomes

$$\log\left[\frac{f(z)}{(1+h)^\alpha z}\right] = \frac{1}{1+h} \int_0^z \frac{p(x)-1}{x}\, dx. \tag{7.4.5}$$

Using the following Herglotz representation (§2.7.1) of functions $p(z)$, which satisfies (7.4.2),

$$p(z) = \int_{-\pi}^{\pi} \frac{1+ze^{it}}{1-ze^{it}}\, dm(t), \quad t \in [-\pi, \pi], \tag{7.4.6}$$

where $m(t)$ is a function of bounded variation with $\int_{-\pi}^{\pi} dm(t) = 1$, we find from (7.4.5) that

$$\log\frac{f(z)}{z} - \int_{-\pi}^{\pi} \log\left[(1+h)^\alpha (1-ze^{it})^{-2/(1+h)}\right] dm(t). \tag{7.4.7}$$

Let $q(z,t) = \log\left[(1+h)^\alpha (1-ze^{it})^{-2/(1+h)}\right]$, $t \in [\pi, \pi]$. Then, since

$$1 + \Re\left\{\frac{zq''(z,t)}{q'(z,t)}\right\} = \Re\left\{\frac{1}{1-ze^{it}}\right\} > \frac{1}{2},$$

the function $q(z,t)$ maps the disk $|z| \leq r < 1$ univalently onto a convex domain E^* which is independent of t. Thus, the relation (7.4.7) means that the points $\log\left(f(z)/z\right)$, for a fixed z, $|z| \le r < 1$, lie in the convex hull of E^*, and therefore, they lie in the image of $|z| \le r$ under the mapping

$$\log\left[(1+h)^\alpha(1-\nu z)^{-2/(1+h)}\right], \quad |\nu| = 1,$$

which yields (7.4.4). ∎

Corollary 7.4.1. The extremal function $f_0(z) \in \mathcal{S}(\alpha,\beta,\gamma)$ has the representation of the form

$$f_0(z) = (1+h)^\alpha z(1-z)^{-2/(1+h)}$$

$$= (1+h)^\alpha\left[z + \sum_{n=2}^{\infty}\prod_{m=0}^{n-2}\frac{\left(m+\dfrac{2}{1+h}\right)}{m+1}z^n\right]. \tag{7.4.8}$$

PROOF. The extremal function is given by

$$\log\frac{f_0(z)}{z} = \log\left[(1+h)^\alpha(1-z)^{-2/(1+h)}\right], \quad |z| = 1,$$

which, in view of (7.4.7), gives (7.4.8). ∎

By evaluating the extrema of the extremal function (7.4.8) for $z = re^{it}$, $|z|\log r < 1$, we get the following two theorems.

Theorem 7.4.2. (Kulshrestha [1973b]) If $f(z) \in \mathcal{S}(\alpha,\beta,\gamma)$, then for $|z| \le r < 1$

$$T(r,\theta_1,\alpha,\beta,\gamma) \le \log\left|\frac{f(z)}{z}\right| \le T(r,\theta_2,\alpha,\beta,\gamma) \quad \text{for } \gamma \ne 0, \tag{7.4.9}$$

where

$$T(r,\theta,\alpha,\beta,\gamma) = (1-\beta)\cos\gamma\left[2\sin\gamma\arctan\frac{r\sin\theta}{1-r\cos\theta}\right.$$

$$\left. - \cos\gamma\log\left(1 - 2r\cos\theta + r^2\right)\right] - \frac{\alpha}{2}\log\left[(1-\beta)\cos\gamma\right],$$

$$\theta_{1,2} = 2\arctan\left\{\frac{-\cot\gamma \mp \sqrt{\csc^2\gamma - r^2}}{1+r}\right\}.$$

For $\gamma = 0$, we have

$$\log\left|\frac{f(z)}{z}\right| \le \frac{\alpha}{2}\log\frac{1}{1-\beta} - 2(1-\beta)\log(1-r). \tag{7.4.10}$$

Theorem 7.4.3. (Kulshrestha [1973b]) If $f \in \mathcal{S}(\alpha,\beta,\gamma)$, then for $|z| = r < 1$

$$T(r,\theta_3,\alpha,\beta,\gamma) \le \arg\left\{\frac{f(z)}{z}\right\} \le T(r,\theta_4,\alpha,\beta,\gamma), \tag{7.4.11}$$

where

$$T(r, \theta, \alpha, \beta, \gamma) = (1 - \beta) \cos \gamma \left[2 \cos \gamma \arctan \frac{r \sin \theta}{1 - r \cos \theta} \right.$$
$$\left. + \sin \gamma \log \left(1 - 2r \cos \theta + r^2 \right) \right] + \alpha \gamma,$$

$$\theta_{3,4} = 2 \arctan \left\{ \frac{\tan \gamma \mp \sqrt{\sec^2 \gamma - r^2}}{1 + r} \right\}.$$

Theorem 7.4.4. (Kulshrestha [1973b]) *Let*

$$\zeta(z) = (1 - \alpha) \frac{z f'(z)}{f(z)} + \alpha \left[1 + \frac{z f''(z)}{f'(z)} \right].$$

Then the images of all circles $|z| \leq r < 1$ under mapping $\zeta(z)$ lie inside the circle $|\zeta - \zeta_0| \leq \rho$, where

$$\zeta_0 = \frac{1 - r^2 + 2(1 - \beta)r^2 \cos^2 \gamma + i(1 - \beta)r^2 \sin 2\gamma}{1 - r^2}$$
$$- \frac{8\alpha\beta(1 - \beta^2)r^2 \cos^4 \gamma \left[1 - 2\beta + 2\beta^2 \right) \cos^2 \gamma - i(1 - 2\beta) \sin \gamma \cos \gamma \right]}{\left(1 - r^2 \right) \left(1 - r^2 + 2\alpha r^2 \cos^2 \gamma \right) \left(1 - 4\beta(1 - \beta) \cos^2 \gamma \right)^2},$$

$$\rho = \frac{2(1 - \beta)r \cos \gamma}{1 - r^2} - \frac{4\alpha r(1 - \beta)^2 \cos^2 \gamma}{\sqrt{1 - 4\beta(1 - \beta) \cos^2 \gamma}}$$
$$\times \left\{ \frac{1}{1 - r^2} + \frac{1}{\sqrt{1 - 4\beta(1 - \beta) \cos^2 \gamma} \left(1 - r^2 + 2\beta r^2 \cos^2 \gamma \right)} \right\}.$$

PROOF. The function $\zeta(z)$ can be written as

$$\zeta(z) = \frac{p(z) + h}{1 + h} + \frac{\alpha z p'(z)}{p(z) + h}.$$

Since $\zeta(z)$ is subordinate to the function $\zeta^*(z)$, we find that

$$\zeta^*(z) = \frac{1}{\mu} \frac{z + \mu}{1 - z} - \frac{4\alpha}{(1 - h)^2} \left[\frac{1}{z - 1} + \frac{\mu}{z + \mu} \right], \quad \mu = \frac{1 + h}{1 - h},$$

and the results of the theorem follow by using superposition of linear mappings. ∎

Corollary 7.4.2. For $\alpha = 0$, we find from Theorem 7.4.4 that for $|z| \leq r < 1$

$$\left| \arg \left\{ \frac{z f'(z)}{f(z)} \right\} \right| \leq \arctan \frac{(1 - \beta)r^2 \sin 2\gamma}{1 - r^2 + 2(1 - \beta)r^2 \cos^2 \gamma}$$
$$+ \arctan \frac{2(1 - \beta)r \cos \gamma}{\sqrt{(1 - r^2) \left[1 - r^2 + 4\beta(1 - \beta)r^2 \cos^2 \gamma \right]}}, \tag{7.4.12}$$

where equality holds for $p(z) = \dfrac{1+z}{1-z} \in P$.

Lemma 7.4.1. *Let $f_1(t)$ and $f_2(t)$ be single-valued continuous complex functions of a real variable $t \in [-\pi, \pi]$ and let $m(t)$ be the function defined (7.4.7). If $\Re\{f_2(t)\} > 0$, then*

$$\left| \frac{\int_{-\pi}^{\pi} f_1(t)\, dm(t)}{\int_{-\pi}^{\pi} f_2(t)\, dm(t)} \right| \leq \max_{t \in [-\pi, \pi]} \frac{|f_1(t)|}{\Re\{f_2(t)\}}.$$

Proof of this lemma can be found in any standard textbook on real and complex analysis.

Theorem 7.4.5. (Kulshrestha [1973b]) *Let $\zeta(z)$ be defined as in Theorem 7.4.4. Then for $|z| \leq r < 1$*

$$|\zeta(z)| \leq 1 + \frac{2(1-\beta)r\cos\gamma}{1-r}\left\{1 + \frac{\alpha}{1-r+2(1-\beta)r\cos^2\gamma}\right\}. \qquad (7.4.13)$$

Equality holds in both Theorems 7.4.4 and 7.4.5 for the function

$$\zeta^*(z) = \frac{p^*(z)+h}{1+h} + \frac{\alpha z p^{*'}(z)}{p^*(z)+h}, \quad p^*(z) = \frac{1+z}{1-z} \in P. \qquad (7.4.14)$$

PROOF. Using (7.4.6), the function $\zeta()$ can be written as

$$\zeta(z) = \int_{-\pi}^{\pi} \left\{1 + \frac{2ze^{it}}{(1+h)(1-ze^{it})}\right\} dm(t)$$

$$+ \frac{\alpha}{1+h} \frac{\displaystyle\int_{-\pi}^{\pi} \frac{2ze^{it}}{(1+h)(1-ze^{it})}\, dm(t)}{\displaystyle\int_{-\pi}^{\pi} \left\{1 + \frac{2ze^{it}}{(1+h)(1-ze^{it})}\right\} dm(t)}. \qquad (7.4.15)$$

Then (7.4.13) follows from (7.4.1) by using Lemma 7.4.1. ∎

A consequence of Theorem 7.4.5 is the following result.

Theorem 7.4.6. (Kulshrestha [1973b]) *Let*

$$F(z) = \log\left\{\left(\frac{zf'(z)}{f(z)}\right)^{\alpha} \frac{f(z)}{z}\right\}.$$

Then for $|z| \leq r < 1$,

$$\Re\{1 - \alpha + zF'(z)\} = \Re\left\{(1-\alpha)\frac{zf'(z)}{f(z)} + \alpha \frac{zf''(z)}{f'(z)}\right\}$$

$$\leq 1 - \alpha + \frac{2(1-\beta)r\cos\gamma}{1-r}\left\{1 + \frac{\alpha}{1-r+2(1-\beta)r\cos^2\gamma}\right\}. \qquad (7.4.16)$$

Corollary 7.4.3. Denoting the right-hand side of (7.4.16) by $\psi(r, \alpha, \gamma)$, we have for $|z| \leq r < 1$

$$1 - \psi(r, \alpha, \gamma) \leq 1 + \Re\{1 - \alpha + zF'(z)\} \leq 1 + \psi(r, \alpha, \gamma). \tag{7.4.17}$$

These inequalities are sharp for

$$r = r_0 \leq [\alpha + (1 + \alpha)(1 - \beta) \cos \gamma - \alpha(1 - \beta) \cos^2 \gamma + \{2\alpha^2(1 - \beta) \cos \gamma$$
$$+ (1 + \alpha^2)(1 - \beta)^2 \cos^2 \gamma + 2\alpha(1 - \alpha)(1 - \beta)^2 \cos \gamma$$
$$+ \alpha^2 \left(1 - \beta^2 \cos^4 \gamma\right)^{1/2}]^{-1}. \tag{7.4.18}$$

PROOF. The inequalities (7.4.17) are obvious. The bound (7.4.18) follows by considering the roots of $1 - \psi(r, \alpha, \gamma) = 0$. ∎

7.4.1 Particular Cases. Note that the subclass $\mathcal{S}(0, \beta, \gamma)$ is the class of spirallike functions of order β, while $\mathcal{S}(0, \beta, 0)$ is the class of starlike functions of order β and $\mathcal{S}(0, 0, 0)$ is the class $\mathcal{S}^\star$ of starlike functions in the unit disk E. Fir these classes the following results hold.

Corollary 7.4.4. If $f \in \mathcal{S}(0, \beta, \gamma)$, the mapping function in Theorem 7.4.1 reduces to $w = \log(1 - z)^{-2c}$, where $c = (1 - \beta)/(1 + i \tan \gamma)$. If $f \in \mathcal{S}(0, \beta, 0)$, this mapping function becomes $w = \log(1 - z)^{-2(1-\beta)}$. If $f \in \mathcal{S}(0, 0, 0) \equiv \mathcal{S}^\star$, the mapping function becomes $w = \log(1 - z)^{-2}$, as found in Marx [1932-33] and Strohhäcker [1923].

Corollary 7.4.5. If $f \in \mathcal{S}(0, 0, 0) \equiv \mathcal{S}^\star$, then Theorems 7.4.2 and 7.4.3 reduce to the following bounds for starlike functions (see Nehari [1952]):

$$\frac{r}{1 + r^2} \leq |f(z)| \leq \frac{r}{1 - r^2}, \quad \text{and} \quad \left|\arg\left\{\frac{f(z)}{z}\right\}\right| \leq 2 \arcsin r, .$$

If $f \in \mathcal{S}(0, \beta, 0)$, then from (7.4.12) and (7.4.13) we get

$$\left|\arg\left\{\frac{zf'(z)}{f(z)}\right\}\right| \leq \arctan \frac{2(1 - \beta)r}{\sqrt{(1 - r^2)(1 - r^2 + 4\beta(1 - \beta)r^2)}}, \tag{7.4.19}$$

$$\left|\frac{zf'(z)}{f(z)}\right| \leq 1 + \frac{2(1 - \beta)r}{1 - r}. \tag{7.4.20}$$

If $f \in \mathcal{S}(0, 0, 0)$, we obtain the well-known results from (7.4.19) and (7.4.20) for the starlike functions in E (see Nehari [1952]).

Using the method developed in Clunie [1959] and applied in Libera [1967], it is easy to find the coefficient bounds for functions $f \in \mathcal{S}(0, \beta, \gamma)$, as presented in the following theorem.

Theorem 7.4.7. (Kulshrestha [1973b]) *If* $f \in \mathcal{S}(0, \beta, \gamma)$ *and* $f(z) = z + \sum\limits_{n=2}^{\infty} a_n z^n$, $z \in E$, *then*

$$|a_n| \leq \prod_{m=0}^{n-2} \frac{\left| m + \dfrac{2}{1+h} \right|}{1+h}, \quad n = 2, 3, \ldots, \tag{7.4.21}$$

where h *is defined in (7.4.1).*

The bounds in (7.4.21) are sharp and equality holds for the extremal function (7.4.8) with $\alpha = 0$. The estimate (7.4.21) and the γ-radius of $f \in \mathcal{S}(0, \beta, \gamma)$, $|\gamma| < \pi/2$, are the same as in Libera [1967].

Corollary 7.4.6. If $f \in \mathcal{S}(1, \beta, \gamma)$ which is the class of convex functions of order β, we find from (7.4.16) that

$$\left| \frac{zf''(z)}{f'(z)} \right| \leq \frac{2(1-\beta)r \cos \gamma}{1-r} \left\{ 1 + \frac{1}{1 - r + 2(1-\beta)r \cos^2 \gamma} \right\}. \tag{7.4.22}$$

This result is sharp for

$$r = r_0 \geq \left[1 + (1-\beta)(2\cos\gamma - \cos^2\gamma) + \{2(1-\beta)\cos\gamma + 4(1-\beta)^2 \cos^2\gamma \right.$$
$$\left. + (1-\beta)^2 \cos^4 \gamma \}^{1/2} \right]^{-1}. \tag{7.4.23}$$

If $f \in \mathcal{S}(1, \beta, 0)$, then the estimate (7.4.22) becomes

$$\left| \frac{zf''(z)}{f'(z)} \right| \leq \frac{2(1-\beta)r}{1-r} \left\{ 1 + \frac{1}{1 + r - 2\beta r} \right\}, \tag{7.4.24}$$

which holds for $r_0 \geq \left[2 - \beta + \sqrt{5\beta^2 - 12\beta + 7} \right]^{-1}$. If $f \in \mathcal{S}(1, 0, 0)$, we get

$$\left| \frac{zf''(z)}{f'(z)} \right| \leq \frac{4r + 2r^2}{1 - r^2}, \tag{7.4.25}$$

which holds for $r > \dfrac{1}{2 + \sqrt{7}} = 0.213\ldots$, as in Nehari [1952].

7.5 Alpha-Convex Functions

A normalized function $f \in \mathcal{S}$ is said to be an α-*convex function*, $\alpha \geq 0$, if the Moçanu angle $\vartheta = (1-\alpha)\phi + \alpha\psi$ is an increasing function of θ for fixed values of r and $\partial\vartheta/\partial\theta \geq 0$ for a given α (see Moçanu [1969]). The class of all α-convex functions is denoted by $\mathcal{M}(\alpha)$, which is equivalent to the class $\mathcal{S}(\alpha, 0, 0)$. Let

$$J(\alpha, f(z)) := (1-\alpha)\frac{zf'(z)}{f(z)} + \alpha\left[1 + \frac{zf''(z)}{f'(z)} \right].$$

Then $f \in \mathcal{M}(\alpha)$ if and only if $\Re\{J(\alpha, f(z))\} \geq 0$ for all $z \in E$. Obviously, $\mathcal{M}(0) = \mathcal{S}^{\star}$ and $\mathcal{M}(1) = \mathcal{K}$. Moreover, $\mathcal{M}(\alpha_1) \subset \mathcal{M}(\alpha_2)$ for $\alpha_1 \geq \alpha_2$. and hence, all members of $\mathcal{M}(\alpha)$ are starlike for $\alpha \geq 0$ and convex for $\alpha \geq 1$ (Moçanu [1969] and Miller et al. [1973]).

Let $\mathcal{B}(1/\alpha)$, $\alpha > 0$ denote the subclass of Bazilevich functions (Bazilevich [1955]) consisting of univalent functions which can be represented in the form

$$f(z) = \left[\frac{1}{\alpha} \int_0^z [g(\zeta)]^{1/\alpha} \zeta^{-1} \, d\zeta \right]^{\alpha}, \quad z \in E, \tag{7.5.1}$$

where $g \in \mathcal{S}^{\star}$, $\alpha > 0$, and the powers appearing in (7.5.1) as well as elsewhere are treated as principal values. Let $\mathcal{P}$ denote the class of functions p regular in E and such that $p(0) = 0$ and $\Re\{p(z)\} > 0$ for $z \in E$. It is known that $f \in \mathcal{S}^{\star}$ if and only if $zf'(z)/f(z) \in \mathcal{P}$. Likewise, functions in the class $\mathcal{M}(\alpha)$ have a representation in terms of the members of $\mathcal{P}$ which, in view of the Herglotz theorem (§2.7.1), possess the following integral representation: if $f \in \mathcal{S}^{\star}$, then there exists a positive measure $\mu(t)$, $-\infty < t < \infty$, such that $\int_{-\pi}^{\pi} d\mu(t) = 1$, and

$$\frac{zf'(z)}{f(z)} = p(z) = \int_{-\pi}^{\pi} \frac{e^{it} + z}{e^{it} - z} \, d\mu(t), \quad z \in E. \tag{7.5.2}$$

Lemma 7.5.1. it If $f \in \mathcal{M}(\alpha)$ and if we choose the branch of $\left(\frac{zf'(z)}{f(z)} \right)^{\alpha}$ which is 1 at $z = 0$ for $\alpha > 0$, then

$$g(z) = f(z) \left(\frac{zf'(z)}{f(z)} \right)^{\alpha}, \quad z \in E$$

belongs to the class $\mathcal{S}^{\star}$.

PROOF. An easy calculation shows that $\Re\left\{ \frac{zg'(z)}{g(z)} \right\} = \Re\{J(\alpha, f(z))\} \geq 0$. The result follows since $g(0) = 0 = g'(0) - 1$. ∎

Theorem 7.5.1. (Kulshrestha [1974a,b]) $f \in \mathcal{M}(\alpha)$ *if and only if* $f \in \mathcal{B}(1/\alpha)$.

PROOF. The necessity part follows directly from (7.5.1), which on differentiation gives

$$J(\alpha, f(z)) = \frac{zg'(z)}{g(z)}, \quad g \in \mathcal{S}^{\star}.$$

To prove sufficiency, the condition $\Re\{J(\alpha, f(z))\} \geq 0$, $z \in E$, can be written in terms of the members of $\mathcal{P}$ as $J(\alpha, f(z)) = p(z)$, which on integration from 0 to z gives

$$\log \left[\frac{f(z)}{z} \left(\frac{zf'(z)}{f(z)} \right)^{\alpha} \right] = \int_0^z \frac{p(x) - 1}{x} \, dx. \tag{7.5.3}$$

Using the representation (7.5.2) in (7.5.3) we get

$$\log\left[\frac{f(z)}{z}\left(\frac{zf'(z)}{f(z)}\right)^{\alpha}\right] = 2\int_{-\pi}^{\pi}\log\left(1 - z\,e^{-it}\right)^{-1} d\mu(t). \qquad (7.5.4)$$

According to Pflatzgraft [1971], the principal values og $\log\left[\frac{f(z)}{z}\left(\frac{zf'(z)}{f(z)}\right)^{\alpha}\right]$ lie, for each $z \in E$, in a closed convex domain bounded by the curve $\Gamma_\rho = \{\log\left(1 - ze^{-it}\right)^{-1}: -\pi \le t < \pi, 0 < \rho < 1\}$ under the mapping $\log\left(1 - \varepsilon z\right)^{-2}$, $|\varepsilon| = 1$. Let $\log\left(f(z)/z\right)$, $log\left(zf'(z)/f(z)\right)$ and $\log(1-z)$ be regular in E and possess branches which have the value 0 at $z = 0$. Assuming that the measure $\mu(t)$ has value 0 everywhere on Γ_ρ except at one point where it has value 1, we get from (7.5.4)

$$\log\left[\frac{f(z)}{z}\left(\frac{zf'(z)}{f(z)}\right)^{\alpha}\right] = 2\log\left(1 - \varepsilon z\right)^{-1}, \quad |\varepsilon| = 1. \qquad (7.5.5)$$

Since $g(z) = z(1 - \varepsilon z)^{-2} \in S^\star$, Eq (7.5.5) is justified in view of Lemma 7.5.1. Set $w = \log(f(z)/z)$ in (7.5.5), and note that $zf'(z)/f(z) = 1 + zdw/dz$. Then Eq (7.5.5) becomes

$$w + \alpha\log\left(1 + z\frac{dw}{dz}\right) = \log(1 - \varepsilon z)^2, \quad |\varepsilon| = 1. \qquad (7.5.6)$$

A formal solution of the differential equation (7.5.6) with the initial condition $f(0) = 0$ is

$$f(z) = \left[\frac{1}{\alpha}\int_0^z \zeta^{1/\alpha - 1}(1 - \varepsilon\zeta)^{-2/\alpha}\,d\zeta\right]^{-1}, \quad |\varepsilon| = 1. \qquad (7.5.7)$$

The function $f(z)$ defined by (7.5.7) is well-defined and regular in E and that $f'(0) = 1$. Since the function $z(1 - \varepsilon z)^{-2} \in S^\star$, it is clear from (7.5.1) that the function f defined by (7.5.7) belongs to the class $\mathcal{B}(1/\alpha)$. ∎

Taking the principal values of the powers in (7.5.7), we can write

$$f_*(z) = zH(z), \qquad (7.5.8)$$

where

$$H(z) = \left\{1 + \sum_{n=1}^{\infty} c_n z^n\right\}^{\alpha}, \qquad (7.5.9)$$

$$c_n = \frac{1}{n!\,\alpha^n(1 + n\alpha)}\prod_{k=0}^{n-1}(2 + k\alpha), \qquad (7.5.10)$$

Then if $f(z) = z + \sum\limits_{\nu=2}^{\infty} a_\nu z^\nu \in \mathcal{M}(\alpha)$, it is easy to see that $|a_{n+1}| \le \dfrac{H^{(n)}(0)}{n!}$, $n = 1, 2, \ldots$.

Theorem 7.5.2. (Kulshrestha [1974]) *Let* $f(z) = z + \sum\limits_{\nu=2}^{\infty} a_\nu z^\nu \in \mathcal{M}(\alpha)$. *Let* $S(n)$ *be the set of all n-tuples* $(x_1, x_2, \cdots, x_n)$ *of non-negative integers for which* $\sum\limits_{j=1}^{n} j x_j = n$, *and for each such n-tuple define q by* $\sum\limits_{j=1}^{n} x_j = q$. *If* $\gamma(\alpha, q) = \alpha(\alpha - 1) \cdots (\alpha - q)$ *with* $\gamma(\alpha, 0) = \alpha$, *then for* $n = 1, 2, \ldots$,

$$|a_{n+1} \le \sum \frac{\gamma(\alpha, q-1)\, c_1^{x_1} c_2^{x_2} \cdots c_n^{x_n}}{x_1!\, x_2! \cdots x_n!}, \tag{7.5.11}$$

where summation is carried over all n-tuples in $S(n)$, *and* c_n *are defined by* (7.5.10).

The proof is given as Exercise 7.6.4. The technique of Goodman [1972] was used to get the bounds in (7.5.11) in a compact form. These bounds are sharp and for $\alpha > 0$ attained by the function $f_*(z)$ defined by (7.5.7). For $\alpha = 0$, we find from (7.5.11) that $|a_n| \le n$ for $n = 2, 3, \ldots$ and the bounds are attained by $f(z) = z(1-z)^{-2}$. For $\alpha = 1$, we have $|a_n| \le 1$ for $n = 2, 3, \ldots$, the bound being attained by $f(z) = z(1-z)^{-1}$.

The formula (7.5.11) is readily computable. Thus, we have, e.g.,

$$|a_2| \le \frac{2}{1+\alpha},$$

$$|a_3| \le \frac{3 + 8\alpha + \alpha^2}{(1+\alpha)^2(1+2\alpha)},$$

$$|a_4| \le \frac{4(3 + 19\alpha + 38\alpha^2 + 11\alpha^3 + \alpha^4)}{3(1+\alpha)^3(1+2\alpha)(1+3\alpha)},$$

$$|a_5| \le \frac{30 + 394\alpha + 2024\alpha^2 + 5284\alpha^3 + 6386\alpha^4 + 2638\alpha^5 + 488\alpha^6 + 36\alpha^7}{6(1+\alpha)^4(1+2\alpha)^2(1+3\alpha)(1+4\alpha)},$$

and so on. Moreover, $\sup|a_{n+1}| < \sup|a_n|$ for $\alpha \ge 2$, $n = 2, 3, \ldots$; and for a given n $(n = 2, 3, \ldots)$, $\sup|a_n|$ is a decreasing function of α, $\alpha \ge 0$.

7.5.1 Radius of α-Convexity. The *radius of α-convexity* for the function f is defined by

$$r_\alpha(f) = \sup\{r : \Re\{J(\alpha, f; z) > 0, |z| \le r\}. \tag{7.5.12}$$

The following results hold.

Theorem 7.5.2. (Cernikov [1972]) *If* $\coth(1) - 1 \approx 0.313 \le \alpha \le 1$, *then* $r_\alpha(\mathcal{S}) = 1 + \alpha - \sqrt{\alpha(\alpha + 2)}$.

Miller et al. [1974] proved that the above result also holds for $\alpha > 1$. Another result is:

Theorem 7.5.3. (Miller et al. [1974]) *The radii of α-convexity for the class $\mathcal{S}^\star$ are*

$$r_\alpha(\mathcal{S}^\star) = \begin{cases} 1 + \alpha - \sqrt{\alpha(\alpha+2)} & \text{if } \alpha \geq 0, \\[2mm] \sqrt{\dfrac{2 - \sqrt{-\alpha}}{2 + \sqrt{-\alpha}}} & \text{if } -3 < \alpha < 0, \\[2mm] -(1+\alpha) - \sqrt{\alpha(\alpha+2)} & \text{if } \alpha \leq -3. \end{cases} \tag{7.5.13}$$

Let $\mathcal{M}_\alpha$ denote the class of univalent functions f in E with $f(0) = 0$ and $f'(0) = 1$ and α-convex in E such that $\Re\{J(\alpha, f; z)\} > 0$ for $z \in E$. Note that $\mathcal{M}_0 = \mathcal{S}^\star$ and $\mathcal{M}_1 = \mathcal{K}$.

Let $p(z) = \dfrac{zf'(z)}{f(z)}$. Then $J(\alpha, f; z) = p(z) + \alpha \dfrac{zp'(z)}{p(z)}$, and thus (e.2) can be written as

$$\Re\left\{p(z) + \alpha \frac{zp'(z)}{p(z)}\right\} > 0, \quad z \in E, \tag{7.5.14}$$

or

$$p(z) + \alpha \frac{zp'(z)}{p(z)} \prec \frac{1+z}{1-z}. \tag{7.5.15}$$

If the condition (7.5.14) holds, then $p(z) = \dfrac{zf'(z)}{f(z)}$ is analytic in E and $p(z) \neq 0$ for $z \in E$, which implies that the condition $\dfrac{f(z)f'(z)}{z} \neq 0$ for $z \in E$ also holds.

Theorem 7.5.4. (Moçanu et al. [1999]) *For $\alpha, \beta \in \mathbb{R}$ such that $0 \leq \beta/\alpha \leq 1$, we have $\mathcal{M}_\alpha \subset \mathcal{M}_\beta$.*

Corollary 7.5.1. *For all $\alpha \in [0, 1]$, we have $\mathcal{K} \subset \mathcal{M}_\alpha \subset \mathcal{S}^\star$.*

Theorem 7.5.5. (Moçanu et al. [1999]) *If $\alpha > 0$, then $f \in \mathcal{M}_\alpha$ if and only if $F(z) \equiv \left(\dfrac{zf'(z)}{f(z)}\right)^\alpha \in \mathcal{S}^\star$.*

This theorem leads to the following result:

Theorem 7.5.6. (Moçanu et al. [1999]) *If $\alpha > 0$, then $f \in \mathcal{M}_\alpha$ if and only if there exists a function $F \in \mathcal{S}^\star$ such that*

$$f(z) = \left[\frac{1}{\alpha} \int_0^z \frac{F^{1/\alpha}(\zeta)}{\zeta} d\zeta\right]^\alpha, \quad z \in E. \tag{7.5.16}$$

A function $f \in \mathcal{M}_\alpha$ is called α-convex of order γ, $0 \leq \gamma < 1$, if

$$\Re\{J(\alpha, f; z)\} > \gamma, \quad z \in E. \tag{7.5.17}$$

All functions that satisfy (7.5.17) will be denoted by $\mathcal{M}_\alpha(\gamma)$. Details about subclasses of α-convex can be found in Acu [2008 : ch. 3].

7.6 Exercises

7.6.1 Consider a function p in the class $\mathcal{P}$, for which $\Re p(z) > 0$, and $p(0) = 0$. Then $\dfrac{1+z}{1-z} \in \mathcal{P}$. By the Schwarz lemma, if w is a regular function such that $|w(z)| < 1$, $w(0) = 0$, then $|w(z)| \leq |z|$ for all $z \in E$. Also,

$$|1 + w(z)| \leq 1 + |w(z)| \leq 1 + |z|,$$
$$|1 - w(z)| \geqq |1 - |w(z)|| = 1 - |w(z)| \geqq 1 - |z|,$$

so that

$$\frac{1}{|1 - w(z)|} \leq \frac{1}{1 - |z|}, \quad \text{and} \quad \left|\frac{1 + w(z)}{1 - w(z)}\right| \leq \left|\frac{1+z}{1-z}\right|.$$

Since $f \in \mathcal{S}^\star$, we have $|Re\left\{\dfrac{zf'(z)}{f(z)}\right\} > 0$, where $\dfrac{zf'(z)}{f(z)} \in \mathcal{P}$. Hence,

$$\frac{zf'(z)}{f(z)} = \frac{1 + w(z)}{1 - w(z)} \in \mathcal{P}.$$

7.6.2. A comparison between the properties of univalent functions f in the class $\mathcal{S}^\star$ and $\mathcal{K}$ is provided in Table 7.6.1.

Table 7.6.1 Properties of f in $\mathcal{S}^\star$ and $\mathcal{K}$.

If $f \in \mathcal{S}^\star$, then	If $f \in \mathcal{K}$, then
$\Re\left\{\dfrac{zf'(z)}{f(z)}\right\} > 0$	$1 + \Re\left\{\dfrac{re^{i\theta} f''\left(re^{i\theta}\right)}{f'\left(re^{i\theta}\right)}\right\} > 0$
$\arg\left\{f\left(re^{i\theta}\right)\right\} \uparrow$	$\arg\left\{T(r,\theta)\right\} = \dfrac{\pi}{2} + \theta + \arg\left\{f'\left(re^{i\theta}\right)\right\}$
$\dfrac{\partial}{\partial \theta} \arg\left\{f\left(re^{i\theta}\right)\right\} > 0$	$\dfrac{\partial}{\partial \theta} \arg\left\{T(r,\theta)\right\} > 0$
$\dfrac{\partial}{\partial \theta} \Im\left\{\log f\left(re^{i\theta}\right)\right\} > 0$	$\dfrac{\partial}{\partial \theta} \left\{\dfrac{\pi}{2} + \theta + \arg\left\{f'\left(re^{i\theta}\right)\right\}\right\} > 0$
$\Im\left\{\dfrac{\partial}{\partial \theta} \log f\left(re^{i\theta}\right)\right\} > 0$	$1 + \dfrac{\partial}{\partial \theta} \arg\left\{f'\left(re^{i\theta}\right)\right\} > 0$
$\Im\left\{\dfrac{re^{i\theta} f'\left(re^{i\theta}\right)}{f\left(re^{i\theta}\right)}\right\} > 0$	$1 + \Im\left\{\dfrac{\partial}{\partial \theta} \log f'\left(re^{i\theta}\right)\right\} > 0$

7.6.3. Let $\alpha \in \mathbb{R}$ and $\beta \geq 1$. Then prove that $\mathcal{M}_{\alpha,\beta} \subset \mathcal{S}^\star$. This proof is due to Oros and Oros [2008]. To prove, we will need the following lemma:

Lemma 7.6.1. Let $\psi : \mathbb{C}^2 \mapsto \mathbb{C}$ satisfy the condition $\Re\{\psi(is, \sigma)\} \leq 0$, $z \in E$, for $s, \sigma \in \mathbb{R}$, and $\sigma \leq -(1 + s^2)/2$. If $p(z) = 1 + \sum\limits_{n=1}^{\infty} p_n z^n$ satisfies $\Re\{\psi(p(z), zp'(z))\} > 0$, then $\Re\{p(z)\} > 0$ for $z \in E$.

This Lemma is a particular case of a general lemma found in Miller and Moçanu [1978]. Now, going back to the proof, let $\alpha, \beta \in \mathbb{R}$ and $f \in \mathcal{S}$ with $\dfrac{f(z)f'(z)}{z} \neq 0$ for $z \in E$. Then function f belongs to the class $\mathcal{M}_{\alpha,\beta}$ if the function $F : E \mapsto \mathbb{C}$, defined by

$$F(z) = z\left(\frac{f(z)}{z}\right)^{\beta}\left[\frac{zf'(z)}{f(z)}\right]^{\alpha} \tag{7.6.1}$$

is a starlike function in E. Let $f \in \mathcal{S}$ with the series expansion $f(z) = z + \sum\limits_{n=2}^{\infty} a_n z^n$, $z \in E$, and

$$\frac{zf'(z)}{f(z)} = p(z) = 1 + \sum_{n=1}^{\infty} p_n z^n, \quad z \in E. \tag{7.6.2}$$

If $f \in \mathcal{M}_{\alpha,\beta}$, then

$$\Re\left\{\frac{zF'(z)}{F(z)}\right\} > 0, \quad z \in E. \tag{7.6.3}$$

By differentiating (7.6.1) with respect to z and using (7.6.2), we get

$$\frac{zF'(z)}{F(z)} = 1 - \beta + \beta p(z) + \alpha\frac{zp'(z)}{p(z)}, \quad z \in E. \tag{7.6.4}$$

Let $\psi(p(z), zp'(z))$ denote the expression $1 - \beta + \beta p(z) + \alpha\dfrac{zp'(z)}{p(z)}$ in (7.6.4). Then inequality (7.6.3) becomes $\Re\{\psi(p(z), zp'(z))\} > 0$, $z \in E$. We now use Lemma 7.6.1 and calculate $\Re\{\psi(is, \sigma)\} = \Re\left\{1 - \beta + i\beta s + \alpha\dfrac{\sigma}{is}\right\} = 1 - \beta \leq 0$ for any $s, \sigma \in \mathbb{R}$ and $\sigma \leq -(1 - s^2)/2$. Then Lemma 7.6.1 implies that $\Re\{p(z)\} > 0$, $z \in E$, which implies that $\Re\left\{\dfrac{zf'(z)}{f(z)}\right\} > 0$ for $z \in E$. Hence, f is starlike in E, i.e., $\mathcal{M}_{\alpha,\beta} \subset \mathcal{S}^\star$. $\blacksquare$

7.6.4. To prove Theorem 7.5.2, let $\mathfrak{h}(z) = 1 + \sum\limits_{n=1}^{\infty} c_n z^n$. Then, in view of (7.5.7) and (7.5.8), we have $H(z) = [\mathfrak{h}(z)]^{\alpha} = 1 + \sum\limits_{n=1}^{\infty} a_{n+1} z^n$, which on differentiating with respect to z gives

$$H'(z) = \alpha\frac{\mathfrak{h}'(z)}{\mathfrak{h}(z)}H(z) = \sum_{n=1}^{\infty} na_{n+1}z^{n-1}.$$

This, after using the power series expansions for $\mathfrak{h}, \mathfrak{h}'$ and H, gives

$$\left(\sum_{n=1}^{\infty} n a_{n+1} z^{n-1}\right)\left(1 + \sum_{n=1}^{\infty} c_n z^n\right) = \alpha \left(\sum_{n=1}^{\infty} n c_n z^{n-1}\right)\left(1 + \sum_{n=1}^{\infty} a_{n+1} z^n\right).$$

$$(7.6.5)$$

For fixed integers $n \geq 1$, we equate the coefficients of z^{n-1} in (7.6.5) and find that

$$\sum_{k=0}^{n} [k - \alpha(n-k)] \, c_{n-k} a_{k+1} = 0, \quad (c_0 = a_1 = 1). \qquad (7.6.6)$$

Since $c_0 = 1$, we solve (7.6.6) for a_{n+1} and get

$$a_{n+1} = -\frac{1}{n} \sum_{k=0}^{n-1} [k - \alpha(n-k)] \, c_{n-k} a_{k+1}, \qquad (7.6.7)$$

Eq (7.6.7) is a recursion formula that allows us to compute a_{n+1} from those with smaller index, and as such determines a sequence of a_{n+1} in a unique manner. Thus, to prove this theorem it would suffice to show that, for each integer n, the coefficients a_{n+1} defined by the equality in (7.5.11) do indeed satisfy (7.6.7). Thus, proceeding by induction, we assume for each $k = 1, 2, \ldots, n-1$,

$$a_{n+1} = \sum \frac{\gamma(\alpha, m-1) \, c_1^{x_1} c_2^{x_2} \cdots c_k^{x_k}}{x_1! \, x_2! \cdots x_k!}, \qquad (7.6.8)$$

where $m = \sum_{j=1}^{k} x_j$, and the sum is taken over $S(k)$, the set of all non-negative k-tuples $(x_1, x_2, \ldots, x_k)$ for which $\sum_{j=1}^{k} j x_j = k$. Next, if $k < n$, we can enlarge the k-tuple to an n-tuple by adjoining suitably many zeros. Then the solution of

$$\sum_{j=1}^{n} j x_j = k, \quad k \leq n, \qquad (7.6.9)$$

in non-negative integers must give $x_j = 0$ for $j = k+1, k+2, \ldots, n$, and the inclusion of the factors $c_j^{x_j}/x_j!$ in (7.6.8) does not change the value because these factors are 1 for $j = k+1, k+2, \ldots, n$. Hence, (7.6.8) can be replaced by

$$a_{n+1} = \sum \frac{\gamma(\alpha, m-1) \, c_1^{x_1} c_2^{x_2} \cdots c_n^{x_n}}{x_1! \, x_2! \cdots x_n!}, \quad k \leq n, \qquad (7.6.10)$$

where $m = \sum_{j=1}^{n} x_j$, and the sum is taken over the set $S(k)$ of all non-negative integer solutions of (7.6.9). Using (7.6.10) in the right hand side of (7.6.7), we get

$$R := -\frac{1}{n}\sum_{k=0}^{n-1}\sum_{S(k)}\frac{[k-\alpha(n-k)]\gamma(\alpha, m-1)\, c_{n-k}\, c_1^{x_1}c_2^{x_2}\cdots c_n^{x_n}}{x_1!\,x_2!\cdots x_n!}. \qquad (7.6.11)$$

Now let $(y_1, y_2, \ldots, y_n)$ be any fixed n-tuple in $S(n)$, such that

$$\sum_{j=1}^{n} j y_j = n, \quad \sum_{j=1}^{n} y_j = q. \qquad (7.6.12)$$

We will determine the coefficient C of $c_1^{y_1}c_2^{y_2}\cdots c_n^{y_n}$ in (7.6.11). This coefficient may arise from combining several terms from the sum and in fact such terms may arise if and only if $c_{n-k}c_1^{x_1}c_2^{x_2}\cdots c_n^{x_n} = c_1^{y_1}c_2^{y_2}\cdots c_n^{y_n}$. To be specific, let a be an index for which $y_a \geq 1$, and let $x_j = y_j$ if $j \neq a$, and let $x_a = y_a - 1$. For this fixed a, we have $m = \sum_{j=1}^{n} x_j = q - 1$. Set $n - k = a$ in (7,5,18). If A is the set of a for which $y_a \neq 0$, then

$$C = -\frac{1}{n}\sum_{a\in A}\frac{(n - a - a\alpha)\gamma(a, q-2)}{x_1!x_2!\cdots x_n!}. \qquad (7.6.13)$$

Inserting the factor y_a in the numerator and denominator of (7.6.13), we get

$$C = -\sum_{a\in A}\frac{y_a(n - a - a\alpha)\gamma(a, q-2)}{n\, y_1!y_2!\cdots y_n!} = \frac{\gamma(\alpha, q-2)}{n\, y_1!y_2!\cdots y_n!}\sum_{a\in A} y_a(a\alpha + a - n).$$

If $y_a = 0$, then the corresponding term in the sum is zero. Hence, using (7.6.12), we have

$$C = \frac{\gamma(\alpha, q-2)}{n\, y_1!y_2!\cdots y_n!}\sum_{a=1}^{n}(a\alpha y_a + a y_a - n y_a) = \frac{\gamma(\alpha, q-2)}{n\, y_1!y_2!\cdots y_n!}\, n[\alpha - (q-1)]$$

$$= \frac{\gamma(\alpha, q-2)}{n\, y_1!y_2!\cdots y_n!},$$

which is precisely the coefficient of $c_1^{y_1}c_2^{y_2}\cdots c_n^{y_n}$ on the right hand side of (7.5.11). Since the argument holds for each fixed $(y_1, y_2, \ldots, y_n)$, the proof is complete. ∎

8

Coefficients Estimates

Littlewood [1925] defined the mean of the modulus of a univalent function $f \in \mathcal{S}$, and found an estimate for the nth coefficient in its Taylor series expansion (6.1.1). This estimation spree continued until 1977 using various conformal inequalities, like Grunsky, Goluzin, Lebedev-Milin, and especially Fitzgerald inequalities which were obtained by exponentiation of Goluzin inequalities. During this period a number of useful conjectures evolved, all aiming at the Bieberbach conjecture of 1916. We will discuss both of these topic, and show how the Bieberbach conjecture was finally proved in 1984.

8.1 Mean Modulus

Littlewood [1925] defined the mean of the modulus $|f(z)|$ for $0 < r < 1$ by

$$M_p(r, f) = \left\{ \frac{1}{2\pi} \int_0^{2\pi} |f(re^{i\theta})|^p \, d\theta \right\}^{1/p}, \quad 0 < p < \infty. \quad (8.1.1)$$

This definition and the related method were used to obtain estimates of the coefficients $|a_n|$ of univalent functions $f \in \mathcal{S}$. For example,

Landau [1926] showed that $|a_n| < \left(\frac{1}{2} + \frac{1}{\pi} \right) en$;

Bazilevich [1948] showed that $|a_n| < \frac{9}{4} \left(\frac{1}{\pi} \int_0^{\pi} \frac{\sin x}{x} \, dx + 0.2649 \right) n$;

Goluzin [1948] showed that $|a_n| < \frac{3}{4} en$; and

Milin in 1951 (see Milin [1977]) and Bazilevich [1951] proved independently that

$$|a_n| < \tfrac{1}{2} en + 1.8 \text{ and } |a_n| < \tfrac{1}{2} en + 1.51, \text{ respectively.}$$

Lemma 8.1.1. For any $f \in \mathcal{S}$,

$$M_1(r, f) \leq \frac{r}{1 - r}, \quad 0 < r < 1. \quad (8.1.2)$$

PROOF. The function $f(z)/z$, $f \in \mathcal{S}$, is analytic and $f(z)/z \neq 0$ for $z \in E$.

Define $g(z) = \{f(z)/z\}^{1/2}$, $g(0) = 1$, and then define $h(z) = z\left(g\left(z^2\right)\right)$. The function $h(z)$ is univalent, because $h(z_1) = h(z_2)$ implies $f\left(z_1^2\right) = f\left(z_2^2\right)$. Thus, $z = 1 = z_2$ or $z_1 = -z_2$. But the latter case is ruled out because h is odd and $h \neq 0$ for $z \neq 0$. Let

$$h(z) = \sqrt{f\left(z^2\right)} = \sum_{n=1}^{\infty} c_{2n-1} z^{2n-1}, \tag{8.1.3}$$

Then $h(z) \in \mathcal{S}$. Using the definition of $h(z)$ and Theorem 3.4.2, we find that $|h(z)| \leq \dfrac{r}{1 - r^2}$, $|z| = r < 1$. Since h maps $z| < r$ into a domain D_r contained on the disk $|w| < \dfrac{r}{1 - r^2}$. The area A_r of D_r is no greater than $\pi r^2 (1 - r^2)^{-2}$, i.e., no greater than the area of the disk $C\left(0, r(1 - r^2)^{-1}\right)$. Also, we have

$$A_r = \int_0^{2\pi} \int_0^r |h'(\rho e^{i\theta}|^2 \rho\, d\theta\, d\rho = \pi \sum_{n=1}^{\infty} n|c_n|^2\, r^{2n}. \tag{8.1.4}$$

Note that

$$\sum_{n=1}^{\infty} n|c_n|^2\, r^{2n} \leq \frac{r}{(1 - r^2)^2}, \quad 0 \leq r < 1.$$

which , upon integration from 0 to r, yields

$$\sum_{n=1}^{\infty} n|c_n|^2\, r^{2n} \leq \frac{r^2}{1 - r^2}.$$

If we substitute r for r^2 in this inequality, we get (8.1.2). ∎

Historically, Löwner used the differential equation (4.2.2) and proved that $|a_3| \leq 3$ (Theorem 4.4.1). Later, Schiffer and other developed a variation method for injective analytic functions (Chapter 5). During the years 1955–1972 these methods provided rather laborious proofs for the special cases $n = 4, 6$ and 5 of the Bieberbach conjecture. In the general case n, the upper bound for $|f(z)|$ in the distortion inequalities (3.4.2) and Cauchy's inequality for the coefficients of a power series show that

$$|a_n| < en^2. \tag{8.1.5}$$

Littlewood [1925] found the correct upper bound for $|a_n|$ as $n \to \infty$ for $f \in \mathcal{S}$, as follows.

Theorem 8.1.2. (Littlewood [1925]) If $f \in \mathcal{S}$ with the series expansion (6.1.1), then

$$|a_n| < en, \quad n = 2, 3, \ldots. \tag{8.1.6}$$

PROOF. Since

$$a_n = \frac{1}{2\pi i} \int_{|z|=r} \frac{f(z)}{z^{n+1}}\, dz = \frac{1}{2\pi} \int_0^{2\pi} e^{-in\theta} r^{-n} f\left(re^{i\theta}\right) d\theta,$$

we get

$$|a_n| r^n \leq \frac{1}{2\pi} \int_0^{2\pi} |f(re^{i\theta}| \, d\theta, \quad \text{or} \quad |a_n| \leq r^{-n} M_1(r, f),$$

which, using (8.1.2), gives

$$|a_n| \leq \frac{r^{-n+1}}{1 - r}.$$

The right-hand side of this inequality has minimum when $r = 1 - \dfrac{1}{n}$. Hence,

$$|a_n| \leq n\left(1 + \frac{1}{n-1}\right)^{n-1} < en. \;\blacksquare$$

If $f \in \mathcal{S}$, then the function $h(z)$ defined by (8.1.3) is an *odd* function. Littlewood and Paley [1932] proved that $|c_n| < 14$ and conjectured that $|c_n| \leq 1$. Since

$$a_n = c_1 c_{2n-1} + c_2 c_{2n-3} + \cdots + c_{2n-1} c_1, \tag{8.1.7}$$

which implies that the Bieberbach conjecture would be true if the Littlewood-Paley conjecture were true. But Fekete and Szegö [1933] proved the Littlewood-Paley conjecture to be wrong by showing that

$$|c_5| \leq \frac{1}{2} + e^{-2/3} \approx 1.013.$$

Also, Chen [1933] proved that $|c_n| < e^2$, and Chen [1935] proved that $|c_n| < 2^{1/4} 3^{1/2} e^{1/2} \approx 3.39$. Schaeffer and Spencer [1943] proved that for any $n \geq 5$ there exists a univalent odd function with real coefficients such that $|c_n| > 1$. Leeman [1976] proved that for univalent odd functions with real coefficients, the sharp bound for $|c_7| \leq \dfrac{1090}{1083} \approx 1.006$. Gong [1955] prove that $|c_n| < 2^{-1/6} 3^{1/2} e^{1/2} \approx 2.54$. Milin [1977] showed that $|c_n| < 1.17$, and improved it to $|c_n| < 1.14$ (Milin [1980]). The best estimate by Hu [1986] has been $|c_n| < 1.1305$. A detailed discussion of some of these estimates is given in §8.3.

8.2 Hayman Index

The Haymann index $\alpha = \lim\limits_{n \to \infty} \dfrac{|a_n(f)|}{n}$ provides much information about the class $\mathcal{S}$, but it is not obvious that the limit exists for every $f \in \mathcal{S}$. On one hand, this index does not indicate that the Bieberbach conjecture is true for large $n \in \mathbb{N}$ since the rate of convergence depends considerably on f. On the other hand, it does provide some truth about the Bieberbach conjecture, however fallacious it may sound, since his regularity theorem (Theorem 8.2.1) does imply that for odd univalent functions $h(z) = z\sqrt{\dfrac{f(z^2)}{z^2}} = \sum\limits_{n=1}^{\infty} c_n z^n \in \mathcal{S}_{\text{odd}}$, the limit

$$\lim_{n \to \infty} |c_{2n+1}| = \sqrt{\alpha} \le 1,$$

even though the Littlewood-Paley conjecture is false! We prove the following results.

Lemma 8.2.1. If $f \in \mathcal{S}$, and $M_\infty(r, f) = \max\limits_{|z|=r} |f(z)|$, then

$$\lim_{r \to 1} (1-r)^2 M_\infty(r, f) = \alpha < 1. \tag{8.2.1}$$

If f is not a Koebe function or one of its rotations, then $\dfrac{(1-r)^2}{r} M_\infty(r, f)$ is strictly decreasing on the interval $(0, 1)$ and $\alpha < 1$.

PROOF. Fix $z \in E$. Then for any $f \in \mathcal{S}$, since $\dfrac{\zeta + z}{1 + \bar{\zeta}}$ is an automorphism of E,

$$F(\zeta) = \frac{f\left(\dfrac{\zeta + z}{1 - \bar{z}\zeta}\right) - f(z)}{(1 - |z|^2)\, f'(z)} \in \mathcal{S}.$$

Using (3.4.2), we have

$$\frac{|\zeta|}{(1 + |\zeta|)^2} \le |F(\zeta)| \le \frac{|\zeta|}{(1 - |\zeta|)^2}.$$

Set $\zeta = -z$. Then

$$\frac{|z|}{(1 + |z|)^2} \le \left| \frac{f(z)}{(1 - |z|^2)} f'(z) \right| \le \frac{|z|}{(1 + |z|)^2},$$

which yields

$$\frac{1-r}{1+r} \le \left| \frac{z f'(z)}{f(z)} \right| \le \frac{1+r}{1-r}, \quad |z| = r < 1,$$

where equality holds if and only if $f(z)$ is the Koebe function or one of its rotations. Hence,

$$\frac{\partial}{\partial r} \log |f(re^{i\theta})| \le \left| \frac{f'(re^{i\theta})}{f(re^{i\theta})} \right| \le \frac{1+r}{r(1-r)}.$$

Integrating this inequality from r_1 to r_2, $(0 < r_1 < r_2 < 1)$, we get

$$\log\left|\frac{f\left(r_2 e^{i\theta}\right)}{f\left(r_1 e^{i\theta}\right)}\right| \leq \int_{r_1}^{r_2} \frac{1+r}{r(1-r)}\, dr = \log\frac{r_2(1-r_1)^2}{r_1(1-r_2)^2}.$$

Thus after exponentiating the above inequality, and for any θ, $0 < r_1 < r_2 < 1$, we have

$$\frac{(1-r_2)^2}{r_2}\left|f\left(r_2 e^{i\theta}\right)\right| < \frac{(1-r_1)^2}{r_1}\left|f\left(r_1 e^{i\theta}\right)\right|.$$

Now, we choose θ so that $\left|f\left(r_2 e^{i\theta}\right)\right| = M_\infty(r_2, f)$. Then the above inequality becomes

$$\frac{(1-r_2)^2}{r_2} M_\infty(r_2, f) < \frac{(1-r_1)^2}{r_1}\left|f\left(r_1 e^{i\theta}\right)\right| \leq \frac{(1-r_1)^2}{r_1} M_\infty(r_1, f).$$

This shows that $\dfrac{(1-r)^2}{r} M_\infty(r, f)$ is a decreasing function of r, and it approaches the limit $\alpha \geq 0$ as $r \to 1$. By using (3.4.2), we get $\dfrac{(1-r)^2}{r} M_\infty(r, f) \leq 1$, whence $\alpha \leq 1$. If f is the Koebe function or one of its rotations, then equality holds, giving $\alpha = 1$; otherwise, the decreasing is strict and $\alpha < 1$ ∎

The limit α is called the *Hayman index* of the function $f \in \mathcal{S}$. The following theorem is based on Lemma 8.2.1.

Lemma 8.2.2. *If the Hayman index $\alpha > 0$ for $f \in \mathcal{S}$, then there exists a unique direction $e^{i\theta_0}$ such that*

$$\lim_{r \to 1}(1-r)^2|f(re^{i\theta_0}| = \alpha. \tag{8.2.2}$$

PROOF. Let $\{r_n\}$ be an increasing sequence with limit 1. Take θ_n, $0 \leq \theta_n \leq 2\pi$ such that $|f(r_n\theta_n)| = M_\infty(r_n, f)$ for $n - 1, 2, \ldots$. Then, by Lemma 8.2.1, we have for $r < r_n$

$$\alpha \leq \frac{(1-r_n)^2}{r_n}|f(r_n\theta_n)| \leq \frac{(1-r)^2}{r} M_\infty(r, f). \tag{8.2.3}$$

Note that $\dfrac{(1-r)^2}{r} M_\infty(r, f) \to \alpha$ as $r \to 1$, Then from (8.2.3) we have $\lim_{r \to 1}(1- r)^2|f(r\theta_0)| = \alpha$. To show that θ_0 is unique, choose $N = 1, \lambda_1 = \mu_1 = 1, \zeta_1 = \zeta_1$, $z_1 = \zeta_2$, $|\zeta_1| = |\zeta_2| = \rho > 1$ in (g.6), which gives

$$1 - \rho^{-2} \leq \left|\frac{g(\zeta_1) - g)\zeta_2)}{\zeta_1 - \zeta_2}\right| \leq \frac{1}{1-\rho^{-2}}. \tag{8.2.4}$$

Recall that $f(z) = \dfrac{1}{g(1/z)}$, $\rho^{-1} = r$, $\zeta_1^{-1} = z$, and $\zeta_2^{-1} = z_2 = re^{i\theta_0}$, so the inequality (8.2.4) becomes

$$\frac{1-r^2}{r}|e^{i\theta} - e^{i\theta_0}| \le \left| \frac{1}{f(re^{i\theta})} - \frac{1}{f(re^{i\theta_0})} \right| \le \frac{1}{|f(re^{i\theta})|} + \frac{1}{|f(re^{i\theta_0})|}.$$

Since $\lim\limits_{r\to 1}(1-r)^2|f(re^{i\theta})| \ne 0$, then if $e^{i\theta} \ne e^{i\theta_0}$ and $\alpha > 0$, we arrive at a contradiction in that $e^{i\theta}$ must be equal to $e^{i\theta_0}$. ∎

The direction $e^{i\theta_0}$ is called the *Hayman direction*. Using these two Lemmas, we have

Theorem 8.2.1. (Hayman regularity theorem, Hayman [1955]) If $f \in \mathcal{S}$, then

$$\lim_{n\to\infty} \frac{|a_n|}{n} = \alpha < 1. \tag{8.2.5}$$

Equality holds if and only if f is the Koebe function or one of its rotations.

The above proofs are from Gong [1999]. There are two other proofs of the Hayman regularity theorem, by Hayman [1955, 1994] and Milin [1970].

8.3 Conformal Inequalities

Before we study the Goluzin inequalities and the subsequent development by Fitzgerald [1972, 1977] and Fitzgerald and Horn [1977] of exponentiating these inequalities leading to the *Fitzgerald inequality*, we will introduce two classes of univalent functions, as follows.

8.3.1 Class $\mathcal{S}_t$. Let $\mathcal{S}_t$ denote the set of all functions in $\mathcal{S}$ which map the disk E onto the domain obtained by excluding from the plane a generalized simple arc extending to infinity. This class is dense in $\mathcal{S}$ in the topology of uniform convergence on compacta, and consist of univalent functions f having representation of the form

$$f(z) = \lim_{t\to\infty} e^t\, f(z,t), \tag{8.3.1}$$

where $f(z,t)$ is a solution of the *Löwner equation* (4.2.2) with $f(z,0) = z$, where k and f are defined as before, and $f(z,t)$, for a fixed t, is a regular function for $z \in E$ such that (i) $|f(z,t)| < 1$ for $z \in E$ and $0 \le t < \infty$; (ii) $f(0,t) = 0$; (iii) $f_z'(0,t) > 0$; and (iv) and $k = k(t)$ is an arbitrary complex-valued piecewise continuous function such that $|k(t)| = 1$ for $0 \le t < \infty$.

Take arbitrary points z_ν, $\nu = 0, 1, 2, \dots, n$, in E and write $f_\nu \equiv f(z_\nu, t)$. Then Eq (4.2.2) becomes

$$\frac{\partial f_\nu}{\partial t} = -f_\nu \frac{1+k(t)}{1-k(t)}, \quad \nu = 1, 2, \dots, n.$$

Theorem 8.3.1. *We have*

$$\frac{\partial}{\partial t} \log \left[\frac{e^{-it}}{f_\nu \, f_\mu} \frac{f_\nu - f_\mu}{z_\nu - z_\mu} \right] = -2 \frac{k \, f_\nu}{1 - k \, f_\nu} \frac{k \, f_\mu}{1 - k \, f_\mu}. \tag{8.3.2}$$

and

$$\frac{\partial}{\partial t} \log \left(1 - f_\nu \, \bar{f}_\mu \right) = 2 \frac{k \, f_\nu}{1 - k \, f_\nu} \overline{\left(\frac{k \, f_\mu}{1 - k \, f_\mu} \right)}. \tag{8.3.3}$$

PROOF. Note that

$$\frac{\partial}{\partial t} \log \left[\frac{e^{-t}}{f_\nu f_\mu} \cdot \frac{f_\nu - f_\mu}{z_\nu - z_\mu} \right]$$

$$= \frac{\partial}{\partial t} \left[\log e^{-t} + \log \frac{f_\nu - f_\mu}{z_\nu - z_\mu} + \log \frac{1}{z_\nu z_\mu} \right]$$

$$= -1 + \frac{z_\nu - z_\mu}{f_\nu - f_\mu} \left(\frac{\dfrac{\partial f_\nu}{\partial t} - \dfrac{\partial f_\mu}{\partial t}}{z_\nu - z_\mu} \right) - \frac{1}{z_\nu z_\mu} \left\{ f_\nu \frac{\partial f_\mu}{\partial t} + f_\mu \frac{\partial f_\nu}{\partial t} \right\}$$

$$= -1 + \frac{1}{f_\nu - f_\mu} \left\{ - f_\nu \frac{1 + k f_\nu}{1 - k f_\nu} + f_\mu \frac{1 + k f_\mu}{1 - k f_\mu} \right\}$$

$$\quad - \frac{1}{f_\nu f_\mu} \left\{ - f_\nu f_\mu \frac{1 + k f_\mu}{1 - k f_\mu} - f_\mu f_\nu \frac{1 + k f_\nu}{1 - k f_\nu} \right\}$$

$$= -1 + \frac{1}{f_\nu - f_\mu} \left\{ \frac{-f_\nu \left(1 + k f_\nu \right) \left(1 - k f_\mu \right) + f_\mu \left(1 + k f_\mu \right) \left(1 - k f_\nu \right)}{\left(1 - k f_\nu \right) \left(1 - k f_\mu \right)} \right\}$$

$$\quad + \frac{\left(1 + k f_\mu \right) \left(1 - k f_\nu \right) + \left(1 + k f_\nu \right) \left(1 - k f_\nu \right)}{\left(1 - k f_\nu \right) \left(1 - k f_\mu \right)}$$

$$= -1 + \frac{k^2 f_\nu f_\mu - k \left(f_\mu + f_\nu - 1 \right)}{\left(1 - k f_\nu \right) \left(1 - k f_\mu \right)} + \frac{2 \left(1 - k^2 f_\nu f_\mu \right)}{\left(1 - k f_\nu \right) \left(1 - k f_\mu \right)}$$

$$= -1 + \frac{k^2 f_\nu f_\mu - k \left(f_\nu + f_\mu \right) - 1 + 2 - 2 k^2 f_\nu f_\mu}{\left(1 - k f_\nu \right) \left(1 - k f_\mu \right)}$$

$$= \frac{-2 k^2 f_\nu f_\mu}{\left(1 - k f_\nu \right) \left(1 - k f_\mu \right)} = -2 \left(\frac{k f_\nu}{1 - k f_\nu} \right) \left(\frac{k f_\mu}{1 - k f_\mu} \right).$$

Eq (8.3.3) can be similarly obtained. ∎

Note that if $z_\nu = z_\mu$, then $\dfrac{f_\nu - f_\mu}{z_\nu - z_\mu} = f'(z_\nu)$, since

$$\lim_{z_\mu \to z_\nu} \frac{f_\nu - f_\mu}{z_\nu - z_\mu} = \lim_{z_\mu \to z_\nu} \frac{f(z_\nu, t) - f(z_\mu, t)}{z_\nu - z_\mu} = f'_z(z_\nu, t),$$

and $\lim_{t \to \infty} e^t f(z_\nu, t) = f'(z_\nu)$, and $f(z) = \lim_{t \to \infty} e^t f(z, t)$. Thus, to summarize, $\mathcal{S}_t \subset \mathcal{S}$ such that (i) $f(z) = \lim_{t \to \infty} e^t f(z, t)$; (ii) $f(z, 0) = z$; (iii) $f(0, t) = 0$; (iv)

$f'_z(0,t) > 0$ for $0 \leq t < \infty$; (v) Löwner Eq: $\dfrac{\partial f}{\partial t} = -f \dfrac{1+kf}{1-kf}$, where $|k(t)| = 1$,

which (4.2.2) and (8.3.2) hold; and (vi) If $z_\nu = z_\mu$, then $\dfrac{f_\nu - f_\mu}{z_\nu - z_\mu} = f'(z_\nu)$.

On integrating Eqs (8.3.2) and (8.3.3) with respect to t from 0 to ∞, we obtain

Theorem 8.3.2. *The following results hold:*

$$\left[\log \frac{e^{-t}}{f_\nu f_\mu} \frac{f_\nu - f_\mu}{z_\nu - z_\mu} \right]_0^\infty = -2 \int_0^\infty \frac{kf_\nu}{1 - kf_\nu} \frac{kf_\mu}{1 - kf_\mu}\, dt, \qquad (8.3.4)$$

$$\left[\log \left(1 - f_\nu \bar{f}_\mu \right) \right]_0^\infty = 2 \int_0^\infty \frac{kf_\nu}{1 - kf_\nu} \left(\overline{\frac{kf_\mu}{1 - kf_\mu}} \right) dt. \qquad (8.3.5)$$

8.3.2 Class Σ. Consider the class Σ of regular univalent functions defined for $|z| > 1$ (§3.1). If $f \in \mathcal{S}$, $z \in E$, then $\dfrac{1}{f(1/\zeta)} \in \Sigma$, $|\zeta| > 1$. Put $\zeta_\nu = 1/z_\nu$, and let $F(\zeta) = \dfrac{1}{f(1/\zeta)}$, $F \in \Sigma$, $\zeta| > 1$. Thus, $f_\nu = f(z_\nu, t) = f\left(\dfrac{1}{f(1/\zeta)}, t \right)$, and $e^t f_\nu \equiv e^t f(z_n u, t)$, which yields the condition for $\mathcal{S}_t$ as

$$\lim_{t \to \infty} e^t f_\nu = f(z_\nu) \equiv f_\nu = f\left(\frac{1}{f(1/\zeta)} \right). \qquad (8.3.6)$$

Results analogous to (8.3.4) for $F \in \Sigma$ are given in the following theorem.

Theorem 8.3.3. *For $F \in \Sigma$,*

$$\log \frac{F(\zeta_\mu) - F(\zeta_\nu)}{\zeta_\mu - \zeta_\nu} = -2 \int_0^\infty \frac{kf_\nu}{1 - kf_\nu} \frac{kf_\mu}{1 - kf_\mu}\, dt. \qquad (8.3.7)$$

$$\log \left(1 - \frac{1}{\zeta_\nu \bar{\zeta}_\mu} \right) = 2 \int_0^\infty \frac{kf_\nu}{1 - kf_\nu} \left(\overline{\frac{kf_\mu}{1 - kf_\mu}} \right) dt. \qquad (8.3.8)$$

PROOF. By taking reciprocals of Eq (8.3.6), we get

$$\frac{e^{-t}}{f_\nu} \to \frac{1}{f(1/\zeta_\nu)} = F(\zeta_\nu) \quad \text{as } t \to \infty.$$

The left hand side of (8.3.4) is equal to

$$\log \left[\frac{e^{-t}}{f_\nu f_\mu} \frac{f_\nu - f_\mu}{z_\nu - z_\mu} \right] = \log \left\{ \frac{e^{-t}}{z_\nu - z_\mu} \left(\frac{1}{f_\mu} - \frac{1}{f_\nu} \right) \right\}$$

$$\to \log \left[\frac{F(\zeta_\mu) - F(\zeta_\nu)}{1/\zeta_\nu - 1/\zeta_\mu} \right] \quad \text{as } t \to \infty$$

$$= \log \frac{\zeta_\nu \zeta_\mu \left[F(\zeta_\mu) - F(\zeta_\nu) \right]}{\zeta_\mu - \zeta_n u}.$$

Also, as $t \to 0$, we find that

$$\log \left[\frac{e^{-t}}{z_\nu - z_\mu} \left(\frac{1}{f_\mu} - \frac{1}{f_\nu} \right) \right] \to \log \left[\frac{1}{z_\nu - z_\mu} \left(\frac{1}{z_\mu} - \frac{1}{z_\nu} \right) \right]$$

$$= \log \frac{1}{z_\mu z_\nu} = \log \zeta_\nu \zeta_\mu.$$

Note that $f_\mu \equiv f(z_\mu, t) \to z_\mu$ as $t \to \infty$. Then (8.3.4) becomes

$$\log \frac{\zeta_\nu \zeta_\mu [F(\zeta_\mu) - F(\zeta_\nu)]}{\zeta_\mu - \zeta_\nu} - \log \zeta_\nu \zeta_\mu = -2 \int_0^\infty \frac{k f_\nu}{1 - k f_\nu} \frac{k f_\mu}{1 - k f_\mu} \, dt,$$

which implies (8.3.7). Also,

$$\text{left side of (8.3.5)} = \log \left(1 - f_\nu \bar{f}_\mu \right) = \log \left(1 - e^{-t} f_\nu e^{-t} \bar{f}_\mu \right)$$

$$= \log \left(1 - e^{2-t} f(z_\nu) \overline{f(z_\mu)} \right) \to \log 1 = 0 \text{ as } t \to \infty;$$

and

$$\log \left(1 - f_\nu \bar{f}_\mu \right) = \log \left(1 - f(z_\nu, t) \bar{f}(z_\mu, t) \right)$$

$$\to \log \left(1 - z_\nu z_\mu \right) = \log \left(1 - \frac{1}{\zeta_\nu \zeta_\mu} \right) \quad \text{as } t \to 0.$$

Thus, Eq (8.3.5) becomes (8.3.8). ∎

We will introduce the notation:

$$\log \frac{F(\zeta_\mu) - F(\zeta_\nu)}{\zeta_\mu - \zeta_\nu} = \sum_{m,n=1}^\infty a_{m,n} \zeta_\nu^m \zeta_\mu^n,$$

where $|\zeta_\nu| > 1$, $|\zeta_\mu| > 1$. Then

$$\frac{f(z) - f(\zeta)}{z - \zeta} = \sum_{n=1}^\infty A_n(\zeta) z^n = \sum_{n=1}^\infty \sum_{m=1}^\infty B_{nm} \zeta^m z^n$$

$$= \sum_{n,m=1}^\infty B_{nm} \zeta^m z^n, \quad |z| < 1, |\zeta| < 1.$$

Now,

$$\frac{kf}{1 - kf} = \frac{k(t) f(z,t)}{1 - k(t) f(z,t)} = \sum_{m=1}^\infty b_m(t) z^m, \quad |z| < 1,$$

so that

$$\overline{\left(\frac{kf}{1 - kf} \right)} = \sum_{m=1}^\infty \bar{b}_m(t) \bar{z}^m.$$

Then (8.3.7) and (8.3.8) become

$$\sum_{m,n=1}^{\infty} a_{m,n}\, \zeta_\nu^{-m}\, \zeta_\mu^{-n} = -2 \int_0^{\infty} \left(\sum_{m=1}^{\infty} b_m(t)\, z_\nu^m \right) \left(\sum_{n=1}^{\infty} b_n(t)\, z_\mu^m \right) dt,$$

which yields

$$a_{m,n} = -2 \int_0^{\infty} b_m(t) b_n(t)\, dt, \quad m,n = 1,2,\dots .$$

Since $\log(1-x) = -\sum_{k=0}^{\infty} \dfrac{x^{2k+1}}{2k+1} = -\sum_{k\,\text{odd}} \dfrac{x^k}{k}$, Eq (8.3.8) gives

$$-\sum_{\substack{m=1 \\ m\,\text{odd}}}^{\infty} \frac{1}{m} \left(\frac{1}{\zeta_\nu \bar{\zeta}_\mu} \right)^m = -2 \int_0^{\infty} \frac{kf_\nu}{1-kf_\nu} \cdot \frac{kf_\mu}{1-kf_\mu}\, dt,$$

$$\sum_{\text{odd }m=1}^{\infty} \frac{1}{m} \left(\zeta_\nu\, \bar{\zeta}_\mu \right)^{-m} = -2 \int_0^{\infty} \left(\sum_{m=1}^{\infty} b_m(t)\, z_\nu^m \right) \left(\sum_{n=1}^{\infty} \bar{b}_n(t)\, z_\mu^n \right) dt$$

$$= -2 \int_0^{\infty} \left(\sum_{m=1}^{\infty} b_m(t)\, \zeta_\nu^{-m} \right) \left(\sum_{n=1}^{\infty} \bar{b}_n(t)\, \zeta_\mu^{-n} \right) dt.$$

Now, by the orthogonality property of $b_n(t)$, we have

$$\int_0^{\infty} b_m(t)\, \overline{b_m(t)}\, dt = \begin{cases} 0, & m=n, \text{ i.e., even powers,} \\ \dfrac{1}{2m}, & m \neq n, \text{ i.e., odd powers.} \end{cases}$$

If we set $\dfrac{kf_\nu}{1-kf_\nu} = X_\nu + i\, Y_\nu$, $\nu = 1,\dots,n$, then Eqs (8.3.7) and (8.3.8), after separating the real and imaginary parts on both sides, yield

$$\log \left| \frac{F(\zeta_\nu) - F(\zeta_\mu)}{\zeta_\nu - \zeta_\mu} \right| = -2 \int_0^{\infty} \Re\left\{ (X_\nu + iY_\nu)(X_\mu + iY_\mu) \right\} dt,$$

or

$$\log \left| \frac{F(\zeta_\nu) - F(\zeta_\mu)}{\zeta_\nu - \zeta_\mu} \right| = -2 \int_0^{\infty} (X_\nu X_\mu - Y_\nu Y_\mu)\, dt; \tag{8.3.9}$$

and

$$\log \left| 1 - \frac{1}{\zeta_\nu - \zeta_\mu} \right| = -2 \int_0^{\infty} \Re\left\{ (X_\nu + iY_\nu) \overline{(X_\mu + iY_\mu)} \right\} dt$$

$$= -2 \int_0^{\infty} \Re\left\{ (X_\nu + iY_\nu)(X_\mu - iY_\mu) \right\} dt,$$

or

$$\log \left| 1 - \frac{1}{\zeta_\nu - \zeta_\mu} \right| = -2 \int_0^\infty (X_\nu X_\mu + Y_\nu Y_\mu) \, dt. \tag{8.3.10}$$

Adding (8.3.9) and (8.3.10) gives

$$\log \left| \frac{F(\zeta_\nu) + F(\zeta_\mu)}{\zeta_\nu - \zeta_\mu} \right| + \log \left| 1 - \frac{1}{\zeta_\nu - \zeta_\mu} \right| = -4 \int_0^\infty X_\nu X_\mu \, dt,$$

while subtracting (8.3.10) from (8.3.9) gives

$$\log \left| \frac{F(\zeta_\nu) - F(\zeta_\mu)}{\zeta_\nu - \zeta_\mu} \right| - \log \left| 1 - \frac{1}{\zeta_\nu - \zeta_\mu} \right| = 4 \int_0^\infty Y_\nu Y_\mu \, dt.$$

Hence, we have proved:

Theorem 8.3.4. *For $F \in \Sigma$, we have*

$$\log \left| \frac{F(\zeta_\nu) - F(\zeta_\mu)}{\zeta_\nu - \zeta_\mu} \right| = -\log \left| 1 - \frac{1}{\zeta_\nu - \zeta_\mu} \right| - 4 \int_0^\infty Y_\nu Y_\mu \, dt. \tag{8.3.11}$$

$$\log \left| \frac{F(\zeta_\nu) - F(\zeta_\mu)}{\zeta_\nu - \zeta_\mu} \right| = \log \left| 1 - \frac{1}{\zeta_\nu - \zeta_\mu} \right| + 4 \int_0^\infty Y_\nu Y_\mu \, dt. \tag{8.3.12}$$

Let $\alpha_{\nu,\mu}$, $(\nu, \mu = 1, \dots, n)$ be real numbers such that $\sum \alpha_{\nu,\mu} X_\nu X_\mu$ is a positive quadratic form, i.e., $\sum \alpha_{\nu,\mu} X_\nu X_\mu \geq 0$. In our case we will then have the quadratic forms $\sum \alpha_{\nu,\mu} X_\nu X_\mu$ and $\sum \alpha_{\nu,\mu} Y_\nu Y_\mu$. Multiply (8.3.11) and (8.3.12) by these $\alpha_{\nu,\mu}$ and sum over ν, μ, and we get

$$\sum_{\nu,\mu} \log \left| \frac{F(\zeta_\nu) - F(\zeta_\mu)}{\zeta_\nu - \zeta_\mu} \right| \alpha_{\nu,\mu} = \sum_{\nu,\mu} \frac{1}{\log \left| 1 - \frac{1}{\zeta_\nu - \zeta_\mu} \right|} \alpha_{\nu,\mu} - 4 \int_0^\infty \alpha_{\nu,\mu} X_\nu X_\mu \, dt.$$

The second term on the right hand side is ≤ 0. So for (8.3.11) we get

$$\prod_{\nu,\mu}^n \left| \frac{F(\zeta_\nu) - F(\zeta_\mu)}{\zeta_\nu - \zeta_\mu} \right|^{\alpha_{\nu,\mu}} \leq \frac{1}{\prod_{\nu,\mu}^n \left| 1 - \frac{1}{\zeta_\nu - \zeta_\mu} \right|}, \tag{8.3.13}$$

and for (8.3.12) we get

$$\sum_{\nu,\mu} \log \left| \frac{F(\zeta_\nu) - F(\zeta_\mu)}{\zeta_\nu - \zeta_\mu} \right| \alpha_{\nu,\mu} = \sum_{\nu,\mu} \frac{1}{\log \left| 1 - \frac{1}{\zeta_\nu - \zeta_\mu} \right|} \alpha_{\nu,\mu} + 4 \int_0^\infty \alpha_{\nu,\mu} X_\nu X_\mu \, dt,$$

which gives

$$\prod_{\nu,\mu}^{n} \left| \frac{F(\zeta_\nu) - F(\zeta_\mu)}{\zeta_\nu - \zeta_\mu} \right|^{\alpha_{\nu,\mu}} \geq \prod_{\nu,\mu}^{n} \frac{1}{\left| 1 - \dfrac{1}{\zeta_\nu - \zeta_\mu} \right|}. \tag{8.3.14}$$

Combining (8.3.13) and (8.3.14) we have proved the following theorem:

Theorem 8.3.5. *The following inequalities hold:*

$$\prod_{\nu,\mu=1}^{n} \left| 1 - \frac{1}{\zeta_\nu - \zeta_\mu} \right|^{\alpha_{\nu,\mu}} \leq \prod_{\nu,\mu=1}^{n} \left| \frac{F(\zeta_\nu) - F(\zeta_\mu)}{\zeta_\nu - \zeta_\mu} \right|^{\alpha_{\nu,\mu}} \leq \frac{1}{\prod_{\nu,\mu=1}^{n} \left| 1 - \dfrac{1}{\zeta_\nu - \zeta_\mu} \right|^{\alpha_{\nu,\mu}}}. \tag{8.3.15}$$

Note that these inequalities, known as *Grunsky-type inequalities*, hold throughout the domain $|\zeta| > 1$, and if $\zeta_\nu = \zeta_\mu$, then $\dfrac{F(\zeta_\nu) - F(\zeta_\mu)}{\zeta_\nu - \zeta_\mu}$ is understood to be $f'(\zeta_\nu)$. Also, recall that if $f \in \Sigma$, then $f(\zeta) = \dfrac{1}{f(1/\zeta)}$, or $f(\zeta) = \dfrac{1}{f(1/\zeta)} + C$, $f \in \mathcal{S}$.

Let γ_ν be arbitrary complex numbers (ζ_ν, ζ_μ are also complex numbers such that $|\zeta_\nu| > 1$, $|\zeta_\mu| > 1$). Then

$$\log \frac{F(\zeta_\nu) - F(\zeta_\mu)}{\zeta_\nu - \zeta_\mu} = -2 \int_0^\infty \frac{k f_\nu}{1 - k f_\nu} \frac{k f_\mu}{1 - k f_\mu}\, dt,$$

$$\sum_{\nu,\mu=1}^{\infty} \gamma_\nu \gamma_\mu \log \frac{F(\zeta_\nu) - F(\zeta_\mu)}{\zeta_\nu - \zeta_\mu} = -2 \int_0^\infty \sum_{\nu,\mu=1}^{\infty} \gamma_\nu \gamma_\mu \frac{k f_\nu}{1 - k f_\nu} \frac{k f_\mu}{1 - k f_\mu}\, dt$$

$$= -2 \int_0^\infty \left(\sum_{\nu=1}^{\infty} \gamma_\nu \frac{k f_\nu}{1 - k f_\nu} \right)^2 dt. \tag{8.3.16}$$

Now, from (8.3.8)

$$\sum_{\nu,\mu=1}^{\infty} \gamma_\nu \gamma_\mu \log \left(1 - \frac{1}{\zeta_\nu} \frac{1}{\zeta_\mu} \right) = -2 \int_0^\infty \sum_{\nu,\mu=1}^{\infty} \gamma_\nu \gamma_\mu \frac{k f_\nu}{1 - k f_\nu} \overline{\left(\frac{k f_\mu}{1 - k f_\mu} \right)}\, dt. \tag{8.3.17}$$

Then

$$\left| \sum_{\nu,\mu=1}^{\infty} \gamma_\nu \gamma_\mu \log \frac{F(\zeta_\nu) - F(\zeta_\mu)}{\zeta_\nu - \zeta_\mu} \right| \le 2 \int_0^{\infty} \left| \sum_{\nu=1}^{\infty} \gamma_\nu \frac{k f_\nu}{1 - k f_\nu} \right|^2 dt$$

$$= 2 \int_0^{\infty} \left(\sum_{\nu=1}^{\infty} \gamma_\nu \frac{k f_\nu}{1 - k f_\nu} \right) \overline{\left(\sum_{\nu=1}^{\infty} \gamma_\nu \frac{k f_\nu}{1 - k f_\nu} \right)} dt$$

$$= 2 \int_0^{\infty} \left(\sum_{\nu,\mu=1}^{\infty} \gamma_\nu \bar\gamma_\mu \frac{k f_\nu}{1 - k f_\nu} \right) \left(\sum_{\nu=1}^{\infty} \gamma_\nu \frac{k f_\nu}{1 - k f_\nu} \right) dt$$

$$= - \sum_{\nu,\mu=1}^{\infty} \gamma_\nu \bar\gamma_\mu \log \left(1 - \frac{1}{\zeta_\nu \bar\zeta_\mu} \right) \quad \text{from (8.3.17).}$$

Instead of the range of ν, μ from 1 to ∞, let $\nu = 1, 2, \dots, n$ and $\mu = 1, 2, \dots, n'$, where n and n' are positive integers (not essentially the same). Then, proceeding as before, we get

$$\left| \sum_{\nu=1}^{n} \sum_{\mu=1}^{n'} \gamma_\nu \gamma_\mu \log \frac{F(\zeta_\nu) - F(\zeta_\mu)}{\zeta_\nu - \zeta_\mu} \right|$$

$$\le + \sqrt{\sum_{\nu=1}^{n} \sum_{\mu=1}^{n'} \left\{ \gamma_\nu \log \left(1 - \frac{1}{\zeta_\nu \bar\zeta_\mu} \right) \right\} \left\{ \bar\gamma_\mu \log \left(1 - \frac{1}{\zeta_\nu \bar\zeta_\mu} \right) \right\}},$$

$$\sum_{\nu,\mu=1}^{n} \gamma_\nu \gamma_\mu \log \frac{F(\zeta_\nu) - F(\zeta_\mu)}{\zeta_\nu - \zeta_\mu} = -2 \int_0^{\infty} \left(\sum_{\nu=1}^{n} \gamma_\nu \frac{k f_\nu}{1 - k f_\nu} \right)^2 dt,$$

$$\sum_{\nu,\mu=1}^{n'} \gamma_\nu \gamma_\mu \log \frac{F(\zeta_\nu) - F(\zeta_\mu)}{\zeta_\nu - \zeta_\mu} = -2 \int_0^{\infty} \left(\sum_{\nu=1}^{n'} \bar\gamma_\mu \frac{k f_\mu}{1 - k f_\mu} \right)^2 dt.$$

Thus, we have proved the following theorem:

Theorem 8.3.6. *For $F \in \Sigma$, we have*

$$\left| \sum_{\nu,\mu} \gamma_\nu \gamma_\mu \log \frac{z_\nu z_\mu f(z_\nu) - f(z_\mu)}{f_\nu f_\mu (z_\nu - z_\mu)} \right| \le - \sum_{\nu,\mu} \gamma_\nu \bar\gamma_\mu \log (1 - z_\nu \bar z_\mu), \tag{8.3.18}$$

$$\log \frac{z_\nu z_\mu f(z_\nu) - f(z_\mu)}{f_\nu f_\mu (z_\nu - z_\mu)} = -2 \int_0^{\infty} \frac{k f_\nu}{1 - k f_\nu} \frac{k f_\mu}{1 - k f_\mu} dt, \tag{8.3.19}$$

$$\log (1 - z_\nu \bar z_\mu) = -2 \int_0^{\infty} \frac{k f_\nu}{1 - k f_\nu} \left(\frac{k f_\mu}{1 - k f_\mu} \right) dt, \tag{8.3.20}$$

where $|z_\nu| < 1$ and $|z_\mu| < 1$.

8.3.3 Goluzin inequalities. We will use the Löwner functions $f(z) \in \mathcal{S}$ for which there exists an analytic and univalent function $f(z,t)$ for $z \in E$ with $|f(z,t)| < 1, f(0,t) = 0, f'(0,t) > 0$, satisfies the differential equation (L.2), and $\lim_{t \to \infty} = f(z)$. Set $\zeta = 1/z$. Then let $F(\zeta) = 1/f(1/z) \in \Sigma$, $z_1, \ldots, z_n$ be n arbitrary points in E, and $\zeta_\nu = 1/z_\nu$, $f_\nu = f(z_\nu, t)$. By direct calculations using

$$\frac{\partial f_\nu}{\partial t} = -f_\nu \frac{1 + k(t) f_\nu}{1 - k(t) f_\nu}, \quad \nu = 1, \ldots, n, \tag{8.3.21}$$

we obtain for $\nu, \mu = 1, \ldots, n$,

$$\frac{\partial}{\partial t} \log \left\{ \frac{e^{-t}}{f_\nu f_\mu} \frac{f_\nu - f_\mu}{z_\nu - z_\mu} \right\} = -2 \frac{k f_\nu}{1 - k f_\nu} \frac{k f_\mu}{1 - k f_\mu}, \tag{8.3.22}$$

$$\frac{\partial}{\partial t} \log \left(1 - f_\nu \bar{f}_\mu \right) = -2 \frac{k f_\nu}{1 - k f_\nu} \left(\overline{\frac{k f_\mu}{1 - k f_\mu}} \right). \tag{8.3.23}$$

Note that integrating (8.3.22) and (8.3.23) from 0 to ∞ and using $f(z,0) = z$, we get Eqs (8.3.7) and (8.3.8); also by taking the real parts of (8.3.7) and (8.3.8) and writing $F(\zeta_{\{\nu,\mu\}}) = X_{\{\nu, \mu\}} + iY_{\{\nu, \mu\}}$, respectively, we obtain Eqs (8.3.13) and (8.3.14). As we have seen before, these results lead to Theorem 8.3.4. Let $\alpha_{\nu,\mu}$, $(\nu, \mu = 1, \ldots, n)$ be real numbers such that $\sum \alpha_{\nu,\mu} X_\nu X_\mu$ is a positive quadratic form, i.e., $\sum \alpha_{\nu,\mu} X_\nu X_\mu \geq 0$. In our case we will then have the quadratic forms $\sum \alpha_{\nu,\mu} X_\nu X_\mu$ and $\sum \alpha_{\nu,\mu} Y_\nu Y_\mu$. Then from (8.3.11) and (8.3.12) we obtain the *first Goluzin inequality*:

$$\sum_{\nu,\mu=1}^n \alpha_{\nu\mu} \log \left| 1 - \frac{1}{\zeta_\nu - \bar{\zeta}_\mu} \right| \leq \sum_{\nu,\mu=1}^n \alpha_{\nu\mu} \log \left| \frac{F(\zeta_\nu) - F(\zeta_\mu)}{\zeta_\nu - \zeta_\mu} \right|$$

$$\leq - \sum_{\nu,\mu=1}^n \alpha_{\nu\mu} \log \left| 1 - \frac{1}{\zeta_\nu - \bar{\zeta}_\mu} \right|. \tag{8.3.24}$$

Next, we find from (8.3.7) that

$$\sum_{\nu,\mu=1}^n \gamma_\nu \gamma_\mu \log \frac{F(\zeta_\nu) - F(\zeta_\mu)}{\zeta_\nu - \zeta_\mu} = -2 \int_0^\infty \sum_{\nu,\mu=1}^n \gamma_\nu \gamma_\mu \frac{k f_\nu}{1 - k f_\nu} \frac{k f_\mu}{1 - k f_\mu} \, dt$$

$$= -2 \int_0^\infty \left(\sum_{\nu=1}^n \gamma_\nu \frac{k f_\nu}{1 - k f_\nu} \right)^2 dt,$$

which in absolute values yields

$$\left| \sum_{\nu,\mu=1}^n \gamma_\nu \gamma_\mu \log \frac{F(\zeta_\nu) - F(\zeta_\mu)}{\zeta_\nu - \zeta_\mu} \right| \leq 2 \int_0^\infty \left| \sum_{\nu=1}^n \gamma_\nu \frac{k f_\nu}{1 - k f_\nu} \right|^2 dt$$

$$= 2 \int_0^\infty \sum_{\nu,\mu=1}^n \gamma_\nu \gamma_\mu \frac{k f_\nu}{1 - k f_\nu} \left(\overline{\frac{k f_\mu}{1 - k f_\mu}} \right) dt. \tag{8.3.25}$$

In view of (8.3.8) the right-hand side of the inequality (8.3.25) is equal to

$$-\sum_{\nu,\mu=1}^{n} \gamma_\nu \bar{\gamma}_\mu \log\left\{1 - \frac{1}{\zeta_\nu - \bar{\zeta}_\mu}\right\}.$$

Thus, we obtain the *second Goluzin inequality*:

$$\left|\sum_{\nu,\mu=1}^{n} \gamma_\nu \gamma_\mu \log \frac{F(\zeta_\nu) - F(\zeta_\mu)}{\zeta_\nu - \zeta_\mu}\right| \leq -\sum_{\nu,\mu=1}^{n} \gamma_\nu \bar{\gamma}_\mu \log\left\{1 - \frac{1}{\zeta_\nu - \bar{\zeta}_\mu}\right\}. \qquad (8.3.26)$$

8.4 Exponentiation of Inequalities

The method is as follows: For arbitrary complex numbers γ_ν, $\nu = 1, \ldots, n$, let m be a positive integer. Consider

$$\sum_{\nu,\mu=1}^{n} \gamma_\nu\,\gamma_\mu\left\{\log \frac{z_\nu z_\mu (f_\nu - f_\mu)}{f_\nu f_\mu (z_\nu - z_\mu)}\right\}^m$$

$$= \sum_{\nu,\mu=1}^{n} \gamma_\nu\,\gamma_\mu\left\{-2\int_0^\infty \frac{k f_\nu}{1 - k f_\nu}\frac{k f_\mu}{1 - k f_\mu}\,dt\right\}^m, \quad \text{by (8.3.19)}$$

$$= \sum_{\nu,\mu=1}^{n} \gamma_\nu\,\gamma_\mu\left\{-2\int_0^\infty \frac{k(t) f(z_\nu,t)}{1 - k(t) f(z_\nu,t)}\frac{k(t) f(z_\mu,t)}{1 - k(t) f(z_\mu,t)}\,dt\right\}^m$$

$$= \sum_{\nu,\mu=1}^{n} \gamma_\nu\,\gamma_\mu\,(-2)^m \int_0^\infty \frac{k(\tau_1) f(z_\nu,\tau_1) k(\tau_1) f(z_\mu,\tau_1)\,d\tau_1}{\left[1 - k(\tau_1) f(z_\nu,\tau_1)\right]\left[1 - k(\tau) f(z_\mu,\tau_1)\right]} \cdots$$

$$\cdots \int_0^\infty \frac{k(\tau_m) f(z_\nu,\tau_m) k(\tau_m) f(z_\mu,\tau_m)\,d\tau_m}{\left[1 - k(\tau_m) f(z_\nu,\tau_m)\right]\left[1 - k(\tau_m) f(z_\mu,\tau_m)\right]}$$

$$= (-2)^m \int_0^\infty \cdots \int_0^\infty \sum_{\nu,\mu=1}^{n} \gamma_\nu\,\gamma_\mu \frac{k(\tau_1) f(z_\nu,\tau_1) k(\tau_1) f(z_\mu,\tau_1)}{\left[1 - k(\tau_1) f(z_\nu,\tau_1)\right]\left[1 - k(\tau_1) f(z_\mu,\tau_1)\right]}$$

$$\cdots \frac{k(\tau_m) f(z_\nu,\tau_m) k(\tau_m) f(z_\mu,\tau_m)}{\left[1 - k(\tau_m) f(z_\nu,\tau_m)\right]\left[1 - k(\tau_m) f(z_\mu,\tau_m)\right]}\,d\tau_1 \cdots d\tau_m.$$

Then

$$\left|\sum_{\nu,\mu=1}^{n} \gamma_\nu\,\gamma_\mu\left\{\log \frac{z_\nu z_\mu (f_\nu - f_\mu)}{f_\nu f_\mu (z_\nu - z_\mu)}\right\}^m\right|$$

$$\leq 2^m \int_0^\infty \cdots \int_0^\infty \left|\sum_{\nu=1}^{n}\gamma_\nu \frac{k(\tau_1) f(z_\nu,\tau_1)}{\left[1 - k(\tau_1) f(z_\nu,\tau_1)\right]} \cdots \frac{k(\tau_m) f(z_\mu,\tau_m)}{\left[1 - k(\tau_m) f(z_\mu,\tau_m)\right]}\right|^2 d\tau_1 \cdots d\tau_m$$

$$= 2^m \int_0^\infty \cdots \int_0^\infty \left(\sum_{\nu=1}^{n}\gamma_\nu \frac{k(\tau_1) f(z_\nu,\tau_1)}{\left[1 - k(\tau_1) f(z_\nu,\tau_1)\right]} \cdots \frac{k(\tau_m) f(z_\mu,\tau_m)}{\left[1 - k(\tau_m) f(z_\nu,\tau_m)\right]}\right) \times$$

$$\left(\sum_{\nu=1}^{n} \gamma_\nu \frac{k\left(\tau_1\right) f\left(z_\nu, \tau_1\right)}{\left[1 - k\left(\tau_m\right) f\left(z_\nu, \tau_m\right)\right]} \cdots \frac{k\left(\tau_m\right) f\left(z_\mu, \tau_m\right)}{\left[1 - k\left(\tau_m\right) f\left(z_\nu, \tau_m\right)\right]} \right) d\tau_1 \cdots d\tau_m$$

$$= 2^m \int_0^\infty \cdots \int_0^\infty \left(\sum_{\nu,\mu=1}^{n} \gamma_\nu \bar\gamma_\mu \left\{ \frac{k\left(\tau_1\right) f\left(z_\nu, \tau_1\right)}{\left[1 - k\left(\tau_1\right) f\left(z_\nu, \tau_1\right)\right]} \cdot \overline{\left(\frac{k\left(\tau_1\right) f\left(z_\mu, \tau_1\right)}{\left[1 - k\left(\tau_1\right) f\left(z_\mu, \tau_1\right)\right]} \right)} \right\} \right.$$

$$\left. \cdots \left\{ \frac{k\left(\tau_m\right) f\left(z_\nu, \tau_m\right)}{\left[1 - k\left(\tau_m\right) f\left(z_\nu, \tau_m\right)\right]} \cdot \overline{\left(\frac{k\left(\tau_m\right) f\left(z_\mu, \tau_m\right)}{\left[1 - k\left(\tau_m\right) f\left(z_\mu, \tau_m\right)\right]} \right)} \right\} \right\} d\tau_1 \cdots d\tau_m$$

$$= \sum_{\nu,\mu=1}^{n} \gamma_\nu \bar\gamma_\mu \left\{ 2 \int_0^\infty \frac{k(t) f\left(z_\nu, t\right)}{\left[1 - k(t) f\left(z_\nu, t\right)\right]} \cdot \overline{\left(\frac{k(t) f\left(z_\mu, t\right)}{\left[1 - k(t) f\left(z_\mu, t\right)\right]} \right)} dt \right\}^m$$

$$= \sum_{\nu,\mu=1}^{n} \gamma_\nu \bar\gamma_\mu \left\{ -\log\left(1 - z_\nu z_\mu\right) \right\}^m, \quad \text{by (8.3.20)}.$$

For a sequence of non-negative numbers $\left\{c_m\right\}_{m=0}^{\infty}$ the indicated summations converge, and the above equality gives

$$\left| \sum_{\nu,\mu=1}^{n} \gamma_\nu \gamma_\mu \sum_{m=0}^{\infty} c_m \left\{ \log \frac{z_\nu z_\mu}{f_\nu f_\mu} \cdot \frac{f_\nu - f_\mu}{z_\nu - z_\mu} \right\}^m \right|$$

$$\leq \sum_{\nu,\mu=1}^{n} \gamma_\nu \bar\gamma_\mu \sum_{m=0}^{\infty} |c_m| \left\{ \log \frac{1}{1 - z_\nu \bar z_\mu} \right\}^m. \qquad (8.4.1)$$

Taking $c_m = 1/(m!)$, $m = 0, 1, \ldots$, we note that $\sum_{m=0}^{\infty} c_m X^m = \sum_{m=0}^{\infty} \dfrac{X^m}{m!} = e^X$.

Let $X = \log \dfrac{z_\nu z_\mu}{f_\nu f_\mu} \cdot \dfrac{f_\nu - f_\mu}{z_\nu - z_\mu}$. Then (8.4.1) yields

$$\left| \sum_{\nu,\mu=1}^{n} \gamma_\nu \gamma_\mu \exp\left\{ \log \frac{z_\nu z_\mu}{f_\nu f_\mu} \cdot \frac{f_\nu - f_\mu}{z_\nu - z_\mu} \right\} \right| = \left| \sum_{\nu,\mu=1}^{n} \gamma_\nu \gamma_\mu \frac{z_\nu z_\mu}{f_\nu f_\mu} \cdot \frac{f_\nu - f_\mu}{z_\nu - z_\mu} \right|$$

$$\leq \sum_{\nu,\mu=1}^{n} \gamma_\nu \bar\gamma_\mu \exp\left\{ \log \frac{1}{1 - z_\nu z_\mu} \right\} = \sum_{\nu,\mu=1}^{n} \gamma_\nu \bar\gamma_\mu \left(1 - z_\nu z_\mu\right)^{-1}.$$

Thus,

$$\left| \sum_{\nu,\mu=1}^{n} \gamma_\nu \gamma_\mu \frac{z_\nu z_\mu}{f_\nu f_\mu} \cdot \frac{f_\nu - f_\mu}{z_\nu - z_\mu} \right| \leq \sum_{\nu,\mu=1}^{n} \gamma_\nu \bar\gamma_\mu \left(1 - z_\nu z_\mu\right)^{-1}. \qquad (8.4.2)$$

If, instead, we choose $c_m = (-1)^m/(m!)$, $m = 0, 1, \ldots$, then, since $\sum_{m=0}^{\infty} c_m X^m = \sum_{m=0}^{\infty} \dfrac{(-1)^m}{m!} X^m = e^{-X}$, we find from (8.4.1) that

$$\left| \sum_{\nu,\mu=1}^{n} \gamma_\nu \gamma_\mu \frac{z_\nu z_\mu}{f_\nu f_\mu} \cdot \frac{f_\nu - f_\mu}{z_\nu - z_\mu} \right| \leq \sum_{\nu,\mu=1}^{n} \gamma_\nu \bar\gamma_\mu \left(1 - z_\nu z_\mu\right)^{-1}. \qquad (8.4.3)$$

We have proved the following result:

Theorem 8.4.1. $f \in S$ *if and only if the inequalities (8.3.18), (8.4.1), (8.4.2), or (8.4.3) hold for any complex numbers* γ_ν, $\nu = 1, 2, \ldots, n$ *and for* z_ν *such that* $|z_\nu| < 1$.

8.4.1 Fitzgerald Inequality. Fitzgerald [1977] dropped the log terms from Goluzin inequalities, and considered the following set of inequalities for $f \in S$:

$$\frac{\partial}{\partial t} \left[\log \frac{z_\nu \bar{z}_\mu \left(1 - f_\nu \bar{f}_\mu\right)(f_\nu - f_\mu)\, e^{-t}}{f_\nu \bar{f}_\mu \left(1 - z_\nu \bar{z}_\mu\right)(z_\nu - \bar{z}_\mu)} \right]$$

$$= \frac{\partial}{\partial t} \left[\log \frac{z_\nu}{f_\nu} \frac{z_\mu}{f_\mu} + \log \frac{1 - f_\nu \bar{f}_\mu}{1 - z_\nu \bar{z}_\mu} + \log \frac{f_\nu - f_\mu}{z_\nu - z_\mu} - t \right]$$

$$= \frac{f_\nu}{z_\nu} \frac{\bar{f}_\mu}{z_\mu}(z_\nu \bar{z}_\mu) \left(- \frac{f_\nu \dfrac{\partial \bar{f}_\mu}{\partial t} + \dfrac{\partial f_\nu}{\partial t} \bar{f}_\mu}{\left[f_\nu \bar{f}_\mu\right]^2} \right) + \frac{1 - z_\nu \bar{z}_\mu}{1 - z_\nu \bar{z}_\mu}\left(- \frac{f_\nu \dfrac{\partial \bar{f}_\mu}{\partial t} - \dfrac{\partial f_\nu}{\partial t} \bar{f}_\mu}{\left[f_\nu \bar{f}_\mu\right]^2} \right)$$

$$+ \frac{z_\nu}{f_\nu} \frac{z_\mu}{f_\mu} \left(\frac{\partial f_\nu}{\partial t} - \frac{\partial \bar{f}_\mu}{\partial t} \right) \quad \text{by Löwner's Eq (4.2.2) and with } f_\nu = f\,(z_\nu, t),$$

$$= -\frac{1}{f_\nu \bar{f}_\mu} \left\{ - f_\nu \bar{f}_\mu \cdot \frac{1 - \overline{kf_\mu}}{1 - \overline{kf_\mu}} - f_\nu \bar{f}_\mu \frac{1 + kf_\nu}{1 - kf_\nu} \right\}$$

$$- \frac{1}{f_\nu \bar{f}_\mu} \left\{ - f_\nu \bar{f}_\mu \cdot \frac{1 - \overline{kf_\mu}}{1 - \overline{kf_\mu}} - f_\nu \bar{f}_\mu \frac{1 + kf_\nu}{1 - kf_\nu} \right\}$$

$$+ \frac{1}{f_\nu \bar{f}_\mu} \left\{ - f_\nu \bar{f}_\mu \cdot \frac{1 - \overline{kf_\mu}}{1 - \overline{kf_\mu}} - f_\nu \bar{f}_\mu \frac{1 + kf_\nu}{1 - kf_\nu} \right\} - 1$$

$$= -1 + \frac{f_\nu f_\mu}{1 - f_\nu \bar{f}_\mu} \left[\frac{1 - k\bar{k}f_\nu \bar{f}_\mu}{(1 - kf_n u)\left(1 - \overline{kf_\mu}\right)} + \frac{1 - k\bar{k}f_\nu \bar{f}_\mu}{(1 - kf_n u)\left(1 - \overline{kf_\mu}\right)} \right]$$

$$\times \frac{f_\nu \bar{f}_\mu}{f_\nu \bar{f}_\mu} \left[\frac{1 - k\bar{k}f_\nu \bar{f}_\mu + 1 - k\bar{k}f_\nu \bar{f}_\mu}{(1 - kf_n u)\left(1 - \overline{kf_\mu}\right)} \right]$$

$$\times \frac{1}{f_\nu - \bar{f}_\mu} \left\{ - f_\nu \frac{1 + kf_\nu}{1 - kf_\nu} + f_\mu \frac{1 + kf_\nu}{1 - kf_\nu} \right\} \quad \text{(Recall } k\bar{k} = |k|^2 = 1.\text{)}$$

$$= -1 + \frac{2 f_\nu \bar{f}_\mu}{(1 - kf_\nu)\left(1 - \overline{kf_\mu}\right)} - \frac{(f_\nu - f_\mu)\left[1 + k\,(f_\nu - f_\mu) - k^2 f_\nu f_\mu\right]}{(f_\nu - f_\mu)\,(1 - kf_\nu)\,(1 - kf_\mu)}$$

$$- 2\frac{\left(1 - f_\nu \bar{f}_\mu\right)}{(1 - kf_\nu)\left(1 - \overline{kf_\mu}\right)}$$

$$= \frac{2}{(1 - kf_\nu)\left(1 - \overline{kf_\mu}\right)} - \frac{2}{(1 - kf_\nu)\,(1 - kf_\mu)},$$

$$= 2\left[\,(a_\nu + ib_\nu)\,(a_\mu - ib_\mu) - (a_\nu + ib_\nu)\,(a_\mu + ib_\mu)\,\right]$$

$$= 4\left[\,b_\nu\, b_\mu - i a_\nu b_\nu\,\right],$$

where we have set $\dfrac{1}{(1 - k f_\nu)} = a_\nu + i\, b_\nu$, so that $\dfrac{1}{(1 - k f_\mu)} = a_\nu\, b_\nu$. Similarly,

$$\frac{\partial}{\partial t}\left[\log \frac{z_\nu \bar{z}_\mu \left(1 - f_\nu \bar{f}_\mu\right)\left(f_\nu - f_\mu\right) e^{-t}}{f_\nu \bar{f}_\mu \left(1 - z_\nu \bar{z}_\mu\right)\left(z_\nu - \bar{z}_\mu\right)}\right] = 4\left[b_\nu\, b_\mu + i a_\nu b_\nu\right].$$

Hence,

$$\frac{\partial}{\partial t}\left[\log \frac{z_\nu \bar{z}_\mu \left(1 - f_\nu \bar{f}_\mu\right)\left(f_\nu - f_\mu\right) e^{-t}}{f_\nu \bar{f}_\mu \left(1 - z_\nu \bar{z}_\mu\right)\left(z_\nu - \bar{z}_\mu\right)}\right]$$

$$+ \frac{\partial}{\partial t}\left[\overline{\log \frac{z_\nu \bar{z}_\mu \left(1 - f_\nu \bar{f}_\mu\right)\left(f_\nu - f_\mu\right) e^{-t}}{f_\nu \bar{f}_\mu \left(1 - z_\nu \bar{z}_\mu\right)\left(z_\nu - \bar{z}_\mu\right)}}\right] = 8\, b_\nu b_\mu. \quad (8.4.4)$$

Let m be a positive integer. Then, denoting the first expression within square brackets on the right hand side of (8.4.4) by $[\cdots]$ and the second one with the square brackets by $\overline{[\cdots]}$, we get

$$\sum_{\nu,\mu=1}^{n} \gamma_n u\, \bar{\gamma}_\mu \int_0^\infty \frac{\partial}{\partial t}\left\{[\cdots] + \overline{[\cdots]}\right\} dt \Big\}^m$$

$$= \sum_{\nu,\mu=1}^{n} \gamma_n u\, \bar{\gamma}_\mu \left\{\int_0^\infty \left\{8 n_\nu b_\mu\, dt\right\}^m \quad \text{by (8.4.4)}\right.$$

$$= 8^m \sum_{\nu,\mu=1}^{n} \gamma_n u\, \bar{\gamma}_\mu \int_0^\infty b_\nu(\tau_1)\, d\tau_1 \cdots \int_0^\infty b_\nu(\tau_m)\, d\tau_m;$$

since b_ν and b_μ are real, i.e., $\bar{b}_\mu = b_\mu$, $\bar{b}_\nu = b_\nu$, and since the series are uniformly convergent,

$$= 8^m \int_0^\infty \cdots \int_0^\infty \left(\sum_{\nu=1}^{n} \gamma_n u\, b_\nu(\tau_1) \cdots b_\nu(\tau_m)\right)$$

$$\left(\sum_{\nu=1}^{n} \gamma_m u\, b_\mu(\tau_1) \cdots b_\mu(\tau_m)\right) d\tau_1 \cdots d\tau_m$$

$$= 8^m \int_0^\infty \cdots \int_0^\infty \left|\sum_{\nu=1}^{n} \gamma_n u\, b_\nu(\tau_1) \cdots b_\nu(\tau_m)\right|^2 d\tau_1 \cdots d\tau_m \geq 0.$$

Hence for a sequence of arbitrary positive constants $\left\{C_m\right\}_{m=1}^{\infty}$, we get

$$\sum_{\nu,\mu=1}^{n} \gamma_n u\, \bar{\gamma}_\mu \sum_{m=1}^{\infty} C_m \left\{\int_0^\infty \frac{\partial}{\partial t}\left\{[\cdots]\right\} + \left\{\overline{[\cdots]}\right\} dt\right\}^m \geq 0. \quad (8.4.5)$$

Let $C_m = \dfrac{1}{2^m m!}$, $m = 1, 2, \ldots ,$; and let $X = \{ \ \}$. Then

$$\sum_{m=1}^{\infty} C_m = \sum_{m=1}^{\infty} \frac{1}{2^m m!} X^m = \sum_{m=1}^{\infty} \frac{1}{m!} \left(\frac{X}{2}\right)^m = e^{X/2} - 1.$$

Now, let

$$\frac{X}{2} = \frac{1}{2}\left\{ [\ ] + \overline{[\ ]}\Big|_0^{\infty} \right\} = \frac{1}{2}\left\{ 2\Re\left\{ [\ ]\right\}\Big|_0^{\infty} \right\}$$

$$= \Re\left\{ \left[\log \frac{z_\nu z_\mu \left(1 - f_\nu \bar{f}_\mu\right)\left(f_\nu - f_\mu\right) e^{-t}}{f_\nu \bar{f}_\mu \left(1 - z_\nu \bar{z}_\mu\right)\left(z_\nu - z_\mu\right)} \right]_0^{\infty} \right\}.$$

Also, from the definition of the class $\mathcal{S}_t$ (see §4.3.1), we have (i) $\Re\left\{\log w\right\} = \log |w|$; (ii) $\lim\limits_{t\to\infty} e^t f_\nu \equiv \lim\limits_{t\to\infty} e^t f\left(z_\nu, t\right) = f\left(z_\nu\right) = f_\nu$; and (iii) $f\left(z_\nu, 0\right) = z_\nu$ so that

$$\frac{X}{2} = \Re\left\{ \left[\log \frac{z_\nu z_\mu \left[f\left(z_\nu\right) - f\left(z_\mu\right)\right]}{f\left(z_\nu\right)\overline{f\left(z_\mu\right)}\left(1 - z_\nu \bar{z}_\mu\right)\left(z_\nu - z_\mu\right)} \right] - \right.$$

$$\left. \left[\log \frac{z_\nu z_\mu \left(1 - z_\nu \bar{z}_\mu\right)\left(z_\nu - z_\mu\right)}{z_\nu z_\mu \left(1 - z_\nu \bar{z}_\mu\right)\left(z_\nu - z_\mu\right)} \right\} \right.$$

$$= \log\left| \frac{z_\nu \bar{z}_\mu \left[f\left(z_\nu\right) - f\left(z_\mu\right)\right]}{f\left(z_\nu\right)\overline{f\left(z_\mu\right)}\left(1 - z_\nu \bar{z}_\mu\right)\left(z_\nu - z_\mu\right)} \right|.$$

Then from (8.4.5) we get

$$\sum_{\nu,\mu=1}^{n} \gamma_\nu \bar{\gamma}_\mu \left\{ \exp \log \left| \frac{z_\nu \bar{z}_\mu \left[f\left(z_\nu\right) - f\left(z_\mu\right)\right]}{f\left(z_\nu\right)\overline{f\left(z_\mu\right)}\left(1 - z_\nu \bar{z}_\mu\right)\left(z_\nu - z_\mu\right)} \right| - 1 \right\} \geq 0,$$

or

$$\sum_{\nu,\mu=1}^{n} \gamma_\nu \bar{\gamma}_\mu \left| \frac{z_\nu \bar{z}_\mu \left[f\left(z_\nu\right) - f\left(z_\mu\right)\right]}{f\left(z_\nu\right)\overline{f\left(z_\mu\right)}\left(1 - z_\nu \bar{z}_\mu\right)\left(z_\nu - z_\mu\right)} \right|$$

$$\geq \sum_{\nu,\mu=1}^{n} \gamma_\nu \bar{\gamma}_\mu \geq \sum_{\nu}^{n} |\gamma_\nu|^2 = \left| \sum_{\nu=1}^{n} \gamma_\nu \right|^2. \tag{8.4.6}$$

Put $\gamma_\nu = \alpha_\nu \left| \dfrac{f\left(z_\nu\right)}{z_\nu} \right|$ in (8.4.6). Then

$$\sum_{\nu,\mu=1}^{n} \alpha_\nu \left| \frac{f\left(z_\nu\right)}{z_\nu} \right| \bar{\alpha}_\nu \left| \overline{\frac{f\left(z_\nu\right)}{z_\nu}} \right| \left| \frac{z_\nu \bar{z}_\mu \left[f\left(z_\nu\right) - f\left(z_\mu\right)\right]}{f\left(z_\nu\right)\overline{f\left(z_\mu\right)}\left(1 - z_\nu \bar{z}_\mu\right)\left(z_\nu - z_\mu\right)} \right|$$

$$\geq \left| \sum_{\nu=1}^{n} \alpha_\nu \left| \frac{f\left(z_\nu\right)}{z_\nu} \right|^2 \right|,$$

which leads to the following result:

Theorem 8.4.1. (Fitzgerald inequality) Let $f \in \mathcal{S}$. If $\alpha_1, \ldots \alpha_n$ are n arbitrary complex numbers, and $z_1, \ldots, z_n$ are n arbitrary points in E, then

$$\sum_{\nu=1}^{n} \sum_{\mu=1}^{n} \alpha_\nu \bar{\alpha}_\mu \left| \frac{f(z_\nu) - f(z_\mu)}{z_\nu - z_\mu} \frac{1}{1 - z_\nu \bar{z}_\mu} \right|^2$$

$$\geq \sum_{\nu=1}^{n} \sum_{\mu=1}^{n} \alpha_\nu \bar{\alpha}_\mu \left| \frac{f(z_\nu) f(z_\mu)}{z_\nu z_\mu} \right|^2 . \qquad (8.4.7)$$

The inequality (8.4.7) gives the following Fitzgerald coefficient inequality:

Theorem 8.4.2. Let $f \in \mathcal{S}$ with the power series expansion (6.1.1). For $\nu \leq \mu$, define $\beta_n(\nu, \mu)$ by

$$\beta_n(\nu, \mu) = \begin{cases} \nu - |n - \mu| & \text{if } |n - \mu| < \nu, \\ 0 & \text{if } |n - \mu| \geq \nu, \end{cases} \qquad (8.4.8)$$

and $\beta_n(\nu, \mu) = \beta(\mu, \nu)$. Denote

$$a_{\nu\mu}(f) = \sum_{k=1}^{\nu+\mu-1} \beta_k(\nu, \mu) |a_k|^2 - |a_\nu|^2 |a_\mu|^2 .$$

Then the matrix

$$(a_{\nu\mu}) \geq 0. \qquad (8.4.9)$$

PROOF. Set $n = Nm$ in (8.4.4). Let z_ν, a_ν be given by $z_{\nu\mu} = r_p \, e^{2\pi i \nu/m}$, $a_{p\mu} = m^{-1} r_p$, respectively, where $0 < r_p < 1$. As $m \to \infty$, the sum in (8.4.4) reduces to an integral, and

$$\left(\sum_{p=1}^{N} \frac{\gamma_p}{2\pi} \int_0^{2\pi} \left| \frac{f(r_p e^{it})}{r_p e^{it}} \right|^2 dt \right)^2$$

$$\leq \sum_{p=1}^{N} \sum_{q=1}^{N} \frac{\gamma_p \gamma_q}{(2\pi)^2} \int_0^{2\pi} \int_0^{2\pi} \left| \frac{f(r_p e^{is}) - f(r_q e^{it})}{(r_p e^{is} - r_q e^{it})(1 - r_p r_q e^{i(s-t)})} \right|^2 ds \, dt. \qquad (8.4.10)$$

To calculate the integrals in (8.4.10), note that since

$$\frac{f(z) - f(\zeta)}{z - \zeta} = \sum_{j=0}^{\infty} \sum_{k=0}^{\infty} a_{j+k+1} z^j \zeta^k, \qquad \frac{1}{1 - z\bar{\zeta}} = \sum_{l=0}^{\infty} z^l \bar{\zeta}^l,$$

the integrand on the right-hand side of (8.4.10) is

$$\sum_{j,k,l \geq 0} \sum_{j',k',l' \geq 0} a_{j+k+1} \bar{a}_{j'+k'+1} r_p^{j+k+j'+k'} r_q^{k+l+k'+l'} e^{i(j+k-j'-k')s} e^{i(k+l-k'-l')t}.$$

Clearly the integrands of all terms except those with $j + k = j' + k'$, $k - l = k' - l'$ are zero. Set

$$n = j + k + 1 = j' + k' + 1,$$
$$p = j + l + 1 = j' + l' + 1,$$
$$\nu = k + l' + 1 = k' + l + 1.$$

Then the integrand on the right hand side of (8.4.10) becomes

$$\sum_{nu=1}^{\infty} \sum_{\mu=1}^{\infty} \sum_{n=1}^{\infty} |a_n|^2 r_p^{2(\nu-1)} r_q^{2(\mu-1)} \beta_n(\nu, \mu),$$

where $\beta_n(\nu, \mu)$ is the number of integers l which satisfy the following conditions:

$$0 \leq l \leq \nu + \mu - n - 1,$$
$$\nu - n \leq l \leq \nu - 1,$$
$$\mu - n \leq l \leq \mu - 1.$$

When $\nu \leq mu$, these conditions become: $\begin{cases} \mu - n \leq l \leq \nu - 1, & \text{when } n \leq \mu, \\ 0 \leq l \leq \nu + \mu - n - 1 & \text{when } n > \mu. \end{cases}$

But this is precisely the definition (8.4.8). Next, the integral on the left hand side of (8.4.10) is equal to $\sum_{n=1}^{\infty} |a_n|^2 r_p^{2(n-1)}$. Thus, the inequality (8.4.10) reduces to

$$\left(\sum_{n=1}^{\infty} |a_n|^2 \lambda_n \right)^2 \leq \sum_{nu=1}^{\infty} \sum_{\mu=1}^{\infty} \sum_{n=1}^{\infty} |a_n|^2 \lambda_\nu \lambda_\mu \beta_n(\nu, \mu), \qquad (8.4.11)$$

where $\lambda_\mu = \sum_{p=1}^{N} \gamma_p r_p^{2(p-1)}$, $\mu = 1, 2, \ldots$. For any $n = 1, \ldots, N$, let $0 < s_1 < \ldots < s_N$. Then we can choose real numbers a_{np}, $p = 1, \ldots, N$, such that for any real numbers $x_1, \ldots, x_N$,

$$\sum_{p=1}^{N} s_p^{2(p-1)} a_{np} = \begin{cases} x_n & \text{if } \nu = n, \\ 0 & \text{if } \nu = 1, \ldots, N, \nu \neq n. \end{cases} \qquad (8.4.12)$$

Since the coefficient determinant of the system of equations (8.4.12) is a Vandermonde determinant, this system has a unique solution. Let $0 < \delta < 1$, and let

$$r_p = \delta s_p, \quad \gamma_p = \sum_{p=1}^{N} s_p^{2(p-1)} \delta^{-2(p-1)} a_{np}, \quad p = 1, \ldots, N.$$

Then

$$\lambda_p = \lambda_p(\delta) = \sum_{n=1}^{N} \delta^{2(\nu-n)} \sum_{p=1}^{N} s_p^{2(p-1)}, \quad \nu = 1, \ldots, N.$$

Since (8.4.12) gives $\lambda_p = x_\nu$ for $1 \leq \nu \leq N$, and from the above equation we get $\lambda_n = O\left(\delta^{2(\nu-n)}\right)$ for $\nu > N$. Let $\delta \to 0$, we get for any real numbers $x_1, \ldots, x_n$,

$$\left(\sum_{n=1}^{N} |a_n|^2 x_n\right)^2 \leq \sum_{\nu=1}^{N} \sum_{\mu=1}^{N} \sum_{n=1}^{\nu+\mu-1} \beta_n(\nu, \mu) |a_n|^2 x_\nu x_\mu, \qquad (8.4.13)$$

which is precisely (8.4.9). ∎

Since the main diagonal of (8.4.9) is non-negative, we have

Corollary 8.4.1. If $f(z) \in S$ has the series expansion (6.1.1), then

$$|a_n|^4 \leq \sum_{k=1}^{N} k|a_k|^2 + \sum_{k=n+1}^{N} (2n-k)|a_k|^2, \quad n = 2, 3, \ldots. \qquad (8.4.14)$$

This corollary implies the following result.

Corollary 8.4.2. (Fitzgerald inequality) If $f(z) \in S$ has the series expansion (6.1.1), then

$$|a_n| < \sqrt{\frac{7}{6}}\, n < 1.081\, n, \quad n = 2, 3, \ldots. \qquad (8.4.15)$$

PROOF. Since $\sum_{k=1}^{N} k^2 = \frac{1}{6}n(n_1)(2n+1)$, and $\sum_{k=1}^{N} k^3 = \frac{1}{4}n^2(n+1)^2$, the substitution of $|a_k|$ by k in the right hand side of inequality (8.4.15) gives

$$\sum_{k=1}^{N} k^3 + \sum_{k=n+1}^{N} (2n-k)\, k^2 = \frac{7}{6}\, n^4 - \frac{1}{6}\, n^2.$$

Let $c = \sup_n \sup_{f \in S} \dfrac{|a_n|}{n}$, which means that c is the smallest constant such that $|a_n| < cn$ for all $f \in S$ and for all n. Then by the Littlewood theorem (Theorem 8.1.1), we have $1 \leq c \leq e$. For any $\varepsilon > 0$, there exists an integer n and a function $f \in S$ such that $|a_n| \geq (c-e)\, n$. Thus, by (8.4.14),

$$(c-\varepsilon)^4 n^4 < |a_n|^4 \leq c^2 \left(\frac{7}{6}\, n^4 - \frac{1}{6}\, n^2\right),$$

i.e.,

$$(c-\varepsilon)^4 n^4 < \frac{7}{6}\, n^4 - \frac{1}{6}\, n^2,$$

which on letting $\varepsilon \to 0$ we get $c = \sqrt{7/6}$. ∎

Corollary 8.4.3. (Horowitz's estimate [1978]) If $f(z) \in \mathcal{S}$ has the series expansion (6.1.1), then

$$|a_n| \le \left(\frac{1659164137}{681080400}\right)^{1/12} n < 1.0657\, n, \quad n = 2, 3, \ldots. \tag{8.4.16}$$

PROOF. we start from the main result (8.4.9) of Theorem 8.3.2. Let $\lambda = (\lambda_2, \ldots, \lambda_{2n})$, where $\lambda_\nu = n - |n - \nu|$, $\nu = 1, 2, \ldots, 2n$. Then we write the inequality $\lambda\,(a_{\nu\mu}) \ge 0$ as

$$\left|\sum_{\nu=1}^{2n} \lambda_\nu |a_\nu|^2\right|^2 \le \sum_{\nu=1}^{2n} |\lambda_\nu|^2 \left\{\sum_{k=n+1}^{N} (2n-k)\,|a_k|^2\right\} + 2 \sum_{1 \le \nu_1 < \nu_2 \le 2n} \lambda_{\nu_1} \lambda_{\nu_2},\, \times$$

$$\times \left\{ \sum_{k=\nu_2-\nu_1}^{\nu_2} (\nu_1 - \nu_2 + k)\,|a_k|^2 + \sum_{k=\nu_2+1}^{\nu_1+\nu_2} (\nu_1 + \nu_2 - k)\,|a_k|^2\right\}.$$

The left hand side of this inequality is the square of the right hand side of (8.4.14). Thus,

$$|a_n|^8 \le \sum_{\nu=1}^{2n} \lambda_\nu^2 \left\{\sum_{k=1}^{\nu} k|a_k|^2 + \sum_{k=\nu+1}^{2\nu} (2\nu - k)\,|a_k|^2\right\}$$

$$+ 2 \sum_{j=2}^{2n} \sum_{m=1}^{j-1} \lambda_j \lambda_m \left\{\sum_{k=j-m}^{j} (m - j + k)|a_k|^2\right.$$

$$+ \left.\sum_{k=j+1}^{j+m} (m + j - k)|a_k|^2\right\}. \tag{8.4.17}$$

Again, let c be defined as in Corollary 8.4.2. Then for any $\varepsilon > 0$, there exists an integer n and a function $f \in \mathcal{S}$ such that $|a_n| > (c - e)\, n$. Using the same argument as in the proof of Corollary 8.4.2, we get

$$n^8 (c - \varepsilon)^8 \le c^2 \left[\sum_{\nu=1}^{2n} \lambda_\nu^2 \left\{\sum_{k=1}^{\nu} k^2 + \sum_{k=\nu+1}^{2\nu} (2\nu - k)k^2\right\}\right.$$

$$+ 2 \sum_{j=2}^{2n} \sum_{m=1}^{j-1} \lambda_j \lambda_m \left\{\sum_{k=j-m}^{j+m} (j)(m - j + k)k^2 + \sum_{k=j+1}^{j+m} (m + j - k)k^2\right\}\right]. \tag{8.4.18}$$

Using the summation formulas for $\sum_{k=1}^{N} k^j$, $j = 1, 2, \ldots, 7$, and after a lengthy calculation we find that the right hand side of the inequality (8.4.18) is equal to

$$c^2 \left[\frac{1}{1260}\left(1881n^8 - 602n^6 + 49n^4 - 68n^2\right)\right] < c^2 \frac{1881}{1260}\, n^8.$$

Since ε is arbitrary, let $\varepsilon \to 0$, and this yields $c^6 \leq \dfrac{209}{140}$, or $c \leq \left(\dfrac{209}{140}\right)^{1/6} <$ 1.0691. Thus, for $f \in \mathcal{S}$, this results $|a_n| \leq \left(\dfrac{209}{140}\right)^{1/6} n < 1.069n$ for $n = 2, 3, \ldots$. The improved result (8.4.16) was obtained by Horowitz [1978] by using Fitzgerald inequality (8.4.15) repeatedly. ∎

This was the best estimate so far. However, by that time it was obvious that the Bieberbach conjecture could not be proved by pursuing this line of estimating the coefficients.

Based on Fitzgerald's inequality, Pommerenke [1975] proved that

$$\limsup_{n \to \infty} \frac{|a_n|}{n} < 1$$

unless f is the Koebe function. A proof is available in Gong [1999: §3.3].

Let $\phi(z) = \sum_{k=1}^{\infty} \beta_k z^k$ be an arbitrary power series expansion with a positive radius of convergence, and $\phi(0) = 0$. Define

$$e^{\phi(z)} = \sum_{k=0}^{\infty} \beta_k z^k, \quad \beta_0 = 1. \tag{8.4.19}$$

Lebedev and Milin [1965] and Milin [1971] used the Cauchy-Schwarz inequality (1.8.4) to prove that if $\psi(z) = \sum_{k=0}^{\infty} \beta_k z^k$, $\beta_0 = 1$, has positive radius of convergence, then the same is true for $\phi(z) = \log \psi(z) = \sum_{k=1}^{\infty} \alpha_k z^k$. In fact, Lebedev and Milin [1965] proved the following three inequalities, which are totally independent of univalent functions, although they helped solve some interesting function-theoretic problems, e.g., as in Aharonov [1970], Leung [1978; 1979], and Nehari [1970], where these inequalities were used in their application to the Bieberbach conjecture.

Theorem 8.4.3. (Lebedev-Milin inequalities) *We have*

$$\sum_{k=0}^{\infty} |\beta_k|^2 \leq \exp\left\{\sum_{k=0}^{\infty} k|\alpha_k|^2\right\}, \tag{8.4.20}$$

$$\frac{1}{n+1} \sum_{k=0}^{\infty} |\beta_k|^2 \leq \exp\left\{\frac{1}{n+1} \sum_{m=1}^{n} \sum_{k=1}^{m} \left(k|\alpha_k|^2 - \frac{1}{k}\right)\right\}, \tag{8.4.21}$$

$$|\beta_k|^2 \leq \exp\left\{\sum_{k=1}^{n} \left(k|\alpha_k|^2 - \frac{1}{k}\right)\right\}. \tag{8.4.22}$$

A proof of these inequalities is given in Exercise 8.5.2.

If we set $\psi(z) = \dfrac{f(z)}{z} = \sum\limits_{n=0}^{\infty} a_{n+1} z^n$, then $\phi(z) = \log \dfrac{f(z)}{z} = \sum\limits_{n=1}^{\infty} c_n z^n$.

If $f \in \mathcal{S}$, then c_n are the logarithmic coefficients of f. Let $h(z) = \sum\limits_{n=1}^{\infty} c_n z^n$ be the square root transform of f (see Exercise 6.8.3). Assume that for some $n \in \mathbb{N}$, the *Milin conjecture*

$$\sum_{k=1}^{n} (n+1-k)\left(k|c_k|^2 - \frac{4}{k} \right) \leq 0, \tag{8.4.23}$$

holds. For its proof, see §10.1. Then, defining $\log \dfrac{h(z)}{z} = \tfrac{1}{2}\phi(z^2)$, it follows from the inequality (8.4.21) that

$$\frac{1}{n+1} \sum_{k=1}^{n+1} |c_{2k-1}|^2 \leq 1, \tag{8.4.24}$$

which implies the Robertson conjecture and therefore, the Bieberbach conjecture for $|a_{n+1}|$.

The Milin conjecture is a statement about a weighted quadratic mean of the logarithmic coefficients or of the coefficients of the function

$$\frac{zf'(z)}{f(z)} = 1 + z\left(\log \frac{f(z)}{z} \right)' = 1 + \sum_{n=1}^{\infty} n c_n z^n, \tag{8.4.25}$$

whose modulus is bounded by 2 for $f \in \mathcal{S}^*$, so that $|c_k| \leq \dfrac{2}{k}$, $(k \in \mathbb{N})$, for starlike functions, and therefore, the Milin conjecture is valid.

Note that for the Koebe function $K_0(z)$, $c_n \leq 2/n$; for the case of the image domain $f(E)$ that are star-shaped relative to the origin, we have $|c_n| \leq 2/n$, and this inequality already implies the Bieberbach conjecture for starlike functions f. This result was proved by Nevanlinna [1920].

For starlike functions f, a geometric argument at the boundary shows that

$$\Re\left\{ \frac{zf'(z)}{f(z)} \right\} = \Re\left\{ 1 + \sum_{n=1}^{\infty} n c_n z^n \right\} > 0, \tag{8.4.26}$$

so that, using the Carathéodory inequality for functions with positive real part (§6.5), we obtain $n|c_n| \leq 2$. Note that the inequality $|c_n| \leq 2/n$, or $n|c_n|^2 - \dfrac{4}{n} \leq 0$, does not hold for every $f \in \mathcal{S}$, but Lebedev and Milin [1971] conjectured that the latter inequality is true in the following average sense:

$$\Omega_n := \sum_{m=1}^{n-1}\sum_{k=1}^{m}\left(k|c_k|^2 - \frac{4}{k}\right) = \sum_{k+1}^{n-1}\left(k|c_k|^2 - \frac{4}{k}\right)(n+1-k) \le 0 \qquad (8.4.27)$$

for $n = 2, 3, \ldots$ and for all $f \in \mathcal{S}$. This conjecture became known in English only in Milin [1977]; it implies the Robertson conjecture and hence the Bieberbach conjecture are true.

The Lebedev-Milin inequality in the general case for the coefficients of $\sum_{n=0}^{\infty}\beta_n z^n = \exp\left(\sum_{n=0}^{\infty}\gamma_n z^n\right)$ implies that (see Duren [1983])

$$\sum_{k=0}^{n-1}|\beta_k|^2 \le n\exp\left\{\frac{1}{n}\sum_{m=1}^{n-1}\sum_{k=1}^{m}\left(k|\gamma_k|^2 - \frac{1}{k}\right)\right\}, \qquad n = 1, 2, \ldots. \qquad (8.4.28)$$

If we apply the inequality (8.4.28) to the identity

$$z + b_3 z + \cdots + b_{2n-1}z^{n-1} + \cdots = \frac{f_1(\sqrt{z})}{\sqrt{z}}$$

$$= \sqrt{\frac{f(z)}{z}} = \exp\left\{\frac{1}{2}\log\frac{f(z)}{z}\right\} = \exp\left\{\frac{1}{2}\sum_{n=1}^{\infty}c_n z^n\right\}, \qquad (8.4.29)$$

we get

$$|b_1|^2 + \cdots + |b_{2n-1}|^2 \le n\exp\left\{\frac{\Omega_n}{4n}\right\}. \qquad (8.4.30)$$

Thus, if the Lebedev-Milin conjecture (8.4.27) holds for $f \in \mathcal{S}$ and for a certain n, then the Robertson conjecture (3.1.10) holds for the corresponding f_1 and the same n, so also the Bieberbach conjecture must hold for the same n. Moreover, if $\Omega_n < 0$ for some n, then there is the strict inequality in (8.4.27) and also in the Bieberbach conjecture.

8.4.2 Bazilevich Inequality. Let $f \in \mathcal{S}$, and let

$$\log\frac{f(z)}{z} = 2\sum_{n=1}^{\infty}\gamma_n z^n, \qquad z \in E. \qquad (8.4.31)$$

If $f(z)$ is the Koebe function, then $\gamma_n = 1/n$. Bazilevich [1967] proved the following result.

Theorem 8.4.4. (Bazilevich inequality) Let $f \in \mathcal{S}$, γ_n be defined by (8.4.31), and let $e^{i\theta_0}$ be the Hayman direction of f. Then

$$\sum_{n=1}^{\infty}n\left|\gamma_n - \frac{1}{n}e^{-in\theta_0}\right|^2 \le \frac{1}{2}\log\frac{1}{\alpha'}, \qquad (8.4.32)$$

where $\alpha > 0$ is the Hayman index of the function f.

It is clear from the inequality (8.4.32) that the closer α gets to 1, the 'closer' f gets to the Koebe function. However, the inequality $|\gamma_n| \le 1/n$ is, in general, not true; it is true only for some sub-families of $\mathcal{S}$, such as the starlike univalent functions. For proof of this theorem, see Exercise 8.4.1.

Lemma 8.4.1. (Milin's Lemma) For any $f \in \mathcal{S}$, the following inequality holds:

$$\sum_{k=1}^{n} k |\gamma_k|^2 \le \sum_{k=1}^{n} \frac{1}{k} + \delta, \quad \delta < 0.312. \tag{8.4.33}$$

For a proof, based on Gong [1999], see Exercise 8.5.4. Lemma 8.4.1 implies the following theorem.

Theorem 8.4.5. Let $h(z)$ be defined by (8.1.3). Then

$$|c_n| < e^{\delta/2} < 1.17 \quad \text{for } n = 2, 3, \ldots. \tag{8.4.34}$$

PROOF. Using the definition of $h(z)$, we have

$$\log \frac{h(\sqrt{z})}{\sqrt{z}} = \frac{1}{2} \log \frac{f(z)}{z} = \sum_{n=1}^{\infty} \gamma_n z^n,$$

or

$$\sum_{n=0}^{\infty} c_{2n+1} z^n = \exp\left\{ \sum_{n=1}^{\infty} \gamma_n z^n \right\}, \quad c_1 = 1.$$

But from inequality (8.4.21) we get

$$|c_{2n+1}|^2 \le \exp\left\{ \sum_{k=1}^{n} k |\gamma_k|^2 - \sum_{k=1}^{n} \frac{1}{k} \right\}.$$

Hence, using Milin's Lemma, we find that $|c_{2n+1}| \le e^{\delta/2} < e^{0.146} < 1.17$ for $n = 2, 3, \ldots$. ∎

Even after determining the best estimate, it was soon realized that the Bieberbach conjecture could not be proved by pursuing this line of estimating the coefficients. In the mean time Pommerenke [1975] proved that

$$\limsup_{n \to \infty} \frac{|a_n|}{n} < 1..$$

It was a fervent hope during the 1970s that such conjectures and inequalities, leading to the numerical estimates for the coefficients, may eventually help find the solution of the Bieberbach conjecture.

We summarize the numerical estimate in Table 8.4.1.

Table 8.4.1 Coefficient Estimates.

Author	Year	$|a_n| < C\,n$
Littlewood	1925	$|a_n| < e\,n \approx 2.7183\,n$
Landau	1926	$|a_n| < \left(\dfrac{1}{2} + \dfrac{1}{\pi}\right) e\,n \approx 2.2244\,n$
Goluzin	1946	$|a_n| < \frac{3}{4} e\,n \approx 2.0388\,n$
Bazilevich	1948	$|a_n| < \dfrac{9}{4}\left(\dfrac{1}{\pi}\displaystyle\int_0^\pi \dfrac{\sin x}{x}\,dx + 0.2649\right) n \approx 1.924\,n$
Milin*	1949	$|a_n| < \frac{1}{2} e\,n + 1.8 \approx 1.3592\,n + 1.8$
Bazilevich	1949	$|a_n| < \frac{1}{2} e\,n + 1.51 \approx 1.3592\,n + 1.51$
Milin	1964	$|a_n| < \dfrac{e^{1.6} - 1}{1.6}\,n \approx 1.2427\,n$
Fitzgerald	1977	$|a_n| < \left(\dfrac{209}{140}\right)^{1/6} n \approx 1.0691\,n$
Horowitz	1978	$|a_n| < \left(\dfrac{1659164137}{681080400}\right)^{1/14} \approx 1.0657\,n$

* See Milin [1977].

8.5 Exercises

8.5.1. A generalization of the mean modulus (8.1.1) is known as the Prawitz theorem (see Gong [1999: 18]) which states that if $f(z) \in \mathcal{S}$, then for arbitrary p, $0 < p < \infty$, we have

$$M_p^p(r, f) \le p \int_0^r \frac{1}{\xi} M_\infty^p(\xi, f)\,d\xi, \qquad (8.5.1)$$

where $0 < r < 1$, and $M_\infty(r, f) = \max_{|z|=r}\{f(z)\}$. To prove this theorem, first note that if Γ_1 and Γ_2 are two smooth Jordan curves containing the origin, and if Γ_1 is in the interior of Γ_2, then

$$\int_{\Gamma_1} r^p\,d\theta \le \int_{\Gamma_2} r^p\,d\theta, \qquad 0 < p < \infty, \qquad (8.5.2)$$

where $z = re^{i\theta}$. Let $w = f\left(re^{i\theta}\right) = Re^{i\phi}$. Using $\dfrac{1}{R}\dfrac{\partial R}{\partial r} = \dfrac{1}{r}\dfrac{\partial \phi}{\partial \theta}$, which is one of the Cauchy-Riemann equations for $\log f$, and taking C_r as the image of $|z| = r$ under f, we get

$$2\pi \frac{d}{dr} M_p^p(r, f) = \int_0^{2\pi} \frac{\partial}{\partial r} R^p \, d\theta = \frac{p}{r} \int_0^{2\pi} R^p \frac{\partial \phi}{\partial \theta} \, d\theta = \frac{p}{r} \int_{C_r} R^p \, d\phi.$$

Note that the circle $\Gamma_r : |w| = R = M_\infty(r, f) + \varepsilon$, where $\varepsilon > 0$ is arbitrary, contains the curve $C - r$, Thus, by (8.5.2), we have

$$\int_{C_r} R^p \, d\phi \leq \int_{\Gamma_r} R^p \, d\phi = 2\pi |M_\infty(r, f) + \varepsilon|^p,$$

which, on letting $\varepsilon \to 0+$, gives

$$\frac{d}{dr} M_p^p(r, f) \leq \frac{p}{r} M_\infty^p(r, f).$$

By integrating both sides of this inequality with respect to r, we obtain (8.5.1). ∎

8.5.2. Prove the Lebedev-Milin inequalities (8.4.20)–(8.4.22). PROOF. *First Inequality:* Differentiate $\psi(z) = e^{\phi(z)}$, where $\phi(z)$ is defined by (8.4.19). This gives $\psi'(z) = \psi(z)\phi'(z)$. Thus, the coefficients β_n are given by

$$\beta_n = \frac{1}{n} \sum_{k=0}^{n-1} (n - k) \alpha_{n-k} \beta_k, \quad \beta_0 = 1, \tag{8.5.3}$$

which, using Cauchy-Schwarz inequality (1.8.4), gives

$$|\beta_n|^2 \leq \frac{1}{n} \sum_{k=0}^{n-1} (n - k)^2 |\alpha_{n-k}|^2 |\beta_k|^2. \tag{8.5.4}$$

Let $k|\alpha_k|^2$ be denoted by a_k, and define b_k recursively by

$$b_k = \frac{1}{n} \sum_{k=0}^{n-1} (n - k) \alpha_{n-k} b_k, \quad b_0 = 1. \tag{8.5.5}$$

Then the proof of the first inequality (8.4.20) will be complete if we prove $|\beta_n|^2 \leq b_n$ for $n = 0, 1, 2, \ldots$. This is done by induction. For $n = 0$, we have $b_0 = 1$ and $\beta_0 = 1$, and the inequality is true in this trivial case. Now assume that the inequality holds for $n \leq m$, and prove it for $n = m + 1$. In fact, by (8.5.4)

$$|\beta_{m+1}|^2 \leq \frac{1}{m+1} \sum_{k=0}^{m} (m + 1 - k)^2 |\alpha_{m+1-k}|^2 |\beta_k|^2$$

$$\leq \frac{1}{m+1} \sum_{k=0}^{m} (m + 1 - k)^2 |\alpha_{m+1-k}|^2 b_k$$

$$= \frac{1}{m+1} \sum_{k=0}^{m} (m + 1 - k) a_{m+1-k} b_k = b_{m+1}.$$

Comparing (8.5.5) and (8.5.3) we find that $\sum_{k=0}^{\infty} b_k z^k = \exp\left\{\sum_{k=1}^{\infty} a_k z^k\right\}$, where both $a_k \geq 0$ and $b_k \geq 0$. Hence, we get the inequality

$$\sum_{k=0}^{\infty} |\beta_k|^2 \leq \sum_{k=0}^{\infty} b_k = \exp\left\{\sum_{k=1}^{\infty} a_k\right\} = \exp\left\{\sum_{k=1}^{\infty} k|\alpha_k|^2\right\},$$

which proves (8.4.20). Note that the equality in (8.4.20) holds if and only if $|\beta_k|^2 = b_k$ for all k, i.e., equality in (8.5.4) holds for all n. In fact, the equality in (8.5.4) holds if $(n-k)\alpha_{n-k}\beta_k = \lambda_n$ for $k = 0, 1, \ldots, n-1$, where λ_n, $n = 1, 2, \ldots$, are complex constants. Thus, $\lambda_n = n\alpha_n$, because $\beta_0 = 1$. Substituting it into (8.5.3) we get $\lambda_n = \beta_n$, which implies that $\lambda_n = \lambda_{n-k}\lambda_k$, i.e., $\lambda_2 = \lambda_1^2, \cdots, \lambda_n = \lambda_1^n$. Let $\lambda_1 = \gamma$, where γ is a complex number such that $|\gamma| = 1$. Then $\alpha_n = \gamma^n/n$, $\beta_n = \gamma^n$. Hence, the equality in (8.4.20) holds if and only if $\phi(z) = -\log(1 - \gamma z)$, $\psi(z) = (1 - \gamma z)^{-1}$. The condition $\gamma| < 1$ guarantees convergence of $\sum k|\alpha_k|^2$.

Second Inequality: Using Cauchy-Schwarz inequality on (8.5.3) we get

$$n^2|\beta_n|^2 \leq \sum_{k=1}^{n} k^2|\alpha_k|^2 \sum_{k=0}^{n-1} |\beta_k|^2. \tag{8.5.6}$$

Set $A_n = \sum_{k=1}^{n} k^2|\alpha_k|^2$ and $B_n = \sum_{k=0}^{n} |\beta_k|^2$. Then, by (8.5.6),

$$B_n = B_{n-1} + |\beta_n|^2 \leq \left\{1 + \frac{1}{n^2}A_n\right\}B_{n-1} = \frac{n+1}{n}\left\{1 + \frac{A_n - n}{n(n+1)}\right\}B_{n-1}$$

$$\leq \frac{n+1}{n}B_{n-1}\exp\left\{\frac{A_n - n}{n(n+1)}\right\},$$

which, after using formulas for $B_{n-1}, B_{n-2}, \ldots$, yields

$$B_n \leq (n+1)\exp\left\{\sum_{k=1}^{n}\frac{A_k - k}{k(k+1)}\right\} = (n+1)\exp\left\{\sum_{k=1}^{n}\frac{A_k}{k(k+1)} + 1 - \sum_{k=1}^{n+1}\frac{1}{k}\right\}.$$

Set $s_n = \sum_{k=1}^{n}\frac{1}{k(k+1)}$. Then, using $s_n = \sum_{k=1}^{n}\frac{1}{k(k+1)} = 1 - \frac{1}{n+1}$, we get

$$\sum_{k=1}^{n} A_k \frac{1}{k(k+1)} = \sum_{k=1}^{n} A_k(s_k - s_{k-1}) = A_n s_n - \sum_{k=1}^{n}(A_k - A_{k-1})s_{k-1}$$

$$= A_n s_n - \sum_{k=1}^{n} k^2|\alpha_k|^2 s_{k-1} = \sum_{k=1}^{n} k|\alpha_k|^2 - \frac{1}{n+1}\sum_{k=1}^{n} k^2|\alpha_k|^2$$

$$= \sum_{k=1}^{n}\frac{k(n+1-k)}{n+1}|\alpha_k|^2,$$

Hence, we have

$$B_n \le (n+1) \exp\left\{ \frac{1}{n+1} \sum_{k=1}^{n} (n+1-k)\left(k|\alpha_k|^2 - \frac{1}{k}\right) \right\}$$

$$= (n+1) \exp\left\{ \frac{1}{n+1} \sum_{m=1}^{n} \sum_{k=1}^{m} \left(k|\alpha_k|^2 - \frac{1}{k}\right) \right\},$$

which proves (8.4.21). Equality holds in (8.4.21) under the same condition as in (8.4.20).

Third Inequality: In view of (8.5.6) and (8.4.21) we have

$$|\beta_n|^2 \le \frac{1}{n} \sum_{k=1}^{n} k^2 |\alpha_k|^2 \exp\left\{ \frac{1}{n+1} \sum_{m=1}^{n} \sum_{k=1}^{m} \left(k|\alpha_k|^2 - \frac{1}{k}\right) \right\}$$

$$= \frac{1}{n} \sum_{k=1}^{n} k^2 |\alpha_k|^2 \exp\left\{ \frac{1}{n} \sum_{k=1}^{n-1} (n-k)\left(k|\alpha_k|^2 - \frac{1}{k}\right) \right\}$$

$$= e \frac{A_n}{n} exp\left\{ -\frac{A_n}{n} + \sum_{k=1}^{n-1} \left(k|\alpha_k|^2 - \frac{1}{k}\right) \right\}, \qquad (8.5.7)$$

where we have used the inequality $xe^{-x} \le 1/e$ which holds for all x. Taking $x = A_n/n$, we obtain (8.4.22), in which equality holds under the same condition as in the inequality (8.4.21), which is the same condition as in the first inequality (8.4.20). ∎

8.5.3. Prove Milin's Lemma 8.3.1. PROOF. Let $f \in \mathcal{S}$, and consider the expansion of $\log \dfrac{f(z)}{z} = 2 \sum\limits_{n=1}^{\infty} \gamma_n z^n$, $z \in E$. In view of (9.3.3), we find that $2\gamma_n = \dfrac{1}{n}\Phi_n(0)$, which gives $4 \sum\limits_{k=1}^{n} k|\gamma_k|^2 = \sum\limits_{k=1}^{n} \dfrac{1}{k}|\Phi_k(0)|^2$, where $\Phi_n(w)$ is the nth Faber polynomial. Using the inequality $(a+b)^2 \le 2\left(a^2 + b^2\right)$ and the equality (8.3.17), we find that

$$\frac{1}{k}|\Phi_k(g(w))|^2 \le 2k\left|A_k\left(\frac{1}{w}\right)\right|^2 + \frac{2}{k}|w|^{2n}.$$

Then using (8.3.18), we have

$$\sum_{k=1}^{n} \frac{1}{k}|\Phi_k(g(w))|^2 \le 2k \sum_{k=1}^{n} k\left|A_k\left(\frac{1}{w}\right)\right|^2 + 2\sum_{k=1}^{n} \frac{1}{k}|w|^{2k}$$

$$\le -2\log\left(1 - \frac{1}{|w|^2}\right) + 2\sum_{k=1}^{n} \frac{1}{k}|w|^{2k}. \qquad (8.5.8)$$

Set $g(w) = \left(f(1/w)\right)^{-1}$ for $|w| > 1$. Then g is in Σ, and therefore, maps $|w| = \rho > 1$ onto a Jordan curve Γ_ρ which contains the origin. The function $2 \sum_{k=1}^{n} |\Phi_k(w)|^2$ is subharmonic in the domain bounded by Γ_ρ. Then, using the maximum modulus principle of subharmonic functions, we get

$$4 \sum_{k=1}^{n} k|\gamma_k|^2 = \sum_{k=1}^{n} \frac{1}{k} |\Phi_k(0)|^2 \leq \max_{w \in \Gamma_\rho} \sum_{k=1}^{n} \frac{1}{k} |\Phi_k(w)|^2$$

$$\leq -2 \log \left(1 - \frac{1}{|w|^2} \right) + 2 \sum_{k=1}^{n} \frac{1}{k} |w|^{2k}, \qquad (8.5.9)$$

which on division by 2 gives

$$2 \sum_{k=1}^{n} k|\gamma_k|^2 \leq \sum_{k=1}^{n} \frac{1}{k} \rho^{2k} - \log \left(1 - \rho^{-2} \right), \quad \rho > 1. \qquad (8.5.10)$$

Taking $\rho^2 = e^t$, $t = \dfrac{2x}{2n+1}$, $t > 0$, in (8.5.10), and using the inequality $\log \left(n + \frac{1}{2} \right) < \sum_{k=1}^{n} \frac{1}{k} - \gamma$, where $\gamma \approx 0.577$ is the Euler's constant, we get

$$-\log \left(1 - \rho^{-2} \right) = -\log \left(1 - e^{-t} \right) = \frac{t}{2} - \log \left(e^{x/2} - e^{-t/2} \right) < \frac{t}{2} - \log t$$

$$= \frac{x}{2n+1} - \log x + \log \left(n + \frac{1}{2} \right)$$

$$< \frac{x}{2n+1} - \log x + \sum_{k=1}^{n} \frac{1}{k} - \gamma. \qquad (8.5.11)$$

Hence,

$$\sum_{k=1}^{n} \frac{1}{k} \rho^{2k} = \sum_{k=1}^{n} \frac{1}{k} \sum_{m=0}^{\infty} \frac{(kt)^m}{m!} = \sum_{m=0}^{\infty} \frac{t^m}{m!} \sum_{k=1}^{n} k^{m-1}$$

$$< \sum_{k=1}^{n} \frac{1}{k} + nt + \sum_{m=2}^{\infty} \frac{\left(n + \frac{1}{2} \right) t^m}{m \, (m!)}$$

$$= \sum_{k=1}^{n} \frac{1}{k} + \frac{2nx}{2n+1} + \sum_{m=2}^{\infty} \frac{x^m}{m \, (m!)}, \qquad (8.5.12)$$

where we have used $m \sum_{k=1}^{n} k^{m-1} < \left(n + \frac{1}{2} \right)^m$, $m = 1, 2, \ldots$. Substituting all these results into (8.5.10) we get

$$2 \sum_{k=1}^{n} k|\gamma_k|^2 \leq \int_0^x \frac{e^s - 1}{s} \, ds - \log x + 2 \sum_{k=1}^{n} \frac{1}{k} - \gamma \equiv G_n(x).$$

The minimum of $G_n(x)$ is obtained from $G'(x) = \dfrac{e^x - 1}{x} - \dfrac{1}{x} = \dfrac{e^x - 2}{x} = 0$, which solves to $x = \log 2$. Hence, from (8.5.12)

$$\sum_{k=1}^{n} \frac{1}{k} \rho^{2k} \leq \frac{1}{2} G_n(\log 2) = \sum_{k=1}^{n} \frac{1}{k} + \delta, \qquad (8.5.13)$$

where

$$\delta = \frac{1}{2} \left[\int_0^{\log 2} \frac{e^s - 1}{s} \, ds - \log \log 2 - \gamma \right] < 0.312 \qquad (8.5.14)$$

is known as *Milin's constant.* ∎

8.5.4. For each odd function $h \in \mathcal{S}$, $h(z) = \sqrt{f(z^2)} = \sum\limits_{n=1}^{\infty} c_{2n-1} z^{2n-1}$, $c_1 = 1$ (see §3.1.2), prove that $|c_n| < e^{\delta/2} < 1.17$ for $n = 3, 5, \ldots$, where δ is the Milin's constant. PROOF. From the function h, defined above, we get

$$\log \frac{h(\sqrt{z})}{\sqrt{z}} = \frac{1}{2} \log \frac{f(z)}{z} = \sum_{n=1}^{\infty} \gamma_n z^n,$$

or

$$\sum_{n=0}^{\infty} c_{2n+1} z^n = \exp \left\{ \sum_{n=1}^{\infty} \gamma_n z^n \right\}, \qquad \gamma_1 = 1.$$

Hence, by the third Lebedev-Milin inequality (8.4.22),

$$|c_{2n+1}|^2 \leq \exp \left\{ \sum_{k=1}^{n} k |\gamma_n|^2 - \sum_{k=1}^{n} \frac{1}{k} \right\}.$$

Thus, Milin's lemma 8.3.1 gives

$$|c_{2n+1}| \leq e^{\delta/2} < e^{0.156} < 1.17, \quad n = 1, 2, \ldots . \blacksquare$$

8.5.5. Prove Bazilevich theorem: Let $f \in \mathcal{S}$ have the Hayman index $\alpha > 0$ and direction of maximum growth $e^{i\theta_0}$. Then

$$\sum_{n=1}^{\infty} n \left| \gamma_n - \frac{1}{n} e^{-in\theta_0} \right| \leq \frac{1}{2} \log \frac{1}{\alpha}. \qquad (8.5.15)$$

PROOF. Let $g(z) = 1/f(1/\zeta)$, and consider the expansion $\log \dfrac{g(z) - g(w)}{z - w} =$
$-\sum\limits_{n=1}^{\infty} A_n(z) w^{-n}$, $z = 1/\zeta$. Since $\sum\limits_{n=1}^{\infty} n |A_n(z)|^2 \leq \sum\limits_{n=1}^{\infty} \frac{1}{n} |w|^{2n} = -\log\left(1 - |w|^2\right)$, we get

$$\sum_{n=1}^{\infty} n |A_n(z)|^2 \leq -\log\left(1 - |z|^2\right), \quad z \in E.$$

Hence,

$$\sum_{n=1}^{\infty} n|A_n(z) - \frac{1}{n}\bar{z}^n|^2 \le \sum_{n=1}^{\infty} n|A_n(z)|^2 - 2\Re\left\{\sum_{n=1}^{\infty} A_n(z)z^n\right\} + \sum_{n=1}^{\infty} \frac{1}{n}|z|^{2n}$$

$$\le -2\log\left(1 - |z|^2\right) - 2\Re\left\{\sum_{n=1}^{\infty} A_n(z)z^n\right\}.$$

On the other hand, $-\sum_{n=1}^{\infty} A_n(z)z^n = \log g'(z) = \log \dfrac{z^2 f(z)}{[f(z)]^2}$. Thus, for $r = |z| < 1$,

$$\sum_{n=1}^{\infty} |A_n(z) - \frac{1}{n}\bar{z}^n|^2 \le 2\log \frac{r^2|f'(z)|}{(1-r^2)|f(z)|^2|} \le -2\log\left\{\frac{(1-r)^2}{r}|f(z)|\right\}.$$

Take $z = re^{i\theta_0}$. Then

$$\sum_{n=1}^{\infty} n\left|A_n(re^{i\theta_0}) - \frac{1}{n}r^n e^{in\theta_0}\right|^2 \le -2\log\alpha, \quad r < 1. \qquad (8.5.16)$$

Since the function A_n are closely related to the Faber polynomials such that $nA_n(re^{i\theta_0} + r^{-n}e^{-in\theta_0}) = \Phi_n\left(1/f(re^{i\theta_0})\right)$, where Φ_n are the Faber polynomials of g (see §9.3), and since $f\left(r^{i\theta_0}\right) \to \infty$ as $r \to 1$, then $A_n\left(re^{i\theta_0}\right)$ has a limit, i.e.,

$$A_n\left(e^{i\theta_0}\right) = \lim_{r \to 1} A_n\left(re^{i\theta_0}\right) = \frac{1}{n}\Phi_n(0) - \frac{1}{n}e^{in\theta_0} = 2\gamma_n - \frac{1}{n}e^{in\theta_0},$$

and then the inequality (8.5.16) yields the (8.5.15). ∎

9

Polynomials

We will provide data for certain orthogonal polynomials, such as Gegenbauer (ultraspherical), Jacobi, and Legendre polynomials, as well as the Faber polynomials and the hypergeometric series, which are all needed later in the proof of de Branges theorem in the next chapter. Also, the presentation is kept to essential formulas and representations; details can, however, be found in the literature cited and Brychkov [2008].

9.1 Orthogonal Polynomials

A set of polynomials $\{f_i\}$ with degree i and such that $\langle f_i, f_j \rangle = 0$ for $i \neq j$ is called a set of orthogonal polynomials with respect to the inner product $\langle f_i, f_j \rangle$. Let $w(x)$ be an admissible weight function on a finite or infinite interval $[a, b]$. If we orthonormalize the powers $1, x, x^2, \ldots$, we obtain a unique set of polynomials $p_n(x)$ of degree n and leading coefficient positive, such that

$$\int_a^b w(x) p_n(x) p_m(x)\, dx = \delta_{mn} = \begin{cases} 0 & \text{if } m \neq n, \\ 1 & \text{if } m = n, \end{cases} \qquad (9.1.1)$$

where δ_{mn} is known as the Kronecker delta. Table 9.1.1 lists some of the useful classical polynomials together with their specific weights and intervals.

Table 9.1.1

Name	Symbol	Interval	$w(x)$
Jacobi	$P_n^{(\alpha,\beta)}(x)$	$[-1,1]$	$(1-x)^\alpha(1+x)^\beta$, $\alpha, \beta > 1$
Gegenbauer (ultraspherical)	$C_n^\mu(x)$	$[-1,1]$	$(1-x^2)^{\mu-1/2}$, $\mu > -1/2$
Legendre	$P_n(x)$	$[-1,1]$	1

We provide certain important data for these polynomials; additional information is available in Abramowitz and Stegun [1965]. Zeros of these polynomials can be easily computed by Mathematica®.

Orthogonal polynomials satisfy the differential equation

$$g_1(x)\, y'' + g_2(x)\, y' + a_n\, y = 0, \tag{9.1.2}$$

where $g_1(x)$ and $g_2(x)$ are independent of n, and a_n are constants that depend on n only.

9.1.1. Jacobi Polynomials. These polynomials, denoted by $P_n^{\alpha,\beta}(x)$, satisfy Eq (9.1.2) over the interval $[-1, 1]$ with $g_1(x) = 1 - x^2$, $g_2(x) = \beta - \alpha - (\alpha + \beta + 2)x$, and $a_n = n(n + \alpha + \beta + 1)$, and with the orthogonality condition

$$\int_{-1}^{1} P_n^{\alpha,\beta}(x) P_m^{\alpha,\beta}(x)(1-x)^\alpha (1+x)^\beta\, dx = \delta_{n,m},$$

and the normalization condition $P_n^{\alpha,\beta}(1) = \binom{n+\alpha}{n}$. Thus, $P_n^{\alpha,\beta}(x)$ satisfies the differential equation

$$(1 - x^2)y'' + [(\beta - \alpha) - (\alpha + \beta + 2)x]\, y' + n(n + \alpha + \beta + 1)y = 0. \tag{9.1.3}$$

For proof, see Exercise 9.4.1. Other relevant data are:

Norm: $\displaystyle \int_{-1}^{1} (1-x)^\alpha (1+x)^\beta \left[P_n^{\alpha,\beta}(x) \right]^2 dx$

$$= \frac{2^{\alpha+\beta+1}\, \Gamma(n+\alpha+1)\Gamma(n+\beta+1)}{(2n+\alpha+\beta)n!\,\Gamma(n+\alpha+\beta+1)}$$

Series form: $\displaystyle P_n^{\alpha,\beta}(x) = \frac{1}{2^n} \sum_{k=0}^{[n/2]} \binom{n+\alpha}{k} \binom{n+\beta}{n-k} (x-1)^{n-k}(x+1)^k,$

Inequality: $\displaystyle \max_{-1 \le x \le 1} \left| P_n^{\alpha,\beta}(x) \right| = \begin{cases} \binom{n+q}{n} \sim n^q & \text{if } q = \max(\alpha, \beta) \ge -1/2, \\ \left| P_n^{\alpha,\beta} \right|(x') \sim n^{-1/2} & \text{if } q < -1/2 \end{cases}$

where x' is one of the two maximum points nearest $(\beta - \alpha)/(\alpha + \beta + 1)$.

Rodrigues' formula: $\displaystyle P_n^{\alpha,\beta}(x) = \frac{(-1)^n}{2^n n!(1-x)^\alpha (1+x)^\beta}$

$$\times \frac{d^n}{dx^n} \left\{ (1-x)^{n+\alpha}(1+x)^{n+\beta} \right\}.$$

Theorem 9.1.1. For a fixed $\alpha > -1$ and $\beta > -1$, if $p_n(x) = \sum_{k=0}^{n} a_k x^n$ is an arbitrary polynomial orthogonal to $1, x, \ldots, x^{n-1}$, then $p_n = c\, P_n^{\alpha,\beta}(x)$, where c is a constant.

PROOF. Write $p_n(x) = \sum_{k=0}^{n} a_k x^n$ as $\sum_{k=0}^{n} c_k P_n^{\alpha,\beta}(x)$. Then p_n is orthogonal to $P_0^{\alpha,\beta}(x), \ldots, P_{n-1}^{\alpha,\beta}(x)$ since p_n is orthogonal to $1, x, \ldots, x^{n-1}$. Thus, $\langle p_n, P_m \rangle = c_m = 0$ when $0 \leq m \leq n - 1$, and therefore, $p_n = c P_n^{\alpha,\beta}(x)$. ∎

See Theorem 9.2.1 for expressing Jacobi polynomials in terms of hypergeometric functions.

9.1.2. Gegenbauer Polynomials. Let

$$G^{\mu}(x, w) = \frac{1}{(1 - 2xw + w^2)^{\mu}} = \sum_{n=0}^{\infty} C_n^{\mu}(x) w^n, \qquad (9.1.4)$$

if $\mu > -\frac{1}{2}$, $-1 \leq x \leq 1$, and $|w| < 1$. The polynomials $C_n^{\mu}(x)$ defined over the interval $[-1, 1]$ such that $C_n^{\mu}(1) = \binom{n+2\mu-1}{n}$, are called the Gegenbauer (or Ultraspherical) Polynomials. These polynomials satisfy Eq (9.1.2) with $g_1(x) = 1 - x^2$, $g_2(x) = -(2\mu+1)x$, and $a_n = n(n+2\alpha)$, i.e., the Gegenbauer polynomials satisfy the differential equation

$$(1 - x^2)y'' - (2\mu + 1)xy' + (n + 2\mu)y = 0. \qquad (9.1.5)$$

See Exercise 9.4.2 for a proof. In particular, for $x = 1$, we have

$$G^{\mu}(1, w) = \frac{1}{(1 - w)^{2\mu}} = \sum_{n+0}^{\infty} C_n^{\mu}(1) w^n, \qquad |w| < 1,$$

which implies that $C_n^{\mu}(1) = \binom{-2\mu}{n} = \frac{(2\mu)_n}{n!}$. Other relevant data are the following:

Norm: $\displaystyle\int_{-1}^{1} (1 - x^2)^{\mu - 1/2} \left[C_n^{\mu}(x)\right]^2 dx = \frac{\pi 2^{1-2\mu} \, \Gamma(n + 2\mu)}{n! \, (n + \mu) \left[\Gamma(\mu)\right]^2}$

Series form: $\displaystyle C_n^{\mu}(x) = \frac{1}{\Gamma(\mu)} \sum_{k=0}^{[n/2]} (-1)^k \frac{\Gamma(\mu + n - k)!}{k!(n - 2k)!} (2x)^{n-2k}$

Inequality: $\displaystyle \max_{-1 \leq x \leq 1} |C_n^{\mu}(x)| = \begin{cases} \binom{n+2\mu-1}{n}, & \text{if } \mu > 0, \\ |C_n^{\mu}(x')|, & \text{if } -1/2 < \mu < 0 \end{cases}$

where $x' = 0$ if $n = 2k$; $x' = $ maximum point nearest zero if $n = 2k + 1$.

Rodrigues' formula: $\displaystyle C_n^{\mu}(x) = \frac{(-1)^n 2^n n! \, \Gamma(\mu + n + 1/2)}{\Gamma.(\mu + 1/2) \, \Gamma(n + 2\mu) \, (1 - x^2)^{\mu - 1/2}} \times$

$$\times \frac{d^n}{dx^n} \left\{ (1 - x^2)^{n + \mu - 1/2} \right\}.$$

See Theorem 9.2.2 for expressing the Gegenbauer polynomials in terms of hypergeometric functions.

9.1.3. Legendre Polynomials. These polynomials, denoted by $P_n(x)$, are the special case of $P_n^{\alpha, \beta}$ for $\alpha = \beta = 0$; they satisfy Eq (9.1.2) over the interval $[-1, 1]$ with $g_1(x) = 1 - x^2$, $g_2(x) = -2x$, and $a_n = n(n+1)$, where $P_n(1) = 1$. Let x be a real number and z a complex number such that $|2xz - z^2| < 1$. Then we can expand $\left(1 - 2xz + z^2\right)^{-1/2}$ into a series of ascending powers of $2xz - z^2$ and then expand it into a power series in powers of z, we get

$$\left(1 - 2xz + z^2\right)^{-1/2} = P_0(x) + zP_1(x) + z^2 P_2(x) + z^3 P_3(x) + \cdots, \quad (9.1.6)$$

where $P_0(x) = 1$, $P_1(x) = x$, $P_2(x) = \frac{1}{2}\left(3x^2 - 1\right)$, $P_3(x) = \frac{1}{2}\left(5x^3 - 3x\right), \ldots$, and, in general,

$$P_n(x) = \sum_{k=0}^{[n/2]} (-1)^k \frac{(2n - 2k)!}{2^n k! \, (n - k)! \, (n - 2k)!} \, x^{n-2k}.$$

Obviously, for integer n, the nth derivative is given by

$$\frac{d^n}{dx^n} \left(x^2 - 1\right)^n = \sum_{k=0}^{[n/2]} (-1)^k \frac{n!}{k! \, (n - k)!} \frac{(2n - 2k)!}{(n - 2k)!} \, x^{n-2k}.$$

Thus, from (9.1.6) we get Rodrigues' formula, which is given below.

If $x_{n,m}$ denotes the mth zero of $P_n(x)$, where $x_{n,1} > x_{n,2} > \cdots > x_{n,n}$, then

$$x_{n,m} = \left(1 - \frac{1}{8n^2} + \frac{1}{8n^3}\right) \cos \frac{(4m - 1)\pi}{4n + 2} + O\left(n^{-4}\right).$$

Other relevant data are:

Norm: $\displaystyle \int_{-1}^{1} \left[P_n(x)\right]^2 dx = \frac{2}{2n + 1}$,

Series form: $\displaystyle P_n(x) = \frac{1}{2^n} \sum_{k=0}^{[n/2]} (-1)^k \binom{n}{k} \binom{2n - 2k}{n} x^{n-2k}$,

Indefinite Integral: $\displaystyle \int P_n(x) \, dx = \frac{1}{2n + 1} \left[P_{n+1}(x) - P_{n-1}(x)\right]$,

Inequality: $|P_n(x)| \le 1, \quad -1 \le x \le 1$,

Rodrigues' formula: $\displaystyle P_n(x) = \frac{(-1)^n}{2^n \, n!} \frac{d^n}{dx^n} \{(1 - x^2)^n\}$.

The Fejér's sum $\sum\limits_{k=0}^{n} P_k(x) \geq 0$ for $-1 \leq x \leq 1$, is obtained by using Mehler's formula:

$$P_n(\cos\theta) = \frac{2}{\pi} \int_{-\pi}^{\pi} \frac{\sin(n+\frac{1}{2})\phi}{[2\cos\theta - 2\cos\phi]^{1/2}}\, d\phi.$$

The integral representation for $P_n(x)$ is

$$P_n(x) = \frac{1}{2\pi i} \int_{\gamma} \frac{(t^2-1)^n}{2^n(-x)^{n+1}}\, dt, \tag{9.1.7}$$

where γ is the counter around the point x once counter-clockwise. This is known as *Schläfli formula* for Legendre polynomials.

Legendre polynomials can be expressed in terms of hypergeometric functions: Suppose $|1-x| \leq 2(1-\delta)$, $0 < \delta < 1$, and let the contour γ be the circle $|1-t| = 2-\delta$. Note that $\left|\dfrac{1-x}{1-t}\right| \leq \dfrac{2-2\delta}{2-\delta} < 1$. Then we expand $(t-x)^{-n-1}$ in the uniformly convergent series

$$(t-x)^{-n-1} = (t-1)^{-n-1}$$
$$\times \left\{ 1 + (n+1)\frac{x-1}{t-1} + \frac{(n+1)(n+2)}{2!}\left(\frac{x-1}{t-1}\right)^2 + \cdots \right\}.$$

Substituting this series expansion into Schläfli formula (9.1.7) and integrating term-by-term, we get

$$P_n(x) = \sum_{k=0}^{\infty} \frac{(x-1)^k}{2^{n+1}\pi i} \frac{(n+1)(n+2)\cdots(n+k)}{k!} \int_{\gamma} \frac{(t^2-1)^n}{(t-1)^{n+k+1}}\, dt$$

$$= \sum_{k=0}^{\infty} \frac{(x-1)^k(n+1)(n+2)\cdots(n+k)}{2^n(k!)^2} \left[\frac{d^k}{dt^k}(t+1)^n\right]_{t=1}$$

$$= \sum_{k=0}^{\infty} \frac{(n+1)(n+2)\cdots(n+k)(-n)(1-n)\cdots(k-1-n)}{(k!)^2} \left(\frac{1}{2} - \frac{1}{2}x\right)^k$$

$$= {}_2F_1\left(n+1, -n; 1; \frac{1}{2} - \frac{1}{2}x\right), \quad \text{for } |1-x| \leq 2(1-\delta) < 2, \tag{9.1.8}$$

where we have used $\left[\dfrac{d^k}{dt^k}(t+1)^n\right]_{t=1} = 2^{n-k}n(n-1)\cdots(n-k+1)$. The following results can be derived from (9.1.8):

$$P_n(x) = P_{-n-1}(x), \tag{9.1.9}$$

$$P_n^k(x) = (1-x^2)^{k/2}\frac{d^k P_n(x)}{dx^k} \quad \text{for } k > 0 \text{ integer and } -1 < x < 1, \tag{9.1.10}$$

$$P_n^k(x) = \frac{(n+1)(n+2)\cdots(n+k)}{2^{n+1}\pi i}(1-x^2)^{k/2}\int_{\gamma}(t^2-1)^n(t-x)^{-n-k-1}\, dt. \tag{9.1.11}$$

where the formula (9.1.9) is called the *Ferrer associated Legendre function of degree n and order k*, and formula (9.1.11) is obtained from (9.1.7) and (9.1.10).

Let $t = x + (x^2 - 1)^{1/2}e^{i\phi}$, i.e., we take the contour γ as a circle with center at x and radius $|(x^2 - 1)^{1/2}|$. Then $dt = (x^2 - 1)^{1/2} e^{i\phi} i \, d\phi$, $t^2 = 2(x^2-1)^{1/2}e^{i\phi} \left((x^2 - 1)\cos\phi + x \right)$. Substituting these quantities into (9.1.11), we get

$$P_n^k(x) = \frac{(n+1)(n+2)\cdots(n+k)}{2\pi}(-1)^{k/2}\int_{-\pi}^{\pi} \left[(x^2-1)^{1/2}\cos\phi + x\right]^n e^{-ik\phi} \, d\phi$$

$$= \frac{(n+1)(n+2)\cdots(n+k)}{2\pi}(-1)^{k/2}\int_{-\pi}^{\pi} \left[(x^2-1)^{1/2}\cos\phi + x\right]^n \cos k\phi \, d\phi, \tag{9.1.12}$$

since $(x^2-1)\cos\phi+x$ is an even function of ϕ, and thus, $\int_{-\pi}^{\pi} \left[(x^2-1)^{1/2}\cos\phi + x\right]^n \sin k\phi \, d\phi = 0$.

Another formula for Legendre polynomials for $x = uv - (u^2 - 1)^{1/2}(v^2 - 1)^{1/2}\cos\theta$ is

$$P_n(x) = P_n(u)P_n(v) + 2\sum_{k=1}^{n} \frac{(n-k)!}{(n+k)!}P_n^k(u)P_n^k(v)\cos k\theta. \tag{9.1.13}$$

For a fixed v, $v > 0$, the quantity $\left|\dfrac{u + (u^2 - 1)^{1/2}\cos(\theta - \varphi)}{v + (v^2 - 1)^{1/2}\cos\varphi}\right|$ is a bounded function of φ. Let M be an upper bound of this quantity and let $|z| < 1/M$. Then the series

$$\sum_{n=0}^{\infty} \frac{\left[u + (u^2 - 1)^{1/2}\cos(\theta - \varphi)\right]^n}{\left[v + (v^2 - 1)^{1/2}\cos\varphi\right]^{n+1}} z^n$$

converges uniformly in φ, and thus,

$$\sum_{n=0}^{\infty} z^n \int_{-\pi}^{\pi} \frac{\left[u + (u^2 - 1)^{1/2}\cos(\theta - \varphi)\right]^n}{\left[v + (v^2 - 1)^{1/2}\cos\varphi\right]^{n+1}} \, d\varphi$$

$$= \int_{-\pi}^{\pi} \sum_{n=0}^{\infty} z^n \frac{\left[u + (u^2 - 1)^{1/2}\cos(\theta - \varphi)\right]^n}{\left[v + (v^2 - 1)^{1/2}\cos\varphi\right]^{n+1}} \, d\varphi$$

$$= \int_{-\pi}^{\pi} \left[v - zu + ((v^2 - 1)^{1/2} - z(u^2 - 1)^{1/2}\cos\theta)\cos\varphi\right.$$

$$\left. - z(u^2 - 1)^{1/2}\sin\theta\sin\varphi\right]^{-1} d\varphi. \tag{9.1.14}$$

The next formula is

$$P_n(x) = \frac{1}{2\pi}\int_{-\pi}^{\pi} \frac{\left[u + (u^2 - 1)^{1/2}\cos(\theta - \varphi)\right]^n}{\left[v + (v^2 - 1)^{1/2}\cos\varphi\right]^{n+1}} \, d\varphi, \tag{9.1.15}$$

which is obtained by using the formula

$$\int_{-\pi}^{\pi} \frac{d\varphi}{A + B \cos \varphi + C \sin \varphi} = \frac{2\pi}{(A^2 - B^2 - C^2)^{1/2}}, \tag{9.1.16}$$

where the value of the radical is taken so that $|A - (A^2 - B^2 - C^2)^{1/2}| < |(B^2 + c^2)^{1/2}|$. Then noting that the value of the integral on the right hand side of (9.1.14) is equal to

$$2\pi \left[(v - zu)^2 - \left[(v^2 - 1)^{1/2} - z(u^2 - 1)^{1/2} \cos \theta \right]^2 - \left[z(u^2 - 1)^{1/2} \sin \theta \right]^2 \right]^{-1/2}$$

$$= \frac{2\pi}{(1 - 2zx + z^2)},$$

which when compared with the definition of Legendre polynomials gives (9.1.16). Finally, the Legendre polynomials $P_n(x)$ can be expanded as a Fourier cosine series

$$P_n(x) = \frac{1}{2} A_0 + \sum_{k=1}^{n} A_k \cos k\theta, \tag{9.1.17}$$

where

$$A_k = \frac{1}{\pi} \int_{-\pi}^{\pi} P_n(x) \cos k\theta \, d\theta$$

$$= \frac{1}{2\pi^2} \int_{-\pi}^{\pi} \int_{-\pi}^{\pi} \frac{\left[u + (u^2 - 1)^{1/2} \cos(\theta - \varphi) \right]^n \cos k\theta}{\left[v + (v^2 - 1)^{1/2} \cos \varphi \right]^{n+1}} \, d\theta \, d\varphi$$

$$= \frac{1}{2\pi^2} \int_{-\pi}^{\pi} \int_{-\pi}^{\pi} \frac{\left[u + (u^2 - 1)^{1/2} \cos \psi \right]^n \cos k(\varphi + \psi)}{\left[v + (v^2 - 1)^{1/2} \cos \varphi \right]^{n+1}} \, d\psi \, d\varphi,$$

setting $\psi = \theta - \varphi$,

$$= \frac{1}{2\pi^2} \int_{-\pi}^{\pi} \int_{-\pi}^{\pi} \frac{\left[u + (u^2 - 1)^{1/2} \cos \psi) \right]^n \cos k\varphi \cos k\psi)}{\left[v + (v^2 - 1)^{1/2} \cos \varphi \right]^{n+1}} \, d\psi \, d\varphi, \tag{9.1.18}$$

because $\int_{-\pi}^{\pi} \left[u + (u^2 - 1)^{1/2} \cos \psi) \right]^n \sin k\psi \, d\psi = 0$. Hence, by (9.1.9) and (9.1.12), we get

$$A_k = 2 \frac{(n - k)!}{(n + k)!} P_n^k(u) P_n^k(v),$$

which completes the proof of (9.1.13). These formulas will be used in §10.3.1 to establish Weinstein's proof of the de Branges theorem.

9.2 Hypergeometric Functions

A hypergeometric function is defined as follows: For $|t| < 1$, $c \neq 0, -1, -1, \ldots$,

$$_2F_1(a, b; c; t) = f(a, b; c; t) = \sum_{k=0}^{\infty} \frac{(a)_k (b)_k}{(c)_k} \frac{t^k}{k!}, \qquad (9.2.1)$$

$$_3F_2(a, b; c; d, e; t) = \sum_{k=0}^{\infty} \frac{(a)_k (b)_k (c)_k}{(d)_k (e)_k} \frac{t^k}{k!}, \qquad (9.2.2)$$

where $(n)_k = \dfrac{\Gamma(n+k)}{\Gamma(n)}$. The functions $_2F_1(a, b; c; t)$ and $_3F_2(a, b; c; d, e; t)$, respectively, satisfy the differential equations

$$t^2(1 - t)y'' + [c - (a + b + 1)t]\, y' - aby = 0, \qquad (9.2.3)$$

and

$$t^2(1 - t)z''' - [(3 + a + b + c)t^2 - (1 + d + e)t]\, z'' + [de - (1 + a + b + c$$
$$+ ab + ac + bc)t]z' - abcz = 0. \qquad (9.2.4)$$

Define an operator T by $T = t\dfrac{d}{dt}$, so that $Tt^k = kt^k$. Set $y = {}_2F_1(a, b; c; t)$. Then

$$T(T + c - 1)y = \sum_{k=1}^{\infty} \frac{k(k + c - 1)(a)_k (b)_k}{k!\,(c)_k} t^k = \sum_{k=1}^{\infty} \frac{(a)_k (b)_k}{(k - 1)!\,(c)_{k-1}} t^k$$

$$= \sum_{m=0}^{\infty} \frac{(a)_{m+1}(b)_{m+1} t^{m+1}}{m!\,(c)_m}$$

$$= \sum_{m=0}^{\infty} \frac{(a + m)(b + m)(a)_m (b)_m t^{m+1}}{m!\,(c)_m} = T(T + a)(T + b)y, \qquad (9.2.5)$$

which shows that the function $_2F_1(a, b; c; t)$ is a solution of the differential equation $[T(T + c - 1) - t(T + a)(T + b)]\, y = 0$, with $y(0) = 1$, $y'(0) = ab/c$, i.e., the function $_2F_1(a, b; c; t)$ satisfies the equation (9.2.3). Similarly it can be shown that $_3F_2(a, b; c; d, e; t)$ satisfies the equation (9.2.4) with $y(0) = 1$, $y'(0) = ab/d$.

Jacobi polynomials $P_n^{\alpha,\beta}(x)$ can be expressed in term of the hypergeometric functions, as the following result shows.

Theorem 9.2.1. For $\alpha > -1$, $\beta > -1$ and $n \geq 1$,

$$P_n^{\alpha,\beta}(x) = \binom{n + \alpha}{n} {}_2F_1\left(-n, n + \alpha + \beta + 1; \alpha + 1; \frac{1 - x}{2}\right). \qquad (9.2.6)$$

PROOF. Since $_2F_1(a, b; c; t)$ satisfies Eq (9.2.3), substitute $t = (1 - x)/2$ in that equation, and write $G(x)$ for $y(t)$. This gives

$$(1 - x^2)G''(x) - [2c - (a + b + 1) + (a + b + 1)x] G'(x) - abG(x) = 0.$$

A comparison of this equation with (9.1.3) shows that

$$ab = -n(n + a + \beta + 1), \ a + b = \alpha + \beta + 1, \ a + b + 1 - 2c = \beta - \alpha,$$

and the solutions of the equation

$$(u+a)(u+b) = u^2 + (a+b)u + ab = u^2 + (\alpha + beta + 1)u - n(n+\alpha+\beta+1) = 0$$

are $u_1 = -n - \alpha - \beta - 1$, $u_2 = n$. If we solve the equation

$$-(\beta - \alpha) + a + b + 1 = 2c = -(\beta - \alpha) + \alpha + \beta + 2 = 2\alpha + 2,$$

we obtain $c = \alpha + 1$. The function $_2F_1(a, b; c; t)$ is the solution of Eq (9.1.5) as well as of Eq (9.1.3) when $a = -n, b = n + \alpha + \beta + 1$, and $c = \alpha + 1$. The solution of (9.1.3) being unique, $_2F_1 \left(-n, n + \alpha + \beta + 1; \alpha + 1; (1 - x)/2\right)$ is linearly dependent on $P_n^{\alpha, \beta}(x)$, i.e.,

$$_2F_1\left(- n, n + \alpha + \beta + 1; \alpha + 1; \frac{1 - x}{2}\right) = cP_n^{\alpha,\beta}(x),$$

where the constant c is determined by $_2F_1(a, b; c; 0) = 1$ and $P_n^{\alpha,\beta} = \binom{n+\alpha}{n}$. ∎

Gegenbauer polynomials $C_n^\mu(x)$ can be expressed in terms of hypergeometric functions:

Theorem 9.2.2. If $\mu > -\frac{1}{2}, n \geq 1$, and $-1 \leq x \leq 1$, then

$$C_n^\mu(x) = \frac{(2\mu)_n}{n!} \frac{P_n^{\mu-1/2,\mu-1/2}(x)}{P_n^{\mu-1/2,\mu-1/2}(1)}$$

$$= \frac{(2\mu)_n}{n!} {}_2F_1\left(- n, n + 2\mu; \mu + \tfrac{1}{2}; \frac{1 - x}{2}\right). \tag{9.2.7}$$

PROOF. Take $\alpha = \beta = \mu - \frac{1}{2}$ in Theorem 9.2.1. This gives

$$\frac{P_n^{\mu-1/2,\mu-1/2}(x)}{P_n^{\mu-1/2,\mu-1/2}(1)} = {}_2F_1\left(- n, n + 2\mu; \mu + \tfrac{1}{2}; \frac{1 - x}{2}\right).$$

Then $P_n^{\mu-1/2,\mu-1/2}(x)$ satisfies Eq (9.1.3) for $\alpha = \beta = \mu - 1/2$, i.e,

$$(1 - x^2)y'' - (2\mu + 1)xy' + n(n + 2\mu)y = 0. \tag{9.2.8}$$

But this is Eq (9.1.5), and Eq (9.1.3) has a unique polynomial solution. Thus, by Theorem 9.1.1, $C_n^\mu(x) = AP_n^{\mu-1/2,\mu-1/2}(x)$, where the constant A is determined by $C_n^\mu(1) = \dfrac{(2\mu)_n}{n!}$ and $P_n^{\mu-1/2,\mu-1/2}(1) = \binom{n+\mu-1/2}{n}$. $\blacksquare$

Theorem 9.2.3. The following result holds between the hypergeometric functions:

$$
{}_3F_2(a,b,c;d,e;t) = \frac{\Gamma(e)t^{1-e}}{\Gamma(c)\Gamma(e-c)} \int_0^t {}_2F_1(a,b;d;y)(t-y)^{e-c-1} y^{e-1} \, dy,
$$
(9.2.9)

where $e - c > 0$.

PROOF. Set $y = tz$. Then $dy = t \, dz$, and

$$
\int_0^t {}_2F_1(a,b;d;y)(t-y)^{e-c-1} y^{e-1} \, dy = \int_0^1 {}_2F_1(a,b;d;tz)(t-tz)^{e-c-1}(tz)^{e-1} t \, dz
$$

$$
= t^{e-c-1+c-1+1} \int_0^1 (1-z)^{e-c-1} z^{c-1} \sum_{k=0}^\infty \frac{(a)_k (b)_k}{(d)_k} \frac{(tz)^k}{k!} \, dz
$$

$$
= t^{e-1} \sum_{k=0}^\infty \frac{(a)_k (b)_k}{k!\,(d)_k} t^k \int_0^1 (1-z)^{e-c-1} z^{c+k-1} \, dz
$$

$$
= t^{e-1} \sum_{k=0}^\infty \frac{(a)_k (b)_k}{k!\,(d)_k} t^k \frac{\Gamma(e-c)\Gamma(c+k)}{\Gamma(e+k)},
$$

which, after multiplying its right hand side by $\dfrac{\Gamma(e)t^{1-e}}{\Gamma(e)\Gamma(e-c)}$ yields

$$
\frac{\Gamma(e)t^{1-e}}{\Gamma(e)\Gamma(e-c)} t^{e-1} \sum_{k=0}^\infty \frac{(a)_k (b)_k}{k!\,(d)_k} \frac{\Gamma(e-c)\Gamma(c+k)}{\Gamma(e+k)} t^k
$$

$$
= \sum_{k=0}^\infty \frac{(a)_k (b)_k (c)_k}{k!\,(d)_k (e)_k} t^k = {}_3F_2(a,b,c;d,e;t). \quad \blacksquare
$$

Theorem 9.2.4. (Clausen formula):

$$
\left[{}_2F_1\left(\alpha,\beta;\alpha+\beta+\tfrac{1}{2};t\right) \right]^2
$$
$$
= {}_3F_2\left(2\alpha,2\beta,\alpha+\beta;2\alpha+2\beta,\alpha+\beta+\tfrac{1}{2};t\right).
$$
(9.2.10)

PROOF. Let $y = {}_2F_1(\alpha,\beta;\gamma;t)$ and

$$
L[y] = t(1-t)y'' + [\gamma - (\alpha+\beta+1)t]\,y' - aby, \quad \gamma = \alpha+\beta+\tfrac{1}{2}.
$$

Then $L[y] \equiv 0$. By differentiating $tL[y]$ we get

$$
\begin{aligned}
\left(tL[y]\right)' &\equiv t^2(1-t)y''' + \left[(\gamma+2)t - (\alpha+\beta+4)t^2\right]y'' \\
&\quad + \left[\gamma - (2\alpha+2\beta+\alpha\beta+2)t\right]y' - \alpha\beta y = 0.
\end{aligned}
$$

Set $z = {}_3F_2(a,b,c;d,e;t)$, and let

$$
\begin{aligned}
M[z] &= t^2(1-t)z''' - \left[(3+a+b+c)t^2 - (1+d+e)t\right]z'' \\
&\quad + \left[dc - (1+a+b+c+ab+ac+bc)t\right]z' - abcz.
\end{aligned}
$$

$$(9.2.11)$$

Note that $M[z] = 0$. If $z = y^2$, then

$$
z' = zyy', \quad z'' = 2yy' + 2(y')^2, \quad z''' = 2yy''' + 6y'y''. \tag{9.2.12}
$$

After replacing z, z', z'' and z''' in (9.2.11) by their values in (9.2.12), we get $M[y^2]$. Then we try to find the values of a, b, c, d, e and A, B such that

$$
M[y^2] = (2Ay + Bty')L[y] + 2y(tL[y])'. \tag{9.2.13}
$$

Then $M[y^2] = 0$ only if (9.2.13) is an identity, which implies that

$$
{}_3F_2(a,b,c;d,e;t) = c_0 \left[{}_2F_1(\alpha,\beta;\gamma;t)\right]^2,
$$

where $\gamma = \alpha + \beta + \frac{1}{2}$, and c_0 is a constant. Since (9.2.13) is an identity, and its both sides are polynomials in y, y', y'' and y''', the coefficients in powers of y, y', y'' and y''' on both sides must be equal. This yields the system of equations:

$$
\begin{aligned}
&3 + r = A + \alpha + \beta + 4, \\
&1 + v = A + \gamma + 2, \\
&1 + r + s = A(\alpha+\beta+1) + \tfrac{1}{2}B\alpha\beta + 2(\alpha+\beta+1) + \alpha\beta, \\
&w = \gamma(A+1), \\
&u = 2\alpha\beta(A+1), \\
&6 = B, \\
&2(3+r) = B(\alpha+\beta+1),
\end{aligned}
$$

where $r = a+b+c$, $s = ab+ac+bs$, $u = abc$, $v = d+e$, and $w = de$. Solving these equations, we get

$$
\begin{aligned}
&B = 6, \ A = 2(\alpha+\beta) - 1, \ r = 3(\alpha+\beta), \ v = 3r - 1, \ w = \gamma(2\gamma-1), \\
&u = 4\alpha\beta(\alpha+\beta), \ s = 2(\alpha+beta)^2 + 4\alpha\beta,
\end{aligned}
$$

and a, b, c are the three roots of the cubic equation $x^3 - rx^2 + sx - u = 0$, which, using the above values, is

$$x^3 - 3(\alpha + \beta)x^2 + \left(2(\alpha + \beta)^2 + 4\alpha\beta\right)x - 4(\alpha + \beta)\alpha\beta = 0.$$

It is easy to see that the three roots of this equation are $2\alpha, 2\beta$ and $\alpha + \beta$. Thus, we have $(a, b, c) = (2\alpha, 2\beta, \alpha + \beta)$. Similarly, d and e are the roots of the equation $x^2 - vx + w = 0$, which, using the above values, is $x^2 - (3\gamma - 1)x + (2\gamma - 1)\gamma = 0$, and its two roots are γ and $2\gamma - 1$. Thus, we have $(d, e) = (\gamma, 2\gamma - 1)$. Hence,

$$_3F_2(2\alpha, 2\beta, \alpha + \beta; \gamma, 2\gamma - 1; t) = c_0 \left[{}_2F_1(\alpha, \beta; \gamma; t)\right]^2, \quad c_0 = 1. \blacksquare$$

Theorem 9.2.5. (Gegenbauer-Hua formula, Hua [1963]):

$$C_n^\mu(x) = \sum_{k=0}^{[n/2]} c_k C_{n-2k}^\lambda(x), \tag{9.2.14}$$

where $\mu > \lambda > -\frac{1}{2}$, $[n/2]$ is the greatest integer $\leq n/2$, and

$$c_k(\mu, \lambda) = \frac{(n - 2k + \lambda)T(\lambda)(\mu - \lambda)_k \Gamma(n + \mu - k)}{k! \Gamma(\mu) \Gamma(n + \lambda - k + 1)}.$$

A proof, based on Gong [1999: 114 ff], is given as Exercise 9.4.3.

Lemma 9.2.1. The following results hold:

$$(2a)_{2j} = 2^{2j}(a)_j \left(a + \tfrac{1}{2}\right)_j, \tag{9.2.15}$$

$$\sum_{k=0}^{n} \frac{(a)_k}{k!} = \frac{(a + 1)_n}{n!}. \tag{9.2.16}$$

PROOF. Formula (9.2.15) is obvious. To prove (9.2.16), note that it holds for $n = 0$. If $\sum_{k=0}^{n-1} \frac{(a)_k}{k!} = \frac{(a + 1)_{n-1}}{(n - 1)!}$ is true, then

$$\sum_{k=0}^{n-1} \frac{(a)_k}{k!} + \frac{(a)_n}{n!} = \frac{(a + 1)_{n-1}}{(n - 1)!} + \frac{(a)_n}{n!} = \frac{(a + 1)_{n-1}}{(n - 1)!}\left(1 + \frac{a}{n}\right) = \frac{(a + 1)_n}{n!},$$

so that formula (9.2.16) is true by induction. $\blacksquare$

Theorem 9.2.6. (Askey-Gasper theorem, Askey and Gasper [1985]):

$$\sum_{k=0}^{n} P_k^{\alpha,0}(x) = 0, \quad \text{for } \alpha > -1 \text{ and } -1 \leq x \leq 1. \tag{9.2.17}$$

PROOF. By Theorem 9.2.1, we have

$$P_k^{\alpha,0}(x) = \frac{(\alpha+1)_k}{k!} \, {}_2F_1\left(-k, k+\alpha+1; \alpha+1; \frac{1-x}{2}\right).$$

Set $a_k = \dfrac{(\alpha+1)_k}{k!}$, and let

$$b_{kj} = \begin{cases} \dfrac{(-k)_j(k+\alpha+1)_j\,((1-x)/2)^j}{j!\,(\alpha+1)_j} & \text{if } 0 \leq j \leq k, \\ 0 & \text{if } j > k. \end{cases}$$

Then

$$\sum_{k=0}^{n} P_k^{\alpha,0}(x) = \sum_{k=0}^{n} a_k \sum_{j=0}^{n} b_{kj} = \sum_{j=0}^{n} \sum_{k=0}^{n} a_k b_{kj}$$

$$= \sum_{j=0}^{n} \left(\sum_{k=j}^{n} a_k b_{kj}\right) = \sum_{j=0}^{n} \sum_{k=0}^{n-j} a_{k+j} b_{k+j,j}$$

$$= \sum_{j=0}^{n} \sum_{k=0}^{n-j} \frac{(\alpha+1)_{k+j}}{(k+j)!} \frac{(-k-j)_j(k+j+\alpha+1)_j}{j!\,(\alpha+1)_j} \left(\frac{1-x}{2}\right)^j$$

$$= \sum_{j=0}^{n} \left(\frac{x-1}{2}\right)^j \frac{(\alpha+1)_{2j}}{j!\,(\alpha+1)_j} \sum_{k=0}^{n-j} \frac{1}{k!} \frac{(\alpha+1)_{k+j}(k+j+\alpha+1)_j}{(\alpha+1)_{2j}}.$$

Note that

$$\frac{(\alpha+1)_{k+j}(\alpha+1+k+j)_j}{k!\,(\alpha+1)_{2j}} = \frac{(\alpha+1+2j)_k}{k!}.$$

Using this equation and formula (9.2.16), we get

$$\sum_{k=0}^{n} P_k^{\alpha,0}(x) = \sum_{j=0}^{n} \left(\frac{1-x}{2}\right)^j \frac{(\alpha+1)_{2j}(\alpha+2j+2)_{n-j}}{j!\,(\alpha+1)_j(n-j)!}. \tag{9.2.18}$$

Take $a = \dfrac{\alpha+1}{2}$ in formula (9.2.15). Then $(\alpha+1)_{2j} = 2^{2j}\left(\dfrac{\alpha+1}{2}\right)_j\left(\dfrac{\alpha+2}{2}\right)_j$.
Substituting these into (9.2.18), we get

$$\sum_{k=0}^{n} P_k^{\alpha,0}(x) = \sum_{j=0}^{n} \frac{\left(\frac{x-1}{2}\right)^j 2^{2j}\left(\frac{\alpha+1}{2}\right)_j\left(\frac{\alpha+2}{2}\right)_j(\alpha+2j+2)_{n-j}}{j!\,(\alpha+1)_j(n-j)!}$$

$$= \sum_{j=0}^{n} \frac{[2(x-1)]^j \left(\frac{\alpha+1}{2}\right)_j\left(\frac{\alpha+2}{2}\right)_j(\alpha+2j+2)_{n-j}}{j!\,(\alpha+1)_j(n-j)!}. \tag{9.2.19}$$

Next, we prove that

$$\sum_{k=0}^{n} P_k^{\alpha,0}(x) = \frac{(\alpha+2)_n}{n!} \, {}_3F_2\left(-n, n+\alpha+2, \frac{\alpha+1}{2}; \alpha+1, \frac{\alpha+3}{2}; \frac{1-x}{2}\right),$$

$$(9.2.20)$$

i.e., we must show that

$$\frac{(2(x-1))^j \left(\frac{\alpha+1}{2}\right)_j \left(\frac{\alpha+2}{2}\right)_j (\alpha+2j+2)_{n-j}}{j! \, (\alpha+1)_j (n-j)!}$$

$$= \frac{(\alpha+2)_n}{n!} \frac{(-n)_j (n+\alpha+2)_j \left(\frac{\alpha+1}{2}\right)_j}{(\alpha+1)_j \left(\frac{\alpha+3}{2}\right)_j} \left(\frac{1-x}{2}\right)^j.$$

This amounts to proving the following result:

$$\frac{1}{(n-j)!} 2^{2j} (-1)^j \left(\frac{\alpha+2}{2}\right)_j (\alpha+2j+2)_{n-j} = \frac{(\alpha+2)_n (-n)_j (n+\alpha+2)_j}{n! \left(\frac{\alpha+3}{2}\right)_j}.$$

$$(9.2.21)$$

Take $a = \dfrac{\alpha+2}{2}$ in formula (9.2.15); then $(\alpha+2)_{2j} = 2^{2j} \left(\dfrac{\alpha+2}{2}\right)_j \left(\dfrac{\alpha+3}{2}\right)_j$, which makes Eq (9.2.21) equivalent to

$$\frac{1}{(n-j)!} (-1)^j (\alpha+2)_{2j} (\alpha+2j+2)_{n-j} = \frac{(\alpha+2)_n (-n)_j (n+\alpha+2)_j}{n!}. \quad (9.2.22)$$

This shows that formulas $(a)_{2j}(a+2j)_{n-j} = (a)_n(a+n)_j$ and $(-1)^j(-n)_j = \dfrac{n!}{(n-j)!}$ are true. Using these formulas for $a = \alpha+2$, it can be verified that (9.2.22) is true. The proof of (9.2.19) is complete.

Next, take $c = \dfrac{\alpha+1}{2}$, and $e = 2c = \alpha+1$ in Theorem 9.2.3. Then

$$\frac{\Gamma(e) \, t^{1-e}}{\Gamma(c)\Gamma(e-c)} = \frac{\Gamma(2c) \, t^{1-e}}{(\Gamma(c))^2} = \frac{\Gamma(\alpha+1) \, t^{-\alpha}}{\left[\Gamma\left(\frac{\alpha+1}{2}\right)\right]^2}, \qquad e - c = \frac{\alpha+1}{2},$$

and so (9.2.9) becomes

$$ {}_3F_2\left(a, b, \frac{\alpha+1}{2}; d, \alpha+1; t\right)$$

$$= t^{-\alpha}\Gamma(\alpha+1)\left[\Gamma\left(\frac{\alpha+1}{2}\right)\right]^{-2} \int_0^t [(t-y)y]^{(\alpha+1)/2} \, {}_2F_1(a, b; d; y) \, dy.$$

The integrand in this equality exists since $\alpha > -1$. Now, define a linear operator L by

$$L[g] \equiv t^{-\alpha}\Gamma(\alpha+1)\left[\Gamma\left(\frac{\alpha+1}{2}\right)\right]^{-2} \int_0^t [(t-y)y]^{(\alpha-1)/2} \, g(y) \, dy, \quad \alpha > -1. \, t > 0,$$

Then

$$_3F_2\left(a, b, \frac{\alpha+1}{2}; d, \alpha+1; t\right) = L\left[_2F_1(a, b; d; t)\right]. \tag{9.2.23}$$

Let $2\mu = \alpha + 2$ and $x = 1 - 2t$ in Theorem 9.2.2; then (9.2.7) becomes

$$C_n^\mu(x) = C_n^{(\alpha+2)/2}(1 - 2t) = \frac{(\alpha+2)_n}{n!} {}_2F_1\left(-n, n+\alpha+2; \frac{\alpha+3}{2}; t\right). \tag{9.2.24}$$

Let $\nu = \dfrac{\alpha+2}{2}$ and $\mu = \dfrac{\alpha+1}{2}$ in Theorem 9.2.5 (Gegenbauer-Hua theorem).
Then

$$C_n^{(\alpha+2)/2}(1 - 2t) = \sum_{j=0}^{[n/2]} \frac{\left(\frac{\alpha}{2}\right)_j \left(\frac{\alpha+2}{2}\right)_{n-j} \left(\frac{\alpha+3}{2}\right)_{n-2j}}{j! \left(\frac{\alpha+3}{2}\right)_{n-j} \left(\frac{\alpha+1}{2}\right)_{n-2j}} C_{n-2j}^{(\alpha+1)/2}(1 - 2t). \tag{9.2.25}$$

From Theorem 9.2.8 we have

$$C_{n-2j}^{(\alpha+1)/2}(1-2t) = \frac{(\alpha+1)_{n-2j}}{(n-2j)!} {}_2F_1\left(-n+2j, n-2j+\alpha+1; \frac{\alpha+2}{2}; t\right). \tag{9.2.26}$$

Combining (9.2.24), (9.2.25) and (9.2.26), we get

$$\frac{(\alpha+2)_n}{n!} {}_2F_1\left(-n, n+\alpha+2; \frac{\alpha+3}{2}; t\right)$$
$$= \sum_{j=0}^{[n/2]} \frac{\left(\frac{\alpha}{2}\right)_j \left(\frac{\alpha+2}{2}\right)_{n-j} \left(\frac{\alpha+3}{2}\right)_{n-2j} (\alpha+1)_{n-2j}}{j! \left(\frac{\alpha+3}{2}\right)_{n-j} \left(\frac{\alpha+1}{2}\right)_{n-2j} (n-2j)!}$$
$$\times {}_2F_1\left(-n+2j, n-2j+\alpha+1; \frac{\alpha+2}{2}; t\right). \tag{9.2.27}$$

Using the definition of the (linear) operator L and (9.2.23) we find that

$$_3F_2\left(-n, n+\alpha+2, \frac{\alpha+1}{2}; \frac{\alpha+3}{2}, \alpha+1; t\right) = L\left[_2F_1\left(-n, n+\alpha+2; \frac{\alpha+3}{2}; t\right)\right],$$
$$_3F_2\left(-n+2j, n-2j+\alpha+1, \frac{\alpha+1}{2}; \frac{\alpha+2}{2}, \alpha+1; t\right)$$
$$= L\left[_2F_1\left(-n+2j, n-2j+\alpha+1; \frac{\alpha+2}{2}; t\right)\right].$$

In view of (9.2.20) and (9.2.27) and the above two formulas, we get

$$\sum_{k=0}^n P_k^{\alpha,\beta}(x) = \frac{(\alpha+2)_n}{n!} L\left[_2F_1\left(-n, n+\alpha+2; \frac{\alpha+3}{2}; t\right)\right]$$
$$= \sum_{j=0}^{[n/2]} \frac{\left(\frac{\alpha}{2}\right)_j \left(\frac{\alpha+2}{2}\right)_{n-j} \left(\frac{\alpha+3}{2}\right)_{n-2j} (\alpha+1)_{n-2j}}{j! \left(\frac{\alpha+3}{2}\right)_{n-j} \left(\frac{\alpha+1}{2}\right)_{n-2j} (n-2j)!}$$
$$\times {}_3F_2\left(-n+2j, n-2j+\alpha+1, \frac{\alpha+1}{2}; \frac{\alpha+2}{2}, \alpha+1; t\right). \tag{9.2.28}$$

Now, set $2a = -n+2j$ and $2b = n-2j+\alpha+1$ in the Clausen formula (9.2.10). This yields

$$\left[{}_2F_1\left(a, b; a+b+\tfrac{1}{2}; t\right)\right]^2 = {}_3F_2\left(2a, 2b, a+b; 2a+2b, a+b+\tfrac{1}{2}; t\right),$$

which gives

$$\left[{}_2F_1\left(\frac{-n+2j}{2}, \frac{n-2j+\alpha+1}{2}; \frac{\alpha+2}{2}; t\right)\right]^2$$
$$= {}_3F_2\left(-n+2j, n-2j+\alpha+1, \frac{\alpha+1}{2}; \alpha+1, \frac{\alpha+2}{2}; t\right). \tag{9.2.29}$$

Since $\alpha > -1$, the coefficients of each term in (9.2.28) are positive, thus implying that $\sum\limits_{k=0}^{n} P_k^{\alpha,\beta}(x) \geq 0$. ∎

9.3 Faber Polynomials

Let $g(\zeta) = \zeta + b_0 + b_1\zeta^{-1} + \cdots \in \Sigma$. The Faber polynomials $\Phi_n(w)$ of $g(\zeta)$ are given by

$$\frac{\zeta g'(\zeta)}{g(\zeta) - w} = \sum_{n=0}^{\infty} \Phi_n(w)\zeta^{-n} = 1 + \sum_{n=1}^{\infty} \Phi_n(w)\zeta^{-n}, \quad w \in \mathbb{C}, \tag{9.3.1}$$

at a neighborhood of $\zeta = \infty$. Let $\Phi_m(b_0, b_1, \ldots, b_m) = \Phi_m(0)$, $m \geq 0$. The function

$$\Phi_n(w) = w^n + \sum_{m=1}^{n} a_{nm}w^{n-m}, \tag{9.3.2}$$

is an nth degree polynomial in w and is called the *Faber polynomial* of the function $g(\zeta)$. By direct calculation we find that

$$\Phi_0(w) = 1,$$
$$\Phi_1(w) = w - b_0,$$
$$\Phi_2(w) = (w - b_0)^2 - 2b_1 = w^2 - 2b_0w + (b_0^2 - 2b_1),$$
$$\Phi_3(w) = (w - b_0)^3 - 3(w - b_0) - 3b_1$$
$$\quad = w^3 - 3b_0w^2 + 3(b_0^2 - b_1)w + (b_0^3 + 3b_1b_0 - 3b_2),$$
$$\Phi_4(w) = (w - b_0)^4 - 4b_1(w - b_0)^2 - 4b_2(w - b_0) + (2b_1^2 - 4b_3)$$
$$\quad = w^4 - 4b_0w^3 + (6b_0^2 - 4b_1)w^2 + (-4b_0^3 + 8b_0b_1 - 4b_2)w$$
$$\quad + (b_0^4 - 4b_0^2b_1 + 4b_0b_2 + 2b_1^2 - 4b_3),$$
$$\cdots \cdots$$

The generating function for the Faber polynomials is $\log \dfrac{g(\zeta) - w}{\zeta}$, which yields (see Schiffer [1948])

$$\log \frac{\zeta}{g(\zeta) - w} = \sum_{n=1}^{\infty} \frac{1}{n} \Phi_n(w) \zeta^{-n}, \tag{9.3.3}$$

where the series expansion is valid in a neighborhood of $\zeta = \infty$. Let

$$\log\left[g(\zeta) - g(w)\right] - \log(\zeta - w) = \sum_{n,m=1}^{\infty} c_{nm} \zeta^{-n} w^{-m}. \tag{9.3.4}$$

Then from (9.3.3) and (9.3.4) we find that

$$-\sum_{n=1}^{\infty} \frac{1}{n} \Phi_n\left[g(w)\right] \zeta^{-n} = \log \frac{g(\zeta) - g(w)}{\zeta - w} + \log\left(1 - \frac{w}{\zeta}\right)$$
$$- \sum_{n=1}^{\infty} \frac{1}{n}\left(w^n - \sum_{n,m=1}^{\infty} n c_{nm} \zeta^{-n} w^{-m}\right) \zeta^{-n}. \tag{9.3.5}$$

Comparing equal powers of ζ on both sides of (9.3.5), we obtain

$$\Phi_n\left[g(w)\right] = w^m - \sum_{n=1}^{\infty} n c_{nm} w^{-m}, \tag{9.3.6}$$

which shows that the coefficients of $\Phi_m(w)$ are the Faber polynomials as defined above, with $c_{nm} = a_{nm}$. If we differentiate (9.3.4) with respect to w, we get

$$\frac{1}{g(\zeta) - w} = \sum_{n=1}^{\infty} \frac{1}{n} f'_n(w) \zeta^{-n}, \tag{9.3.7}$$

which is a generating function for the derivatives of Faber polynomials. Further, if we differentiate (9.3.7) withy respect to w, we find that

$$\frac{1}{(g(\zeta) - w)^2} = \sum_{n=1}^{\infty} \frac{1}{n} f''_n(w) \zeta^{-n} = \sum_{n,m=1}^{\infty} \frac{1}{nm} \Phi'_n(w) \Phi'_m(w) \zeta^{-(n+m)}, \tag{9.3.8}$$

which yields

$$\frac{1}{k} \Phi''_k(w) = \sum_{n+m=k} \frac{1}{nm} \Phi'_n(w) \Phi'_m(w). \tag{9.3.9}$$

9.3.1. Variation Formula. Let $g(\zeta) \in \Sigma$. Then g maps a domain D in the ζ-plane, which contains the point at infinity and is bounded by a finite number of proper continua, onto a domain G in the w-plane. If w_0 is an arbitrary point in the w-plane which does not belong to G, and ρ any positive constant, then according to Schiffer [1938a], there exist infinitely many functions which are univalent in G and have in the domain $|w - w_0| > 4\rho$ a series expansion of the form

$$v(w) = w + \frac{a\rho^2}{w - w_0} + \frac{b\rho^3}{(w - w_0)^2} + \frac{c\rho^4}{(w - w - 0)^3} + \cdots , \qquad (9.3.10)$$

where $|a| \leq 4^2$, $|b| \leq 4^3$, $|c| \leq 4^4, \dots$. The function

$$g^*(\zeta) = v[g(\zeta)] = g(\zeta) + \frac{a\rho^2}{f(\zeta) - w_0} + \cdots \qquad (9.3.11)$$

is again analytic and univalent in D and again has the series expansion (9.3.1). However, it maps D onto a new domain G^* in the w-plane. If ρ is small enough, G^* will be nearly like G. Thus, we can regard the mapping (9.3.10) as a small variation of the domain G. Such variations are used in solving extremal problems in the class of normalized univalent functions.

Now, if we subject $g(\zeta)$ to a variation of the type (9.3.11), it will be interesting to see how the corresponding Faber polynomial will change. Using the generating function (9.3.3) and writing $\Phi_n^*(w)$ as a variation in $\Phi_n(w)$, we have

$$\Phi_n^*(w) = \Phi_n(w) - a\rho^2 \frac{\Phi_n'(w) - \Phi_n'(w_0)}{w - w_0} + o\left(\rho^2\right), \qquad (9.3.12)$$

and for a variation of $\Phi"_n[g^*(\zeta)]$ we have for a fixed point $\zeta \in D$

$$\Phi_n^{*'}[g^*(\zeta)] = \Phi_n'[g(\zeta)] + a\rho^2 \frac{\Phi_n'[g(\zeta)]}{(g(\zeta) - \zeta_0)^2} + o\left(\rho^2\right). \qquad (9.3.13)$$

These variation formulas have application in solving extremal problems in geometric function theory.

Since $g \in \Sigma$ is univalent, the function

$$\frac{\zeta g'(\zeta)}{g(\zeta) - w} - \frac{\zeta}{\zeta - w} = \sum_{n=1}^{\infty} \sum_{m=1}^{\infty} d_{nm} \zeta^{-m} w^{-n} \qquad (9.3.14)$$

is analytic in $|\zeta| > 1$, $|w| > 1$. Then, in view of (9.3.1), we have

$$\sum_{n=0}^{\infty} \Phi_n(g(\zeta)) w^{-n} = 1 + \sum_{n=1}^{\infty} \left\{ w^n + \sum_{m=1}^{\infty} d_{nm} w^{-m} \right\} \zeta^{-n}, \qquad (9.3.15)$$

The Faber polynomials satisfy the condition:

$$\Phi_n\left(g(w)\right) = w^n + \sum_{m=1}^{\infty} d_{nm} w^{-m}, \quad n = 1, 2, \ldots. \tag{9.3.16}$$

The coefficients d_{nm} are called the *Grunsky coefficients* of g. Using the series expansion (g.1) for the function $\log \dfrac{g(\zeta) - g(w)}{\zeta - w}$ which is analytic in $\{\zeta, w : |\zeta| > 1, |w| > 1\}$, and symmetric in ζ and w, we find that $\gamma_{nm} = d_{nm}$. Thus, dividing (9.3.14) by ζ and then integrating with respect to ζ, we get

$$\log \frac{g(\zeta) - g(w)}{\zeta - w} = -\sum_{n=1}^{\infty} \sum_{m=1}^{\infty} \frac{1}{n} d_{nm} w^{-m} \zeta^{-n} = -\sum_{n=1}^{\infty} A_n\left(\frac{1}{w}\right) \zeta^{-n}, \tag{9.3.17}$$

where $A_n(w) = \sum_{k=1}^{\infty} \gamma_{nk} w_k$. Thus, $d_{nm} = n\gamma_{nm}$, $md_{nm} = n\gamma_{nm}$, and $md_{nm} = nd_{mn}$, $m, n = 1, 2, \ldots$.

The matrix (d_{nm}) is an $n \times m$ matrix, and its elements are obtained by direct computation as follows:

$$d_{11} = b_1, \quad d_{12} = b_2, \quad d_{13} = b_3, \quad d_{14} = b_4, \quad \ldots,$$

$$d_{21} = 2b_2, \quad d_{22} = 2b_3 + b_1^2, \quad d_{23} = 2(b_4 + b_1 b_2), \quad d_{24} = 2b_5 + 2b_1 b_3 + b_2^2, \quad \ldots,$$

$$d_{31} = 3b_3, \quad d_{32} = 3(b_4 + b_1 b_2), \quad d_{33} = 3(b_5 + b_1 b_3 + b_2^2) + b_1^3,$$

$$d_{34} = 3(b_0 + b_1 b_4 + 2b_2 b_3 + b_1^2 b_2), \quad \ldots,$$

$$d_{41} = 4b_4, \quad d_{42} = 4(b_5 + b_1 b_3), \quad d_{43} = 4(b_6 + b_1 b_4 + 2b_2 b_3 + b_1^2 b_2),$$

$$d_{44} = 4(b_7 + b_1 b_3 + b_1^2 b_3) + 8(b_2 b_4 + b_1 b_2^2) + 6b_3^2 + b_1^4, \quad \ldots$$

$$\ldots \ldots.$$

Notice that if we take $\lambda_k = w^k$ in the strong Grunsky inequality (3.5.6) and use (9.3.17), we get

$$\sum_{n=1}^{\infty} n|A_n(w)|^2 \le \sum_{n=1}^{\infty} \frac{1}{n} |w|^{2n} = -\log\left(1 - |w|^2\right). \tag{9.3.18}$$

9.4 Exercises

9.4.1. Show that the Jacobi polynomials $y = P_n^{\alpha,\beta}(x)$ are solutions of Eq (9.1.3). PROOF. Note that since

$$\left[(1 - x)^{\alpha+1}(1 - x)^{\beta+1} y'\right]'$$
$$= (1 - x)^{\alpha}(1 + x)^{\beta} \left[(1 - x^2)y'' + (\beta - \alpha)y' - (\alpha + \beta + 2)x\right],$$

Eq (9.1.3) is equivalent to

$$\left[(1-x)^{\alpha+1}(1-x)^{\beta+1}y'\right]' = -n(n+\alpha+\beta+1)(1-x)^{\alpha}(1+x)^{\beta}y. \quad (9.4.1)$$

Substituting $y = P_n^{\alpha,\beta}(x)$ on the left hand side of Eq (9.4.1), we obtain an expression of the form $(1-x)^{\alpha}(1+x)^{\beta}Q(x)$, where $Q(x)$ is a polynomial of degree n. Therefore, if we prove that $Q(x) = -n(n+\alpha+\beta+1)P_n^{\alpha,\beta}(x)$, then the proof is complete. First, we prove that $Q(x)$ is orthogonal to $P_m^{\alpha,\beta}(x)$, $0 \le m \le n-1$. For $\alpha > -1, \beta > -1$, we have

$$\int_{-1}^{1}(1-x)^{\alpha}(1+x)^{\beta}x^{m}Q(x)\,dx = \int_{-1}^{1}\left[(1-x)^{\alpha+1}(1+x)^{\beta+1}y'\right]'x^{m}\,dx.$$

On integrating by parts on the right side of this equation, we get

$$-\int_{-1}^{1}(1-x)^{\alpha+1}(1+x)^{\beta+1}mx^{m-1}y'\,dx,$$

which, on integrating by parts again, yields

$$m\int_{-1}^{1}y\frac{d}{dx}\left[(1-x)^{\alpha+1}(1+x)^{\beta+1}x^{m-1}\right]dx = \int_{-1}^{1}y(1-x)^{\alpha}(1+x)^{\beta}p_m(x)\,dx,$$

$$(9.4.2)$$

where $p_m(x)$ is a polynomial of degree m. Since $P_n^{\alpha,\beta}(x)$ is orthogonal to $1, x, \dots, x^{n-1}$, the integral on the right hand side of Eq (9.4.2) is equal to 0 when $0 \le m \le n-1$. This means that $Q(x)$ is orthogonal to x^m, $0 \le m \le n-1$. Thus, by Theorem 9.1.1, $p(x) = cP_n^{\alpha,\beta}(x)$, where c is a constant, which is determined as follows: Since

$$\left[\frac{d}{dx}P_n^{\alpha,\beta}(x)(1-x)^{\alpha+1}(1+x)^{\beta+1}\right]' = (1-x)^{\alpha}(1+x)^{\beta}Q(x)$$

$$= (1-x)^{\alpha}(1+x)^{\beta}cP_n^{\alpha,\beta}(x),$$

we find that

$$cP_n^{\alpha,\beta}(x) = \frac{d^2}{dx^2}P_n^{\alpha,\beta}(x)(1-x^2)\left[(\beta-\alpha)-(\alpha+\beta+2)x\right]\frac{d}{dx}P_n^{\alpha,\beta}(x).$$

If we set $P_n^{\alpha,\beta}(x) = \sum\limits_{k=0}^{n}a_k x^k$ in this equation and compare the coefficients of x^n on both sides, we get $c = -n(n+\alpha+\beta+1)$. Thus, $Q(x) = -n(n+\alpha+\beta+1)P_n^{\alpha,\beta}(x)$. ∎

9.4.2. Show that the Gegenbauer polynomials $y = C_n^{\mu}(x)$ satisfy Eq (9.1.5). PROOF. Since, by definition of C_n^{μ}, we have $G \equiv G^{\mu}(x,t) = \sum\limits_{n=0}^{\infty}C_n^{\mu}(x)t^n$, we find that

$$G_x = \sum_{n=0}^{\infty} [C_n^\mu(x)]' \, t^n, \qquad G_{xx} = \sum_{n=0}^{\infty} [C_n^\mu(x)]'' \, t^n,$$

$$tG_t = \sum_{n=1}^{\infty} nC_n^\mu(x) t^n, \qquad t\,(tG_t)_t = \sum_{n=1}^{\infty} n^2 C_n^\mu(x) t^n.$$

Substituting these expressions into Eq (9.1.4), we get

$$\sum_{n=0}^{\infty} \left\{ (1 - x^2) \, [C_n^\mu(x)]'' - (2\mu + 1)x \, [C_n^\mu(x)]' + n^2 C_n^\mu(x) + 2\mu n C_n^\mu(x) \right\} t^n = 0,$$

which completes the proof. ∎

9.4.3. Proof of Gegenbauer-Hua Theorem (Theorem 9.2.5): For sufficiently small t such that $|2xt - t^2| < 1$, we have

$$(1 - 2xt + t^2)^{-\mu} = \sum_{n=0}^{\infty} \frac{(\mu)_n}{n!} \left(2xt - t^2 \right)^n$$

$$= \sum_{n=0}^{\infty} \sum_{k=0}^{\infty} \frac{(-1)^k (\mu)_n (2x)^{n-k} t^{n+k}}{k! \, (n - k)!}$$

$$= \sum_{n=0}^{\infty} t^n \sum_{k=0}^{[n/2]} \frac{(-1)^k (\mu)_{n-k} (2x)^{n-2k}}{k! \, (n - 2k)!} \qquad \text{by binomial theorem.}$$

Thus,

$$C_n^\mu(x) = \sum_{k=0}^{[n/2]} \frac{(-1)^k (\mu)_{n-k} (2x)^{n-2k}}{k! \, (n - 2k)!}. \qquad (9.4.3)$$

Differentiating both sides of (9.4.3) with respect to x, we get

$$(1 - 2xt + t^2)^{-\lambda} = \sum_{n=0}^{\infty} C_n^\lambda(x) t^n, \qquad \text{for } \mu > \lambda > -\tfrac{1}{2},$$

and thus,

$$\frac{2\lambda t}{(1 - 2xt + t^2)^{\lambda+1}} = \sum_{n=1}^{\infty} \left(\frac{d}{dx} C_n^\lambda(x) \right) t^n = 2\lambda \sum_{m=0}^{\infty} C_m^{\lambda+1}(x) t^{m+1}.$$

If we compare the corresponding coefficients of t^n, we get

$$\frac{d}{dx} C_n^\lambda(x) = 2\lambda C_{n-1}^{\lambda+1}(x). \qquad (9.4.4)$$

However, from (9.4.3) we have

$$\frac{(2x)^n}{n!} = \sum_{k=0}^{[n/2]} a_{k,n}(\lambda) C_{n-2k}^{\lambda}(x). \tag{9.4.5}$$

Differentiating with respect to x on both sides of the equality

$$\frac{(2x)^{n+1}}{(n+1)!} = \sum_{k=0}^{[(n+1)/2]} a_{k,n+1}(\lambda) C_{n+1-2k}(x),$$

we get

$$\frac{2(2x)^n}{n!} = \sum_{k=0}^{[(n+1)/2]} a_{k,n+1}(\lambda) \frac{d}{dx} C_{n+1-2k}^{\lambda}(x).$$

By (9.4.5) the left hand side of this equality is equal to $\sum_{k=0}^{[n/2]} 2a_{k,n}(\lambda+1) C_{n-2k}^{\lambda+1}(x)$,

while by (9.4.4) its right hand side is equal to $\sum_{k=0}^{[(n+1)/2]} 2a_{k,n+1}(\lambda)\lambda C_{n-2k}^{\lambda+1}$.

Thus,

$$a_{k,n}(\lambda+1) = \lambda a_{k,n+1}(\lambda). \tag{9.4.6}$$

Next, consider $\dfrac{1}{1 - 2t\cos\theta + t^2} = \sum\limits_{n=0}^{\infty} C_n^1(\cos\theta) t^n$. Since

$$\frac{1}{1 - 2t\cos\theta + t^2} = \frac{1}{1 - e^{-i\theta} t} \frac{1}{1 - e^{i\theta} t} = \sum_{k=0}^{\infty} e^{-ik\theta} t^k \sum_{m=0}^{\infty} e^{im\theta} t^m$$

$$= \sum_{n=0}^{\infty} \frac{\sin(n+1)\theta}{\sin\theta} t^n,$$

then, in view of (9.1.5), $C_n^\lambda(x)$ is, in fact, the Jacobi polynomial $P_n^{\lambda-1/2,\lambda-1/2}(x)$ multiplied by a constant, where the system $\{P_n^\lambda\}$ is orthogonal with weight function $(1-x^2)^{\lambda-1/2}$, $\mu > -\frac{1}{2}$. Now, since

$$\int_{-1}^{1} C_n^1(x) C_m^1(x)(1-x^2)^{1/2}\,dx = \int_{-\pi/2}^{\pi/2} \sin\left((n+1)\theta\right) \sin\left((m+1)\theta\right)\,d\theta = \frac{\pi}{2}\delta_{n,m},$$

$$\int_{-1}^{1} \frac{(2x)^n}{n!} C_{n-2k}^1(x)(1-x^2)^{1/2}\,dx = \int_{-\pi/2}^{\pi/2} \frac{(2\cos\theta)^n}{n!} \sin\left((n-2k+1)\theta\right)\sin\theta\,d\theta$$

$$= \frac{\pi}{2}\frac{n-2k+1}{k!\,(n-2k)!},$$

and since $a_{k,n}(1) = \dfrac{n - 2k + 1}{k! \, (n - k + 1)!}$ from (9.4.5), and $a_{k,n}(\lambda) = \dfrac{n - 2k + \lambda}{k! \, (\lambda)_{n-k+1}}$ from (9.4.6), then substituting it into (9.4.5) we have

$$\frac{(2x)^n}{n!} = \sum_{k=0}^{[n/2]} \frac{n - 2k + \lambda}{k!(\lambda)_{n-k+1}} C_{n-2k}^{\lambda}(x), \tag{9.4.7}$$

while substitution into (9.4.3) gives

$$C_n^{\mu}(x) = \sum_{k=0}^{[n/2]} c_k C_{n-2k}^{\lambda}(x),$$

where

$$c_k = \frac{(-1)(\mu)_{n-k}}{k!} \sum_{j=0}^{[n/2]-k} \frac{\lambda + n - 2k - 2j}{j! \, (\lambda)_{n-2p-j+1}},$$

or

$$\frac{\Gamma(\mu)}{\Gamma(\lambda)} c_k = \frac{\lambda + n - 2k}{k!} \sum_{j=0}^{k} (-1)^j \binom{k}{j} \frac{\Gamma(\mu + n - j)}{\Gamma(\lambda + n - j + k + 1)}.$$

Let $\Delta(x) = f(x + 1) - f(x)$. Then (Hua [1963])

$$\Delta^k \left[\frac{f(a+x)}{f(b+x)} \right] = \sum_{j=0}^{k} (-1)^j \binom{k}{j} \frac{f(a + x + k - j)}{f(b + x + k - j)}.$$

If we let $x = n, a = \mu - k$ and $b = \lambda - 2k + 1$, then

$$\frac{\Gamma(\mu)}{\Gamma(\lambda)} c_k = \frac{\lambda + n - 2k}{k!} \Delta^k \left[\frac{\Gamma(\mu - k + n)}{\Gamma(\lambda - 2k + 1 + n)} \right].$$

Using $\Delta \left[\dfrac{\gamma(a+n)}{\Gamma(b+n)} \right] = \dfrac{(a-b)\Gamma(a+n)}{\Gamma(b+n+1)}$ and

$$\Delta^k \left[\frac{\Gamma(a+n)}{\Gamma(b+n)} \right] = \frac{\Gamma(a - b + 1)\Gamma(a+n)}{\Gamma(a - b - k + 1)\Gamma(b + nk)},$$

we find that

$$\frac{\Gamma(\mu)}{\Gamma(\lambda)} c_k = \frac{\lambda + n - 2k}{k!} \frac{\Gamma(\mu - \lambda + k)\gamma(\mu - l + n)}{\Gamma(\mu - \lambda)\Gamma(\lambda - k + n + 1)},$$

which proves the theorem. ∎

9.4.4. Derive the formulas:

(a) $\ln(1+z) = z \, {}_2F_1(1,1;2;-z)$;

(b) $(1-z)^\alpha = {}_2F_1(\alpha,1;1,z)$;

(c) $\arcsin(z) = z \, {}_2F_1\left(\frac{1}{2},\frac{1}{2};\frac{3}{2};z^2\right)$.

HINT. Use Taylor series expansion and the definition (9.2.1) of hypergeometric function.

9.4.5. Show that the circle of convergence of the hypergeometric function ${}_2F_1(a,b;c;z)$ is the unit circle $|z| = 1$. The hypergeometric series diverges when $\Re\{c - a - b\} \leq -1$, converges absolutely when $\Re\{c - a - b\} \geq 0$, and converges conditionally when $-1 < \Re\{c - a - b\} \leq 0$, the point $z = 1$ excluded. This series reduces to a polynomial of degree n in z when a or b is equal to $-n$, $(n = 0, 1, \dots)$. The hypergeometric series is not defined when $c = -m$, $(m = 0, 1, \dots)$, provided a or b is not a negative integer n with $n < m$.

The integral representation

$$
{}_2F_1(a,b;c;z) = \frac{\Gamma(c)}{\Gamma(b)\Gamma(c-b)} \int_0^1 t^{b-1}(1-t)^{c-b-1}(1-tz)^{-a}\, dt
$$

is a single-valued analytic function in the z-plane cut along the real axis from 1 to ∞.

10

---

de Branges Theorem

A discussion of all conjectures that developed just to prove a single conjecture, namely the Bieberbach conjecture, is important to appreciate how it points to the only way to settle this problem. Many researchers in the field saw it but de Branges [1984] succeeded in his attempt. We will first discuss all the conjectures, then prove de Branges theorem, as he himself proceeded, and then provide additional historical information on his accomplishment.

10.1 Conjectures

A number of conjectures developed during the pursuit of proving the Bieberbach conjecture. The main conjectures, which are the asymptotic Bieberbach conjecture, Littlewood conjecture, Robertson conjecture, and Milin conjecture, have two related conjectures known as Rogosinski conjecture and Sheil-Small conjecture. A detailed account of all these conjectures follows.

Asymptotic Bieberbach Conjecture. If $f \in \mathcal{S}$ has the series expansion (6.1.1), i.e., if $f(z) = z + \sum_{n=2}^{\infty} a_n z^n$ and if $A_n = \max_{f \in \mathcal{S}} |a_n|$, then $\lim_{n \to \infty} \dfrac{A_n}{n} = 1$.

Littlewood Conjecture. If $f \in \mathcal{S}$ has the series expansion (6.1.1) and if $f \neq w$, then $|a_n| \leq 4|w| n$ for $n = 2, 3, \ldots$. Since $|w| \geq 1/4$, the Bieberbach conjecture implies the Littlewood conjecture. The asymptotic Bieberbach conjecture implies the Littlewood conjecture, as Nehari [1957] proved that if $f \in \mathcal{S}$ with series expansion $f(z) = z + \sum_{n=2}^{\infty} a_n z^n$ and if $f \neq w$,, and if w is not a value of $f(z)$ for any $z \in E$, then $|a_n| \leq 4|w| \lambda n$, $n = 2, 3, \ldots$, where $\lambda = \lim_{n \to \infty} \dfrac{a_n}{n}$. Later, Hamilton [1982] showed that the Littlewood conjecture implies the asymptotic Bieberbach conjecture, i.e.,

$$\text{Littlewood conjecture} \Longleftrightarrow \text{asymptotic Bieberbach conjecture.} \qquad (10.1.1)$$

Robertson conjecture. For any odd function $h(z) = z + c_3 z^3 + \cdots \in \mathcal{S}$, the following inequalities hold:

$$1 + |c_3|^2 + \cdots + |c_{2n-1}|^2 \leq n, \quad n = 2, 3, \ldots . \tag{10.1.2}$$

From (10.1.2) it is obvious that

$$\text{Robertson conjecture} \Longrightarrow \text{Bieberbach conjecture}. \tag{10.1.3}$$

For $n = 2$, Robertson conjecture yields $|a_2| \leq 2$, and then Robertson [1936] proved that the conjecture is true for $n = 3$ by using the Löwner method. Friedland [1970], using the Grunsky inequalities, proved that the Robertson conjecture is true for $n = 4$. This conjecture remained open for $n \geq 5$ until de Branges [1984] proved the Milin conjecture (see below), which, in view of (10.1.3), leads to the following truth-chain:

$$\text{Milin conjecture} \Longrightarrow \text{Robertson conjecture} \Longrightarrow \text{Bieberbach conjecture}. \tag{10.1.4}$$

Note that since the Littlewood-Paley conjecture is not true, the value of δ cannot be zero. However, the Milin conjecture (8.4.23) asserts that $\delta = 0$ in some average sense.

Theorem 10.1.1. (Milin's conjecture [1971]) *For any $f \in \mathcal{S}$, let c_k be the logarithmic coefficients. Then the following inequality holds:*

$$\sum_{k=1}^{n} (n + 1 - k)\left(k\,|c_k|^2 - \frac{4}{k}\right) \leq 0, \quad n = 1, 2, \ldots . \tag{10.1.5}$$

PROOF. The inequality (10.1.5) is the same as (8.4.23). Assume that for some $n \in \mathbb{N}$, the Milin conjecture (10.1.5) is valid. Then, with $\log \dfrac{h(z)}{z} = \frac{1}{2} f(z^2)$, and using the second Lebedev-Milin inequality (8.4.21), we find that

$$\frac{1}{n+1} \sum_{k=1}^{n+1} |c_{2k-1}|^2 \leq 1. \tag{10.1.6}$$

Hence, if the Milin conjecture is true, then $\sum_{k=0}^{n} |c_{2k+1}|^2 \leq n+1$, which implies that the Robertson conjecture is true., i.e., the proposition (10.1.4) would hold if the Milin conjecture were true. That is what de Branges [1984] did; he proved the Milin conjecture, and thereby proved the Bieberbach conjecture.

Milin conjecture is about a weighted quadratic mean of the logarithmic coefficients d_k or of the coefficients of the function $\dfrac{z f'(z)}{f(z)} = 1 + z\left(\log \dfrac{f(z)}{z}\right)' =$

$1 + \sum\limits_{n=1}^{\infty} n d_n z^n$, whose modulus is bounded by 2 for $f \in \mathcal{S}^\star$, so that for these functions $|d_k| \leq 2/k$ for $k \in \mathbb{N}$, and therefore, the Milin conjecture is valid, although it has yet to be proved. ∎

10.1.1 Some Related Conjectures. Let $g(z) = b_1 z + b_2 z^2 + \cdots$ be an analytic function in E, and let $f \in \mathcal{S}$. If $g(E) \subset f(E)$, then g is said to be *subordinate* to f, denoted by $g \prec f$, i.e., if $g \prec f$, then there exists a Schwarz function $S(z)$ such that $g(z) = f(S(z))$ (see Exercise 2.8.10).

Littlewood [1925] proved the following result.

Theorem 10.1.2. *If $g \prec f$, then for $0 < r < 1$ and $0 < p < \infty$,*

$$M_p(r, g) \leq M_p(r, f), \tag{10.1.7}$$

where $M_p(r, g)$ is defined by (8.1.1).

Using this definition, Rogosinski [1939, 1943] proved the following result.

Theorem 10.1.3. (Rogosinski [1939,1943]) *If $g \prec f$, then*

$$\sum_{n=1}^{N} |b_n|^2 \leq \sum_{n=1}^{N} |a_n|^2, \quad N = 1, 2, \ldots. \tag{10.1.8}$$

The inequality

$$\sum_{n=1}^{N} |b_n|^p \leq \sum_{n=1}^{N} |a_n|^p \tag{10.1.9}$$

is not true when $p \neq 2$. Moreover, $g \prec f$ does not imply $|b_n| \leq |a_n|$, for which the counter-example is $z^2 \prec z$.

Theorem 10.1.4. (Rogosinski conjecture) *If $g \prec f$, and $f \in \mathcal{S}$, then $|b_n| < n$ for $n = 1, 2, \ldots.$*

Since $f \prec f$ is trivial, we see that

$$\text{Rogosinski conjecture} \Longrightarrow \text{Bieberbach conjecture.} \tag{10.1.10}$$

However, the Robertson conjecture implies the Rogosinski conjecture, as shown by the following result.

Theorem 10.1.5. *Let c_{2k-1} be defined as in (8.1.3). If $g \prec f$, and $f \in \mathcal{S}$, then*

$$|b_n| \leq \sum_{k=1}^{n} |c_{2k-1}|^2. \tag{10.1.11}$$

PROOF. Let $h(z)$ be defined by (8.1.3). Since $g \prec f$, we have $g(z) = f(w(z))$. Let

$$\phi(z) = \frac{h(\sqrt{z})}{\sqrt{z}} = 1 + c_3 z + c_5 z^2 + \cdot\,,$$

i.e., $\phi^2(z) = f(z)/z$. Then

$$g(z) = w(z) \left[1 + c_3 w(z) + c_5 w^2(z) + \cdots \right]^2.$$

Denote the partial sum of the first n terms of ϕ by $S_n(z) = \sum\limits_{k=1}^{n} c_{2k-1} z^{k-1}$. Since $w(0) = 0$, we find that

$$b_n = \frac{1}{2\pi i} \int_{|z|=r} \frac{w(z) \left[S_n(w(z)) \right]^2}{z^{n+1}} \, dz.$$

Since $S_n(w(z)) \prec S_n(z)$, by Theorem 10.1.2, we get

$$|b_n| \leq \frac{[M_2(r, S_n(w(z)))]^2}{r^n} \leq \frac{[M_2(r, S_n(z))]^2}{r^n} = \frac{\sum\limits_{k=1}^{n} |c_{2k-1}|^2 r^{2k-2}}{r^n},$$

which, on letting $r \to 1$ yields (10.1.11). ∎

This shows that

$$\text{Robertson conjecture} \Longrightarrow \text{Rogosinski conjecture.} \qquad (10.1.12)$$

Let $f(z) \in S$ with series expansion (6.1.1), and $g(z) = \sum\limits_{n=1}^{\infty} b_n z^n$ be two power series. Then $h(z) = \sum\limits_{n=1}^{\infty} a_n b_n z^n$ is called the *convolution* or *Hadamard product* of f and g, and is denoted by $h = f \star g$.

Theorem 10.1.6. (Sheil-Small conjecture [1973]) *For any $f \in S$ and any polynomial $P(z)$ of degree n, the following inequality holds:*

$$\|P \star f\|_\infty \leq n \|P\|_\infty, \qquad (10.1.13)$$

where $\| \, \|_\infty$ denotes the maximum modulus in E.

Note that for $P(z) = z^n$ the Sheil-Small conjecture becomes the Bieberbach conjecture. The Sheil-Small conjecture lies between the Robertson conjecture and the Rogosinski conjecture.

Hence, the relationship between all these conjectures follows the logical chain:

Milin conjecture $\Longrightarrow$ Robertson conjecture $\Longrightarrow$ Sheil-Small conjecture $\Longrightarrow$
Rogosinski conjecture $\Longrightarrow$ Bieberbach conjecture $\Longrightarrow$
Asymptotic Bieberbach conjecture $\Longleftrightarrow$ Littlewood conjecture.

10.2 de Branges Theorem

Before we prove de Branges theorem, we introduce the *de Branges function system* $\{\tau_{m,n}(t)\}_{m=1,\ldots,n}$, where t, $0 \leq t < \infty$, is a parameter. Let $m = 1, 2, \ldots$, and for a fixed m, define

$$\tau_{m,n}(t) = n \sum_{j=0}^{m-n} (-1)^j \frac{(2n+j+1)_j (2n+2j+2)_{m-n-j}}{(n+j)j!\,(m-n-j)!} e^{-(j+n)t},$$
$$\tau_{m,m+1}(t) \equiv 0, \quad n = 1, 2, \ldots, m. \tag{10.2.1}$$

This system consists of functions $\tau_{m,n} : \mathbb{R}^+ \mapsto \mathbb{R}$ specified below in properties (10.2.4)–(10.2.6). Let $c_m(t)$ denote the logarithmic coefficients of $e^{-t} f(z,t)$, where $f(z,t)$ is a Löwner chain of $f \in S$. To prove the Milin conjecture, de Branges defined a function $\phi(t)$ as

$$\phi(t) = \sum_{m=1}^{n} \left(m\,|c_m(t)|^2 - \frac{4}{m} \right) \tau_{m,n}(t). \tag{10.2.2}$$

Then the inequality $\phi(0) \leq 0$ becomes equivalent to the Milin conjecture if we set

$$\tau_{m,n}(0) = n + 1 - m, \quad m = 1, \ldots, n. \tag{10.2.3}$$

The success of de Branges proof of the Milin conjecture for $n \geq 1$ is true if the function system $\{\tau_{m,n}(t)\}_{m=1,\ldots,n}$ has the following properties:

$$\tau_{m,n}(t) - \tau_{m+1,n}(t) = -\frac{\dot{\tau}_{m,n}}{m} - \frac{\dot{\tau}_{m+1,n}(t)}{m+1}, \quad m = 1, \ldots, n; \tag{10.2.4}$$

$$\lim_{t \to \infty} \tau_{m,n}(t) = 0, \quad m = 1, \ldots, n; \quad \text{and} \tag{10.2.5}$$

$$\dot{\tau}_{m,n} \leq 0 \quad \text{for } t \in \mathbb{R}^+, \tag{10.2.6}$$

where the 'dot' over τ denotes its derivative with respect to t (as in time-derivative). Using these properties and the Löwner differential equation

$$\Re\left\{ \frac{\partial_t f(z,t)}{z f'(z,t)} \right\} \geq 0, \quad z \in E,$$

de Branges, after a lengthy computation, succeeded in obtaining the inequality $\dot{\phi}(t) \geq 0$ and thus, in proving that $\phi(0) = -\int_0^\infty \dot{\phi}(t)\,dt \leq 0$. The

Bieberbach conjecture was reduced to the existence of the de Branges function system $\{\tau_{m,n}\}_{m=1,\dots,n}$ with the properties (10.2.4)–(10.2.6). Note that the coupled system of first-order differential equations (10.2.4) together with the initial conditions (10.2.3) has a unique solution, and therefore, the properties (10.2.4)–(10.2.5) must also be satisfied simultaneously. The property (10.2.5) is easily fulfilled using the theory of ordinary differential equations, but (10.2.6) is a deep theorem, for which de Branges gave the explicit representations in terms of the hypergeometric functions (§9.2) as

$$\tau_{m,n}(t) = e^{-mt} \binom{n+m+1}{2m+1} {}_4F_3\left(\begin{matrix} m+\frac{1}{2}, & n_m+2, & m, & m-n \\ m+1, & 2m+1, & m+\frac{3}{2} \end{matrix} \middle| e^{-t}\right),$$

$$\dot{\tau}_{m,n}(t) = -m\, e^{-mt} \binom{n+m+1}{1+2m} {}_3F_2\left(\begin{matrix} m+\frac{1}{2}, & m-n, & n+m+2 \\ 2m+1, & m+\frac{3}{2} \end{matrix} \middle| e^{-t}\right).$$

Note that $\dot{\tau}_{m,n}(t)$ are polynomials with respect to the variable $y = e^{-t}$.

Lemma 10.2.1. If $P_j^{(\alpha,\beta)}(x)$ are Jacobi polynomials (§9.1.1), then

$$\dot{\tau}_{m,n}(t) = -ne^{-nt} \sum_{j=0}^{m-n} P_j^{(2n,0)}\left(1-2e^{-t}\right). \tag{10.2.7}$$

PROOF. By definition (10.2.1), we have

$$-\frac{\dot{\gamma}_{m,n}(t)\, e^{nt}}{n} = \sum_{j=0}^{m-n}(-1)^j \frac{(2n+j+1)_j(2n+j+2)_{m-n-j}}{j!(m-n-j)!}\, e^{-jt}. \tag{10.2.8}$$

Set $\alpha = 2n, x = 1 - 2e^{-t}$ in (9.2.28), and replace m by $m - n$. Then we have

$$\sum_{j=0}^{m-n} P_j^{(2n,0)}\left(1-2e^{-t}\right) = \sum_{j=0}^{m-n}(-1)^j \frac{2^{2j}\left(\frac{2n+1}{2}\right)_j \left(\frac{2n+2}{2}\right)_j (2n+2j+2)_{m-n-j}}{(2n+1)_j\, j!\,(m-n-j)!}\, e^{-jt}. \tag{10.2.9}$$

Since, by (9.2.15), $2^{2j}\left(\frac{2n+1}{2}\right)_j \left(\frac{2n+2}{2}\right)_j = (2n+1)_{2j}$, and since $\dfrac{(2n+1)_{2j}}{(2n+1)_j} = (2n+j+1)_j$, Eq (10.2.9) becomes

$$\sum_{j=0}^{m-n} P_j^{(2n,0)}\left(1-2e^{-t}\right) = \sum_{j=0}^{m-n}(-1)^j \frac{(2n+j+1)_j(2n+2j+2)_{m-n-j}}{j!(m-n-j)!}\, e^{-jt},$$

which after comparing with (10.2.8) gives (10.2.7). ∎

As a result of this Lemma and the Askey-Gasper theorem (Theorem 9.2.6), we have two additional results.

Theorem 10.2.1. $\dot{\tau}_{m,n}(t) \geq 0$ for $n = 1, 2, \ldots, m$, and $0 \leq t < \infty$.

Proof of this theorem is obvious from (9.2.29).

Theorem 10.2.2. The de Branges function system $\{\tau_{m,n}(t)\}_{m=1,\ldots,n}$ satisfies the relations (10.2.4)–(10.2.6).

PROOF. Since $\tau_{m,m+1}(t) \equiv 0$ for each $m = 1, 2, \ldots$, the initial value problem (10.2.4)-(10.2.5) has a unique solution, and therefore, it defines a unique system. We will solve this initial value problem first for $\tau_{m,m}(t)$ and then for $\tau_{m,m-1}(t)$, and so on. Since $\tau_{m,m+1}(t) \equiv 0$ for the de Branges system and satisfies the initial condition (10.2.5), we will show that Eq (10.2.4) is satisfied, which amount to showing that $\tau_{m,n}(t) + \dfrac{\dot{\tau}_{m,n}(t)}{n} = \tau_{m,n+1}(t) - \dfrac{\dot{\tau}_{m,n+1}(t)}{n+1}$. Denote

$$V_n = \frac{\tau_{m,n}(t)\,e^{nt}}{n}, \quad \text{and} \quad W_n = \frac{\tau_{m,n}(t)\,e^{-nt}}{n}, \quad 1 \leq n \leq m+1, \quad (10.2.10)$$

Then

$$\dot{V}_n = \left(\frac{\dot{\tau}_{m,n}(t)}{n} + \tau_{m,n}(t)\right)e^{nt}, \quad \dot{W}_n = \left(\frac{\dot{\tau}_{m,n}(t)}{n} - \tau_{m,n}(t)\right)e^{-nt}.$$

Thus, we should show that

$$\dot{V}_n(t)\,e^{-nt} = -\dot{W}_{n+1}(t)\,e^{(n+1)t} \quad \text{for } 1 \leq n \leq m. \quad (10.2.11)$$

By (10.2.1) and (10.2.10), we have

$$V_n = \sum_{j=0}^{m-n}(-1)^j \frac{(2n+j+1)_j(2n+2j+2)_{m-n-j}}{(n+j)\,j!(m-n-j)!}\,e^{-jt},$$

$$W_n = \sum_{j=0}^{m-n}(-1)^j \frac{(2n+j+1)_j(2n+2j+2)_{m-n-j}}{(n+j)\,j!(m-n-j)!}\,e^{-jt-2nt}.$$

Then (10.2.11) is obtained simply by evaluating $\dot{V}_n$ and $\dot{W}_{n+1}$ from these two formulas.

As a consequence, it is easy to prove that $P_n^{(\alpha,\beta)}(-1) = (-1)^n P_n^{(\beta,\alpha)}(1)$, and thus, $P_j^{(\alpha,0)}(-1) = (-1)^j$. Since $\dot{\tau}_{n,0}(0) \equiv \dot{\tau}_n(0) = -n\sum_{j=0}^{m-n}(-1)^j$ by Lemma 10.2.1, we find that

$$-\frac{\dot{\tau}_n(0)}{n} = \sum_{j=0}^{m-n}(-1)^j = \begin{cases} 1 & \text{if } m-n \text{ is even,} \\ 0 & \text{if } m-n \text{ is odd.} \end{cases} \quad (10.2.12)$$

Thus, to prove (10.2.4), note that by (10.2.3) and (10.2.12) we have $\tau_{m,n}(0) -$
$\tau_{m,n+1}(0) = 1$ for $n = 1, 2, \ldots, m$. Then the initial condition (10.2.4) follows
if we add the above two formulas. ∎

Since the Milin conjecture implies Robertson conjecture which in turn
implies Bieberbach conjecture, de Branges [1984] proved his famous theorem
which is as follows.

Theorem 10.2.3. (de Branges theorem) *The Milin conjecture (10.1.5) is*
true, where equality holds if and only if $f(z)$ is the Koebe function or one of
its rotations.

PROOF. The proof,[1] based on de Branges [1984, 1985, 1987] and Fitzgerald
and Pommerenke [1985] as presented in Gong [1999], is divided into two parts:
the first part shows that the inequality (10.1.5) holds, and the second proves
the later part of the theorem. Consider a function $f \in S$ such that it maps
the unit disk E onto the complex plane with a silt which is a Jordan curve
extending to the point at infinity. It is known from §4.5 that such slit mappings
are dense in S. We will prove the theorem only for such functions. From §4.1.1
we know that there exists a chain, the Löwner chain, of functions of the form
$f(z,t) = e^t z + \cdots + a_n(t) z^n + \cdots$, $|z| < 1$, $0 \le t < \infty$, such that $f(z,t)$
satisfies the Löwner differential equation (4.2.2) which can be written as

$$\frac{\partial f(z,t)}{\partial t} = \frac{1 + k(t)z}{1 - k(t)z} z \frac{\partial f(z,t)}{\partial z}, \quad f(z,0) = f(z), \tag{10.2.13}$$

where $k(t)$ is a continuous function defined on $0 \le t < \infty$ such that $|k(t)| = 1$.
Let

$$\log \left(\frac{f(z,t)}{e^t z} \right) = \sum_{n=1}^{\infty} c_n(t) z^n, \quad |z| < 1, \tag{10.2.14}$$

for $0 \le t < \infty$, and $c_n(0) = 2\gamma_n$, where γ_n are the coefficients in the power
series expansion of $\log \dfrac{f(z)}{z}$ given by (3.5.4) with $d_{0n} = \gamma_n$, such that for the
Koebe function $\gamma_n = \dfrac{1}{n}$. Differentiating (10.2.14) with respect to t and z
respectively, we find that

$$\frac{\partial_t f(z,t)}{f(z,t)} - 1 = \sum_{n=1}^{\infty} \dot{c}_n(t) z^n, \tag{10.2.15}$$

$$\frac{\partial_z f(z,t)}{f(z,t)} - \frac{1}{z} = \sum_{n=1}^{\infty} n c_n(t) z^{n-1}, \tag{10.2.16}$$

---

[1] For another approach to the proof, see Exercise 10.5.3.

From (10.2.13), (10.2.15) and (10.2.16), we get

$$1 + \sum_{n=1}^{\infty} \dot{c}_n(t) z^n = \frac{1 + k(t)z}{1 - k(t)z} \left(1 + \sum_{n=1}^{\infty} n c_n(t) z^n \right)$$

$$= \left(1 + 2k(t)z + 2k^2(t)z^2 + \cdots \right) \left(1 + \sum_{n=1}^{\infty} n c_n(t) z^n \right).$$

If we compare the coefficients of z^n on both sides of the above equality, we get

$$\dot{c}_n(t) = 2k^n(t) + n c_n(t) + 2 \sum_{j=1}^{n-1} k^{n-j}(t) j c_j(t). \tag{10.2.17}$$

Let

$$b_0 \equiv 0, \quad b_n(t) = \sum_{j=1}^{n} j c_j(t) k^{-j}(t), \quad n = 1, 2, \ldots, \tag{10.2.18}$$

Then from (10.2.17) we get

$$\dot{c}_n(t) = 2k^n(t) - n c_n(t) + 2k^n(t) b_n(t), \quad n = 1, 2, \ldots. \tag{10.2.19}$$

Now, for a fixed m, define

$$\phi(t) = \sum_{n=1}^{m} \left(n |c_n(t)|^2 - \frac{4}{n} \right) \tau_{m,n}(t). \tag{10.2.20}$$

Then the Milin conjecture (10.1.5) reduces to

$$\phi(0) = \sum_{n=1}^{m} \left(n |c_n(0)|^2 - \frac{4}{n} \right) (m - n + 1) \leq 0. \tag{10.2.21}$$

The proof of (10.2.21) is equivalent to proving that

$$\dot{\phi}(t) = - \sum_{n=1}^{m} |b_{n-1}(t) + b_n(t) + 2|^2 \frac{\dot{\tau}_{m,n}(t)}{n} \quad \text{for } t \geq 0, \tag{10.2.22}$$

because if (10.2.22) is true, then $\dot{\phi}(t) \geq 0$ by Theorem 10.2.1, and therefore, $\phi(t)$ is a monotone increasing function. Also, since S is compact and by (10.2.2) we have $\phi(\infty) = 0$, we find that $\phi(0) \leq 0$, which is precisely the Milin conjecture.

To prove (10.2.22), note that $k^{-1}(t) = \overline{k(t)}$ since $|k(t)| = 1$. Also, from (10.2.18),

$$\overline{b}_n(t) - \overline{b}_{n-1}(t) = n\overline{c}_n(t)k^n(t),$$

$$b_n(t) - b_{n-1}(t) = nc_n\overline{k^n(t)}, \tag{10.2.23}$$

$$\left|b_n(t) - b_{n-1}(t)\right|^2 = n^2|c_n|^2.$$

Thus, from (10.2.20) and (10.2.23) we have

$$\phi(t) = \sum_{n=1}^{m} \left(|b_n(t) - b_{n-1}(t)|^2 - 4\right) \frac{\tau_{m,n}(t)}{n}. \tag{10.2.24}$$

Now, let us write b_n for $b_n(t)$ for brevity, and let $nc_n(t) = k^n (b_n - b_{n-1}) = u(t)$. Then $|u(t)|^2 = |b_n - b_{n-1}|^2$, and

$$\frac{\partial|u(t)|^2}{\partial t} = \frac{\partial u(t)}{\partial t}\bar{u}(t) + u(t)\frac{\partial \bar{u}(t)}{\partial t} = 2\Re\left\{\frac{\partial u(t)}{\partial t}\bar{u}(t)\right\} = 2\Re\{n\dot{c}_n(t)\bar{u}(t)\}$$

$$= 2\Re\{n\dot{c}_n(t)k^{-n}(t)\left[\overline{b}_n - \overline{b}_{n-1}\right]\},$$

which yields

$$\frac{\partial}{\partial t}\left(\frac{|b_n - b_{n-1}|^2}{n}\right) = 2\Re\{\dot{c}_n(t)k^{-n}(t)\left[\overline{b}_n - \overline{b}_{n-1}\right]\}. \tag{10.2.25}$$

Substitute (10.2.19) into (10.2.25); this gives

$$\frac{\partial}{\partial t}\left(\frac{|b_n - b_{n-1}|^2}{n}\right) = 2\Re\{\left[2 - nc_n(t)k^{-n}(t) + 2b_n\right]\left(\overline{b}_n - \overline{b}_{n-1}\right)\},$$

or, since from the first equality in (10.2.23) $nc_n k^{-n}(t) = b_n - b_{n-1}$, the above equality becomes

$$\frac{\partial}{\partial t}\left(\frac{|b_n - b_{n-1}|^2}{n}\right) = 2\Re\{2\left(\overline{b}_n - \overline{b}_{n-1}\right) - |b_n - b_{n-1}|^2 + 2b_n\left(\overline{b}_n - \overline{b}_{n-1}\right)\}$$

$$= -2|b_n - b_{n-1}|^2 + 4\Re\{(1 + b_n)\left(\overline{b}_n - \overline{b}_{n-1}\right)\}. \tag{10.2.26}$$

Differentiating (10.2.24) and using (10.2.26), we get

$$\dot{\phi}(t) = \sum_{n=1}^{m} \left(|b_n - b_{n-1}|^2 - 4\right) \frac{\dot{\tau}_{m,n}(t)}{n}$$

$$+ \sum_{n=1}^{m} \tau_{m,n}(t)\left[-2|b_n - b_{n-1}|^2 + 4\Re\{(1 + b_n)\left(\overline{b}_n - \overline{b}_{n-1}\right)\}\right]. \tag{10.2.27}$$

Since

$$-\sum_{n=1}^{m} \left(2|b_n|^2 + 4\Re\{b_n\}\right) \tau_{m,n+1} = -\sum_{n=2}^{m+1} \left(2|b_{n-1}|^2 + 4\Re\{b_{n-1}\}\right) \tau_{m,n}$$

$$= -\sum_{n=1}^{m} \left(2|b_{n-1}|^2 + 4\Re\{b_{n-1}\}\right) \tau_{m,n},$$

we get

$$\sum_{n=1}^{m} \left(2|b_n|^2 + 4\Re\{b_n\}\right) \left(\tau_{m,n} - \tau_{m,n+1}\right)$$

$$= \sum_{n=1}^{m} \left(2|b_n|^2 + 4\Re\{b_n\} - 2|b_{n-1}|^2 - 4\Re\{b_{n-1}\}\right) \tau_{m,n}.$$

Note that

$$-2|b_n - b_{n-1}|^2 + 4\Re\{1 + b_n\}\left(\bar{b}_n - \bar{b}_{n-1}\right)$$
$$= 2|b_n|^2 - 2|b_{n-1}|^2 + 4\Re\{b_n\} - 4\Re\{b_{n-1}\}.$$

Since $\tau_{m,m+1} = 0$ and $b_0 = 0$, so from (10.2.27) and (10.2.3) we get

$$\dot{\phi}(t) = \sum_{n=1}^{m} \left(|b_n - b_{n-1}|^2 - 4\right) \frac{\dot{\tau}_{m,n}}{n}$$

$$+ \sum_{n=1}^{m} \left(2|b_n|^2 + 4\Re\{b_n\}\right) \left(\tau_{m,n} - \tau_{m,n+1}\right)$$

$$= \sum_{n=1}^{m} \left(|b_n - b_{n-1}|^2 - 4\right) \frac{\dot{\tau}_{m,n}}{n}$$

$$+ \sum_{n=1}^{m} \left(2|b_n|^2 + 4\Re\{b_n\}\right) \left(-\frac{\dot{\tau}_{m,n}}{n} - \frac{\dot{\tau}_{m,n+1}}{n+1}\right)$$

$$= \sum_{n=1}^{m} \left[\left(|b_n - b_{n-1}|^2 - 4\right) - 2|b_n|^2 - 4\Re\{b_n\}\right] \frac{\dot{\tau}_{m,n}}{n}$$

$$- \sum_{n=1}^{m} \left(2|b_{n-1}|^2 + 4\Re\{b_{n-1}\}\right) \frac{\dot{\tau}_{m,n}}{n}$$

$$= \sum_{n=1}^{m} \left[|b_n - b_{n-1}|^2 - 4 - 2|b_n|^2\right.$$

$$\left. - 4\Re\{b_n\} - 2|b_{n-1}|^2 - 4\Re\{b_{n-1}\}\right] \frac{\dot{\tau}_n}{n}$$

$$= -\sum_{n=1}^{m} |b_n + b_{n+1} + 2|^2 \frac{\dot{\tau}_{m,n}}{n},$$

which is (10.2.22). This proves first part of the theorem.

To prove that equality in the Milin conjecture holds if and only if $f(z)$ is the Koebe function or one of its rotations, let us choose a sequence of functions $\{f_m(z) \in \mathcal{S}\}$ such that f_m maps the unit disk E onto the complex plane minus a Jordan curve which tends to infinity, and the sequence $\{f_m\}$ converges uniformly to f on any compact subsets of E. Corresponding to $f_n(z)$ there exists a function $f_m(z,t)$, whose coefficients $a_{m,n}(t)$ in the series expansion of $f_m(z,t)$ correspond to the coefficients $c_{n,m}(t)$ which tend to $c_n(t)$, defined in (10.2.14), as $m \to \infty$. For sufficiently large m there exists an α such that

$$|c_{1,m}(0)| = |a_{2,m}(0)| < \alpha < 2. \tag{10.2.28}$$

In view of (10.2.19) we have $|\dot{c}_{1,m}(t)| = |c_{1,m}(t) + 2k(t)| \le |a_{2,m}(t)| + 2 \le 4$. Thus, $|c_{1,m}(t)| \le \alpha + 4t$ for $t \ge 0$. Since $\alpha + 4t \le 2$ for $0 \le t \le \dfrac{2-\alpha}{4}$, and $\dot{\tau}_m(t) < 0$ (recall that $\tau_{m,0}(t) \equiv \tau_m(t)$), we get from (10.2.28) for sufficiently large m,

$$\dot{\phi}_m(t) \ge |c_{1,m}(t)\bar{k}_m(t)+2|^2\,(-\dot{\tau}_1(t)) \ge (2 - \alpha - 4t)^2\,(-\dot{\tau}_1(t)) \text{ for } 0 \le t \le \frac{2-\alpha}{4}.$$

Since $-\int_0^\infty \dot{\phi}_m(t)\,dt = \phi_m(0) = \sum\limits_{k=1}^{n} \left(k|c_{k,m}(0)|^2 - \frac{4}{k}\right)(n+1-k)$, we get

$$\sum_{k=1}^{n} \left(k|c_{k,m}(0)|^2 - \frac{4}{k}\right)(n+1-k) \le - \int_0^{(2-\alpha)/8} \dot{\tau}_m(t)\,dt$$

$$\le \left(\frac{2-\alpha}{2}\right)^2 \int_0^{(2-\alpha)/8} \dot{\tau}_1(t)\,dt$$

$$= \left(\frac{2-\alpha}{2}\right)^2 \left[\tau_1\left(\frac{2-\alpha}{8}\right) - \tau_1(0)\right] < 0,$$

which, as $m \to \infty$, reduces to $\sum\limits_{k=1}^{n} \left(k|c_k|^2 - \frac{4}{k}\right)(n+1-k) < 0$. This completes the second part of the proof. ∎

10.2.1. Historical Notes. de Branges verified the inequality (10.2.5) by hand computation up to $n = 5$, thus proving the Bieberbach conjecture for $|a_6|$. Since these computations were getting larger, he asked his colleague Walter Gautschi of Purdue University to verify his computations for this inequality numerically, which was done in February 1984 up to $n \le 30$. This account is available as Gautschi's reminiscences (Gautschi [1986]). Since the function $\dot{\tau}_{m,n}$ has an oscillatory behavior, the computations had to be carefully done on the computers in those days to obtain correct results. However, nowadays those computation, using a robust computer algebra system, can be performed and checked out within seconds using a PC.

Gautschi also called Dick Askey, and de Branges finally realized that the inequality (10.2.5) had already been proved by Askey and Gasper [1976]. This was enough to finish the proof of the Milin conjecture.

Using the so called *Zeiberger algorithm*, developed by Zeilberger [1990], Koepf proved the Askey-Gasper identity computationally; a Maple program is available in Koepf [2007]; also available at http://www.mathematik.uni-kassel.de/~koepf/Publikationen.

10.3 Alternate Proofs of de Branges Theorem

After de Branges proved the Bieberbach conjecture, the fervor in this topic waned a little bit but not completely. The research still continues in old topics with new directions, especially in proving alternate proofs of the de Branges theorem, and in related topics in slit mappings and applications, quasi-conformal mappings, and the like, which are described in Chapter 11. However, we will discuss in detail the proof by Weinstein [1991], and a generalization of de Branges theorem by Ming-Qin [1997].

10.3.1 Weinstein's proof. Weinstein [1991] proved de Branges theorem on purely classical lines; his proof is simple and impressive. He had the advantage of knowing that the Milin conjecture was true as proved by de Branges (Theorem 10.2.3). Weinstein uses the following representation of the t-derivative of the logarithmic coefficients $c_k(t)$ of the Löwner chain $f(z,t)$ with $z = re^{i\theta}$, $0 < r < 1$:

$$
\begin{aligned}
\dot{c}_k(t) &= \frac{1}{2\pi} \int_0^{2\pi} \frac{\partial_t f(z,t)}{z^k f(z,t)}\, d\theta, \quad \text{by Cauchy's formula} \\
&= \frac{1}{2\pi} \int_0^{2\pi} p(z,t) \frac{z f'(z,t)}{z^k f(z,t)}\, d\theta, \quad \text{by (4.1.2)} \\
&= \frac{1}{2\pi} \int_0^{2\pi} p(z,t)\Big(1 + \sum_{j=1}^{\infty} j c_j(t) z^j\Big) \frac{d\theta}{z^k} \\
&= \lim_{r \to 1} \frac{1}{2\pi} \int_0^{2\pi} p(z,t)\Big(1 + \sum_{j=1}^{\infty} j c_j(t) z^j\Big) \frac{d\theta}{z^k}.
\end{aligned}
\tag{10.3.1}
$$

Let $\phi(z,t) = K^{-1}\left(e^{-t} K(z)\right)$ for fixed $z \in E$. Then the Koebe function is

$$
K(z) = \frac{z}{(1-z)^2} = e^t \frac{\phi(z,t)}{(1-\phi(z,t))^2}, \quad t \geq 0.
\tag{10.3.2}
$$

Weinstein considered the generating function of the inequality (10.1.5). Since

$\mathcal{S}$ is compact, this generating function converges absolutely and locally uniformly in E. Thus, from (10.1.5) define

$$w(z) = \sum_{n=1}^{\infty} \left[\sum_{k=1}^{n} (n+1-k) \left(\frac{4}{k} - k|c_k(0)|^2 \right) \right] z^{n+1} = \frac{z}{(1-z)^2} \sum_{k=1}^{\infty} \left(\frac{4}{k} - k|c_k(0)|^2 \right) z^k.$$

(10.3.3)

Since, in view of (10.3.3) and the fundamental theorem of calculus, we have $\phi(z,\infty) = 0$ and $\phi(z,0) = z$, so (10.3.3) reduces to

$$w(z) = - \int_0^{\infty} \frac{e^t \, \phi(z,t)}{(1-\phi(z,t))^2} \frac{d}{dt} \left\{ \sum_{k=1}^{\infty} \left(\frac{4}{k} - k|c_k(t)|^2 \right) \phi^k(z,t) \right\} dt,$$

which, using (10.3.1) and the fact that $\dot{\phi}(z,t) = -\phi(z,t)\dfrac{1-\phi(z,t)}{1+\phi(z,t)}$, where $z \in E$, $z = re^{i\theta}$, simplifies to

$$w(z) = \int_0^{\infty} \frac{e^t \, \phi(z,t)}{(1-\phi(z,t))^2} \sum_{k=1}^{\infty} A_k(t) \phi^k(z,t) \, dt,$$

(10.3.4)

where

$$A_k(t) = \lim_{r \to 1} \frac{1}{2\pi} \int_0^{2\pi} \Re\{p(z,t)\} \left| 1 + 2 \sum_{j=1}^{k} j c_j(t) z^j - k d_k(t) z^k \right|^2 d\theta.$$ (10.3.5)

For derivation of (10.3.4), see Exercise 10.5.5. Now, let

$$W_k(z,t) = \frac{e^t \phi^{k+1}(z,t)}{1 - \phi^2(z,t)} = \sum_{n=k}^{\infty} \Lambda_{k,n}(t) z^{n+1}.$$

(10.3.6)

Then

$$w(t) = \sum_{n=1}^{\infty} \left(\int_0^{2\pi} \sum_{k=1}^{\infty} \Lambda_{k,n}(t) A_k(t) \, dt \right) z^{n+1}.$$

(10.3.7)

Note that the Milin conjecture is equivalent to the statement that $w(t)$ has non-negative coefficients, whereas from Löwner's equation (4.1.2) we have $\Re\{p(z,t)\} > 0$. Thus, $A_k(t) \geq 0$ for $t \geq 0$, so that the Milin conjecture will be valid if

$$\Lambda_{k,n}(t) \geq 0, \quad \text{for } t \geq 0.$$

(10.3.8)

But this statement is simply a positivity property of special real functions, which are related to the Koebe function.

The range of $\phi(z,t) = K^{-1}(z)\big(e^{-t}K(z)\big)$ is E with a slit on the real axis $u = 0$. Note that the mapping

$$h_\gamma(z) = \frac{z}{1 - 2z \cos\gamma + z^2} \tag{10.3.9}$$

maps E for $\gamma \neq 0 \pmod{\pi}$ onto a domain which is slit on the real axis twice (see Kober [1957]). Hence, we interpret $\phi(z,t)$ as the composite function $\phi = h_\gamma^{-1}(e^{-t} h_\gamma)$ for a suitable pair (θ, γ), which after some computation gives

$$\cos\gamma = \big(1 - e^{-t}\big) + e^{-t} \cos\theta. \tag{10.3.10}$$

Thus,

$$
\begin{aligned}
h_\gamma(z) = e^t h_\theta(\phi) &= \frac{e^t \phi}{1 - \phi^2}\left(\frac{1 - \phi^2}{1 = 2\phi\cos\gamma + \phi^2}\right) \\
&= \frac{e^t \phi}{1 - \phi^2}\Re\left\{\frac{1 + e^{i\theta}\,\phi}{1 - e^{i\theta}\,\phi}\right\} = \frac{e^t \phi}{1 - \phi^2}\left(1 + 2\sum_{k=1}^{\infty}\phi^k \cos k\theta\right) \\
&= \frac{e^t \phi}{1 - \phi^2} + 2\sum_{k=1}^{\infty}\left(\sum_{n=1}^{\infty}\Lambda_{k,n}(t)z^{n+1}\right)\cos k\theta.
\end{aligned}
\tag{10.3.11}
$$

Since

$$\eta(z) = \frac{\sqrt{h_\gamma(z)}}{z} = \frac{1}{\sqrt{1 - 2z\cos\gamma + z^2}} \tag{10.3.12}$$

is the generating function of the Legendre polynomials (§9.1.3), we find from (10.3.10) and (9.1.13) that

$$\eta(z) = \sum_{n=0}^{\infty}\nu_n^2 z^n + 2\sum_{k=1}^{\infty}\left(\sum_{n=k}^{\infty}\frac{(n-k)!}{(n+k)!}\mu_{kn}^2 z^n\right)\cos k\theta, \tag{10.3.13}$$

where ν_n and μ_{kn} are real numbers for $k, n \in \mathbb{N}$. The representation (10.3.13) has non-negative coefficients. After squaring the results, it follows from (10.3.13) that the condition (10.3.8) is satisfied, and therefore, the Milin conjecture is true. ∎

10.3.2 Ming-Qin's Generalization. Recall that the Askey-Gasper theorem [1976] was the backbone in de Branges proof (§6.1.1). However, this generalization of de Branges theorem by Ming-Qin does not use the Askey-Gasper theorem; instead, it replaces de Branges system of functions by a special system $\{s_k(t)\}_{k=1}^{n}$, although de Branges method is used throughout the proof.

Theorem 10.3.2. (Ming-Qin [1997]) *Let $f \in S$ with the series expansion (6.1.1), and let* $\log \dfrac{f(z)}{z} = \sum\limits_{k=1}^{\infty} c_k z^k$, $|z| < 1$. *If for any real number set* $\{\lambda_k\}_{k=1}^{n}$, $\lambda_k \geq 0$ *for* $k = 1, \ldots, n$, *the condition*

$$\lambda_k + 2 \sum_{m=k+1}^{n} (-1)^{m-k} \lambda_k \geq 0 \qquad (10.3.14)$$

is satisfied, then

$$\sum_{k=1}^{n} k \lambda_k |c_k|^2 \leq 4 \sum_{k=1}^{n} \frac{\lambda_k}{k}, \quad n = 1, 2, \ldots, \qquad (10.3.15)$$

where equality holds if and only if $f(z)$ is the Koebe function or one of its rotations.

Note that if $l_k = n - k + 1$ for $k = 1, \ldots, n$, then de Branges theorem becomes a corollary of this theorem.

PROOF. Instead of de Branges' function system, consider the following system of differential equations in $s_k(t)$, $k = 1, \ldots, n$:

$$\frac{\dot{s}_{k+1}(t)}{k+1} + \frac{\dot{s}_k(t)}{k} = s_{k+1}(t) - s_k(t), \qquad (10.3.16)$$

with the initial conditions

$$s_{n+1}(t) \equiv 0, \quad s_k(0) = \lambda_k, \ \lambda_k \geq 0 \text{ for } k = 1, \ldots, n. \qquad (10.3.17)$$

The initial value problem (10.3.16)–(10.3.17) defines a unique system of solutions $\{s_k(t)\}_{k=1}^{n}$. To determine these solutions, first we find from (10.3.16) that

$$\frac{\dot{s}_k(t)}{k} = 2 \sum_{m=k+1}^{n} (-1)^{m-k-1} s_m(t) - s_k(t), \ k = 1, \ldots, n, \qquad (10.3.18)$$

where the characteristic equation of this system is

$$\Delta(\lambda) = \begin{vmatrix} 1+\lambda & -2 \cdot 1 & 2 \cdot 1 & \cdots & (-1)^{n-1} 2 \cdot 1 \\ 0 & 2+\lambda & -2 \cdot 2 & \cdots & (-1)^{n-2} 2 \cdot 2 \\ 0 & 0 & 3+\lambda & \cdots & (-1)^{n-3} 2 \cdot 3 \\ \vdots & \vdots & \vdots & \cdots & \vdots \\ 0 & 0 & 0 & \cdots & n+\lambda \end{vmatrix} = 0. \qquad (10.3.19)$$

The characteristic roots of this equation are

$$\lambda_k = -k, \text{ for } k = 1, 2, \ldots, n. \qquad (10.3.20)$$

Thus, we can assume that

$$s_k(t) = \sum_{j=k}^{n} c_{k,j}\, e^{-jt}, \quad k = 1,\ldots,n. \tag{10.3.21}$$

The proof is completed in the following four steps.

STEP 1. We will show that the functions $s_k(t)$, defined by (11.2.21), have the form

$$s_k(t) = c_{k,k}\, e^{-kt} + \sum_{m=k+1}^{n} \alpha_{k,m} c_{m,m}\, e^{-mt}, \tag{10.3.22}$$

where $\alpha_{k,m}$ and $c_{m,m}$, $m = k, k+1, \ldots, n$, are constants defined by

$$\alpha_{k,m} = \frac{(-1)^{m-k} 2k(2m-1)(2m-2)\cdots(m+k+1)}{(m-k)!}, \quad m = k, k+1, \ldots, n,$$

$$\alpha_{k,k} = 1. \tag{10.3.23}$$

$$\sum_{j=k}^{m} (-1)^{j-k} \alpha_{j,m} = \frac{(-1)^{m-k}(2m-1)(2m-2)\cdots(m+k)}{(m-k)!}. \tag{10.3.24}$$

Note that (10.3.21) when substituted in (10.3.18) gives

$$\dot{s}_k(t) + k s_k(t) = 2k \sum_{\nu=k+1}^{n} (-1)^{\nu-k-1} \sum_{j=\nu}^{n} c_{\nu,j}\, e^{-jt}.$$

This differential equation has the solution

$$s_k(t) = c_{k,k}\, e^{-kt} + 2k \sum_{m=k+1}^{n} \left[\sum_{p=k+1}^{m} (-1)^{p-k} \frac{c_{p,m}}{m-k} \right] e^{-mt}.$$

Hence, for $k = 1, \ldots, n-1$ and $m = k+1, k+2, \ldots, n$,

$$c_{k,m} = \frac{2k}{m-k} \sum_{p=k+1}^{m} (-1)^{p-k} c_{p,m}, \tag{10.3.25}$$

which yields

$$c_{k,m} = \alpha_{k,m} c_{m,m},\ c_{k+1,m} = \alpha_{k+1,m} c_{m,m}, \cdots, c_{m-1,m} = \alpha_{m-1,m} c_{m,m}$$

for $k = 1, 2, \ldots, n-1$ and $m = k+1, k+2, \ldots, n$. Note that (10.3.25) for $k = m-1$ gives $c_{m-1,m} = -2(m-1)c_{m,m}$, and $c_{m-1,m} - c_{m,m} = -(2m-1)c_{m,m}$, which implies that (10.3.23) and (10.3.24) are true if $k = \nu+1, \nu+2, \ldots, m-1$. But then by (10.3.25) we have

$$
\begin{aligned}
c_{\nu,m} &= -\frac{2\nu}{m-\nu}\left(c_{\nu+1,m} - c_{\nu+2,m} + \cdots + (-1)^{m-\nu-1}c_{m,m}\right) \\
&= -\frac{2\nu}{m-\nu}\frac{(-1)^{m-\nu-1}(2m-1)(2m-2)\cdots(m+\nu+1)}{(m-\nu)!} \\
&= \frac{(-1)^{m-\nu}2\nu(2m-1)(2m-2)\cdots(m+\nu+1)}{(m-\nu)!}c_{m,m},
\end{aligned}
$$

which shows that (10.3.23) is true. Again, note that

$$
\begin{aligned}
c_{\nu,m} &- c_{\nu+1,m} + \cdots + (-1)^{m-\nu}c_{m,m} \\
&= \left[\frac{(-1)^{m-\nu}2\nu(2m-1)(2m-2)\cdots(m+\nu+1)}{(m-\nu)!}\right. \\
&\quad \left.- \frac{(-1)^{m-\nu-1}(2m-1)(2m-2)\cdots(m+\nu+1)}{(m-\nu-1)!}\right]c_{m,m} \\
&= \frac{(-1)^{m-\nu}(2m-1)(2m-2)\cdots(m+\nu)}{(m-\nu)!}c_{m,m},
\end{aligned}
$$

which proves that (11.2.24) is true.

STEP 2. We will show that for any positive integers $n = 1, 2, \ldots,$ the identity

$$
\sum_{j=0}^{n}(-1)^j\frac{j^k}{j!\,(n-j)!} = 0 \tag{10.3.26}
$$

holds for $k = 0, 1, 2, \ldots, n-1$. In fact, this result follows by using the binomial theorem $(1-x)^n = n!\sum_{j=0}^{n}(-1)^j\frac{x^j}{j!(n-j)!}$, and applying k-times the operator $x\frac{d}{dx}$ to $(1-x)^n$ and then letting $x = 1$.

STEP 3. Show that the coefficients $c_{m,m}$ in (10.3.22) have the form

$$
c_{m,m} = \sum_{m=k}^{n}(-1)^{m-k}\delta_{k,m}\lambda_m, \quad k = 1, 2, \ldots, n, \tag{10.3.27}
$$

where

$$
\begin{aligned}
\delta_{k,m} &= \begin{vmatrix} \alpha_{k,k+1} & \alpha_{k,k+2} & \cdots & \alpha_{k,m-1} & \alpha_{k,m} \\ 1 & \alpha_{k+1,k+2} & \cdots & \alpha_{k+1,m-1} & \alpha_{k+1,m} \\ \vdots & \vdots & \cdots & \vdots & \cdots \\ 0 & 0 & \cdots & 1 & \alpha_{m-1,m} \end{vmatrix} \\
&= \frac{(-1)^{m-k}2k(2k+1)(2k+2)\cdots(m+k-1)}{(m-k)!}, \quad (m > k),
\end{aligned}
$$

$$
\delta_{k,k} \equiv 1. \tag{10.3.28}
$$

In fact, From (10.3.17) and (10.3.22) we get

$$c_{1,1} + \alpha_{1,2}c_{2,2} + \alpha_{1,3}c_{3,3} + \cdots + \alpha_{1,n}c_{n,n} = \lambda_1,$$
$$c_{2,2} + \alpha_{2,3}c_{3,3} + \cdots + \alpha_{2,n}c_{n.n} = \lambda_2,$$
$$\vdots$$
$$c_{n,n} = \lambda_n,$$

which, when solved for $c_{k,k}$ gives

$$
c_{k,k} = \begin{vmatrix}
\lambda_k & \alpha_{k,k+1} & \cdots & \alpha_{k,n-1} & \alpha_{k,n} \\
\lambda_{k+1} & 1 & \cdots & \alpha_{k+1,n-1} & \alpha_{k+1,n} \\
\vdots & & & & \\
\lambda_{n-1} & 0 & \cdots & 1 & \alpha_{n-1,n} \\
\lambda_n & 0 & \cdots & 0 & 1
\end{vmatrix}
$$
$$
= \sum_{m=k}^{n} (-1)^{m-k} \delta_{k,m} \lambda_m, \quad k = 1, 2, \ldots, n.
$$

After some calculations we find that

$$\delta_{m-1,m} = -\frac{2m-2}{1!}, \; \delta_{m-2,m-1} = -\frac{2m-4}{1!}, \; \delta_{m-2,m} = \frac{(2m-4)(2m-3)}{2!}.$$

So we assume that

$$\delta_{j,m} = \frac{(-1)^{m-j}2j(2j+1)(2j+2)\cdots(m+j-1)}{(m-j)!}$$

for $j = k+1, k+2, \ldots, m-1$, and $j < m \leq n$. Then by (10.3.28) we have

$$
\begin{aligned}
\delta_{k,m} &= \sum_{j=k+1}^{m} (-1)^{j-k-1} \alpha_{k,j} \delta_{j,m} \\
&= \sum_{j=k+1}^{m} (-1)^{j-k-1} \frac{2k(2j-1)(2j-2)\cdots(j+k+1)}{(j-k)!} \\
&\quad \times \frac{2j(2j+1)(2j+2)\cdots(m+j-1)}{(m-j)!} \\
&= \sum_{j=k+1}^{m} (-1)^{j-k-1} \frac{2k(m+j-1)(m+j-2)\cdots(j+k+1)}{(j-k)!\,(m-j)!}.
\end{aligned}
$$
$$\text{(10.3.29)}$$

Let $j - k = \mu$. Then (10.3.29) becomes

$$\delta_{k,m} = \sum_{\mu=1}^{m-k} (-1)^{m-k-\mu-1} \frac{2k(m+k+\mu-1)(m+k+\mu-2)\cdots(2k+\mu+1)}{\mu!\,(m-k-\mu)!}$$

$$= (-1)^{m-k} \frac{2k(2k+1)(2k+2)\cdots(m+k-1)}{(m-k)!}$$

$$+ \sum_{\mu=0}^{m-k} (-1)^{m-k-\mu-1} \frac{2k(2k+\mu+1)(2k+\mu+2)\cdots(m+k+\mu-1)}{\mu!\,(m-k-\mu)!}.$$

Note that for fixed m and k, the numerator in the second term in the above formula is a polynomial in μ of degree $(m-k-1)$. Hence,

$$\delta_{k,m} = (-1)^{m-k} \frac{2k(2k+1)(2k+2)\cdots(m+k-1)}{(m-k)!},$$

which completes this step.

STEP 4. We will show that the functions $s_k(t)$, $k = 1, 2, \ldots, n$, defined by the system of differential equation (10.3.16)–(10.3.17), are the solutions of

$$-\frac{1}{k}\dot{s}_k(t) = \lambda_k + 2 \sum_{m=k+1}^{n} (-1)^{m-k}\lambda_m, \qquad (10.3.30)$$

In fact, from (10.3.16)–(10.3.17) we have

$$-\frac{1}{n}\dot{s}_n(0) = \lambda_n, \quad \text{and} \quad -\frac{1}{n-1}\dot{s}_{n-1}(0) = \lambda_{n-1} - 2\lambda_n.$$

Assume that for $j = k+1, k+2, \ldots, n$,

$$-\frac{1}{j}\dot{s}_j(0) = \lambda_j + 2 \sum_{m=j+1}^{n} (-1)^{m-j}\lambda_m. \qquad (10.3.31)$$

But by (10.3.22) we have

$$-\frac{1}{k+1}\dot{s}_{k+1}(0) = c_{k+1,k+1} + \sum_{m=k+2}^{n} \frac{m}{k+1}\alpha_{k+1,m}\,c_{m,m}$$

$$= c_{k+1,k+1} + \sum_{m=k+2}^{n} \alpha_{k+1,m}c_{m,m} + \sum_{m=k+2}^{n} \frac{m-k-1}{k+1}\alpha_{k+1,m}c_{m,m}$$

$$= \lambda_{k+1} + \sum_{m=k+2}^{n} \frac{m-k-1}{k+1}\alpha_{k1,m}c_{m,m}.$$

Hence,

$$\sum_{m=k+2}^{n} \frac{m-k-1}{k+1} \alpha_{k+1,m} c_{m,m} = 2 \sum_{m=k+2}^{n} (-1)^{m-k-1} \lambda_m. \qquad (10.3.32)$$

However, by (10.3.22)

$$-\frac{1}{k} \dot{s}_k(0) = c_{k,k} + \sum_{m=k+1}^{n} \frac{m}{k} \alpha_{k,m} c_{m,m}$$

$$= c_{k,k} + \sum_{m=k+1}^{n} \alpha_{k,m} c_{m,m} + \sum_{m=k+1}^{n} \frac{m-k}{k} \alpha_{k,m} c_{m,m}$$

$$= \lambda_k + \sum_{m=k+1}^{n} \frac{m-k}{k} \alpha_{k,m} c_{m,m}$$

$$= \lambda_k - 2c_{k+1,k+1} + \sum_{m=k+2}^{n} \frac{m-k}{k} \alpha_{k,m} c_{m,m}. \qquad (10.3.33)$$

Since

$$\frac{m-k}{k} \alpha_{k,m} = \frac{(-1)^{m-k} 2(2m-1)!}{(m-k-1)!(m+k)!} = \frac{m+k+1}{k+1} \alpha_{k+1,m},$$

we find from (10.3.32) and (10.3.33) that

$$-\frac{1}{k} \dot{s}_k(0) = \lambda_k - 2c_{k+1,k+1} - \sum_{m=k+2}^{n} \frac{m+k+1}{k+1} \alpha_{k+1,m} c_{m,m}$$

$$= \lambda_k - 2\left(c_{k+1,k+1} + \sum_{m=k+2}^{n} \alpha_{k+1,m} c_{m,m} \right)$$

$$\qquad - \sum_{m=k+2}^{n} \frac{m-k-1}{k+1} \alpha_{k+1,m} c_{m,m}$$

$$= \lambda_k - 2\lambda_{k+1} - 2 \sum_{m=k+2}^{n} (-1)^{m-k-1} \lambda_m$$

$$= \lambda_k + 2 \sum_{m=k+1}^{n} (-1)^{m-k} \lambda_m,$$

which shows that (10.3.30) is true. ∎

Note that for $\lambda_k = n - k + 1$ $(k = 1, 2, \dots, n)$, this theorem becomes de Branges theorem.

10.4 de Branges and Weinstein Systems of Functions

Some interesting relationships between the de Branges system of functions $\{\tau_{m,n}(t)\}$ and the Weinstein system of functions $\{\Lambda_{m,n}(t)\}$ are as follows.

Theorem 10.4.1. (Todorov [1992], Wilf [1994]) *For $k = 1,\ldots,n$, $n \in \mathbb{N}$,*

$$\dot{\tau}_{m,n}(t) = -m\,\Lambda_{m,n}(t). \tag{10.4.1}$$

PROOF. (Koepf and Schmersau [1996]) Since the functions $\tau_{m,n}(t)$ satisfy the conditions (10.2.1), (10.2.2) and (10.2.3), these functions are obtained by solving the initial value problem consisting of the system of differential equations (10.2.3), which, since $\dot{\tau}_{m,n}(t)$ are polynomials with respect to the variable e^{-t} (§10.2), reduces to the system of equations

$$\dot{\tau}_{m,n}(t) = -ne^{-nt}, \tag{10.4.2}$$

and the initial conditions

$$\dot{\tau}_{m,n}(t) = \begin{cases} -m & \text{if } n-m \text{ is even,} \\ 0 & \text{if } n-m \text{ is odd.} \end{cases} \tag{10.4.3}$$

Define $y_n(t) := -m\Lambda_{m,n}(t)$. Then to prove (10.4.1) we will show that the functions $y_n(t)$ satisfy (10.4.2) and (10.4.3). In view of the Weinstein function $W_m(z,t)$ defined by (10.3.6) as

$$W_m(z,t) = \frac{e^t\,w(z,t)^{m+1}}{1 - w^2(z,t)} = \sum_{n=m}^{\infty} \Lambda_{m,n}(t)z^{n+1}, \tag{10.4.4}$$

we write w for $w(z,t)$ and get

$$W_{m+1}(z,t) = \frac{e^t w^{m+2}}{1 - w^2} = wW_m(z,t). \tag{10.4.5}$$

Then we have

$$W_m(z,t) = \frac{e^t\,w^{m+1}}{1 - w^2} = \frac{e^t\,w}{(1-w)^2}\frac{1-w}{1+w}w^m = K(z)\frac{1-w}{1+w}w^m, \tag{10.4.6}$$

where $K(z)$ is the Koebe function or one of its rotations. From (10.4.5) and (10.4.6) we obtain the equation

$$W_m(z,t) + W_{m+1}(z,t) = (1+w)W_m(z,t) = K(z)(1-w)w^m$$
$$= (z)w^m - K(z)w^{m+1}. \tag{10.4.7}$$

Differentiating this equation with respect to t, we obtain

$$\frac{\partial}{\partial t}W_m(z,t) + \frac{\partial}{\partial t}W_{m+1}(z,t) = K(z)\,mw^{m-1}\dot{w} - K(z)(m+1)w^m\dot{w}$$

$$= -m\,K(z)\frac{1-w}{1+w}w^m + (m+1)K(z)\frac{1-w}{1+w}w^{m+1}$$

$$= (m+1)W_{m+1}(z,t) - mW_m(z,t),$$

$$(10.4.8)$$

which is similar to Eq (4.1.1). Using the definition for $y_n(t)$ and comparing it with the coefficients $\Lambda_{m,n}$ in (10.4.4), we find that the equation

$$y_{m+1}^n(t) - y_m^n(t) = \frac{\dot{y}_m^n(t)}{m} + \frac{\dot{y}_{m+1}^n(t)}{m+1},$$

which precisely the de Branges system (10.2.3). Since, by (104.x),

$$\Lambda_{n,n}(t) = \lim_{z\to 0}\frac{W_n(z,t)}{z^{n+1}} = \lim_{z\to 0}\frac{e^t}{1-w^2(z,t)}\left(\frac{w(z,t)}{z}\right)^{n+1} = e^t\left(e^{-(n+1)t}\right) = e^{-nt},$$

we find that $y_n^n(t) = -ne^{-nt}$, and thus, (10.4.2) is satisfied. Finally, from the initial conditions (10.4.3) we get

$$W_m(z,0) = \frac{z^{m+1}}{1-z^2} = \sum_{j=0}^{\infty}z^{2j+m} = \sum_{n=m}^{\infty}\Lambda_{m,n}(0)\,z^{n+1}.\ \blacksquare$$

10.4.1 Generating Function. The generating function for de Branges system of functions $\{\tau_{m,n}(t)\}_{m=1}^n$ is defined by

$$B_m(z,t) = \sum_{n=m}^{\infty}\tau_{m,n}(t)z^{n+1}. \qquad (10.4.9)$$

Theorem 10.4.2. (Koepf [2003]) *The generating function (10.4.9) has the representation*

$$B_m(z,t) = \sum_{n=m}^{\infty}\tau_{m,n}(t)z^{n+1} = K(z)w(z,t)^m$$

$$= K(z)\left(\frac{4e^{-t}z}{\left(1-z+\sqrt{1-2xz+z^2}\right)^1}\right)^m, \qquad (10.4.10)$$

where $x = 1 - 2e^{-t}$, and $K(z)$ is the Koebe function.

PROOF. Define

$$C_m(z, y) = K(z)w(z, t)^m. \qquad (10.4.11)$$

Then $\dot{C}_m(z, t) = K(z)mw^{m-1}\dot{w}$. Since, in view of (4.5.1), $\dfrac{\dot{w}}{t} = -w\dfrac{1-w}{1+w}$ and since $K(w(z,t)) = e^{-t}K(z)$, we have $\dot{C}_m(z, t) = K(z)mw^{m-1}\dot{w}$. Thus,

$$\frac{\dot{C}_{m+1}(z, t)}{m+1} + \frac{\dot{C}_m(z, t)}{m} = K(z)w^{m-1}(1+w)\dot{w}$$
$$= -K(z)w^m(1-w) = K(z)w^{m+1} - K(z)w^m$$
$$= C_{m+1}(z, t) - C_m(z, t). \qquad (10.4.12)$$

In view of (10.4.11), the Taylor series expansion of $C_m(z, t)$ about $z = 0$ is

$$C_m(z, t) = \sum_{n=m}^{\infty} y_m^n(t) z^{n+1}, \qquad (10.4.13)$$

where $y_m^n(t)$ is defined in Theorem 10.4.1, with $y_{n+1}^n(0)$. Substituting (10.4.13) into (10.4.12), we find from comparing the coefficients of z^{n+1} that the functions $y_m^n(t)$ satisfy de Branges differential equation (10.2.3). The $(n+1)$st coefficient in the Taylor expansion of $C_m(z, 0) = K(z)w(z, 0)^m = \dfrac{z^{m+1}}{(1-z)^2}$ is obviously $n + 1 = m$; thus, the functions $y_m^n(t)$ have the initial value (10.2.2). Hence, $y_m^n(t)$ are de Branges functions $y_m^n(t) = \tau_{m,n}(t)$, and so $B_m(z, t) = C_m(z, t)$. Also, the explicit representation

$$w(z, t) = \left(\frac{4e^{-t} z}{\left(1 - z + \sqrt{1 - 2xz + z^2}\right)^1} \right)^m$$

is on the right hand side of (10.4.10) ∎

10.5 Exercises

10.5.1. Prove the Milin conjecture for $n = 2$ and 3. PROOF. For $n = 2$, this conjecture means that $|c_1|^2 \leq 4$. Since

$$1 + a_2 z + a_3 z^2 + \cdots = \frac{f(z)}{z} = \exp\left\{ c_1 z + c_2 z^2 + \cdots \right\} = 1 + c_1 z + \left(\tfrac{1}{2}c_1^2 + c_2\right)z^2 + \cdots, \qquad (10.5.1)$$

we find that $|a_2| = |c_1| \leq 2$, which is the Bieberbach inequality. For $n = 3$, the Milin conjecture is equivalent to $|c_1|^2 + |c_2|^2 \leq 5$, and using the Löwner inequality (Theorem 4.4.1), we get

$$|a_3| = \left| \tfrac{1}{2}c_1^2 + c_2 \right| \leq \tfrac{1}{2}|c_1|^2 + |c_2| \leq \tfrac{5}{2} - \tfrac{1}{2}|c_2|^2 + |c_2| \leq 3.$$

10.5.2. What theorems, conjectures and known results were used in the proof of de Branges theorem? ANSWER. The following items were used by de Branges to complete his proof:

(i) Löwner's partial differential equation for the Löwner chains $\{f_t(z)\}$ of injective analytic functions from E to $\mathbb{C}$ [1923].

(ii) The Milin conjecture about the logarithmic coefficients of $f \in \mathcal{S}$, i.e., the coefficients in the expansion $\sum_{n=1}^{\infty} c_n z^n$ for a branch of $\log\left\{\dfrac{f(z)}{z}\right\}$ (1965–1970; known to West in 1977 after publication of Milin [1977])

(iii) de Branges' system of functions $\{\tau_n(t)\}$ associated with the Milin conjecture, which varies monotonically along Löwner chains.

(iv) de Branges' introduction and solution of a system of differential equations which he devised so that the above system becomes manageable.

(v) Use of a positivity result for hypergeometric functions which established the monotonicity of the functional, using the work of Askey and Gasper [1976].

10.5.3. Describe the steps used in the proof of de Branges theorem. ANSWER. The following nine steps, in fact, complete the proof. This provides another approach to the proof.

STEP 1. To prove the Milin conjecture (10.1.5), assume that f maps E onto a domain D bounded by a Jordan curve, i.e., for any given $f \in \mathcal{S}$ and $0 < r < 1$, define

$$f^*(z) = \frac{1}{r}f(rz) = z + \sum_{n=1}^{\infty} a_n r^{n-1} z^n.$$

The function f^* maps E onto the domain $\dfrac{1}{r}f(rE)$, which is bounded by the Jordan curve given by $\dfrac{1}{r}$-times the the image of the circle $|z| = r$ under f. Since

$$\log \frac{f^*(z)}{z} = \log \frac{f(rz)}{rz} = \sum_{n=1}^{\infty} c_n^* r^n z^n,$$

the logarithmic coefficients c_n^* for f^* are equal to $c_n r^n$. Hence, if (10.1.5) is proved for the coefficients c_n^*, then it is also proved for c_n by letting $r \to 1$.

STEP 2. To obtain the Löwner chains, construct, as in §4.1.1, a nice continuously increasing family of simply connected domains G_t, $0 \le t < \infty$, such that $G_0 = G$, $G_s \subsetneq G_t$ if $z < t$, and $G_t \to \mathbb{C}$ as $t \to \infty$. According to Pommerenke [1975], this construction can be done for every simply connected domain G. Define $f(z, t)$, $0 \le t < \infty$, as an injective (one-to-one) conformal map of E onto G_t such that $f(0,t) = 0$, $f'(0,t) > 0$. Then the function $w(t) = f'(0,t)$ is a strictly increasing function such that $w(0) = 1$ and $w(t) \to \infty$ as $t \to \infty$. Thus, we can assume that $w(t) = e^t$. Then the corresponding family of injective analytic functions is the Löwner chain:

$$f(z,t) = e^t\left(z + \sum_{n=2}^{\infty} a_n(t)z^n\right), \quad 0 \le t < \infty; \; f(z,0) = f(z), \qquad (10.5.2)$$

which depends continuously on t and starts at $f(z)$. In fact, every $f \in S$ is the starting point of a Löwner chain (Pommerenke [1975 : 159].

STEP 3. Derive Löwner's differential equation (4.1.2). The function $f(z,t)$ of a Löwner chain satisfies this partial differential equation, where $p(z,t)$ is analytic in z, $\Re\{p(z,t)\} > 0$, and $p(0,t) = 1$.

STEP 4. Consider the logarithmic coefficients for $e^{-t} f(z,t)$. Let

$$\log \frac{f(z,t)}{e^t z} = \sum_{n=1}^{\infty} c_n(t)z^n. \qquad (10.5.3)$$

Since $e^{-t} f(z,t) \in S$, there exists, in view of (8.1.5), constants A_n, e.g., $A_n = en^2$, such that $|a_n(t)| \le A_n$ for all t. Hence, by recursion (see Exercise 10.5.1, Eq (10.5.1)), there are constants C_n such that

$$|c_n(t)| \le C_n \quad \text{for all } t. \qquad (10.5.4)$$

Differentiating (10.5.3) with respect to t and with respect to z, substituting the results into the Löwner equation (4.1.2), where we set

$$p(z,t) = 1 + 2\sum_{n+1}^{\infty} d_n(t)z^n, \qquad (10.5.5)$$

and then equating the coefficients of similar powers of z, we obtain the system of differential equations

$$\dot{c}_n(t) = 2d_n(t) + nc_n(t) + 2\sum_{j=1}^{n-1} jc_j(t)d_{n-j}(t), \quad n = 1, 2, \dots, \qquad (10.5.6)$$

where the dot denotes the 't'-derivative.

STEP 5. Fix n in (8.4.23), and introduce an auxiliary functional $\Omega(t)$ by

$$\Omega(t) = \Omega_n(t) = \sum_{k=1}^{n-1}\left(k|c_k(t)|^2 - \frac{4}{k}\right)\sigma_k(t),$$

where $\sigma(t)$ is the weight function to be suitably chosen, with the following properties: (i) The inequality $\Omega \le 0$ must be the same as the Milin conjecture (10.1.5). Since $c_n(0) = c_n$, we impose the initial conditions

$$\sigma_k(0) = n - k \quad \text{for } k = 1, \dots, n - 1. \tag{10.5.7}$$

Then the inequality $\Omega \le 0$ would follow if $\Omega(t)$ were a non-decreasing function of t which vanishes at $t = +\infty$, that is, if

$$\dot{\Omega}(t) \ge 0 \quad \text{for } 0 \le t < \infty, \text{ and } \Omega \to 0 \text{ as } t \to \infty. \tag{10.5.8}$$

Since, in view of (10.5.4), every $c_n(t)$ is bounded, the condition (10.5.8) is satisfied if

$$\lim_{t \to \infty} \sigma_k(t) = 0 \quad \text{for } k = 1, \dots, n - 1. \tag{10.5.9}$$

STEP 6. Determine the conditions on $\sigma_k(t)$, and calculate $\dot{\Omega}(t)$ using the differential equations (10.5.6) for $c_n(t)$. Although the resulting expression gets quite complicated, yet it gets simplified if we impose de Branges' conditions

$$\sigma_k(t) - \sigma_{k+1}(t) = -\left(\frac{\dot{\sigma}_k(t)}{k} - \frac{\dot{\sigma}_{k+1}(t)}{k+1}\right) \quad \text{for } k = 1, \dots, n-1; \ \sigma_n(t) \equiv 0, \tag{10.5.10}$$

which finally gives

$$\dot{\Omega}(t) = -\sum_{k=1}^{n-1} Q_k(c, d)\,\dot{\sigma}_k(t), \tag{10.5.11}$$

where $Q_k(t)$ are non-negative functions of $c_k(t)$ and $d_k(t)$.

The precise form of the functions $Q_k(c, d)$ is obtained as follows: Using the Herglotz representation (2.7.8) for analytic functions on the unit disk with positive part, we have

$$p(z, t) = \int_{-\pi}^{\pi} \frac{e^{i\theta} + z}{e^{i\theta} - z}\, d\mu_t(\theta),$$

where μ_t is a positive Borel measure of total mass equal to $p(0, t) = 1.$[2] Thus, in view of (10.5.5), the coefficients $d_n(t)$ of $p(z, t)$ are of the form

$$d_n(t) = \int_{-\pi}^{\pi} e^{in\theta}\, d\mu_t(\theta).$$

---

[2] An outer Borel measure μ on $\mathbb{R}^n$ is called a regular Borel measure if (i) every Borel set $B \subseteq \mathbb{R}^n$ is μ-measureable in the sense of Carathéodory's criterion which states that $\mu(a) = \mu(A \cap B) + \mu(A \backslash B)$ for any $A \subseteq \mathbb{R}^n$; and (ii) for every set $A \subseteq \mathbb{R}^n$ (which need not be μ measureable) there exists a Borel set $B \subseteq \mathbb{R}^n$ such that $A \subseteq B$ and $\mu(A) = \mu(B)$. If the outer measure satisfies these two conditions, it is called a Borel measure. Lebesgue outer measure on $\mathbb{R}$ is an example of a Borel regular measure.

Let $s_n(t) = \sum_{j=1}^{n} jc_j(t)\, e^{ij\theta}$, $s_0(t) = 0$. Then $nc_n(t) = (s_n(t) - s_{n-1}(t))\, e^{-in\theta}$, and using (10.5.6) we get $\dot{c}_n(t) = \int_{-\pi}^{\pi} (2 + s_{n-1}(t) + s_n(t))\, e^{-in\theta}\, d\mu_t(\theta)$. Then $\dot{\Omega}(t)$ can be written as an integral in terms of μ_t, with the integrand as a sum involving $s_k(t)$, $\sigma_k(t)$ and $\dot{\sigma}(t)$. After summation by parts and using the differential equation (10.5.10), we obtain (10.5.11), where

$$Q_k(c, d) = \frac{1}{k} \int_{-\pi}^{\pi} |2 + s_{k+1}(t) + s_k(t)|^2\, d\mu_t. \qquad (10.5.12)$$

STEP 7. Using $\dot{\Omega}(t)$, defined by (10.5.11), we can establish that $\dot{\Omega}(t) > 0$ provided that we show that

$$\dot{\sigma}(t) \le 0 \quad k = 1, \dots, n - 1. \qquad (10.5.13)$$

Note that the first $\sigma_{n-1}(t)$ and the next $\sigma n - 1(t), \dots, \sigma_1(t)$ are determined completely by the system of differential equations (10.5.10) and the initial conditions (10.5.7). See Exercise 10.5.1 for the cases $n = 2$ and $n = 3$.

For the general case n, de Branges found the solution of the system of differential equations (10.5.10) and initial conditions (10.5.7) for $k = 1, \dots, n-1$ as

$$\sigma_k(t) = k \sum_{m=0}^{n-k-1} (-1)^m \frac{(2k + m + 1)_m (2k + 2m + 2)_{n-k-1-m}}{(k + m)\, m!\, (n - k - 1 - m)!}\, e^{-mt - kt}.$$
$$(10.5.14)$$

Obviously, $\lim_{t \to \infty} \sigma_k(t) = 0$, as in the condition (10.5.9). However, in view of the condition (10.5.13) which implies non-positive values of $\dot{\sigma}(t)$, we find from (10.5.14) that

$$-\frac{\dot{\sigma}(t)}{k} e^{kt} = \sum_{m=0}^{n-k-1} (-1)^m \frac{(2k + m + 1)_m (2k + 2m + 2)_{n-k-1-m}}{m!\, (n - k - 1 - m)!}\, e^{-mt},$$

and thus, the sums (10.5.14) must be non-negative for $k = 1, \dots, n - 1$ and for all $n \ge 2$.

STEP 8. Complete the proof of the Milin conjecture, i.e., verify that the sums (10.5.14) are positive for $0 < t < \infty$ and for all $n \ge 2$. The verification for small n was easy; de Branges verified through calculations the sum (10.5.14) for n up to 5. Then his colleague Gautschi [1986] helped him verify numerically the sums up to $n = 30$. The following result of Askey and Gasper [1976] on generalized hypergeometric functions established that the sums (10.5.14) were all known to be positive:

$$-\frac{\dot{\sigma}(t)}{k} e^{kt} = \binom{n + k}{n - k - 1}\, {}_3F_2\left(\begin{matrix} -n + k + 1,\ k + \frac{1}{2},\ n + k + 1 \\ k + \frac{3}{2},\ 2k + 1 \end{matrix} \middle|\, e^{-t} \right).$$
$$(10.5.15)$$

This establishes the proof of de Branges theorem for positive values of the sums (10.5.15).

STEP 9.   To establish the sums (10.5.15) for the case of equality in (10.5.13), start with any $f \in S$ and an associated Löwner chain of the form (10.5.2). It is obvious that the equality holds in the Lebedev-Milin inequality (8.4.20) for f and for $n \geq 2$ if and only if $\dot{\Omega} \equiv 0$. Since $\dot{\sigma}_k(t) < 0$ for $0 < t < \infty$ and $k = 1, \ldots, n-1$, we must show in (10.5.11) that $Q_k(c, d) \equiv 0$ for $l = 1, \ldots, n-1$. But it is clear from (10.5.11) that $Q_k(c, d) = 0$ is necessary for the case of equality, since by (10.5.12) with $\mu_t > 0$ we get $2 + s_1 = 2 + c_1(t)e^{i\theta} = 0$ a.e. $[\mu_t]$. Thus, the absolutely continuous part of μ_t must be zero. In fact, μ_t must be a point mass at some point θ_t. Thus, we find that $c_1| = 2$, and therefore $|a_2| = 2$ so that f is the Koebe function, for which $\Omega(0) = 0$ for all $n \geq 2$. ∎

10.5.4. Derive Eq (10.3.4). SOLUTION: Eq (4.1.2) implies that

$$\Re\{p(z,t)\} > 0 \quad \text{for } p(z,t) = \frac{\partial_t f(z,t)}{z \partial_z f(z,t)} \tag{10.5.16}$$

is a characteristic univalent function. Then there exists a Loewner chain $\{f(z,t)\}$ for all $f \in S$. Also, the following three known results are used:

$$\dot{c}_k(t) = \frac{1}{2\pi}\int_0^{2\pi} \frac{\partial_t f(z,t)}{f(z,t)}\frac{d\theta}{z^k} = \lim_{r \to 1}\frac{1}{2\pi}\int_0^{2\pi}\frac{\partial_t f(z,t)}{f(z,t)}z^{-k}\,d\theta. \tag{10.5.17}$$

$$\frac{z}{(1-z)^2} = e^t\frac{w}{(1-w)^2}, \quad t \geq 0. \tag{10.5.18}$$

$$\frac{\partial}{\partial t}w = -w\frac{1-w}{1+w}. \tag{10.5.19}$$

Setting $w = phi(z,t)$, from (10.3.3) we get

$$
\begin{aligned}
w(z) &= -\int_0^\infty \frac{z}{(1-z)^2}\frac{\partial}{\partial t}\left(\sum_{k=1}^\infty \left(\frac{4}{k} - k|c_k(t)|^2\right)w^k\right)dt \\
&= -\int_0^\infty \frac{e^t w}{(1-w)^2}\frac{\partial}{\partial t}\left(\sum_{k=1}^\infty \left(\frac{4}{k} - k|c_k(t)|^2\right)w^k\right)dt, \quad \text{using (10.5.18)} \\
&= \int_0^\infty \frac{e^t w}{1-w^2}\left(\frac{1+w}{1-w}\sum_{k=1}^\infty k\frac{d}{dt}\{c_k(t)\overline{c_k(t)}\}w^k \right. \\
&\quad \left. + \sum_{k=1}^\infty \left(4 - k^2|c_k(t)|^2\right)w^k\right)dt, \quad \text{using (10.5.19)} \\
&= \int_0^\infty \frac{e^t w}{1-w^2}\left(\frac{1+w}{1-w}\left(1+\sum_{k=1}^\infty\left(\lim_{r\to 1}\frac{1}{2\pi}\int_0^{2\pi}\frac{\partial_t f(z,t)}{f(z,t)}k\,\overline{c_k(t)}z^k\,d\theta\right)w^k\right)\right)
\end{aligned}
$$

$$+ \frac{1+w}{1-w}\left(1 + \sum_{k=1}^{\infty}\left(\lim_{r\to 1}\frac{1}{2\pi}\int_0^{2\pi}\frac{\partial_t \overline{f(z,t)}}{f(z,t)}\,kc_k(t)z^k\,d\theta\right)w^k\right)$$

$$- 2\frac{1+w}{1-w} + \frac{4w}{1-w} - \sum_{k=1}^{\infty}k^2|c_k(t)|^2 w^k\bigg)\,dt,\quad \text{using (10.5.17)}$$

$$= \int_0^{\infty}\frac{e^t w}{1-w^2}\left(\left(1+\sum_{k=1}^{\infty}2w^k\right)\left(1+\sum_{k=1}^{\infty}\left(\lim_{r\to 1}\frac{1}{2\pi}\int_0^{2\pi}\frac{\partial_t f(z,t)}{f(z,t)}\right.\right.\right.$$

$$\left.\left.\times\,k\,\overline{c_k(t)z^k}\,d\theta\right)w^k\right) + \left(1+\sum_{k=1}^{\infty}2w^k\right)\left(1+\sum_{k=1}^{\infty}\left(\lim_{r\to 1}\frac{1}{2\pi}\right.\right.$$

$$\left.\left.\times\int_0^{2\pi}\frac{\partial_t \overline{f(z,t)}}{f(z,t)}\,kc_k(t)z^k\,d\theta\right)w^k\right) - 2 - \sum_{k=1}^{\infty}k^2|c_k(t)|^2 w^k\bigg)\,dt$$

$$= \int_0^{\infty}\frac{e^t w}{1-w^2}\left(\left(1+\sum_{k=1}^{\infty}\left(\lim_{r\to 1}\frac{1}{2\pi}\int_0^{2\pi}\frac{\partial_t f(z,t)}{f(z,t)}\left(2+2\sum_{j=1}^{k}j\,\overline{c_j(t)z^j}\right.\right.\right.\right.$$

$$\left.\left.- k\,\overline{c_k(t)z^k}\right)d\theta\right)w^k\right) + \left(1+\sum_{k=1}^{\infty}\left(\lim_{r\to 1}\frac{1}{2\pi}\int_0^{2\pi}\frac{\partial_t \overline{f(z,t)}}{f(z,t)}\right.\right.$$

$$\left.\left.\times\left(2+2\sum_{j=1}^{k}j\,c_j(t)z^j - k\,c_k(t)z^k\right)d\theta\right)w^k\right) - 2 - \sum_{k=1}^{\infty}k^2|c_k(t)|^2 w^k\bigg)\,dt$$

$$\text{by (10.5.16)}$$

$$= \int_0^{\infty}\frac{e^t w}{1-w^2}\left(\left(\lim_{r\to 1}\frac{1}{2\pi}\int_0^{2\pi}p(z,t)\left(1+\sum_{j=1}^{\infty}jc_j(t)z^j\right)\left(2+2\sum_{j=1}^{k}j\,\overline{c_j(t)z^j}\right.\right.\right.$$

$$\left.\left.- k\,\overline{c_k(t)z^k}\right)d\theta\right)w^k\right) + \left(\sum_{j=1}^{\infty}\left(\lim_{r\to 1}\frac{1}{2\pi}\int_0^{2\pi}\overline{p(z,t)}\left(1+\sum_{j=1}^{\infty}j\overline{c_j(t)z^j}\right)\right.\right.$$

$$\left.\left.\times\left(2+2\sum_{j=1}^{k}j\,c_j(t)z^j - k\,c_k(t)z^k\right)d\theta\right)w^k\right) - \sum_{k=1}^{\infty}k^2|c_k(t)|^2 w^k\bigg)\,dt$$

$$= \int_0^{\infty}\frac{e^t w}{1-w^2}\left(\sum_{k=1}^{\infty}\left(\lim_{r\to 1}\frac{1}{2\pi}\int_0^{2\pi}\Re\{p(z,t)\}\times\right.\right.$$

$$\left.\left.\left|2+2\sum_{j=1}^{k}j\,c_j(t)z^j - k\,c_k(t)z^k\right|^2 d\theta\right)w^k\right)$$

$$= \int_0^{\infty}\frac{e^t w}{1-w^2}\sum_{k=1}^{\infty}A_k(t)w^k\,dt,$$

where $A_k(t)$ is given by (10.3.5). ∎

11

Epilogue: After de Branges

After de Branges proved the Bieberbach conjecture, the fervor in this topic did wane a little bit but not completely. The research still continues in some circles on old topics, as in, e.g., Acu [2008], and in others with new directions, especially in providing alternate proofs of de Branges theorem, and on topics related to Loewner equation in the Euclidean n-space $\mathbb{C}^n$, slit mappings and applications such as statistical Loewner equation, and quasi-conformal mappings. For example, in the area of quasiconformal mappings, which can be found in Ahlfors [1966] and later in the area of quasiconformal extensions as in Kühnau [1971], Schiffer and Schober [1976] and Semes [1986], the recent research occupies areas like quasiconformal variation of slit domains as in Earle and Epstein [2001], quasiconformal mappings and partial differential equations as in Astala et al. [2009], geodesic curvature and conformal mapping as in Fernández and Granados [1998], Ahlfors-Beurling operator in Petermichl and Volberg [2002], and evolution families as in Bracci et al. [2008]. In the area of several complex variables and multivariate holomorphic mappings, we find extensive work by Rosay and Rudin [1988], Duren et al. [2010], Graham et al. [2002], Graham and Kohr [2003], Hamada [2011], Roper and Suffridge [1995], and Vodă [2011]; in the area of stochastic Loewner equations and random walks in Lawler et al. [2001] and Schramm [2000, 2011]; on biholomorphic mappings in Liu [2006]; and on slit mappings in Lind [2005], Marshall and Rohde [2005] and Wong [2014]. We will, however, not discuss all these topics in this chapter, but confine to recent developments in Loewner equations and Cauchy and Beurling transforms, which refer to the main theme of the book.

11.1 Chordal Löwner Equations

Löwner [1923] developed a method of embedding a univalent map from E onto a slit domain in a continuous family of univalent mappings. He proved that such a family of univalent maps $f(z,t) : E \mapsto \mathbb{C}$ which satisfy the partial differential equation (4.2.2), where $k : [0, +\infty) \mapsto \partial E$ is a continuous map that encodes information about the slit. The Loewner theory had been successfully applied to (i) extremal problems of univalent maps; (ii) univalence criteria for holomorphic maps; (iii) geometric function theoretic properties (spirallike and starlike maps); and (iv) estimates of coefficients of univalent maps. There are two types of Loewner equations: radial and chordal (forward and backward, in the sense of the 'time' variable t). These equations played an important role in the development of Schramm-Loewner evolution theory (Schramm [2000], [2007], [2011]).

The *forward chordal Löwner equation* is defined by (4.2.17). Suppose z_0 is a point such that the denominator of the right hand side of (11.1.2) is zero, i.e., $f(z_0, t) = \lambda(t)$. It will introduce a singularity in the derivative $\partial f(z, t)/\partial t$, which would force us to keep the point z_0 out of our domain. Under certain conditions on λ, to be determined later, the set of all such singular points will produce a curve extending from the real axis. However, the curve may be space filling, depending on λ. Let us denote this curve by γ. If we set $T_0 = \sup\{t_0 \in [0, T] : f(z, t)$ exists on $[0, t_0)\}$, we will obtain the largest possible value for t such that a solution $f(z, t)$ will make sense. Define $G_t = \{z \in \mathbb{U}^+ : t < T_0\}$, which omits only the points in $\mathbb{U}^+$ that cause $\partial f(z, t)/\partial t$ to become singular for some $t < T_0$. Thus, G_t is now the domain for Eq (11.1.2), as shown in Figure 11.1.1, in which the function $f(z, t)$ maps the curve γ onto the real axis, and the complement of γ bounded by real axis in (a) is the domain G_t, i.e, $G_t = \mathbb{U}^+ \backslash \gamma$ which is mapped onto $\mathbb{U}^+$ by $f(z, t)$.

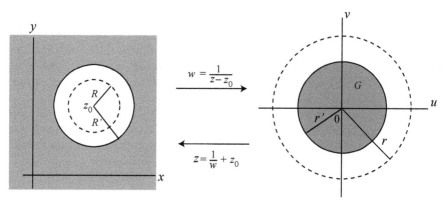

Figure 11.1.1 Slit Domain in the upper half-plane $\mathbb{U}^+$.

The above defined function $\lambda(t)$ is called the *driving term* in the following sense: Since putting different functions in place of λ would result in different functions g, the term λ generates $f(z, t)$ and the corresponding domain G_t. By Riemann mapping theorem, the function $f(z, t)$ is a conformal map from G_t onto $\mathbb{U}^+$. This mapping can be made unique by specifying certain conditions on points in the domain G_t. For this purpose, in view of the Möbius transformation, we choose three degrees of freedom as follows: (i) map ∞ to

∞; (ii) map real line to the real line; and (iii) have the derivative at ∞ to be 1. This process is known as the *hydrodynamic normalization* which states that $\lim_{z \to \infty} f(z,t) - z = 0$. This will make the mapping look like the identity map when z is far away from the origin. Then all coefficients of z^n for $n \geq 2$ and the constant term will be zero in the expression for $f(z,t)$, while the coefficient of z would be 1. Thus, near infinity the mapping $f(z,t)$ has the form

$$f(z,t) = z + \frac{c(t)}{z} + O\left(z^{-2}\right), \tag{11.1.1}$$

where the symbol $O(z^{-2})$ signifies that the rate at which all terms after the first two terms decrease is at least as fast as $1/z^2$, and $c(t)$, which is continuously increasing in t, is called the *half-plane capacity* which plays a role in statistical Loewner equation (SLE).[1] In fact, c is the residue of $f(z,t)$. Also, $c(0) = 0$ because $f(z,0) = z$. To parametrize γ so that $c(t)$ is linear in t, we choose $c(t) = 2t$. Although it is an arbitrary choice, the reasoning behind it is that in the radial version of Loewner equation, the version used by Loewner himself, it is natural to parametrize γ such that $c(t) = e^t$, while in the chordal version $c(t)$ is taken as $2t$.

A *hull* in the half-plane $\mathbb{U}^+$ is defined as a compact set $K \in \overline{\mathbb{U}}^+$ so that $\mathbb{U}^+\backslash K$ is simply connected. Here $K = \overline{K \cap \mathbb{U}^+}$, which guarantees that K contains no intervals of $\mathbb{R}$ that are 'sticking out' to the left or right. The hull can be thought of as the generated curve γ although the hull i not necessarily the generated curve γ. The hull depends on time, so that $G_t = \mathbb{U}^+\backslash K$; sometimes K is written as K_t.

The *backward chordal Löwner equation* is defined by (4.2.18). Just as the forward Loewner equation takes the curve γ and moving it along the real axis as time moves forward, so in the case of backward Loewner equation we start at the time T and move backwards to time 0, thus resulting in some curve γ from the previously 'empty' (at time T) upper half-plane, as presented in Figure 11.1.2, which shows the bolded interval on the real axis is mapped onto the curve γ under $g(z,t)$. Such functions are generated by Eq (4.2.18).

Although the difference between Eqs (4.2.17) and (4.2.18) is obvious, yet these two equations are related such that if T is the largest possible value for t, then letting $\lambda(T - t) = \xi(t)$ we get one equation from the other. However, it is generally not true that the function $g(z,t)$ generated by Eq (4.2.17)) will be inverse of $g(z,t)$ generated by Eq (4.2.18), i.e., the curve γ generated by (4.2.18) is not necessarily the same curve γ generated by (4.2.17). However, it is true that $g(z,T) = f^{-1}(z,T)$, where T is the final time. The hydrodynamic

---

[1] We will not discuss this topic, but for interested readers the following references are suggested: Gruzberg and Kadanoff [2003] and research papers cited therein; Schramm [2001]; work by Lawler, Schramm and Werner [2001], and Rohde and Schramm [2005].

normalization at infinity for the backward Löwner equation is

$$g(z,t) = z + \frac{-c(t)}{z} + O(z^{-2}). \tag{11.1.2}$$

where $-c(t)$ represents the backward half-plane capacity, because of 'time running backwards'. Note that in the forward case the real axis is mapped to one point on γ under $f(z,t)$. But in the backward case the same phenomenon occurs in the opposite direction under the mapping $g(z,t)$ such that a part of the real axis is mapped to γ. This mapping is two-to-one, is called *welding*, which is easily understood by setting the driving term $\xi(t)$ equal to a constant. Then Eq (4.2.18) can be solved by the method of separation of variables. The solution is

$$g(z,t) = A + \sqrt{(z-A)^2 - 4t}. \tag{11.1.3}$$

This solution ensures that the image stays in the upper half-plane. Also, as time moves forward, the singular point $z = A$ moves upward along the line $\Re\{z\} = A$. Thus, γ is perpendicular to the real axis.

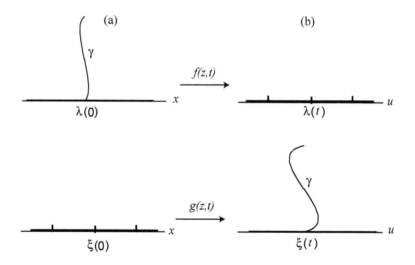

Figure 11.1.2 Slit Domain in the image of the upper half-plane $\mathbb{U}^+$.

The *welding phenomenon* describes how two real-valued curves, one on either side of $z = A$, are 'welded' together to form the curve (slit) γ. In fact, set $g(z,t) = A$, and then solve for z in Eq (11.1.3), which gives $z = A \pm 2\sqrt{t}$. These values of z are on the real axis and are centered symmetrically around A, since $t > 0$. This shows the two-to-one correspondence between real axis and γ.

A very interesting case arises when the driving term causes a spiral to generate. The curve remains simple until it reaches infinity and then the curve bounces back on itself and closes off the disk in the center. When the curve touches back on itself, the points completely enclosed by the curve all to reach a singularity at the same time; thus, the whole enclosed set is no longer a part of the domain, as shown in Figure 11.1.3, where at ∞ the disk inside the spiral becomes closed off from the rest of the upper half-plane; at this point every point within the disk is a singularity.

Chordal equation has been used in extremal problems of univalent maps, and in fluid dynamics. Schramm [2000] introduced a stochastic Loewner equation (SLE), which was used to obtain several results in statistical mechanics and to solve the Mandelbrot conjecture in 2000. For this work Schramm and Werner won the Fields Medal.

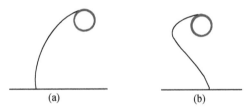

(a)   (b)

Figure 11.1.3 Spiral Curve at $t = \infty$.

11.2 Löwner Curves

Let D be a Jordan domain. Consider a family $\Gamma_{D,a,b}$ of *simple* curves in $\bar{D}$ that begins at a and ends in b (which can be $\pm\infty$). It is required that the curves be simple, i.e., they do not cross themselves. Given $\gamma[0,T]$, define G_t to be the set of all conformal maps $\phi(t) : D\backslash\gamma[0,t] \mapsto D$ with $\phi(0) = a$ and $\phi(T) = b$. Then the notation $\gamma : (0,T) \mapsto (D,a,b)$, $0 < T < \infty$, means that $\gamma[0,T]$ is a simple curve with $\gamma(0,T) \subset D$ and $\gamma(0) = a$. If there is no confusion, we will simply write γ for $\gamma(0,T)$, which is the image of the open 'time'-interval $(0,T)$.

A curve $\gamma : (0,T) \mapsto (D,a,b)$ is said to be *self-similar* if $\gamma \in C^3(0,T)$ and if for each $t \in (0,T)$, there exists a conformal map $\phi \in G_t$ such that $\phi(\gamma(t,T)) = \gamma$. The family of self-similar curves in (D,a,b) is denoted by $S(D,a,b)$. If for a fixed (D_0,a_0,b_0) the family $S(D_0,a_0,b_0)$ is known, then we can use the conformal map $\psi : D_0 \mapsto D$ with $\psi(a_0) = a$ and $\psi(b_0) = b$ and obtain the family $S(D,a,b)$; that is, $S(D,a,b) = \{\psi \circ \gamma : \gamma \in S(D_0,a_0,b_0)\}$. The concept of self-similarity is a generalization of the one introduced by Lind et al. [2010], where they say that a curve γ in the upper half-plane $\mathbb{U}^+$ with finite half-plane capacity is self-similar if $g(\gamma(t,T),t)$ is a transition and dilation of γ for all $t \in (0,T)$. In fact, γ is self-similar if and only if the driving term $\lambda(t) = k + c\sqrt{\tau - t}$ for some $c,k \in \mathbb{R}$ and $\tau > 0$.

In the sequel, we will use the Jordan domain $D = \mathbb{U}^+$, the upper half-plane, with $a_0 = 0$ and $b_0 = \infty$. Let $\lambda : [0, T] \mapsto \mathbb{R}$ be a continuous function of t, and let $z \in \mathbb{U}^+$. Consider the initial value problem (4.2.17) in $\mathbb{U}^+$, which we reformulate as

$$\frac{\partial}{\partial t} g(z, t) = \frac{2}{g(z, t) - \lambda(t)}, \tag{11.2.1}$$

$$g(z, 0) = z, \tag{11.2.2}$$

where the function $\lambda(t)$, known as the *driving function*, determines the unique solution $g(z, t)$. The difficulty in solving this type of initial value problems arises from the denominator in Eq (11.2.1). Unless this problem is resolved, the initial value problem remains ill-posed. Let all well-posed initial value problems of the above type form a set K_t, known as the *Löwner hull*, such that $K_t = \{z \in \mathbb{U}^+ : g(z, s) = \lambda(s) \text{ for some } s \in (0, T]\}$. If $z \notin K_t$, then $g(z, t)$ is well-defined, and $\mathbb{U}^+ \backslash K_t$ is simply connected, i.e., $g(z, t)$ is the unique conformal map from $\mathbb{U}^+ \backslash K_t$ onto $\mathbb{U}^+$ which satisfies the following normalization-at-infinity condition, known as the *hydrodynamic normalization*:

$$g(z, t) = z + \frac{c(t)}{z} + O\left(1/z^2\right). \tag{11.2.3}$$

The quantity $c(t) = 2t$ is called the *half-plane capacity* of K_t (see Lawler [2005]). Some simple cases of the form $\lambda(t) = c - c\sqrt{1 - t}$ in K_1 (i.e., for $t \in (0, 1]$) are discussed in Lind, Marshall and Rohde [2010]. In fact, we have the following result:

Theorem 11.2.1. *Assume that* $\gamma : (0, T) \mapsto (\mathbb{U}^+, 0, \infty)$ *has the driving term* $\lambda(t)$. *Then* $\gamma \in S(\mathbb{U}^+, 0, \infty)$ *if and only if* $\lambda(t)$ *is of one the following forms:* 0, ct, $c\sqrt{\tau} - c\sqrt{(\tau - t)}$, *or* $c\sqrt{\tau_t} - c\sqrt{\tau}$, *where* $c \in \mathbb{R}\backslash\{0\}$ *and* $\tau > 0$. *Moreover,* λ *is completely determined by two real parameters* $\dot{\lambda}(t)$ *and* $\ddot{\lambda}(t)$.

Assume that the Löwner hulls K_t are generated by the driving term $\lambda(t)$. Then some properties of Eq (11.2.1) are:

(i) SCALING. For $r > 0$, the driving term of the scaled hulls $r K_{t/r^2}$ is $r\lambda(t/r^2)$;

(ii) TRANSLATION. For $x \in \mathbb{R}$, the driving term of the shifted hulls $K_t + x$ is $\lambda(t) + x$;

(iii) REFLECTION. If R_I denotes reflection in the imaginary axis, then the driving term of the reflected hulls $R_I(K_t)$ is $-\lambda(t)$; and

(iv) CONCATENATION. For fixed τ, the driving function of the mapped hulls $g_\tau(K_{\tau+t})$ is $\lambda(\tau + t)$.

Theorem 11.2.2. (Earle and Epstein [2001]) *Let* $\lambda(t)$ *be the driving term for* $\gamma : (0, T) \mapsto (\mathbb{U}^+, 0, \infty)$. *If any parametrization of* γ *is a* $C^m(0, T)$

function, then both half-plane capacity parametrization of γ as well as of λ are in $C^{m-1}(0,\tau)$, where τ is the half-plane capacity of γ (which may be infinite.)

Assume that $\lambda \in C^2$. Then the *Loewner curvature* is defined by

$$LC_\gamma(t) = \frac{[\dot{\lambda}(t)]^3}{\ddot{\lambda}(t)}. \tag{11.2.4}$$

Let $\gamma \in S(\mathbb{U}^+, 0, \infty)$. Then (i) if γ is generated by $\lambda(t) = 0$, then $LC_\gamma \equiv 0$; (ii) if γ is generated by $\lambda(t) = ct$, then $LC_\gamma \equiv \infty$; (iii) if γ is generated by $\lambda(t) = c\sqrt{\tau} - c\sqrt{\tau - t}$, then $LC_\gamma \equiv c^2/2$; and (iv) if γ is generated by $\lambda(t) = c\sqrt{\tau + t} - c\sqrt{\tau}$, then $LC_\gamma \equiv -c^2/2$. Thus, for $\gamma \in S(\mathbb{U}^+, 0, \infty)$, the Loewner curvature is a scaling invariant; for self-similar curves $\gamma \in S(D, a, b)$, the Loewner curvature satisfies conformal invariance; and for a C^3-curve γ : $(0,T) \mapsto (D, a, b)$, the Loewner curvature is defined by comparing it to curves of constant curvature, as LC_{γ^*}, where $\gamma^* \in S\left(D\backslash\gamma(0,T], \gamma(t), b\right)$ is the unique best fit to the curve $\gamma(t,T)$ at $\gamma(t)$.

Theorem 11.2.3. (Lind and Rohde [2014]) *Let $\gamma : (0,T) \mapsto (D, a, b)$ be a C^3-curve. If $LC_\gamma(t) < 8$ for all $t \in (0,T)$, then $\gamma(0,T)$ is a simple curve in $D \cup \{b\}$.*

The constant 8 is the best possible and corresponds to the constant 4 in the condition $\|\lambda\|_{1//2} < 4$ for simple Loewner traces More details and further results can be found in Lind [2005], Lind et al. [2010], Lind and Rohde [2014], and Wong [2014]. For exact solutions of Loewner evolutions, see Krager et al. [2004]. The definition (11.2.4) can be compared with the classical definition of curvature (B.2) for a parametrized $C^2(a,b)$-curve $\gamma(t) = x(t) + iy(t)$.

11.3 Loewner Chains in $\mathbb{C}^n$

Recent work on Loewner chains and extension operators is available in Arosio [2013], Chirilă [2013], Duren et al. [2010], Graham et al. [2002, 2003, 2005, 2012], Hamada [2011], Hamada and Kohr [2000], Liu [2006], Muir [2008], Roper and Suffridge [1995], Vodă [2011], and other references found in them. Loewner chains in the Euclidean unit ball in $\mathbb{C}^n$ have also been studied extensively, More details can be found in Bracci et al. [2008], Contreras et al. [2006, 2010], Lawler [2005, 2006], Lawler, Schramm and Werner [2001], and Marshall and Rohde [2005].

Let $\mathbb{C}^n$ denote the space n complex variables $z = \{z_1, \ldots, z_n\}$ with the Euclidean inner product $\langle z, w \rangle = \sum_{j=1}^n z_j \bar{w}_j$ and the Euclidean norm $\|z\| = \langle z, z \rangle^{1/2}$. For $n \geq 2$, let $\tilde{z} = (z_2, \ldots, z_n) \in \mathbb{C}^{n-1}$. Then $z = (z_1, \tilde{z}) \in \mathbb{C}^n$. Let $B^n(0,r) = \{z \in \mathbb{C}^n : \|z\| < r\}$ denote an open ball in $\mathbb{C}^n$; $B^n(0,1)$ an open unit ball in $\mathbb{C}^n$; and $B^1(0,1)$ the unit disc, same as $C(0,1)$, denoted by E. Let $L(\mathbb{C}^n, \mathbb{C}^m)$ denote the space of linear continuous operators from $\mathbb{C}^n$ into $\mathbb{C}^m$, with the standard operator norm $\|A\| = \sup\{\|A(z)\| : \|z\| = 1\}$.

Let I_n be the identity of $L\left(\mathbb{C}^n, \mathbb{C}^m\right)$. If Ω is a domain in $\mathbb{C}^n$, the set of holomorphic mappings from Ω into $\mathbb{C}^n$ is denote by $H(\Omega)$. If $f \in H(B^n)$, we say that f is *normalized* if $f(0) = 0$ and $\partial f(0) = I_n$, where ∂f denotes the derivative of f. The function $f \in H(B^n)$ is said to be *locally biharmonic* on B^n if the complete Jacobian matrix $\Delta_z f(z)$ is nonsingular at each $z \in B^n$. A holomorphic mapping $f : B^n \mapsto \mathbb{C}^n$ is said to be *biholomorphic* if the inverse f^{-1} exists and is holomorphic on the open set $f(B^n)$. Also, a univalent mapping on B^n (holomorphic and injective on B^n) is also biharmonic. Let $\mathcal{L}\mathcal{S}_n$ denote the set of normalized locally biholomorphic mappings on B^n, and let $S(B^n)$ be the set of normalized biholomorphic mappings on B^n. Moreover, let $S^\star(B^n)$ (respectively, $\mathcal{K}(B^n)$) be the subsets of $S(B^n)$, which consists of starlike mappings with respect to zero (respectively, convex mappings). Note that for $n = 1$, $\mathcal{L}\mathcal{S}_1 = \mathcal{L}\mathcal{S}$, $S(B^1) = \mathcal{S}$, $\mathcal{K}(B^1) = \mathcal{K}$, and $S^\star(B^1) = \mathcal{S}^\star$.

Let $f : B^n \mapsto \mathbb{C}^n$ be a normalized locally biholomorphic mapping and let $0 \leq \gamma < 1$. Then the mapping f is said to be *starlike of order γ* if

$$\Re\left\{\frac{\|z\|^2}{\langle[\Delta_z f(z)]^{-1} f(z), z\rangle}\right\} > \gamma, \quad z \in B^n \backslash \{0\}. \qquad (11.3.1)$$

For $n = 1$, the inequality (11.3.1) reduces to $\Re\{zf'(z)/f(z)\} > \gamma$ for $z \in E$ (class $S(0, \beta, 0)$,§7.4, with $\beta = \gamma$). Moreover, f is starlike of order 0 on B^n if and only if f is starlike, and if $0 < \gamma < 1$, then f is starlike of order *gamma* if and only if

$$\left|\frac{1}{\|z\|^2}\langle[\Delta_z f(z)]^{-1} f(z), \rangle - \frac{1}{2\gamma}\right| < \frac{1}{2\gamma}, \quad z \in B^n \backslash \{0\}. \qquad (11.3.2)$$

Let $S_\gamma^\star(B^n)$ denote the set of starlike mappings of order γ on B^n. In the case $n = 1$, $S_\gamma^\star(B^1) \equiv S_\gamma^\star$. If $f \in S_\gamma^\star(B^n)$, then

$$\Re\{\langle[\Delta_z f(z)]^{-1} f(z), z\rangle\} > 0, \quad z \in B^n \backslash \{0\}. \qquad (11.3.3)$$

11.3.1 Spirallike Functions of Type δ and Order γ. (Hamada and Kohr [2000]) Let $f \in \mathcal{L}\mathcal{S}_n$, $\pi/2 < delta < \pi/2$ and $0 \leq \gamma < 1$. Then f is *spirallike of type δ and order γ* if

$$\Re\left\{\frac{1}{(1 - i\tan\delta)\frac{1}{\|z\|^2}\langle[\Delta_z f(z)]^{-1} f(z), z\rangle + i\tan\delta}\right\} > \gamma, \quad z \in B^n \backslash \{0\}.$$
$$(11.3.4)$$

In particular, f is spirallike of type δ and order 0 on B^n if and only if f is spirallike of type δ on B^n. If $0 < \gamma < 1$, then f is spirallike of type δ and order γ if and only if

$$\left|e^{-i\delta}\frac{1}{\|z\|^2}\langle[\Delta_z f(z)]^{-1} f(z), z\rangle + i\sin\delta - \frac{\cos\delta}{2\gamma}\right| < \frac{\cos\delta}{2\gamma}, \quad z \in B^n \backslash \{0\}.$$
$$(11.3.5)$$

The class of spirallike mappings of type δ is denoted by $\hat{S}_\delta(B^n)$, which for $n = 1$ becomes $\hat{S}_\delta(B^1) \equiv \hat{S}_\delta$ (in short).

Pfaltzgraff [1974] introduced the class $\mathcal{M}$ of holomorphic mappings on B^n by

$$\mathcal{M} = \left\{ h \in H(B^n) : h(0) = 0, \Delta_z h(0) = I_n, \Re\{\langle h(z), z \rangle > 0, z \in B^n \backslash \{0\}\} \right\}.$$
(11.3.6)

This class is related to various subclasses of biholomorphic mappings on B^n, such as starlike, spirrallike of type δ, mapping with parametric representations, and others.

Let $0 \le \gamma < 1$ and $g : E \mapsto \mathbb{C}$ defined by $g(\zeta) = \dfrac{1 - \zeta}{1 + (1 - 2\gamma)\zeta}$, $|\zeta| < 1$.

Let $\mathcal{M}_g$ be the subclass of $H(B^n)$ defined by

$$\mathcal{M}_g = \left\{ h : B^n \mapsto \mathbb{C}^n : h \in H(B^n), h(0) = 0, \Delta)z h(0) = I_n, \right.$$

$$\left. \left\langle h(z), \frac{z}{\|z\|^2} \right\rangle \in g(E), z \in B^n \right\},$$
(11.3.7)

where $\left\langle h(z), z/\|z\|^2 \right\rangle \big|_{z=0} = 1$, since h is normalized. Obviously, $\mathcal{M}_g \subseteq \mathcal{M}$, and if $gamma = 0$, then $\mathcal{M}_g \equiv \mathcal{M}$. Also, if $0 < \gamma < 1$, then g maps E onto the open disk $C\left(1/(2\gamma), 1/(2\gamma)\right)$, thus

$$\mathcal{M}_g = \left\{ h \in H(B^n) : h(0) = 0, \Delta_z h(0) = I_n, \right.$$

$$\left. \left| \frac{1}{\|z\|^2} \langle h(z), z \rangle - \frac{1}{2\gamma} \right| < \frac{1}{2\gamma}, z \in B^n \backslash \{0\} \right\}.$$
(11.3.8)

11.3.2 Subordination and Loewner Chains. Let $f, g \in H(B^n)$. Then f is subordinate to g ($f \prec g$) if there is a Schwarz mapping $v = v(z, s, t)$, i.e, $v \in H(B^n)$ and $\|v(z)\| \le \|z\|$, $z \in B^n$ such that $f(z) = g(v(z)), z \in B^n$.[1] A mapping $f : B^n \times [0, \infty) \mapsto \mathbb{C}^n$ is called a *Loewner chain* if $f(\cdot, t)$ is biholomorphic on B^n, $f(0, t) = 0, \Delta_z f(0, t) = e^t I_n$ for $t \ge 0$, and $f(\cdot, s) \prec f(\cdot, t)$ for $0 \le s \le t < \infty$.

Theorem 11.3.1. Pfaltzgraff [1974]) *Let $h = h(z, t) : B^n \times [0, \infty) \mapsto \mathbb{C}^n$ be such that (i) $h(\cdot, t) \in \mathcal{M}$ for $t > 0$, and (ii) $h(z, \cdot)$ is measurable on $[0, \infty)$ for $z \in B^n$. Let the mapping $f = f(z, t) : B^n \times [0, \infty) \mapsto \mathbb{C}^n$ be such that $f(\cdot, t) \in H(B^n)$, $f(0, t) = 0, \Delta_z(0, t) = e^t I_n$ for $t \ge 0$, and $f(z, \cdot)$ is locally absolutely continuous on $[0, \infty)$ locally uniformly with respect to $z \in B^n$. Assuming that*

$$\frac{\partial f}{\partial t}(z, t) = \Delta_z f(z, t) h(z, t), \text{ a.e. } t \ge 0 \text{ for all } z \in B^n,$$
(11.3.9)

---

1 For the subordination principle, see Appendix E.

and that there exists an increasing sequence $\{t_m\}_{m\in\mathbb{N}}$ such that $t_m > 0$, $t_m \to \infty$ and $\lim_{m\to\infty} f(z,t_m) = F(z)$ locally uniformly on B^n. Then $f(z,t)$ is a Loewner chain.

For $n = 1$, it is known (see Chapter 4) that if $f(\zeta,t)$ is a Loewner chain, then $\{e^{-t}f(\cdot,t)\}$, $t \geq 0$, is a normal family on E, and there exists a function $p(\zeta,t)$ such that $p(\zeta,\cdot)$ is measurable on $[0,\infty)$ for $\zeta \in E$ and

$$f t(\zeta,t) = \zeta f'(\zeta,t)p(\zeta,t), \text{ a.e. } t \geq 0 \text{ for all } z \in B^n, \tag{11.3.10}$$

For higher dimensions, Graham and Kohr [2003] have proved that if $f(z,t)$ is a Loewner chain on B^n, then $f(z,\cdot)$ is locally Lipschitz on $[0,\infty)$ locally uniformly with respect to $z \in B^n$; that there exists a mapping $h = h(z,t)$ which satisfies conditions (i) and (ii) of Theorem 11.3.1, such that

$$\frac{\partial f}{\partial z}(z,t) = \Delta_z f(z,t)h(z,t), \text{ a.e. } t \geq 0 \text{ for all } z \in B^n, \tag{11.3.11}$$

The mapping $h = h(z,t)$ in the Loewner differential equation (11.3.11) is unique up to a measurable set of measure zero which does not depend on $z \in B^n$. This means that if there is another mapping $q = q(z,t)$ such that $q(\cdot,t) \in \mathcal{M}$ for a.e. $t \geq 0$, if $q(z,\cdot)$ is measurable on $[0,\infty)$ for $z \in B^n$, and if Eq (11.3.11) holds for $q(z,t)$, then $h(\cdot,t) = q(\cdot,t)$ a.e. $t \geq 0$.

11.3.3 g-Loewner Chain and g-Parametric Representation. As defined in Graham et al. [2002], Let

$$g(\zeta) - \frac{1-\zeta}{1+(1-2\gamma)\zeta}, \ |\zeta| < 1, \text{ and } 0 \leq \gamma < 1.$$

A mapping $f = f(z,t) : B^n \times [0,\infty) \mapsto \mathbb{C}^n$ is called a g-Loewner chain if $f(z,t)$ is a Loewner chain such that $\{e^{-t}f(\cdot,t)\}_{t\geq 0}$ is a normal family on B^n and the mapping $h = h(z,t)$ which occurs in Eq (11.3.11) is such that $h(\cdot,t) \subset \mathcal{M}_g$ for a.e. $t \geq 0$. Moreover, a normalized holomorphic mapping $f : B^n \mapsto \mathbb{C}^n$ has g-representation if there exists a g-Loewner chain $f(z,t)$ such that $f = f(\cdot t)$. Let $S_g^0(B^n)$ denote the set of mappings that have g-parametric representation, with $g(\zeta) = \dfrac{1-\zeta}{1+(1-2\gamma)\zeta}, |\zeta| < 1$, and $0 \leq \gamma < 1$.

If $g(\zeta) = \dfrac{1-\zeta}{1+\zeta}$, then $S_g^0(B^n)$ reduces to the set $S^0(B^n)$ of mappings which have parametric representation. In the case $n = 1$, a g-Loewner chain $f(\zeta,T)$ is a Loewner chain such that the function $p(\zeta,t)$ in Eq (11.3.10) satisfies the condition $p(\cdot,t) \in g(E)$ for a.e. $t \geq 0$. In the case $g(\zeta) = \dfrac{1-\zeta}{1+\zeta}, \zeta| < 1$, any Loewner chain on the unit disk E is also a g-Loewner chain.

Let the open ball $\{z \in \mathbb{C}^n : \|z\| < r\}$ be denoted by B_r^n and the unit ball B_1^n by B_n. Then in the case $n = 1$, B^1 is the same as E.

Theorem 11.3.1. *Assume that $f \in \mathcal{S}$ can be embedded as the first element of a g-Loewner chain, where $g(z) = \dfrac{1-z}{1+(1-2\alpha)z}, |z| < 1$, and $0 < \gamma < 1$. Then $F = \Phi_{n,\alpha,\beta}$ can be embedded as the first element of a g-Loewner chain on $\mathbb{B}^n$ for $0 \leq \alpha \leq 1, 0 \leq \beta \leq \frac{1}{2}, \alpha + \beta \leq 1$.*

This theorem shows that the operator $\Phi_{n,\alpha,\beta}$ preserves the notion of g-Loewner chain for $g(z)$. For a proof, see Chirilă [2013].

Corollary 11.3.1. *If $E \to \mathbb{C}$ has g-parametric representation and $0 \leq \alpha \leq 1, 0 \leq \beta \leq \frac{1}{2}, \alpha + \beta \leq 1$, then $F = \Phi_{n,\alpha,\beta} \subset S_g^0(\mathcal{B}^n)$, where g is the same as in Theorem 11.3.1, an $z \in E$.*

Corollary 11.3.2. *(Liu [2006]) If $f \in S_\gamma^*$ and $0 \leq \alpha \leq 1, 0 \leq \beta \leq \frac{1}{2}$, $\alpha + \beta \leq 1$, the $F = \Phi_{n,\alpha,\beta} \in S_\gamma^*, 0\gamma < 1$. In particular, the Roper-Suffridge extension operator preserves the starlikeness of order γ.*

11.3.4 Extension Operators. The operator $\phi_{n,\alpha,\beta}$ is defined by (see Graham et al. [2002])

$$\Phi_{n,\alpha,\beta}(f)(z) = \left(f(z_1), \tilde{z}\left(\frac{f(z_1)}{z_1}\right)^\alpha (f'(z_1))^\beta \right), \quad z = (z_1, \tilde{z}) \in B^n, \quad (11.3.12)$$

where $\alpha \geq 0, \beta \geq 0$ and f is a locally univalent function on E, normalized by $f(0) = 0, f'(0) = 1$, and is such that $f(z_1) \neq 0$ for $z_1 \in E \backslash \{0\}$, and the branches of the power function are chosen such that

$$\left(\frac{f(z_1)}{z_1}\right)^\alpha \Big|_{z_1=0} = 1 \text{ and } (f'(z_1))^\beta|_{z_1=0} = 1. \quad (11.3.13)$$

The operator $\Phi_{n,0,1/2}$, defined by

$$\Phi_n(f)(z) = \left(f(z_1), \tilde{z}(f'(z_1))^{1/2} \right), \quad z = (z_1, \tilde{z}) \in B^n, \quad (11.3.14)$$

is known as the Roper-Suffridge extension operator (Roper and Suffridge [1995]). Note that $\Phi_n(\mathcal{K}) \subset \mathcal{K}(B^n)$, $\Phi_n(\mathcal{S}^\star) \subset \mathcal{S}^*$, and $\Phi_n(\mathcal{S}) \subset S^0(B^n)$. As determined in Graham et al. [2002], the operator $\Phi_{n,\alpha,\beta}$ preserves starlikeness and parametric representation from one-dimensional to the n-dimensional case whenever $\alpha \in [0,1], \beta \in [0,1/2]$, and $\alpha + \beta \leq 1$. However $\Phi_{n,\alpha,\beta}(\mathcal{K}) \subset \mathcal{K}(B^n)$ if and only if $(\alpha, \beta) = (0, 1/2)$. The following result shows that the operator $\Phi_{n,\alpha,\beta}$ preserves g-Loewner chain for $g(\zeta) = \dfrac{1-\zeta}{1+(1=2\gamma)\zeta}, |\zeta| < 1$, and $0 < \gamma < 1$; a proof is given in Chirilă [2013].

Theorem 11.3.2. *(Graham et al. [2002] for $\alpha = 0$; Chirilă [2013]) Let $f \in \mathcal{S}$ be embedded as the first element of a g-Loewner chain, where $g(\zeta) =$*

$\dfrac{1-\zeta}{1+(1-2\gamma)\zeta}, |\zeta| < 1$, and $0 < \gamma < 1$. Then $F = \Phi_{n,\alpha,\beta}(f)$ can be embedded as the first element of a g-Loewner chain on B^n for $\alpha \in [0,1], \beta \in [0,1/2]$, and $\alpha + \beta \leq 1$.

Some particular cases of this theorem are:

(i) If $F : E \mapsto \mathbb{C}$ has a g-parametric representation and $0 \leq \alpha \leq 1, 0 \leq \beta \leq 1/2$, and $\alpha + \beta \leq 1$, then $F = \phi_{n,\alpha,\beta}(f) \in S_g^0(B^n)$, where $g(\zeta)$ is defined in Theorem 11.3.2 (Graham et al. [2002]);

(ii) If $f \in S_\gamma^\star$ and $0 \leq \alpha \leq 1, 0 \leq \beta \leq 1/2$, and $\alpha + \beta \leq 1$, then $F = \Phi_{n,\alpha,\beta}(f) \in S_\gamma^\star$, where $0 < \gamma < 1$. In particular, the Roper-Suffridge extension operator preserves starlikeness of order γ (Hamada et al. [2005] for $\alpha = 0, \beta = \gamma = 1/2$; Liu [2006] for $0 < \gamma < 1, 0 \leq \alpha \leq 1, 0 << \beta \leq 1/2, \alpha + \beta \leq 1$;

(iii) Since $\mathcal{K} \subset S_{1/2}^\star$, we have $\Phi_{n,\alpha,\beta}(\mathcal{K}) \subset S_{1/2}^\star(B^n)$ for $0 \leq \alpha \leq 1, 0 \leq \beta \leq 1/2$, and $\alpha + \beta \leq 1$, and $\Phi_{n,\alpha,\beta}(\mathcal{K}) \not\subseteq \mathcal{K}(B^n)$ for $(\alpha, \beta) \neq (0, 1/2)$ (Graham et al. [2002]); and

(iv) Let $0 \leq \alpha \leq 1, 0 \leq \beta \leq 1/2$, and $\alpha + \beta \leq 1, -\pi/2 < \delta < \pi/2$, and $0 < \gamma < 1$. If $f : E \mapsto \mathbb{C}$ be a spiralilke function of type δ and order γ on E, and if $F = \Phi_{n,\alpha,\beta}(f)$, then F is also spiralike of type δ and order γ on B^n (Liu [2002]).

11.4 Multivariate Holomorphic Mappings

Examples of biholomorphic functions are provided in Exercises 11.6.10 and 11.6.11. Another example is $w_1 = z_1 e^{az_2}, w_k = z_k$ for $k = 2, \ldots, n$. There are many biholomorphic mappings of $\mathbb{C}^n$ to $\mathbb{C}^n$. There are many coefficients of second order terms: e.g., for each coordinate function there are $n(n + 1)/2$ coefficients, and for the full mapping there are $n^2(n+1)/2$ coefficients of second order terms. The magnitude of these terms may be unbounded, as shown in the preceding exercise. However, the magnitude of some combination of coefficients may be bounded. Although there is no limitation on the choice of coefficients of second order terms, the following result extends the multivariate power series to maps that are defined and biholomorphic on $\mathbb{C}^n$.

Theorem 11.4.1. (Fitzgerald [1994]) *Let $\{P_1, P_2, \ldots, P_n\}$ be a sequence of n homogeneous polynomials of second order in n variables ($n \geq 2$). Then, for each $k = 1, \ldots, n$, there exists a function*

$$f_k(z_1, \ldots, z_n) = z_k + P_k(z_1, \ldots, z_n) + O(|z|^2), \tag{11.4.1}$$

such that $\mathbf{f} = (f_1, \ldots, f_n)$ is a biholomorphic mapping of $\mathbb{C}^n$ into $\mathbb{C}^n$.

For a proof, see Fitzgerald [1994] and Gong [1999].

11.5 Beurling Transforms

We will study the perturbation of the Beurling transform in the complex plane $\mathbb{C}$. First, we define certain transforms of a square-integrable function $f(z)$ in $\mathbb{C}$. Thus,

(i) *Fourier transform* of the function f is defined as

$$\mathcal{F}\{f(\xi)\} = \int_{\mathbb{C}} e^{-2\Re\{z\bar{\xi}\}} f(z)\, dA(z), \quad \xi \in \mathbb{C}, \tag{11.5.1}$$

where $dA(z) = \dfrac{dx\, dy}{\pi}$, $z = x + iy$.

(ii) *Beurling transform* is defined as the singular integral operator

$$\mathcal{B}\{f(z)\} = \mathrm{pv} \int_{\mathbb{C}} \frac{f(\zeta)}{(\zeta - z)^2}\, dA(\zeta), \quad z \in \mathbb{C}, \tag{11.5.2}$$

where 'pv' denotes the principal value. These two transforms are related to each other by

$$\mathcal{F}\mathcal{B}\{f(\xi)\} = -\frac{\bar{\xi}}{\xi}\mathcal{F}\{f(\xi)\}, \quad \xi \in \mathbb{C}. \tag{11.5.3}$$

In view of the Plancherel identity,[1] $\mathcal{F}$ is a unitary transformation on $L^2(\mathbb{C})$, with the norm $\|f\|_{L^2(\mathbb{C})}^2 = \int_{\mathbb{C}} |f(z)|^2\, dA(z)$. Since an operator T on a complex Hilbert space H is unitary if $T^*T = TT^* = I$, where T^* and I are the adjoint and the identity operator, respectively, the Beurling transform $\mathcal{B}_{\mathbb{C}}$ is unitary on $L^2(\mathbb{C})$. Detailed information on Fourier transform, in general, is also available in Kythe et al. [2003] and Kythe [2014].

(iii) *Cauchy transform* $\mathcal{C}$ is defined for square-integrable functions by

$$\mathcal{C}\{f(z)\} = \int_{\mathbb{C}} \frac{f(\zeta)}{\zeta - z}\, dA(\zeta), \tag{11.5.4}$$

and it is related to the Beurling transform by

$$\mathcal{B}\{f(z)\} = \partial_z \mathcal{C}_{\mathbb{C}}\{f(z)\}. \tag{11.5.5}$$

Let $L_\theta^2(\mathbb{C})$, for real θ, denote the Hilbert space H of square-integrable functions on $\mathbb{C}$ with the norm $\|f\|_{L_\theta^2(\mathbb{C})}^2 = \int_E |f(z)|^2 |z|^{2\theta}\, dA(z) < +\infty$. Moreover, let $\mathcal{T}$ denote the operator defined on $\mathbb{C}$ as

$$\mathcal{T}\{h(z)\} = \frac{1}{z}\mathcal{C}\{h(z)\}, \tag{11.5.6}$$

---

[1] Plancherel identity states that the integral of a function's squared modulus is equal to the integral of the squared modulus its frequency spectrum. It corresponds to Parseval's theorem for Fourier series; see Kythe [2014:62].

for square-integrable functions $h(z)$. Then $h \in L^2_\theta(\mathbb{C})$ is well defined for some positive θ for $\mathcal{T}\{h(z)\}$. Also, we define the associated opeartor $\mathcal{T}'$ by

$$\mathcal{T}'\{h(z)\} = \mathcal{C}\left\{\frac{h(z)}{z}\right\}. \tag{11.5.7}$$

Both of these operators are defined on $\mathbb{C}$. Now, the *perturbed Beurling transform* is defined by

$$\mathcal{B}^\theta = \begin{cases} \mathcal{B} + \theta\mathcal{T} & \text{for } 0 \le \theta \le 1, \\ \mathcal{B} + \theta\mathcal{T}' & \text{for } -1 \le \theta \le 0. \end{cases} \tag{11.5.8}$$

Then we have the following theorem, proof of which is available in Hedenmalm [2008].

Theorem 11.5.1. *For* $-1 \le \theta \le 1$*, the perturbed Beurling operator* $\mathcal{B}^\theta$ *behaves as a unitary operator on* $L^2_\theta(\mathbb{C})$*.*

This theorem is also useful in quasiconformal mapping, for which see Hedenmalm and Shimorin [2005] and Petermichl and Volberg [2002].

PROOF. Let α denote an Nth root of unity, $N = 1, 2, \ldots$, i.e., $\alpha^N = 1$, then $\bar{\alpha}$ is also an Nth root of unity. Let $\mathcal{A}_N$ denote the set of all Nth roots of unity. For $n = 1, 2, \ldots N$, consider the closed subspace $L^2_{n,N}(\mathbb{C}) \subset L^2(\mathbb{C})$ consisting of functions f such that $f(\alpha z) = \alpha^n f(z)$ for $z \in \mathbb{C}$. Obviously, $f \in L^2_{n,N}(\mathbb{C})$ if and only if $f \in L^2(\mathbb{C})$ is of the form $f(z) = z^n g(z^N)$ for $z \in \mathbb{C}$, where g is any complex-valued function. Now, for a fixed N, suppose $f \in L^2_{n,N}(\mathbb{C})$, $n = 1, 2, \ldots, N$. Then by change of variables formula

$$\mathcal{B}\{f(z)\} = \text{pv} \int_{\mathbb{C}} \frac{f(\zeta)}{(\zeta - z)^2} \, dA(\zeta) = \text{pv} \int_{\mathbb{C}} \frac{\alpha^n f(\zeta)}{(\alpha\zeta - z)^2} \, dA(\zeta). \tag{11.5.9}$$

If we take the average over all the N roots of unity, we obtain

$$\mathcal{B}\{f(z)\} = \frac{1}{N} \, \text{pv} \int_{\mathbb{C}} \sum_{\alpha \in \mathcal{A}_n} \frac{\alpha^n}{(\alpha\zeta - z)^2} f(\zeta) \, dA(\zeta), \quad z \in \mathbb{C}. \tag{11.5.10}$$

Consider a sum of the form $F(z) = \dfrac{1}{N} \displaystyle\sum_{\alpha \in \mathcal{A}_N} \dfrac{\alpha^n}{1 - \alpha z}$. This sum has the property that $F(\alpha z) = \bar{\alpha}^n f(z)$ for $a \in \mathcal{A}_N$. Thus, $F(z) = z^{N-n} G(z^N)$, where the function G is analytic on $\mathbb{C}$ except a simple pole at 1, and $G(\infty) = 0$, so is $F(\infty) = 0$. Then, obviously, G is of the form $G(z) = \dfrac{C}{1 - z}$, where the constant $C = 1$ can be easily proved. Hence,

$$F(z) = \frac{1}{N} \sum_{\alpha \in \mathcal{A}_N} \frac{\alpha^n}{1 - \alpha z} = \frac{z^{N-n}}{1 - z^N}, \quad z \in \mathbb{C}. \tag{11.5.11}$$

Let $H(z) = F(z) + zF'(z)$. Then

$$H(z) = [zF(z)]' = \frac{1}{N} \sum_{\alpha \in \mathcal{A}_N} \frac{\alpha^n}{(1 - \alpha z)^2} = z^{N-n} \left\{ \frac{N}{(1 - z^N)^2} - \frac{n-1}{1 - z^N} \right\}.$$

This identity then gives the value of the sum in (11.5.10) as

$$\frac{1}{N} \sum_{\alpha \in \mathcal{A}_n} \frac{\alpha^n}{(\alpha \zeta - z)^2} = \frac{1}{z^2} H\left(\frac{\zeta}{z}\right) = z^{n-2} \zeta^{N-n} \left\{ \frac{N z^N}{(z^N - \zeta^N)^2} - \frac{n-1}{z^N - \zeta^N} \right\}.$$

Hence, (11.5.10) can be expressed as

$$\mathcal{B}\{f(z)\} = z^{n-2} \operatorname{pv} \int_{\mathbb{C}} \left\{ \frac{N z^N}{(z^N - \zeta^N)^2} - \frac{n-1}{z^N - \zeta^N} \right\} \zeta^{N-n} f(\zeta) \, dA(\zeta), \quad z \in \mathbb{C}.$$

(11.5.12)

Let $f \in L^2(\mathbb{C})$ such that $f(z) = z^n g(z^N)$ for $z \in \mathbb{C}$. Then (11.5.12) can be expressed in terms of the function g as

$$\mathcal{B}\{f(z)\} = z^{n-2} \operatorname{pv} \int_{\mathbb{C}} \left\{ \frac{N z^N}{(z^N - \zeta^N)^2} - \frac{n-1}{z^N - \zeta^N} \right\} \zeta^N g(\zeta^N) \, dA(\zeta), \quad z \in \mathbb{C},$$

(11.5.13)

and an expression for the Cauchy transform, similar to (11.5.12), is

$$\mathcal{C}\{f(z)\} = z^{n-N-1} \int_{\mathbb{C}} \frac{\zeta^N}{(\zeta^N - z^N)^2} g(\zeta^N) \, dA(\zeta), \quad z \in \mathbb{C}, \qquad (11.5.14)$$

This implies that the operator $\mathcal{B}^\theta$ is unitary on the space $L_\theta^2(\mathbb{C})$ for $0 \le \theta \le 1$. Thus, the relationship between the Beurling transform of f and h is given by

$$\mathcal{B}\{f(z)\} = z^{N+n-2} \mathcal{B}^{(n-1)/N}\{h(z^N)\}, \quad z \in \mathbb{C}. \qquad (11.5.15)$$

Thus, the Beurling transform $\mathcal{B}$ from an isometry becomes the norm identity, i.e,

$$\int_{\mathbb{C}} |h(z)|^2 |z|^{2\theta} \, dA(z) = \int_{\mathbb{C}} |\mathcal{B}^\theta \{h(z)\}|^2 |z|^{2\theta} \, dA(z), \qquad (11.5.16)$$

where we suppose that $\theta = (n-1)/N$. But since such fractional value are θ are dense in the interval $[0, 1]$, the identity is, in fact, valid for all θ, $0 \le \theta \le 1$. This implies that the operator $\mathcal{B}^\theta$ is unitary on the space $L_\theta^2(\mathbb{C})$ for $0 \le \theta \le 1$. To show that $\mathcal{B}^\theta$ is unitary on the space $L_\theta^2(\mathbb{C})$ for $-1 \le \theta \le 0$, note that

$$\frac{1}{(\zeta - z)^2} + \frac{\theta}{\zeta(\zeta - z)} = \frac{z}{\zeta} \left\{ \frac{1}{(\zeta - z)^2} + \frac{\theta + 1}{z(\zeta - z)} \right\},$$

which implies that $\mathcal{B}^{\theta} = \mathfrak{M}_z \mathcal{B}^{\theta+1} \mathfrak{M}_z^{-1}$ for $-1 \leq \theta \leq 0$, where $\mathfrak{M}$ denotes multiplication operator by the variable z. Hence, $\mathcal{B}^{\theta}$ is unitary on $L_{\theta}^2($ for $-1 \leq \theta \leq 0$. ∎

Petermichl and Volberg [2002] have proved that $\mathcal{B}$ is a bounded operator on $L_{\theta}^2(\mathbb{C})$ only for $-1 < \theta < 1$.

11.5.1 Grunsky Identity and Inequality. This is an application of the Beurling transform. Let $\phi : E \mapsto G$, where $G = \phi(E) \subset \mathbb{C}$, be a conformal mapping. Then

$$\mathcal{B}_{\phi}\{f(z)\} = \mathrm{pv} \int_E \frac{\phi'(z)\phi'(\zeta)}{(\phi(\zeta)\phi(z))^2} f(\zeta)\, dA(\zeta), \quad z \in E$$

is a contraction on $L^2(E)$. In fact, $\mathcal{B}$ is unitary on $L^2(\mathbb{C})$. Let $\epsilon(z) := z$, so that

$$\mathcal{B}_{\epsilon}\{f(z)\} = \mathrm{pv} \int_E \frac{f(\zeta)}{(\zeta - z)^2}\, dA(\zeta), \quad z \in E. \tag{11.5.17}$$

Then the *Grunsky identity* is

$$\mathcal{B}_{\phi} - \mathcal{B}_{\epsilon} = \mathfrak{P}\mathcal{B}_{\phi} = \mathcal{B}_{\phi}\overline{\mathfrak{P}} = \mathfrak{P}\mathcal{B}_{\phi}\overline{\mathfrak{P}}, \tag{11.5.18}$$

where $\mathfrak{P}$ is the Bergman projection defined by (6.1.9) and $\overline{\mathfrak{P}}$ is the associated Bergman projection defined by (6.1.10), which are both contractions on $L^2(E)$. Hence, for $f \in L^2(E)$,

$$\| (\mathcal{B}_{\phi} - \mathcal{B}_{\epsilon}) \{f(z)\} \|_{L^2(E)} \leq \|f\|_{L^2(E)}. \tag{11.5.19}$$

This is known as the *Grunsky inequality* in a compact form, which takes the following form for the transferred Beurling transform function ϕ an the unit disk E (Hedenmalm [2008]):

$$\| (\mathcal{B}_{\phi}^{\theta} - \mathcal{B}_E^{\theta}) \{f(z)\} \|_{L_{\theta}^2(E)} \leq \|f\|_{L_{\theta}^2(E)}, \quad z \in E. \tag{11.5.20}$$

11.6 Exercises

11.6.1. Describe the curve $\gamma(t) = x(t) + iy(t)$, where $x(t) = 3\cos t$, $y(t) = 2\sin t$ for $0 \leq t \leq \pi$. ANSWER. The x and y coordinates of the curve $z = \gamma(t)$ satisfy the equation $x^2/9 + y^2/4 = 1$ so that Γ is the part of parabola lying in the upper half-plane. By tracking the motion of $\gamma(t)$ as t increases, we see that $\gamma(t)$ starts at the point $\gamma(0) = 3 + it$ and ends at $\gamma(\pi) = -3 + it$ and moves counterclockwise along the parabolic arc which is the trajectory Γ of the parametrized curve. ∎

11.6.2. Explain why the curve $\gamma(t) = \begin{cases} 0 + it^2 & \text{if } -1 \le t \le 0 \\ t^2 + i0 & \text{if } 0 \le t \le 1 \end{cases}$ is smooth.

ANSWER. The function $\gamma(t)$ is differentiable, even at $t = 0$:

$$\dot{\gamma}(t) = \begin{cases} 0 + 2it & \text{if } -1 \le t \le 0, \\ 2t + i0 & \text{if } 0 \le t \le 1. \end{cases}$$

Obviously, $\dot{\gamma}(t)$ is a continuous function of t, and so is the speed $|\dot{\gamma}(t)| = 2|t|$ of the moving point. Note that although $\dot{\gamma}(0) = 0$, the direction of motion changes abruptly as t passes through the origin. The point $\gamma(t)$ moves downwards from $= i$ to 0, slows to a momentary halt at the origin, and then moves to the right from 0 to 1. The origin ($\gamma(0)$) is known as the *critical point* where the tangent vanishes, and in spite of this sharp point (kink) in the trajectory, $\gamma(t)$ is a smooth curve since $\dot{\gamma}(t)$ varies continuously. ∎

11.6.3. Show that the Loewner curvature is neither local nor reversible. SOLUTION. Although Loewner curvature is similar to other types of curvature, yet there are some significant difference. For example, Loewner curvature is not reversible but it depends on the orientation of the curve, and it is not local but it depends on the 'past' history of the curve. Assume, on the contrary, that Loewner curvature is purely local, and consider, for example, the curve in Figure 11.6.1, where, say, $\gamma(t_0) = 2i$ at time t_0. If Loewner curvature were purely local, then $LC_\gamma(t) = 0$ for $t > t_0$, since in the neighborhood of $\gamma(t)$ (i.e., locally) the curve γ looks like a vertical slit (with zero Loewner curvature). That is, $g_{t_0}(\gamma)$ must be a vertical half-ray, and therefore $\gamma(t_0, \infty)$ would be a hyperbolic geodesic in $\mathbb{U}^+ \backslash \gamma[0, t_0]$, which it is not. Hence, Loewner curvature cannot be reversible. If we trace γ from ∞ to 0, we find that Loewner curvature will be 0 before reaching $2i$. ∎

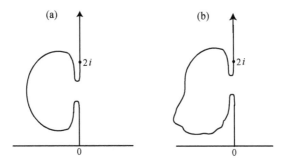

Figure 11.6.1 Exercise 11.6.3.

11.6.4. Set $\xi(t) = A$ (const) in Eq (4.2.18) and solve the resulting differential equation by the method of separation of variable. SOLUTION. Separating the variables, we get $(g - A)\partial g = -2\partial t$, which after integrating and moving al terms to the left side gives $\frac{1}{2}g^2 - Ag + 2t + C = 0$, where C is a constant of integration. Using the quadratic formula, we get $g = A \pm \sqrt{A^2 - 2(2t + C)}$, which, after using the initial value $g(z, 0) = z$ yields $-2C = (z - A)^2 - A^2$, thus giving

$$g(z, t) = A \pm \sqrt{(z - A)^2 - 4t}. \tag{11.6.1}$$

Note that at the singular point $z = A$, $g(A, t) = A + 2i\sqrt{t}$. Thus, in view of continuity of ξ, it is clear that the minus sign in the solution $g(z, t)$ must be discarded. ∎

11.6.5. Use composition property of conformal mapping and discuss the composition of two forcing functions $\xi_1(t)$ and $\xi_2(t)$. SOLUTION. Let the two mappings be $f^A(z, t)$ and $f^B(z, t)$, generated respectively by two forcing functions $\xi^A(t)$ and $\xi^B(t)$ which are defined on the intervals $[0, t_A]$ and $[0, t_B]$, respectively. Suppose that the mapping (11.1.2) generate the maps $f_A = f^A(z, t_A)$ and $f_B = f^B(z, t_B)$ for each forcing function. Then the composite forcing term is given by

$$\xi(t) = \begin{cases} \xi^A(t) & \text{for } 0 < t < t_A, \\ \xi^B(t - t_a) & \text{for } t_A < t < t_A + t_B. \end{cases} \tag{11.6.2}$$

Thus, this composite forcing function generates a map $f(z, t)$ such that by the time $t_A + t_B$ it is given by

$$f(z, t_A + t_B) = (f_B \circ f_A)(z, t) = f_B(f_A(z, t)). \blacksquare \tag{11.6.3}$$

11.6.6. Consider the classical Loewner equation (4.1.2), with $p(z, t)$ independent of t:[1]

$$\partial_t f(z, t) = z f'(z, t) p(z), \tag{11.6.4}$$

where $f' \equiv \partial_z f$. Assume that $f(z, t)$ is of the form $f(z, t) = \phi(e^t \psi(t))$, $f(z, 0) = z$, where $\psi(z) = 1 + c_0 + c_1 z^{-1} + \cdots$, and $\phi = \psi^{-1}$, i.e., $\phi(\psi(z)) = z$. Derive an integral representation for $\psi(z)$. SOLUTION. Differentiate Eq (10.6.4) with respect to t and z, to get

$$\partial_t f(z, t) = e^t \psi(z) \phi'(e^t \psi(z)), \qquad f'(z, t) = e^t \psi'(t) \phi'(e^t \psi(z)).$$

Substituting these into Eq (10.x.1), we obtain $\psi'(z) = \dfrac{1}{z p(z)} \psi(z)$, $z \in E$, which after integration gives

$$\log \psi(z) = \int \frac{dz}{z p(z)}, \quad \text{or} \quad \psi(z) = \exp \left\{ \int \frac{dz}{z p(z)} \right\}.$$

---

[1] Note that $p(z) : E \mapsto \mathbb{C}$ such that $\Re\{p(z)\} > 0$, $p(\infty) = 1$, and $p(z)$ has the integral representation $p(z) = \displaystyle\int_{\partial E} \frac{z + \zeta}{z - \zeta} \, d\mu(z)$.

11.6.7. Substitute $u(t) = r(t)e^{i\theta(t)}$ in the initial value problem $\dot{u}(t) = -u(t)p(u(t), t)$, $u(0) = f(z, T) = z, T > 0$. Let $u(t) = r(t)e^{i\theta(t)}$. Find $\dot{r}(t)$ and $\dot{\theta}(t)$. ANSWER. $\dot{r}(t) = -r(t)\Re\{p(u(t), t)\}$, $r(T) = z$, and $\dot{\theta}(t) = -\Im\{p(u(t), t)\}$, $\theta(T) = \arg\{z\}$. ∎

11.6.8. The Loewner equation associated with the constant point mass is

$$\partial_t f(z, t) = z f'(z, t)\frac{z+1}{z-1},$$

which has as its solutions the conformal maps of the exterior disk to the disk minus a straight slit starting at the point 1 extending to infinity. Determine an explicit formula for such a mapping. SOLUTION. First solve the associated equation $\psi'(z) = \dfrac{z-1}{z(z+1)}\psi(z)$, which has the solution: $\psi(z) = z + \dfrac{1}{z} + 2$, or $\phi(z) = \frac{1}{2}\left(z - 2 + \sqrt{z^2 - 4z}\right)$, where $\phi(\psi(z)) = z$, as in Exercise E.4. Let $s(z, t) = \phi\left(e^t\psi(z)\right)$. Then

$$s(z, t) = \frac{e^t}{2z}\left[z^2 + 2\left(1 - e^t\right)z + 1 + (z+1)\sqrt{z^2 + 2\left(1 - 2e^{-t}\right)z + 1}\right],$$

where $s(-1, t) = -1$ is a fixed point on the boundary, ∎

Exercise 11.6.9. Consider the function $f(z) = |z|^{-2\theta}\left(\dfrac{1}{(1 - \lambda\bar{z})^2} - \dfrac{\theta}{1 - \lambda\bar{z}}\right)$ for $z \in E$, where λ is scalar. Compute the inequality (11.5.20). SOLUTION.

$$(\mathcal{B}_\phi^\theta - \mathcal{B}_E^\theta)\{f(z)\} = \left[\frac{\lambda\phi(z)}{z\phi(z)}\right]^\theta \frac{\phi'(z)\phi'(\lambda)}{(\phi(\lambda) - \phi(z))^2} - \frac{1}{(\lambda - z)^2}$$
$$+ \theta\frac{\phi'(z)}{\phi(z)}\left[\frac{\lambda\phi(z)}{z\phi(z)}\right]^\theta \frac{\phi'(\lambda)}{\phi(\lambda) - \phi(z)} - \frac{\theta}{z(\lambda - z)}.$$

Then (11.5.20) becomes for $0 \le \theta \le 1$

$$\int_E \left|\left[\frac{\lambda\phi(z)}{z\phi(z)}\right]^\theta \frac{\phi'(z)\phi'(\lambda)}{(\phi(\lambda) - \phi(z))^2} - \frac{1}{(\lambda - z)^2}\right.$$
$$\left. + \theta\frac{\phi'(z)}{\phi(z)}\left[\frac{\lambda\phi(z)}{z\phi(z)}\right]^\theta \frac{\phi'(\lambda)}{\phi(\lambda) - \phi(z)} - \frac{\theta}{z(\lambda - z)}\right.$$
$$\le \int_E |f(z)|^2|z|^{2\theta}\, dA(z)$$
$$= \int_E \left|\frac{1}{(1 - \lambda\bar{z})^2} - \frac{\theta}{z(\lambda - z)}\right|^2 |z|^{-2\theta}\, dA(z)$$
$$= \frac{1}{(1 - |\lambda|^2)^2} - \frac{\theta}{1 - |\lambda|^2}.$$

In particular, if $\lambda = 0$, and if we assume $\phi'(0) = 1$, we obtain the *Prawitz inequality* (see Milin [1977], and Hedenmalm and Shimorin [2005]):

$$\int_E \left| \phi'(z) \left[\frac{\phi(z)}{z} \right]^{\theta-2} - 1 \right|^2 |z|^{2\theta} \, dA(z) \leq \frac{1}{1-\theta}. \quad \blacksquare$$

11.6.10. Show that there exists *no* Bieberbach conjecture in the case of several complex variables. ANSWER. The following counter-example, found in Gong [1999], proves this statement. Let $f(\mathbf{z}) = (f_1(\mathbf{z}), f_2(\mathbf{z})) ,\ \mathbf{z} = (z_1, z_2) \in \mathbb{C}^2$, such that

$$f_1(\mathbf{z}) = z_1,$$
$$f_2(\mathbf{z}) = z_2 (1 - z_2)^{-k} = z_2 + k z_1 z_2 + \cdots , \qquad (11.6.5)$$

where k is any positive integer. Then $f(\mathbf{z})$ is a normalized biholomorphic mapping[1] on the unit ball $B(0, r) : \{\mathbf{z} \in \mathbb{C}^2 : \mathbf{z}\bar{\mathbf{z}}^T < 1\}$, where $\bar{\mathbf{z}}^T$ means the conjugate transpose of $\mathbf{z}$ in $\mathbb{C}^2$, i.e., $f(\mathbf{0}) = \mathbf{0}$ and the Jacobian J_f at $\mathbf{z} = \mathbf{0}$ is the identity matrix. It is obvious that the modulus of the coefficient of the second order terms in the Taylor series expansion (11.6.5) of $f(\mathbf{z})$, the growth of $|f(\mathbf{z})|$, and the distortion of $f(\mathbf{z})$ given by $|\det J_f(\mathbf{z})|$ are all unbounded. $\blacksquare$

11.6.11. Let $\mathbf{f} = (f_1, \ldots, f_n) : \mathbb{C}^n \mapsto \mathbb{C}^n$ be a normalized hoomorphic mapping, i.e., $\mathbf{f}(\mathbf{0}) = \mathbf{0}$ and the Jacobian $J_{\mathbf{f}}(\mathbf{0}) = \mathbf{I}$, where $\mathbf{I}$ is the identity matrix. Then each component $f_1, \ldots, f_n$ is a holomorphic function of the variable $\mathbf{z} = (z_1, \ldots, z_n)$ and each has the Taylor series expansion

$$f_k (z_1, \ldots, z_n) = z_k + \sum_{j_m} d^{(k)}_{(j_1, \ldots, j_n)} z_1^{j_1} \cdots z_n^{j_n}, \qquad (11.6.6)$$

for $k = 1, \ldots, n$, where each j_m is a non-negative integer ($m = 1, \ldots, n$) such that $j_1 + \cdots + j_n \geq 2$. Let $\mathbf{a} = (a_1, \ldots, a_n)$ be a complex number in $\mathbb{C}^n$, and let $\mathbf{b} = (b_1, \ldots, b_n) \in \mathbb{C}^n$, $\mathbf{c} = (c_1, \ldots, c_n) \in \mathbb{C}^n$, and define $\mathbf{b} \cdot \mathbf{c} := \sum_{i=1}^{n} b_i c_i$. Let $\mathbf{v} \in \mathbb{C}^n$, $\mathbf{v} \neq \mathbf{0}$. Assume that $\mathbf{A}_k$, $k = 1, \ldots, n$, are vectors from $\mathbb{C}^n$ such that $\mathbf{A}_k \cdot \mathbf{v} = 0$ for each $k = 1, \ldots, n$. Then show that the normalized polynomial mapping defined by $\mathbf{f} = \mathbf{z} + \mathbf{a}\mathbf{v}(\mathbf{A}_1 \cdot \mathbf{z})(\mathbf{A}_2 \cdot \mathbf{z} \cdots (\mathbf{A}_n \cdot \mathbf{z})$ is holomorphic. SOLUTION. To prove, it suffices to obtain the inverse of the mapping. Let $\mathbf{w} = \mathbf{f}(\mathbf{z})$. Since $\mathbf{A}_k \cdot \mathbf{v} = 0$, the equation $\mathbf{A}_k \cdot \mathbf{w} = \mathbf{A}_k \cdot \mathbf{z}$ holds for each $k = 1, \ldots, n$. Hence, $\mathbf{z} = \mathbf{w} - \mathbf{a}\mathbf{v}(\mathbf{A}_k \cdot \mathbf{w})$, which the inverse mapping for $k = 1, \ldots, n$. $\blacksquare$

11.6.12. The SLE (stochastic Loewner evolution) or the recent development known as the Schramm-Loewner evolution (Schramm [2000], Lawler

---

[1]This is one of many normalized biholomorphic mappings that map the space $\mathbb{C}^n$ into itself (see Rosay and Rudin [1988]).

[2005]), include systems that describe the known stochastic models as well as the critical phenomena such as percolation, self-avoiding random walks, spanning trees and so on. The new development is devoted to non-local structures that characterizes a given system, either a boundary of an Ising model or percolation cluster, or loops in the $O(n)$ model. It appears that the questions that are difficult to pose and/or answer using the conformal field theory (CFT) have become easy and natural within the SLE model.

SLE is the study of the Loewner equation with stochastic driving, specially driving by a forcing function $\lambda(t)$ which is a Gaussian random variable, obeying the familiar Langevin equations of Brownian motion:

$$\langle \dot{\lambda}(t), \dot{\lambda}(s) \rangle = \nu \delta(t - s), \qquad (11.6.7)$$

or, equivalently in the more integrated form

$$\langle (\lambda(t) - \lambda(s))^2 \rangle = \nu |t - s|, \qquad (11.6.8)$$

where ν is a dimensionless constant whose significance lies in the fact that an SLE is referred at a particular value of ν as SLE$_\nu$. Schramm [2000] showed that the Loewner equation can be used to describe conformally-invariant random curves by choosing $\lambda(t)$ to be random function that satisfies certain conditions, namely, (i) $\lambda(t)$ must be continuous with probability 1; (ii) to generate a conformally-invariant curve, $\lambda(t)$ must undergo a process to have identically distributed increment, since the required map $g(z, t)$ can be constructed by iterations of some infinitesimal identically-distributed conformal maps; and (iii) the invariance of $g(z, t)$ with respect to the relations $x + iy \to -x_i y$ is needed to make the choice of $\lambda(t)$ unique, i.e., it can only be a scaled version of the Brownian motion without drift of Eq (11.6.7).

If we choose $\lambda(t)$ to be a smooth real-valued function, the solution $g(z, t)$ of the forward Loewner equation

$$\frac{\partial}{\partial t} g(z, t) = \frac{2}{g(z, t) - \lambda(t)}, \quad g(z, 0) = z, \qquad (11.6.9)$$

would give a conformal map from $\mathbb{U}^+$ cut along a segment traced out by a simple, non self-intersecting, curve γ. Similarly, in SLE the function g gives a map from a domain D_t in $\mathbb{U}^+$ onto $\mathbb{U}^+$; this region is $\mathbb{U}^+$ with some part cut away by the singularities of g, and the cut out part may be a simple curve which avoids the real line (for $0 < \nu < 4$), a self-intersecting curve (for $4 < \nu < 8$), or a a filled-in region (for $\nu > 8$); these three cases are shown in Figure 11.6.2. There are theorems and some speculations still that provide direct and useful information about the trace of cut-out singularities, or the curves which surround the self-intersecting trace, or the filled-in region.

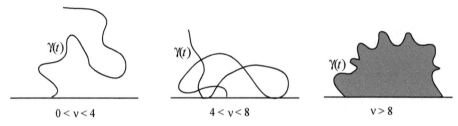

Figure 11.6.2 Exercise 11.6.12.

A

Mappings

Let X and Y be non-empty sets. Let $X \times Y$ denote the *Cartesian product* of X and Y, defined as the set of all ordered pairs whose first element belongs to X and whose second element belongs to Y, i.e., $X \times Y = \{(x,y) : x \in X, y \in Y\}$. A function f from X into Y is a subset of $X \times Y$ such that $(x,y) \in f$. The set X is called the *domain of definition* or simply the *domain* of f, and we say that f is defined on X. The set $\{y \in Y : (x,y) \in f$ for some $x \in X\}$ is called the *range* of f, denoted by $\mathcal{R}(f)$. For each $(x,y) \in f$, we call y the *value* of f at x and denote it by $y = f(x)$. Sometimes we write $f : X \longrightarrow Y$ to denote the function f from X into Y. Since functions are defined as sets, equality of functions must be taken in the sense of equality of sets.

The terms *function, mapping, map, operator,* and *transformation* are often used interchangeability. If the term *mapping* is used, we say 'a mapping of X into Y.' In such statements, the term 'into Y' is important as we sometimes speak of 'onto Y.' Generally, the term 'function' is a *rule* whereby for each $x \in X$ there is a unique $y \in Y$ that is assigned to x. In this sense the term *mapping* becomes very descriptive. Obviously, two mappings f and g of X into Y are equal if and only if $f(x) = g(x)$ for every $x \in X$.

Let f be a function from X into Y. If $\mathcal{R}(f)$ is equal to Y, then f is said to be *surjective* (or a *surjection*), and we say that f *maps* X *onto* Y. If f is a function such that for every $x_1, x_2 \in X$, $f(x_1) = f(x_2)$ implies that $x_1 = x_2$, then f is said to be *injective* or a *one-to-one mapping*, or an *injection*. If f is both injective and surjective, we say that f is *bijective* or *one-to-one and onto*, or a *bijection*.

Let f be an injective mapping of X into Y. Then we say that f *has an inverse*, and we call the function $g \equiv f^{-1}$ the *inverse* of f. Clearly, if f has an inverse, then f^{-1} is a mapping from $\mathcal{R}(f)$ onto X.

Theorem A.1. *Let f be an injective mapping of X into Y. Then (i) f is a one-to-one mapping of X onto $\mathcal{R}(f)$; (ii) f^{-1} is a one-to-one mapping of $\mathcal{R}(f)$ onto X; (iii) for every $x \in X$, $f^{-1}(f(x)) = x$; and (iv) for every $y \in \mathcal{R}(f)$, $f(f^{-1}(y)) = y$.*

Let X, Y and Z be non-empty sets. Let $f : X \longrightarrow Y$ and $g : Y \longrightarrow Z$. For each $x \in X$, we have $f(x) \in Y$ and $g\left(f(x)\right) \in Z$. Since f and g are mappings from X into Y and from Y into Z, respectively, it follows that for each $x \in X$ there is one and only one element $g\left(f(x)\right) \in Z$. Hence, the set $\{(x, z) \in X \times Z : z = g\left(f(x)\right), x \in X\}$ is a function from X into Z. This function is called the *composite function* of g and f and denoted by $g \circ f$, i.e.,

$$(g \circ f)(x) = g \circ f(x) := g\left(f(x)\right).$$

Theorem A.2. *(i) If f is a mapping of a set X onto a set Y and g is a mapping of the set Y onto a set Z, then $f \circ g$ is a mapping to X onto Z; and (ii) If f is a one-to-one mapping of a set X onto a set Y, and if g is a one-to-one mapping of the set Y onto a set Z, then $g \circ f$ is a one-to-one mapping of X onto Z. Moreover, $(g \circ f)^{-1} = \left(f^{-1}\right) \circ \left(g^{-1}\right).$*

For more details, see Ponnusamy [2011].

B

Parametrized Curves

B.1 Curves in the Complex Plane

A curve in the complex plane always implies a *parametrized curve* in the sense that it is determined by specifying a point $\gamma(t) = x(t) + iy(t)$ in the plane for each t in some interval $[a, b]$ of real numbers, where t is a parameter, and the real valued functions $x(t)$ and $y(t)$, assumed to be continuous on the interval $[a, b]$, specify the coordinate of $\gamma(t)$ for each t. Thus, a parametrized curve γ is a mapping on $[a, b]$ such that each value of t corresponds to a well defined point $\gamma(t)$ in the complex plane. The curve traced by $\gamma(t)$ as t varies over the interval $[a, b]$ is denoted by Γ and is called its *trajectory*. A physical interpretation is as follows: if t is regarded as the 'time' variable and $\gamma(t)$ specifies the location of some moving particle in a two-dimensional physical problem, we want to know the trajectory of the particle as well as the motion of the particle along its orbit (parametrization).

The terminology used for a parametrized curve for $t \in [a, b]$ is as follows. The point $\gamma(a)$ is called the *initial point* and the point $\gamma(b)$ the *final point* of Γ; if $\gamma(a) = \gamma(b)$ (i.e., if the curve $\gamma(t)$ returns to its initial point), then Γ is called a *closed curve*; and if $\gamma(t_1) \neq \gamma(t_2)$ for $t_1 \neq t_2$, $(t_1, t_2 \in [a, b])$, so that Γ never crosses itself, we say that Γ is a *simple curve*. Let the interval $[a, b]$ be in a (Jordan) domain D. Then we can assume that there is a family $\Gamma_{D,a,b}$ of *simple* curves in $\bar{D}$ that begins at a and ends in b.

Given a parametrized curve $\gamma(t) = x(t) + iy(t)$, $t \in [a.b]$, the derivative $\dot{\gamma}(t_0)$ at a point $t_0 \in [a, b]$ is defined by

$$\dot{\gamma}(t) = \lim_{t \to t_0} \frac{\Delta \gamma}{\Delta t} = \lim_{t \to t_0} \frac{x(t) - x(t_0)}{t - t_0} + i \lim_{t \to t_0} \frac{y(t) - y(t_0)}{t - t_0} = \frac{dx}{dt}(t_0) + \frac{dy}{dt}(t_0).$$

Note that instead of the usual notation $\gamma'(t_0)$, we have used the "t"-derivative (or, in physical sense, the time-derivative) notation $\dot{\gamma}(t_0)$ for the derivative of $\gamma(t)$ with respect to t. This derivative is a complex number, and most of the curves we discuss will be (continuously) differentiable, twice differentiable, and so on, so that the functions $\dot{\gamma}(t)$, $\ddot{\gamma}(t), \ldots$ are functions of the same kind as $\gamma(t)$.

A parametrized curve $\gamma(t) = x(t) + iy(t)$, $t \in [a, b]$, is said to be *smooth* if it is continuously differentiable. If we think of t as a parameter, then the derivative $\dot{\gamma}(t)$ has the following physical interpretation: the vector corresponding the complex number $\dot{\gamma}(t_0)$ is associated with the point on the trajectory Γ and defines the *velocity* (i.e., speed and direction) of the moving point $\gamma(t)$ when $t = t_0$. The Euclidean length of the *velocity* vector or *tangent* vector is just the absolute value of the corresponding complex number:

$$|\dot{\gamma}(t_0)| = \sqrt{\dot{x}(t_0) + i\dot{y}(t_0)},$$

which is interpreted as the *speed* (a scalar quantity) of the moving point at $t = t_0$.

If $\dot{\gamma}(t) \neq 0$, the derivative $\dot{\gamma}(t)$ determines a well defined direction at $\gamma(t_0)$, in the sense that there is a definite tangent line at this point that is parallel to $\dot{\gamma}(t_0)$. If $\dot{\gamma}(t)$ is continuous and non-vanishing throughout $[a, b]$, the orientation of the tangent line at $\gamma(t)$ varies continuously as t increases, and the trajectory cannot have sharp corners (kinks).

A smooth curve $\gamma(t)$ in the z-plane is transformed into a smooth curve $\eta(t)$ in the w-plane by a mapping $w = f(z)$ whose domain of definition includes the trajectory Γ of $\gamma(t)$. The new curve is given by the formula

$$\eta(t) = f(z)\big|_{z=\gamma(t)} = f(\gamma(t)) \equiv (f \circ \gamma)(t),$$

which is defined on the same interval $[a, b]$ as $\gamma(t)$. The transformed curved is just the composite map $\eta = f \circ \gamma$.

Now, suppose that $f(z) = U(x, y) + iV(x, y)$ is a smooth mapping of the plane. Then $\eta(t) = u(t) + iv(t)$, where $u(t) = U(x(t), y(t))$, $v(t) = V(x(t), y(t))$. Using the chain rule for partial derivatives we can show that these functions of t are continuously differentiable on $[a, b]$, i.e.,

$$\dot{u}(t) = \frac{\partial U}{\partial x}(\gamma(t))\,\dot{x}(t) + \frac{\partial U}{\partial y}(\gamma(t))\,\dot{y}(t),$$
$$\dot{v}(t) = \frac{\partial V}{\partial x}(\gamma(t))\,\dot{x}(t) + \frac{\partial V}{\partial y}(\gamma(t))\,\dot{y}(t),$$

(B.1.1)

which yield $\dot{\eta}(t) = \dot{u}(t) + i\dot{v}(t)$. If f is a smooth analytic function, Cauchy-Riemann equations reduce (B.1.1) to a simpler form

$$\dot{\eta}(t) = \frac{df}{dz}(\gamma(t)) \cdot \dot{\gamma}(t).$$

(B.1.2)

In fact, we have

$$\dot{\eta}(t) = \dot{u}(t) + i\dot{v}(t) = \left(\frac{\partial U}{\partial x} + i\frac{\partial V}{\partial x}\right)\dot{x}(t) + \left(\frac{\partial V}{\partial y} + i\frac{\partial V}{\partial y}\right)\dot{y}(t)$$
$$= \left(\frac{\partial U}{\partial x} + i\frac{\partial V}{\partial x}\right)\dot{x}(t) + i\left(\frac{\partial U}{\partial x} + i\frac{\partial V}{\partial x}\right)\dot{y}(t)$$
$$= \frac{df}{dz}(\dot{x}(t) + i\dot{y}(t)) = \frac{df}{dz}\dot{\gamma}(t). \ \blacksquare$$

B.2 Curvature

According to Coolidge [1952], Cauchy defined the *center of curvature* C as the intersection point of two infinitely close normals to the curve, the *radius of curvature* as the distance from the point to the center C, and the *curvature* itself as inverse of the radius of curvature. The curvature of a plane Γ at a point on it measures the sensitivity of its tangent line as the point is moving towards adjacent points. Among different ways to make this idea precise, the easiest one is to geometrically define the curvature of a straight line to be identically zero, and the curvature of a circle as the reciprocal of its radius. Given any plane curve $\gamma(t)$ at a point P, there is a unique circle or line which most closely approximates the curve near P; this circle is called the *osculating circle* at P (see Figure B.2.1). The curvature of $\gamma(t)$ at P is then defined to be the curvature of that circle or line. The radius of curvature is defined as the reciprocal of the curvature.

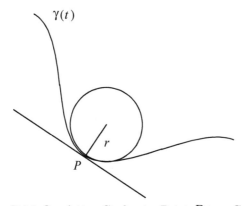

Figure B.2.1 Osculating Circle at a Point P to a Curve Γ.

The physical interpretation is as follows: Suppose that a particle moves along the curve with unit speed. Taking the parameter for $\gamma(t)$ as time t provides a natural parametrization for the curve. The unit tangent vector $\mathbf{T}$ at P is also the velocity vector, since the moving particle also depends on time. Then the curvature κ is the magnitude of the rate of change of $\mathbf{T}$, i.e., $\kappa = \left\| \dfrac{d\mathbf{T}}{dt} \right\|$. This is the magnitude of the acceleration of the particle and the vector $d\mathbf{T}/dt$ is the acceleration vector. The curvature κ measures how fast the unit tangent vector to the curve rotates. If the curve keeps close to the same direction, the unit tangent vector changes very little and the curvature is small, while if the curve undergoes a tight turn, the curvature is large.

The precise definition is as follows: Suppose the $\gamma(t)$ is a $C^2(a, b)$ plane curve, i.e., there exists a parametric representation of $\gamma(t)$ by a pair of functions $\gamma(t) = (x(t), y(t))$ such that the first and second derivatives of both x and y exist and are continuous, and $\|\dot{\gamma}\|^2 = \dot{x}(t)^2 + \dot{y}(t)^2 \neq 0$ throughout the domain D. For such plane curves there exists a reparametrization with respect to the arclength s such that $\|\dot{\gamma}\|^2 = \dot{x}(s)^2 + \dot{y}(s)^2 = 1$ (Kline [1998]). The velocity vector $\mathbf{T}(s)$ is the unit length tangent vector, and the unit normal vector $\mathbf{N}(s)$, the curvature $\kappa(s)$, the oriented or signed curvature $\mathfrak{k}(s)$, and the radius of curvature $R(s)$ are, respectively, given by

$$\mathbf{T}(s) = \dot{\gamma}(s), \ \dot{\mathbf{T}}(s) = k(s)\mathbf{N}(s), \ \kappa(s) = \|\dot{\mathbf{T}}(s)\| = \|\ddot{\gamma}(s)\| = |\mathfrak{k}(s)|, \ R(s) = \frac{1}{\kappa(s)}.$$

For a parametrized plane curve $\gamma(t) \in C^2(a, b)$, the curvature κ is given by

$$\kappa = \frac{|\dot{x}\ddot{y} - \dot{y}\ddot{x}|}{(\dot{x}^2 + \dot{y}^2)^{3/2}}, \tag{B.2.1}$$

which can be expressed in a coordinate-independent form as

$$\kappa = \frac{|\det(\dot{\gamma}, \ddot{\gamma})|}{\|\dot{\gamma}\|^3}. \tag{B.2.2}$$

For a curve defined by $y = f(x)$, the curvature is given by

$$\kappa = \frac{|y''|}{(1 + y'^2)^{3/2}}. \tag{B.2.3}$$

This formula is common in physics and engineering and appears in problems dealing with bending of beams, one-dimensional vibrations of a tout string, approximations of fluid flows around surfaces (in aerodynamics), and free surface boundary conditions in ocean waves. In these applications it is assumed that the slope is small compared with unity, so the approximation $\kappa \approx \left|\dfrac{d^2 y}{dx^2}\right|$ can be used, mainly to obtain linear equation to govern the physical problem. If the curve is defined in terms of polar coordinates (r, θ), then the curvature is

$$\kappa = \frac{|r^2 + 2r'^2 - rr''|}{(r^2 + r'^2)^{3/2}}, \tag{B.2.4}$$

where the prime denotes differentiation with respect to θ. Other examples are:

(i) For the parabola $y = x^2$, with parametric equation $x(t) = t, y(t) = t^2$, the curvature is

$$\kappa = \left|\frac{\dot{x}\ddot{y} - \dot{y}\ddot{x}}{(\dot{x}^2 + \dot{y}^2)^{3/2}}\right| = \frac{2}{(1 + 4t^2)^{3/2}}.$$

(ii) For a curve reparametrized as $\gamma(t) = \begin{pmatrix} \cos(3t) \\ \sin(2t) \end{pmatrix}$, the curvature is

$$\kappa(t) = \frac{\left|6\cos t(8\cos^4 t - 10\cos^2 t + 5)\right|}{(232\cos^4 t - 97\cos^2 t + 13 - 144\cos^6 t)^{3/2}}.$$

C

Green's Theorems

C.1 Green's Identities

Let Ω be a finite domain in $\mathbb{R}^n$ bounded by a piecewise smooth, orientable surface $\partial\Omega$, and let w and F be scalar functions and $\mathbf{G}$ a vector function in the class $C^0(\Omega)$. Then

$$\text{Gradient theorem:} \quad \int_\Omega \nabla F \, d\Omega = \oint_{\partial\Omega} \mathbf{n} F \, dS,$$

$$\text{Divergence theorem:} \quad \int_\Omega \nabla \cdot \mathbf{G} \, d\Omega = \oint_{\partial\Omega} \mathbf{n} \cdot \mathbf{G} \, dS,$$

$$\text{Stokes's theorem:} \quad \int_\Omega \nabla \times \mathbf{G} \, d\Omega = \oint_{\partial\Omega} \mathbf{G} \cdot \mathbf{t} \, dS,$$

where $\mathbf{n}$ is the outward normal to the surface $\partial\Omega$, $\mathbf{t}$ is the tangent vector at a point on $\partial\Omega$, $\oint$ denotes the surface or line integral, and dS (or ds) denotes the surface (or line) element depending on the dimension of Ω. The divergence theorem in the above form is also known as the Gauss theorem. Stokes's theorem in $\mathbb{R}^2$ is a generalization of *Green's theorem* which states that if $\mathbf{G} = (G_1, G_2)$ is a continuously differentiable vector field defined on a region containing $\Omega \cup \partial\Omega \subset \mathbb{R}^2$ such that $\partial\Omega$ is a Jordan contour, then

$$\int_\Omega \left(\frac{\partial G_2}{\partial x_1} - \frac{\partial G_1}{\partial x_2} \right) dx_1 \, dx_2 = \oint_{\partial\Omega} G_1 \, dx_1 + G_2 \, dx_2. \tag{C.1.1}$$

Let the functions $M(x,y)$ and $N(x,y)$, where $(x,y) \in \Omega$, be the components of the vector $\mathbf{G}$. Then, by the divergence theorem

$$\int_\Omega \left(\frac{\partial M}{\partial x} + \frac{\partial N}{\partial y} \right) dx \, dy = \oint_\Gamma [M \cos(\mathbf{n}, x) + N \cos(\mathbf{n}, y)] \, ds,$$

$$= \oint_\Gamma M \, dx + N \, dy, \tag{C.1.2}$$

with the direction cosines $\cos(\mathbf{n}, x)$ and $\cos(\mathbf{n}, y)$, where $\Gamma = \partial\Omega$. If we take $M = f \, g_x$ and $N = f \, g_y$, then (C.1.2) yields

$$\int_\Omega \left(\frac{\partial f}{\partial x} \frac{\partial g}{\partial x} + \frac{\partial f}{\partial y} \frac{\partial g}{\partial y} \right) dx \, dy = \int_\Gamma f \frac{\partial g}{\partial n} \, ds - \int_\Omega f \nabla^2 g \, dx \, dy, \tag{C.1.3}$$

which is known as *Green's first identity*. Moreover, if we interchange f and g in (C.1.2), we get

$$\int_\Omega \left(\frac{\partial f}{\partial x} \frac{\partial g}{\partial x} + \frac{\partial f}{\partial y} \frac{\partial g}{\partial y} \right) dx\, dy = \int_\Gamma g \frac{\partial f}{\partial n}\, ds - \int_\Omega g \nabla^2 f\, dx\, dy. \qquad \text{(C.1.4)}$$

If we subtract (C.1.3) from (C.1.4), we obtain *Green's second identity*:

$$\int_\Omega \left(f \nabla^2 g - g \nabla^2 f \right) dx\, dy = \int_\Gamma \left(f \frac{\partial g}{\partial n} - g \frac{\partial f}{\partial n} \right) ds, \qquad \text{(C.1.5)}$$

which is also known as *Green's reciprocity theorem*. Note that Green's identities are valid even if the domain Ω is bounded by finitely many closed curves. In that case, however, the line integrals must be evaluated over all paths that make the boundary of Ω. If f and g are real and harmonic in $\Omega \subset \mathbb{R}$, then from (C.1.5)

$$\int_\Gamma \left(f \frac{\partial g}{\partial n} - g \frac{\partial f}{\partial n} \right) ds = 0. \qquad \text{(C.1.6)}$$

Let D be a simply connected region in the complex plane $\mathbb{C}$ with boundary Γ. Let z_0 be any point inside D, and let Ω be the region obtained by indenting a disk $B(z_0, \varepsilon)$ from D, where $\varepsilon > 0$ is small (Fig. C.1.1 (a)). Then ∂D consists of the contour Γ together with the contour $\partial B(z_0, \varepsilon) = \Gamma_\varepsilon$.

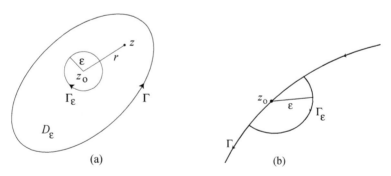

(a)   (b)

Fig. C.1.1 Contour $\partial B(z_0, \varepsilon) = \Gamma_\varepsilon$.

If we set $f = u$ and $g = \log r$ in (C.1.6), where $z \in D$ and $r = |z - z_0|$, then, since $\dfrac{\partial}{\partial n} = -\dfrac{\partial}{\partial r}$ on Γ_ε, we get

$$\int_\Gamma \left(u \frac{\partial}{\partial n} (\log r) - (\log r) \frac{\partial u}{\partial n} \right) ds - \int_{\Gamma_\varepsilon} \left(\frac{u}{r} - (\log r) \frac{\partial u}{\partial r} \right) ds = 0. \quad \text{(C.1.7)}$$

Now, let $\varepsilon \to 0$ in (C.1.7). Then, since

$$\lim_{\varepsilon \to 0} \int_{\Gamma_\varepsilon} \frac{u}{r}\, ds = \lim_{\varepsilon \to 0} \int_0^{2\pi} u(z_0 + \varepsilon\theta)\, \frac{1}{\varepsilon}\, \varepsilon\, d\theta = 0,$$

$$\lim_{\varepsilon \to 0} \int_{\Gamma_\varepsilon} \log r\, \frac{\partial u}{\partial r}\, ds = \lim_{\varepsilon \to 0} \int_0^{2\pi} \log \varepsilon\, \frac{\partial u}{\partial \varepsilon}\, \varepsilon\, d\theta = 0,$$

we obtain

$$2\pi\, u(z_0) = \int_\Gamma \left[u \frac{\partial}{\partial n}(\log r) - (\log r) \frac{\partial u}{\partial n} \right] ds, \qquad (C.1.8)$$

which is known as *Green's third identity*. Note that Eq (C.1.8) gives the value of a harmonic function u at an interior point in terms of the boundary values of u and $\dfrac{\partial u}{\partial n}$. If the contour Γ has no corners and if the point z_0 is on the boundary Γ, then instead of the whole disk $B(z_0, \varepsilon)$ we consider a half disk at the point z_0 deleted from D (Fig. C.1.1(b)), and Green's third identity becomes

$$\pi\, u(z_0) = \text{p.v.} \int_\Gamma \left[u \frac{\partial}{\partial n}(\log r) - (\log r) \frac{\partial u}{\partial n} \right] ds, \qquad (C.1.9)$$

where p.v. denotes the principal value of the integral, i.e., it is the limit , as $r \to 0$, of the integral over the contour Γ obtained by deleting that part of Γ which lies within the circle of radius ε and center z_0. More details can be found in Kythe [1996].

C.2 Exercises

C.1. Derive the Green's formulas:

$$\int_{\partial D} \phi(z)\, dz = 2i \iint_D \bar{\partial}\phi\, dx\, dy, \qquad (C.2.1)$$

$$\int_{\partial D} \phi(z)\, d\bar{z} = -2i \iint_D \partial\phi\, dx\, dy. \qquad (C.2.2)$$

SOLUTION. The gradient theorem implies Green's formula in the complex form

$$\int_{\partial D} \phi(z)\, dz = \int_{\partial D} \phi(z)(dx + i\, dy) = \int_{\partial D} \phi(z)\, dx + i\phi(z)\, dy$$

$$= \iint_D (i\phi_x - \phi_y)\, dx\, dy = 2i \iint_D \bar{\partial}\phi\, dx\, dy,$$

which is (C.2.1). The other formula can be derived similarly.

C.2. Use formula (C.2.1) to derive Cauchy's integral formula. SOLUTION. Using formula (C.2.1) with $\phi(z) = \dfrac{f(\zeta)}{\zeta - z}$ over the domain $D \backslash D_\varepsilon(z)$, we get

$$\int_{D\backslash D_\varepsilon(z)} \frac{f(\zeta)}{\zeta - z}\, d\zeta = 2i \iint_{D\backslash D_\varepsilon(z)} \frac{\bar{\partial} f}{\zeta - z}\, dx\, dy.$$

Let $\varepsilon \to 0$. Then we have *Cauchy integral formula*

$$f(z) = \frac{1}{2i\pi} \int_{\partial D} \frac{f(\zeta)}{\zeta - z}\, d\zeta - \frac{1}{\pi} \int_{\partial D} \frac{\bar{\partial} f(\zeta)}{\zeta - z}\, dx\, dy, \qquad \text{(C.2.3)}$$

where the integral in the formula is the principal value (p.v.) integral, i.e.,

$$\int_{\partial D} \frac{\bar{\partial} f(\zeta)}{\zeta - z}\, dx\, dy = \lim_{\varepsilon \to 0^+} \iint_{D\backslash D_\varepsilon(z)} \frac{\bar{\partial} f}{\zeta - z}\, dx\, dy. \qquad \text{(C.2.4)}$$

D

Two-Dimensional Potential Flows

Let $\mathbf{F} = \mathbf{F}(\mathbf{x}, t)$ denote the velocity field in $\mathbb{R}^n$. The time-dependent (i.e., unsteady, or nonstationary) flow of a perfect fluid is irrotational if the curl $\nabla \times \mathbf{F} = \mathbf{0}$. In the two-dimensional case, there is a potential function $\phi(z)$ such that $\mathbf{F} = -\nabla\phi$. In any simply connected domain there is a conjugate harmonic function $\bar{\phi}(z)$ and an associated *complex potential* for the flow, giving the analytic function $f(z) = \phi(z) + i\bar{\phi}(z)$. The potential ϕ fully determines the velocity field, although it does not have any physical significance. Thus, the velocity vector at a point a, given by

$$\mathbf{F}(a) = -\nabla\phi(a) = \left(-\frac{\partial\phi}{\partial x}\right)\mathbf{i} + \left(-\frac{\partial\phi}{\partial y}\right)\mathbf{j} = -\left[\frac{\partial\phi}{\partial x}\mathbf{i} + \frac{\partial\phi}{\partial y}\mathbf{j}\right], \quad \text{(D.1)}$$

can be identified with the complex number

$$-\left[\frac{\partial\phi}{\partial x} + i\frac{\partial\phi}{\partial y}\right] = -\left[\frac{\partial\phi}{\partial x} - i\frac{\partial\bar{\phi}}{\partial y}\right] = \left(-\frac{\overline{df}}{dz}\right), \quad \text{(D.2)}$$

which is the complex conjugate of $-df/dz$. Let a vector $\mathbf{x} = x\mathbf{i} + y\mathbf{j}$ be identified with the corresponding complex number $z = x + iy$. Then the length (or magnitude) of the vector $\|\mathbf{x}\| = \sqrt{x^2 + y^2}$ is the same as the absolute value of the complex number $\mathbf{x}$. Thus, the speed $\|\mathbf{f}(a)\|$ of the fluid passing through the position a is given in terms of the complex potential by

$$\|\mathbf{F}(a)\| = \left|\frac{df}{dz}(a)\right|. \quad \text{(D.3)}$$

There are certain properties of the flow which may be determined from the complex potential $f(z)$ and its derivative df/dz. For example, the velocity vector at a point a, where $\mathbf{F} \neq \mathbf{0}$, is always tangent to the curve $\bar{\phi}(z) = \text{const} = \bar{\phi}(a)$ which passes through the point a, and is perpendicular to the curve $\phi(z) = \text{const} = \phi(a)$, since the gradient $\mathbf{F} = -\nabla\phi$ is always perpendicular to the curve $\phi = \text{const}$, whenever the gradient vector is nonzero. Thus, $\mathbf{F}$ is perpendicular to the curve $\phi = \text{const}$, and therefore, is parallel (tangent) to the curve $\bar{\phi} = \text{const}$.

Example D.1. Consider the flow associated with the complex potential $f(z) = z$, i.e., $\phi(x + iy) = x$ and $\bar{\phi}(x + iy) = y$. The fluid velocity is given by $-\overline{df/dz} = -1 + i0$, which is interpreted as a vector field. Thus, there is a constant velocity at each point, since the velocity vector is horizontally directed. Similarly, any linear function $f(z) = az + b$, $a \neq 0$, is the complex potential of a constant velocity flow whose velocity vector corresponds to the complex number $-\overline{df/dz} = -\bar{a}$. Note that the choice of b does not affect the velocity field. ∎

Example D.2. Consider a flow with the potential $\phi(z) = \log|z| = \frac{1}{2}\log(x^2+y^2)$ directed on a domain $D = \{z : z \neq 0\}$, i.e., the origin is not in D and will turn out to be a fluid sink ($\nabla \cdot \mathbf{F} < 0$). Since D is not simply connected, the conjugate harmonic function $\bar{\phi}$ is given the multiple-valued function: $\bar{\phi}(z) = \arg\{z\}$. The complex potential $f = \phi + i\bar{\phi} = \log z$ is also multiple valued. However, the single valued $\log z$ can be defined on simply connected subdomains of D. The derivative of the complex potential is single valued, although the potential itself is multiple valued: $df/dz = 1/z$, no matter with which locally defined $f = \log z$ we start with. This derivative determines the velocity field:

$$\mathbf{F} = -\overline{\frac{df}{dz}} = \left(-\overline{\frac{1}{z}}\right) = -\frac{1}{\bar{z}} \quad \text{for } z \neq 0..$$

At every point of the circle $|z| = r$, the magnitude of the velocity is constant: $|df/dz| = |1/z| = 1/r$, and the velocity vector is directed toward the origin. There is an inward flow of fluid from infinity to a sink located at the origin, associated with the potential $\phi(z)$. ∎

E

Subordination Principle

Let f and g be analytic functions defined on E. We say that the function f is subordinate to the function g, and write $f \prec g$, if there exists a function ω which is analytic in E with $\omega(0) = 0$, $|\omega(z)| < 1$, $z \in E$, such that $f = (g \circ \omega)z$, or

$$f(z) = g(\omega(z)) \quad \text{for all } z \in E. \tag{E.1}$$

Theorem E.1. (Subordination principle) *Let* $f, g : E \mapsto \mathbb{C}$ *be analytic functions and let* g *be injective. Then*

(a) $f \prec g$ *is equivalent to* $f(0) = g(0)$ *and* $f(E) \subset g(E)$*; and hence,*

(b) $f(E_r) \subset g(E_r)$ *for all* $0 < r < 1$*, where* $E_r = C(0, r)$*.*

Moreover, $f \prec g$ *implies the following properties for the coefficients of the power series expansions of* $f(z) = a_0 + a_1 z + a_2 z^2 + \cdots$ *and* $g(z) = b_0 + b_1 z + b_2 z^2 cdots$:

(i) $|a_1| \leq |b_1|$ *with equality only for* $f(z) = g\left(e^{i\theta} z\right)$ *for some* $\theta \in \mathbb{R}$;

(ii) $|a_2| \leq max\{|b_1|, |b_2|\}$; *and*

(iii) $\sum_{k=0}^{n} |a_k|^2 \leq \sum_{k=0}^{n} |b_k|^2$ *for all* $n \in \mathbb{N}$ *(Rogosinski [1939]).*

PROOF. Clearly, (a) follows from $f \prec g$; thus, $f(z) = g(\omega(z))$ as in (E.1), and in particular, $f(0) = g(\omega(0)) = g(0)$ and $f(E) = g(\omega(E)) \subset g(E)$. But if $f(0) = g(0)$ and $f(E) \subset g(E)$, using the injective property of g the function $\omega = g^{-1} circ f$ is analytic, and hence, $\omega(0) = g^{-1} \circ f(0) = g^{-1} \circ g(0) = 0$, and $\omega(E) = g^{-1} \circ f(E) \subset E$. (b) follows directly from Schwarz lemma, so that $|\omega(z)| \leq |z|$ implies $\omega(E_r) \subset E_r$. Next, let $\omega(z) = c_1 z + c_2 z^2 + \cdots$. Then $|a_1| = |b_1 \cdot c_1| = |b_1| \cdot |c_1| \leq |b_1|$, where the equality follows from the Schwarz lemma. To prove (ii), consider the bounded function $\phi(z) = \dfrac{\omega(z)}{z} = c_1 + c_2 z + \cdots$. The $\left(1 - |z|^2\right) |\phi'(z)| \leq 1 - |\phi(z)|^2$, and in particular, at $z = 0$ we have $|\phi'(0)| \leq 1$ and $c_2| \leq 1 - |c_1|^2$. Then

$$|a_2| = |b_2 c_1^2 + b_1 c_2| \leq |c_1|^2 |b_2| + \left(1 - |c_1|^2\right) |b_1| \leq max\{|b_1|, |b_2|\}. \tag{E.2}$$

Lastly, to prove (iii) we need the following two lemmas:

Lemma E.1. (Parseval's identity) *For an analytic function* $f : E \mapsto \mathbb{C}$ *with the power series expansion* $f(z) = \sum\limits_{n=0}^{\infty} a_n z^n$, *the following identity holds for* $0 < r < 1$:

$$\frac{1}{2\pi} \int_0^{2\pi} |f\left(re^{i\theta}\right)|^2 \, d\theta = \sum_{n=0}^{\infty} |a_n|^2 r^{2n}. \tag{E.3}$$

PROOF. We have

$$\frac{1}{2\pi} \int_0^{2\pi} |f\left(re^{i\theta}\right)|^2 \, d\theta = \frac{1}{2\pi} \int_0^{2\pi} f\left(re^{i\theta}\right) \overline{f\left(re^{i\theta}\right)} \, d\theta$$

$$= \frac{1}{2\pi} \int_0^{2\pi} \sum_{n=0}^{\infty} a_n r^n e^{in\theta} \sum_{k=0}^{\infty} \overline{a_k} r^k e^{-ik\theta} \, d\theta$$

$$= \frac{1}{2\pi} \sum_{n=0}^{\infty} \sum_{k=0}^{\infty} a_n \overline{a_k} r^{n+k} 2\pi \delta_{n,k} = \sum_{n=0}^{\infty} |a_n|^2 r^{2n}, \tag{E.4}$$

where $\delta_{n,k}$ is the Kronecker delta. ∎

Lemma E.2. (Littlewood [1925]) *Let* $0 < \lambda < \infty$. *If* $f \prec g$, *then for* $0 < r < 1$

$$\frac{1}{2\pi} \int_0^{\infty} |f\left(re^{i\theta}\right)|^{\lambda} \, d\theta \le \frac{1}{2\pi} \int_0^{\infty} |g\left(re^{i\theta}\right)|^{\lambda} \, d\theta. \tag{E.5}$$

This lemma is a result in harmonic functions.

PROOF OF (iii). Let $p_n(z) = \sum\limits_{k=0}^{n} a_k z^k$ and $q_n(z) = \sum\limits_{k=0}^{n} b_k z^k$. If $f(z) = g(\omega(z))$, we can write $p_n(z) + \sum\limits_{k=n+1}^{\infty} a_k z^k - \sum\limits_{k=n+1}^{\infty} b_k \omega(z)^k = q_n\left(\omega(z)\right)$, which, by combining the two power series from $n+1$ to ∞ into one, can be written as

$$p_n(z) + \sum_{k=n+1}^{\infty} d_k z^n = q_n\left(\omega(z)\right). \tag{E.6}$$

The result follows using (E.3). ∎

Theorem E.3. (Littlewood [1925]) *Let* $f, g : E \mapsto \mathbb{C}$ *be analytic*, $f \prec g$ *and* $n = 2, 3, \ldots$, *and let* $\Delta_n := a_{n+1} - a_n$. *Then for the power series of* $f(z) = \sum\limits_{n=0}^{\infty} a_n z^n$ *and of* $g(z) = \sum\limits_{n=0}^{\infty} b_n z^n$, *we have (a) If the coefficients* $b_1, b_2, \ldots$ *are nonnegative, monotone decreasing and convex, i.e.,* $b_n \ge 0$, $b_n - b_{n-1} \le 0$ *and* $\Delta_k^2 b_k = b_{k+2} - 2b_{k+1} + b_k \ge 0$ *for* $k = 1, \ldots, n-2$, *then* $|a_n| \le b_1$; *and (b) If the coefficients* $b_1, b_2, \ldots$ *are nonnegative, monotone increasing and convex, i.e.,* $b_1 \ge 0$, $b_2 - b_1 \ge 0$ *and* $\Delta_k^2 b_k \ge 0$ *for* $k = 3, 4, \ldots, n$, *then* $|a_n| \le b_n$.

Bibliography

(Note: First author or single author is cited with last name first.)

Abramowitz, M. and I. A. Stegun (eds.). 1965. *Handbook of Mathematical Functions.* New York: Dover.

Acu, Mugur A. 2008. *Subclasses of α-Convex Functions.* "Lucian Blaga" University Publishing House.

Aharonov, D. 1970. On Bieberbach-Eilenberg functions. *Bull. Amer. Math. Soc.* 76: 101–104.

Ahlfors, Lars V. 1952. Remarks on the Neumann-Poincaré equation. *Pacific J. Math.* 2: 271–280.

———. 1953. *Complex Analysis.* New York: McGraw-Hill.

———. 1955. Conformality with respect to Riemannian metrics. *Ann. Acad. Sci. Fenn.* Ser. A I. no. 206.

———. 1966. *Lectures on Quasiconformal Mappings.* Van Nostrand.

———. 2010. *Conformal Invariants. Topics in geometric function theory.* Reprint of the 1973 original. With a forward by P. Duren, F. W. Gehring and Brad Osgood. AMS Chelsea Publishing. ISBN 978-0-8218-5270-5.

Airault, Helene. 2008. Remarks on Faber polynomials. *International Mathematics Forum.* 3: 449–456.

Alexander, J. W. 1915-1916. Functions which map the interior of the unit circle upon simple regions. *Ann. of Math. (2).* 17: 12–22.

Arosio, L. 2013. Resonances in Loewner equations. *Adv. Math.* 227:1413–1435.

———. 2013. Loewner equations on complete hyperbolic domains. *J. Math. Anal. Appl.* 398: 609–621.

Askey, R. 1975. *Orthogonal Polynomials and Special Functions.* Philadelphia, PA: SIAM.

——— and G. Gasper. 1976. Positive Jacobi polynomial sums, II. *American Journal of Mathematics,* 98, No. 3:709–737.

——— and G. Gasper. 1985. Inequalities for Polynomials, in *The Bieberbach Conjecture.* West Lafayette, IN; Providence, RI: Amer. Math. Soc., pp. 7–32.

Astala, K., T. Iwaniec and G. Martin. 2009. *Elliptic Partial Differential Equations and Quasiconformal Mappings in the Plane.* Princeton mathematical series 48. Princeton University Press. ISBN 0-691-13777-3.

Baerstein, A. 1982. *Bieberbach's Conjecture for Tourists.* Lecture Notes in Math., vol. 908. Berlin: Springer-Verlag, pp. 48–72.

Bailey, W. 1935. *Generalized Hypergeometric Series.* London: Cambridge University Press.

———, D. Drasin, P. Duren and A. Marden (eds.). 1986. *The Bieberbach Conjecture*, Mathematical Surveys and Monographs, 21. Providence, RI: American Mathematical Society. MR875226, ISBN 978-0-8218-1512-2.

Baranov, A. and H. Hedenmalm. 2008. Boundary properties of Green functions in the plane. *Duke Math. J.* 145: 1–24.

Bazilevich[1], I. E. 1948. Improvement of the estimates of coefficients of univalent functions. *Matem. Sbornik.* 22(64): 381–390 (Russian).

———, 1951. On distortion theorems in the theory of univalent functions. *Matem. Sbornik.* 28(70): 283–292 (Russian).

———. 1955. On a case of integrability by quadratures of the equation of Löwner-Kufarev. *Matem. Sbornik.* 37(79): 471–476 (Russian).

———. 1965. Coefficient dispersion of univalent functions. *Matem. Sbornik.* 68(110): 549–560 (Russian).

———. 1967. On a univalence criterion for regular function and the dispersion of their coefficients. *Matem. Sbornik.* 74(116): 133–146 (Russian).

Beardon, Alan. 1979. *Complex Analysis: the Winding Number Principle in Analysis and Topology.* New York : John Wiley and Sons. ISBN 0-471-99672-6.

Bell, S. R. 1992. *The Cauchy Transform, Potential Theory, and Conformal Mapping.* Studies in Advanced Mathematics. CRC Press. ISBN 0-8493-8270-X.

Bergman, S. and M. Schiffer. 1951. Kernel functions and conformal mapping. *Compositio Mathematica.* 8: 205–249.

Bernardi, S. D. 1952. A survey of the development of the theory of schlicht functions. *Duke Math. J.* 19: 263–287.

———. 1983. *Bibliography of Schlicht Functions.* Tampa, FL: Mariner.

Bieberbach, L. 1914. Zur Theorie und Praxis der konformalen Abbildung. *Rendiconti del Circolo Mathematico di Palermo.* 38: 98–112.

———. 1916. Über die Koeffizienten derjenigen Potenzreihen, welche eine schlichte Abbildung des Einheitskreises vermitteln. *Sitzungber. Preuss. Akad. Wisse. Phys-Math. Kl.* 138: 940–955.

———. 1924. Über die konforme Kreisabbildung nahezu kreisförmig Bereiche. *Sitzungsbereichte der Preuss. Akad. Wissen. Phys-Math. Kl.* 146: 181–188.

Bombieri, E. 1963. Sul problema di Bieberbach per le funkzii univalenti. *Lincei-Red. Sc. fis. mat. e nat.* 35: 469–471.

Bourbaki, N. 1968. *Elements of Mathematics, Theory of Sets.* Reading, MA: Addison-Wesley Publishing Co.

Boyarskiĭ, B. V. 1957. Generalized solutions of systems of differential equa-

---

[1] Also spelled as Bazilevič,

tions of first order and elliptic type with discontinuous coefficients. *Mat. Sb.* 43: 451–503.

Bracci, F., M. D. Contreras and S. Diaz-Madrigal. 2008. Evolution families and the Loewner equation I. Preprint. Available at ArXiv 0807.1594.

————. 2009. Evolution families and the Loewner equation II. Complex hyperbolic manifolds. *Math. Ann.* 344 : 947–962.

Brickman, L. 1970. Extreme points of the set of univalent functions. *Bull. Amer. Math. Soc.* 76: 372–374.

————, T. H. MacGregor and D. R. Wilken. 1971. Convex hulls of some classical families of univalent functions. *Trans. Amer. Math. Soc.* 156: 91–107.

Brill, A. and M. Noether. 1894. Bericht über die Entwicklung der Theorie der analytischen Funktionen in älterer and neuerer Zeit. *Jahresbericht der Deutschen Mathematiker-Vereningung.* 3: 107–566.

Brychkov, Y. A. 2008. *Handbook of Special Functions: Derivatives, Integrals, Series and Other Formulas.* Boca Raton: CRC Press.

Carathéodory, C. 1907. Über den Variabilitätsbereich der Koeffizienten von Potenzreihen, die gegebene Werte nicht annnehmen. *Math. Ann.* 64: 95–115.

————. 1911. Über en Variabilitätsbereich der Fourier'schen Konstanten von positiven harmonischen Funktionen. *Rend. Circ. Mat. Palermo* 32: 193–217.

————. 1912. Untersuchungen über die konformen Abbildungen von festen und veränderlichen Gebieten. *Mat. Ann.* 72: 107–144.

————. 1913a. Über die gegenseitige Beziehung der Ränder bei der konformen Abbildung des Inneren einer Jordanschen Kurve auf einen Kreis. *Mathematische Annalen.* 73: 305–320.

————. 1913b. Über die Begrenzung einfach zusammenhängender Gebiete. *Mathematische Annalen.* 73: 323–370.

————. 1914. Elementarer Beweis für den Fundamentalsatz der konformen Abbildung, in *Schwarz-Festschrift.* Berlin: J. Springer. pages 19–41.

Casey, James. 1996. *Exploring Curvature.* Vieweg+Teubner Verlag.

Cernikov, V. V. 1972. The α-convexity of univalent functions. *Mat. Zametki.* 11 (2) : 227–232.

Charzyński, Z. and M. Schiffer. 1960. A new proof of the Bieberbach conjecture for the fourth coefficient. *Arch. Rational Mech. Anal.* 5: 187–193.

Chen, K.K. 1933. On the theory of schlicht functions. *Proc. Imp. Acad. Japan.* 9: 465–467.

Chirilă, Teodora. 2013. An extension operator associated with certain G-Loewner chains. *Taiwanese J. Math.* 17: 1819–1837. doi: 10.11650/tjm.17. 2013.2966. http://journal.taiwanmathsoc.org.tw

Cima, J. A. and A. Matheson. 1998. Cauchy transforms and other composition operators. *Illinois J. Math.* 42 : 58–69.

————and W. A. Siskakis. 1999. Cauchy transforms and Cesàro averaging

326

operators. *Acta Sci, Math. (Szeged)*. 65 : 505–513.

Clunie, J. 1959. On meromorphic schlicht functions. *J. London Math. Soc.* 34: 215–216.

Coddington, E. A. and N. Levinson. 1955. *Theory of Ordinary Differential Equations*. New York: McGraw-Hill.

Contreras, M. D., S. Diaz-Madrigal, Ch. Pommerenke. 2006. On boundary critical points for semigroups of analytic functions. *Math. Scand.* 98: 125-142.

––––––, S. Diaz-Madrigal, P. Gumenyuk. 2010. Geometry behind Loewner chains. *Complex Anal. Oper. Theory*. 4: 541-587.

Coolidge, J. L. 1952. The unsatisfactory story of curvature. *Amer. Math. Monthly*. 59 : 375–379.

Convey, J. B. 1955. *Functions of One Complex Variable II*. Berlin, New York: Springer-Verlag. ISBN 978-0-387-94460-9.

Davis, P. 1974. *The Schwarz Function and its Applications*. Carus Math. Monographs 17. Mathematical Association of America.

de Branges, L. 1984. A proof of the Bieberbach conjecture. *V. A. Steklov Math. Inst., LOMI, preprint E 5-84, Leningrad*: 1–21.

––––––. 1985. A proof of the Bieberbach conjecture. *Acta Math.* 154: 137–152. MR772434.

––––––. 1987. Underlying concepts in the proof of the Bieberbach conjecture. *Proc. International Congress of Mathematicians, Vols. 1, 2, Berkeley, CA, 1986*: 25–42. Providence, RI: American Mathematical Society. MR934213.

Dieudonné, J. 1931. Sur les fonctions univalentes. *C. R. Acad. Sci. Paris*. 192: 1148–1150.

Drasin, D. 1984. Letters. *Science*. 226: 240.

––––––, P. Duren and A. Marden (eds.) 1986. *Proceedings of the Symposium on the occasion of the proof of the Bieberbach conjecture held at Purdue University, West Lafayette, IN, March 11-14, 1985*. Mathematical Surveys and Monographs, 21. Providence, RI: American Mathematical Society. ISBN 0-8218-1521-0, MR875226.

Dunford, Nelson and Jacob T. Schwarz. 1958. *Linear Operators*. Vol. 1. New York: Wiley-Interscience.

Duren, P. L. 1977. Coefficients of univalent functions. *Bull. Amer. Math. Soc.* 83: 891–911.

––––––. 1983. *Univalent Functions*. New York: Springer-Verlag.

––––––, I. Graham, H. Hamada and G. Kohr. 2010. Solutions for generalized Loewner differential equation in several complex variables. *Math. Ann.* 347: 411–435.

Earle, C. and A. Epstein. 2001. Quasiconformal variation of slit domains. *Proc. Amer. Math. Soc.* 129 : 3363–3372.

Ekhad, S. B. and D. Zellberger. 1994. A high-school algebra, "Formal Calculus", proof of the Bieberbach conjecture [after L. Weinstein]. *Jerusalem*

Combinatorics '93: International Conference in Combinatorics, May 9-17, 1993, Jerusalem, Israel. Barcelo et al. (eds.) Providence, RI: American Mathematical Society. Contemp. Mat. 178: 113–115.

Fejér, L. 1933. Gestaltliches über die Partialsummen und ihre Mittelwerte bei der Fourierreihe und der Potenzreihe. *ZAMP.* 13: 80–88; also in *Gesammelte Arbiten*, II, 1970, pp. 479–492. Basel: Birkhäuser.

———. 1936. Trigonometrische Reihen und Potenzreihen mit mehrfach monotoner Koeffizientenfolge. *Trans. Amer. Math. Soc.* 39: 18–59; also in *Gesammelte Arbiten*, II, 1970, pp. 581–620. Basel: Birkhäuser.

Fekete, M. and G. Szegö. 1933. Eine Bemerkung über ungerade schlichte Funktionen. *J. London Math. Soc.* 8: 85–89.

Fernández, J. L. and A. Granados. 1998. On geodesic curvature and conformal mapping. *St. Petersburg Math. J.* 9 : 615–637.

Fitzgerald, C. H. 1972. Quadratic inequalities and coefficient estimates for schlicht functions. *Arch. Rational Mech. Anal.* 46:356–368.

———. 1977. Quadratic inequalities and analytic continuation. *J. Analyse Math.* — 31: 19–47.

———.1982/83. Univalent functions with large late coefficients. *Journal d'Analyse Mathématique.* 42: 167–174.

———. 1985. The Bieberbach Conjecture: Retrospective. *Notices of the American Mathematical Society*: 2–5.

———. 1994. Geometric Function Theory in One and Several Complex Variables: Parallels and Problems. *Complex Analysis and its Applications.* (C. C. Yang, G. C. Wen, K. Y. Li and Y. M. Chiang, Eds.) Pitman Research Notes in Math. Series 305, pp. 14–25. Longman Scientific and Technical.

———and R. A. Horn. 1977. On the structure of Hermitian-symmetric inequalities. *J. London Math. Soc.* (2), 15: 419–430.

———and Ch. Pommerenke. 1985. The de Branges theorem on univalent functions. *Trans. Amer. Math. Soc.* 290: 683–690.

Friedland, S. 1970. On a conjecture of Robertson. *Arch. Rational Mech. Anal.* 37: 255–261.

Fuchs, I. 1971. Potenzreihen mit mehrfach monotonen Koeffizienten. *Arch. Math.* 22: 275–278.

Gal, Sorin G. 2001. Starlike, convex and alpha-convex functions of the hyperbolic complex and of dual complex variable. *Mathematica, Studia Univ. Babeş-Bolyai.* 46 : 23–39.

Garabedian, P. R. and M. Schiffer. 1955. A proof of the Bieberbach conjecture for the fourth coefficient, *J. Rat. Mech. Anal.*. 4: 427–465.

———, G. G. Ross and M. Schiffer. 1965. On the Bieberbach conjecture for even *n*. *J. Math. Mech.* 14: 975–988.

———and M. Schiffer. 1967. The local maximum theorem for the coefficients of univalent functions. *Arch. Rat. Mech. Anal.* 26: 1–31.

Gautschi, W. 1986. Reminiscences of my involvement in de Branges's proof of the Bieberbach conjecture. In *The Bieberbach Conjecture*, Baerstein, Drasin, Duren and Marden (eds.). Math. Surveys Monograph., 21. Providence, RI: American Mathematical Society. pp. 205–211.

Goluzin, E. G. 2001. Bieberbach Conjecture, in M. Hazewinkel's *Encyclopaedia of Mathematics*. New York: Springer-Verlag. ISBN 978-1556080104.

Goluzin, G. M. 1936. On distortion theorems in the theory of conformal mappings. *Matem. Sbornik.* 1(43): 127–135 (Russian).

_____. 1939. Interior problems of the theory of univalent functions. *Uspekhi Mat. Nauk.* 6: 26–89 (Russian).

_____. 1940. On *p*-valent functions. *Mat. Sb.* 8: 277–284.

_____, 1946. On distortion theorems and coefficients of univalent functions. *Matem. Sbornik.* 19(61): 183–202 (Russian).

_____, 1947. A method of variation in conformal mapping, II. *Matem. Sbornik.* 21(63): 83–117 (Russian).

_____, 1948. On the coefficients of univalent functions. *Matem. Sbornik.* 22(64): 373–380(Russian).

_____, 1951. On the theory of univalent functions. *Matem. Sbornik.* 20(71): 197–208 (Russian).

_____. 1969. *Geometric Theory of Functions of a Complex Variable*. English Translation from Russian. Vol. 26. Providence, RI: American Mathematical Society.

Gong, Sheng. 1955. The coefficient problem. *Scientia Sinica.* 4: 359–373.

_____. 1999. *The Bieberbach Conjecture*. AMS/IP Studies in Advanced Mathematics 12. Providence, RI: American Mathematical Society: International Press. ISBN 0-8218-0655-6.

Goodman, A. W. 1972. Coefficients for the area theorem. *Proc. Amer. Math. Soc.* 33: 438–444;

_____. 1983. *Univalent Functions*, Vols. I, II. Tampa, FL: Mariner.

Graham, I., H. Hamada and G. Kohr. 2002. Parametric representation of univalent mappings in several complex variables. *Canadian J. Math.* 54: 324–351.

_____, I., H. Hamada, G. Kohr and T. J. Suffridge. 2002. Extension operators for locally univalent mappings. *Michigan Math. J.* 50: 37–55.

_____ and G. Kohr. 2003. *Geometric Function Theory in One and Higher Dimensions*. New York: Marcel Dekker Inc.

_____, G. Kohr and M. Kohr. 2005. Parametric representation and extension operators for biholomorphic mappings on some Reinhardt domains. *Complex Variables.* 50: 507–519.

_____, I., H. Hamada and G. Kohr. 2012. Extension operators and subordination chains. *J. Math. Anal. Appl.* 386: 278–289.

Gray, J. 1994. On the history of the Riemann mapping theorem. *Rendiconti del Circolo Matematico di Palermo*. Serie II, Supplemento. 34: 47–94.

Gronwall, T. H. 1914. Some remarks on conformal representation. *Ann. of Math.* 16: 72-76.

Grunsky, H. 1932. Neue Abschätzungen zur konformen Abbildung ein- und mehrfach zusammenhängender Berichte. *Schr. Math. Seminars Inst. Angew. Math. Univ. Berlin.* 1: 93–140.

———. 1934. Zwei Bemerkungen zur konformen Abbildung. *Jahrber. Deutsch. Math.-Verein.* 43: 140–143.

———. 1939. Koeffizientenbedingungen für schlichte abbildende meromorphe Funktionen. *Math. Z.* 45: 29–61.

Gruzberg, Ilya A. and Leo P. Kadanoff. 2003. The Loewner equation: maps and shapes. *arXiv:cond.-mat/0309292v1 [cond-mat.stat-mech]* 11 Sep 2003.

Gustafsson, B. and A. Vasil'ev. 2006. *Conformal and Potential Analysis in Hele-Shaw cells* Berlin: Birkhäuser-Verlag.

Hadamard, J. 1908. *Mémoire sur le problème d'analyse relatif à l'équilibre des plaques élastiques encastrées.* Mémoires presentés par divers savants à l'Académie des Sciences. Vol. 33: 1–128.

Hamada, H. 2011. Polynomially bounded solutions to the Loewner differential equation in several complex variables. *J. Math. Anal. Appl.* 381: 179–186.

——— and G. Kohr. 2000. Subordination chains and the growth theorem of spirallike mappings. *Mathematica (Cluj).* 42(65): 153–161

Hamilton, D. H. 1982. On Littlewood's conjecture for univalent functions. *Proc. Amer. Math. Soc.* 86: 32–36.

———. 1986. Extremal Methods, in *The Bieberbach Conjecture.* Proceedings of the Symposium on the Occasion of the Proof. Mathematical Surveys and Monographs # 21. Providence, RI: American Mathematical Society. 85–94.

Hardy, G. H. 1984. 50 Years Ago (Letter to Nature). *Math. Intelligencer.* 6: 7.

Harnack, A. 1887. *Die Grundlagen der Theorie des logarithmischen Potentiales und der eindeutigen Potentialfunktion in der Ebene.* Leipzig: V. G. Teubner.

Hayman, W. K. 1955. The asymptotic behaviour of p-valent functions. *Proc. London Math. Soc..* Third Series. 5: 257–284. MR0071536.

———. 1994. *Multivalent Functions.* Cambridge Tracts in Mathematics 110. 2nd ed. Cambridge: Cambridge University Press.

——— and J. A. Hummel. 1986. Coefficients of powers of univalent functions. *Complex Variables.* 7: 51–70.

Hedenmalm, H. 2008. Planar Beurling transform and Grunsky inequalities. 2008. *Annales Acad. Scient. Fenn. Math.* 33 : 585–596.

———, H., B. Korenblum and K. Zhu. 2000. *Theory of Bergman Spaces.* Grad. Text in Math. 199. New York: Springer-Verlag.

——— and S. Shimorin. 2005. Weighted Bergman spaces and the integral means spectrum of conformal mappings. *Duke Math. J.* 127 : 341–393.

Hilbert, D. 1904. Über das Diricletsche Prinzip, *Matheatische Annalen.* 59:

161–186; *Gesammelte Abhandlungen.* 3: 15–37.

———. 1905. Über das Diricletsche Prinzip, *Journal für Mathematik.* 129: 63–67; *Gesammelte Abhandlungen.* 3: 10–14.

———. 1909. Zur Theorie der konforme Abbildung. *Göttinger Nachrichten.* 314–323; *Gesammelte Abhandlungen.* 3: 73–80.

Hille, E. 1962. *Analytic Function Theory.* Vol. II. New York: Chelsea.

Holland, F. 1973. The extreme points of a class of functions with positive real part. *Math. Ann.* 202: 85–87.;

Horowitz, D. 1976. A refinement for coefficient estimates of univalent functions. *Proc. Am. Math. Soc.* 54: 176-178.

———. 1978. A further refinement for coefficient estimates of univalent functions. *Proc. Amer. Math. Soc.* 7: 217–221.

Hu, Ke. 1986. Coefficients of odd schlicht functions. *Proc. Amer. Mat. Soc.* 96: 183–186.

Hua, Loo-Keng. 1963. *Harmonic Analysis of Functions of Several Complex Variables in the Classical Domains.* Transl. of Math. Monographs, Vol. 6. Providence, RI: Amer. Math. Soc.

Hurwitz, A. 1897. Über die Entwicklung der allgemeinen Theorie der analytischen Funktionen in neuerer Zeit. *Verhandlungen der ersten Internationalen Mathematiker-Kongresses.* Leipzig: Teubner.

Jabotinsky, E. 1949. Sur les fonctions inverses. *C. R. Acad. Sci. Paris.* 229: 508–509.

———. 1953. Representation of functions by matrices. Application to Faber polynomials. *Proc. Amer. Math. Soc.* 4: 546–553.

Jack, I. S. 1971. Functions starlike and convex of order α. *J. London Math. Soc.* (2) 3 : 469474.

Jenkins, J. A. 1964. Some area theorems and a special coefficient theorem. *Ill. J. Math.* 8: 88–99.

———and M. Ozawa. 1967. On local maximality of the coefficient a_6. *Ill. J. Math.* 11: 596–601.

Juia, G. 1931. *Leçons sur la Representation Conforme des Aires Simplement Connexes.* Paris: Gauthier-Villars.

Kager, W., B. Nienhuis and L. Kandoff. 2004. Exact solutions for Loewner evolutions. *J. Statis. Phys.* 115 : 805–822.

Kazarinoff, N. D. 1988. Special functions and the Bieberbach conjecture. *Amer. Math. Monthly.* 95: 689–696.

Keogh, F. R. and E. P. Merkes. 1969. A coefficient inequality for certain classes of analytic functions. *Proc. Amer. Math. Soc.* 20 : 8 12.

Khavinson, D., M. Putinar and H. S. Shapiro. 2007. Poincaré's variational problem in potential theory. *Arch. Ration. Mech. Anal.* 185: 143–184.

Kline, Morris. 1998. *Calculus: An Intuitive and Physical Approach.* New York: Dover.

Kober, H. 1957. *Dictionary of Conformal Representations.* 2nd ed. New York: Dover.

Koebe, P. 1907. Über die Uniformisierung beliebiger analytischer Kurven. *Nachr. Akad. Wiss. Gottingen, Math.-Phys. Kl.* 191–210.

———. 1909. Über die Uniformisierung der algebraischen Kurven durch automorphe Funktionen mit imagiärer Substitutionsgruppe. *Nachr. Kgl. Ges. Wiss. Göttingen, Math.-Phys. Kl.* 68–76.

———. 1912. Über eine neue Methode der konformen Abbildung und Uniformisierung. *Nachr. Kgl. Ges. Wiss. Göttingen, Math.-Phys. Kl.* 844–848.

———. 1913. Ränderzuordung bei konformer Abbildung. *Göttinger Nachrichten.* 286–288.

———. 1915. Abhandlungen zur Theorie der konformen Abbildung. I. Die Kreisabbildung des allgemeisnten einfach und zweifach zusammenhängenden schlichten Bereichs und die Ränderzuordnung bei konformer Abbildung. *J. Reine Angew. Math.* 145: 177–223.

———. 1918. Abhandlungen zur Theorie der konformen Abbildung. IV. Abbildung mehrfach zusammenhängender schlichter Bereiche auf Schlichtbereiche. *Acta Math.* 41: 304–344

Koepf, W. 1986. On nonvanishing univalent functions with real coefficients. *Math. Z.* 192: 575–579.

———. 1987. Extrempunkte und Stützpunkte in Familien nichverschwindender schlichten Funktionen. *Complex Variables.* 8: 153–171.

———. 1990. *On the Interplay between Geometrical and Analytical Properties of Univalent Functions.* Technical Report (Habilitationsschrift). Berlin: Free University.

———. 1992. Power series in Computer Algebra. *J. Symb. Comp.* 13: 581–603.

———. 1994. Von der Bieberbachschen Vermutung zum Satz von de Branges sowie der Beweisvariante von Weinstein. *Jahrbuch Überlicke Mathematik.* Vieweg, Braunschweig/Wiesbaden. 175–193.

———. 2003. Power series, Bieberbach conjecture and the de Branges and Weinstein functions. *Proceedings of the ISSAC 2003*, Philadelphia, J. R. Sendra (ed.) New York: ACM. pp. 169–175.

———. 2007. Bieberbach's conjecture, the de Branges and Weinstein functions and the Askey-Gasper inequality. *The Ramanujan Journal.* 13: 103–129. (Also http://www.mathematik.uni.kassel.de/~koepf/Publicationen/Koeph_Bexbach2003.pdf)

———and D. Schmersau. 1996. On the de Branges theorem. *Complex Variables.* 31: 213–230.

Kolata, G. 1984. Surprise proof of an old conjecture. *Science.* 225: 1006–1007.

Korevaar, J. 1986. Ludwig Bieberbach's conjecture and its proof by Louis de Branges. *The American Mathematical Monthly.* 93: 505–514. MR856290.

Krager, W., B. Nienhuis and L. Kadanoff. 2004. Exact solutions for Loewner evolutions. *J. Statis. Phys.* 115: 805–822.

Krantz, S. G. 1999. The Bieberbach Conjecture. §12.1.2 in *Handbook of Complex Variables*. Boston, MA: Birkhäuser, pp. 149–150.

Krushkal, Samuel. 2005. Beyond Mosser's conjecture on Grunsky inequalities. *Georgian Math. J.* 12 : 485–492.

Kufarev, P. P. 1943. On one-parameter family of analytic functions. *Mat. Sbornik*. 13: 87–118 (Russian).

Kühnau, R. 1971. Verzerrungssätze und Koeffizientenbedingungen von Grunskyschen Typ für quasikonforme Abbildungen. *Nath. Nachr.* 48 : 77–105.

_____. 1982. Quasikonforme Fortsetzbarkeit, Fredholmsche Eigenwerte und Grunskysche Koeffizientenbedingungen. *Ann. Acad. Sci. Fenn. Ser. A I Math.* 7 : 383–391.

_____. 1947. A remark on the integrals of the Loewner equation. *Dokl. Akad. Nauk SSSR.* 57: 655–656 (Russian).

Kulshrestha, P. K.[1] 1956. On evaluations of the measure of curvature of level curves of schlicht functions. *Ganita.* 7: 123–137.

_____. 1958. On the measure of curvature of level curves and orthogonal trajectories of a class of mean p-valent functions in the unit circle. *Ganita.* 9: 1–4.

_____. 1959. On the growth of derivatives of admissible functions. *Proc. Natl. Inst. Sci. India.* 25 A: 75–76.

_____. 1959. On the question of representation of analytic functions by Faber polynomials. *Bull. Calcutta Math. Soc.* 51: 73–76.

_____. 1973a. Distortion of spiral-like mappings. *Proc. Royal Irish Acad.* 73 A: 1–5.

_____. 1973b. Generalized convexity in conformal mappings. *J. Math. Anal. Appl.* 43: 441–449.

_____. 1974a. Coefficients for alpha-convex univalent functions. *Bull. Amer. Math. Soc.* 80: 341–342.

_____. 1974b. Coefficient problem for alpha-convex univalent functions. *Archive Ratnl. Mech. Anal.* 54: 204–211.

_____. 1976a. Coefficient problem for a class of Moçanu-Bazilevič functions. *Annales Polon. Math.* 31: 291–299.

_____. 1976b. Bounded Robertson functions. *Rendi. di Matem.* 9, Ser. VI: 137–150.

_____. 1984. Convex combinations of some regular locally univalent functions. *Bull. Inst. Math., Acad. Sinica.* 12: 37–50

Kythe, P. K. 1996. *Fundamental Solutions for Differential Operators and Applications*. Boston : Birkhäuser. ISBN 0-8167-3869-5.

_____. 1998. *Computational Conformal Mapping*. Boston, MA: Birkhäuser. ISBN 0-8176-3996-9.

_____. 2011. *Green's Functions and Linear Differential Equations: Theory, Applications, and Computation*. Boca Raton, FL: Taylor & Francis

---

[1] Kulshrestha changed his last name to Kythe at the time of naturalization as U. S. citizen.

Group/CRC Press., ISBN 1-4398-4008-5.

———. 2014. *Sinusoids: Theory and Technological Applications.* Boca Raton, FL : CRC Press. ISBN 978-1-4822-2106-0.

———, P. Puri and M. R. Schäferkotter. 2003. *Partial Differential Equations and Boundary Value Problems with Mathematica.* 2nd ed. Boca Raton, FL: Chapman & Hall/CRC Press.

Landau, E. 1926. Einige Bemerkungen über schlichte Abbildung. *Jahresbericht Deutsch. Math.-Vereinigung.* 34: 239–243.

Lawler, G. 2005. Conformality Invariant Processes in the Plane. *Mathematical Surveys and Monographs,* 114. Providence, RI : American Mathematical Society.

———. 2006. *Introduction to the stochastic Loewner evolution.* 2nd ed. Chapman-Hall. *also online at URL http://www.math.cornell.edu/ ˜lawler /papers.html.*

———, O. Schramm and W. Werner. 2001. The dimension of the planar Brownian frontier is 4/3. *Math. Res. Lett.* 8 : 401–411.

Lebedev, N. A. 1975. *The Area Principle in the Theory of Univalent Functions.* Moscow. (Russian).

———. and I. M. Milin. 1951. On the coefficients of certain classes of analytic functions. *Mat. Sb.* 28(2): 359–400 (Russian).

———. and I. M. Milin. 1965. On an inequality. *Vestnik Leningrad Gos. University. Math.* 19 : 157–158 (Russian).

Leeman, G. B. 1976. The seventh coefficient of odd symmetric univalent functions. *Duke Math. J.* 43: 301–307.

Leung, Y. 1978. Successive coefficients of starlike functions. *Bull. London Math. Soc.* 10: 193–196.

———. 1979. Robertson's conjecture on the coefficients of close-to-convex functions. *Proc. Amer. Math. Soc.* 76: 89–94.

Libera, R. J. 1967. Univalent α-spiral functions. *Can. J. Math.* 19: 449–456.

Lind, J. 2005. A sharp condition for the Loewner equation to generate slits. *Ann. Acad. Sci. Fenn. Math.* 30 : 143–158.

———, D. E. Marshall and S. Rohde. 2010. Collisions and Spirals of Loewner Traces. *Duke Math. J.* 154 : 527–573.

——— and S. Rohde. 2014. Loewner curvature. Preprint, January 15, 2014. pp. 1–18.

Littlewood, J. E. 1925. On inequalities in the theory of functions. *Proc. London Math. Soc.* 23 : 481–519.

———and E. A. C. Paley. 1932. A proof that an odd schlicht function has bounded coefficients. *J. London Math. Soc.* 7: 167–169.

Liu, X. 2006. The generalized Roper-Suffridge extension operator for some biholomorphic mappings. *J. Math. Anal. Appl.* 324 : 604–614.

Löwner, K.[2] 1917. Untersuchungen über die Verzerrung bei konformen Abbildungen des Einheitskreises $|z| < 1$, die durch Funktionen mit nichtver-

---

[2] K. Löwner later changed his name to C. Loewner after immigrating to the USA.

schwindender Ableitung geliefert wereden. *S.-B. Verh. Sächs. Ges. Wiss. Leipzig.* 69: 89–106.

_____. 1923. Untersuchungen über schlichte konforme Abbildungen des Einheitskreises. I. *Math. Ann.* 89: 103–121 (http://dx.doi.org/10.1007/BF 01448091).

MacGregor, T. H. 1962. Functions whose derivative has a positive real part. *Trans. Amer. Math. Soc.* 104 : 532537.

Mardare, Sorin. 2008. On Poincaré and de Rham's Theorems. *Rev. Roumaine Math. Pures Appl.* 53: 523–541.

Marsden, J. E. and M. J. Hoffman. 1987. *Basic Complex Analysis*, 2nd ed. New York: W. H. Freeman and Company.

Marshall, D. and S. Rohde. 2005. The Loewner differential equation and slit mappings. *J. Amer. Math. Soc.* 18: 763–778.

Marx, A. 1932-33. Untersuchungen über schlichten Abbildungen. *Math. Ann.* 107 : 40–67.

Milin, I. M. 1964. The area method in the theory of univalent functions. *Dokl. Akad. Nauk SSSR.* 154: 264–267.

_____. 1965. Estimation of coefficients of univalent functions. *Dokl. Akad. Nauk SSSR.* 160: 769–771 (Russian) = *Soviet Math. Dokl.* 6: 196–198.

_____. 1967. On the coefficients of univalent functions. *Dokl. Akad. Nauk SSSR.* 176: 1015–1018 (Russian) = *Soviet Math. Dokl.* 8: 1255–1258.

_____. 1968. Adjoint coefficients of univalent functions. *Soviet Math. Doklady.* 9: 762–765.

_____. 1969. *Univalent Functions and Orthogonal Series.* English Translation from Russian. Providence, RI: American Mathematical Society.

_____. 1970. Hayman's regularity theorem for the coefficients of univalent functions. *Soviet Math. Doklady.* 11: 724–728.

_____. 1971 . *Univalent Functions and Orthogonal Systems.* Izdat. Nauka, Moskua (Russian). Translations of Mathematical Monographs, vol. 49. Providence, RI: American Mathematical Society (Translated from the Russian 1971 edition).

_____. 1977. *Univalent Functions and Orthogonal Systems.* Trans;. Math. Monogr., (translated from Russian), Vol. 49. Providence, RI: American Mathematical Society.

Milin, V. I. 1980. Estimate of the coefficients of odd schlicht univalent functions, in *Metric Question of the Theory of Functions*, (G. D. Surorov, ed.) Kiev: Naukova Dumka (Russian), pp. 78–80.

Miller, Sanford S. 1973. Distortion properties of alpha-starlike functions. *Proc. Amer. Math. Soc.* 38 : 311–318.

_____, Petru Moçanu and Maxwell O. Reade. 1973. All α-convex functions are univalent and starlike. *Proc. Amer. Math. Soc.* 37 : 553554.

_____, P. T. Moçanu and M. O. Reade. 1974. Vazilevič functions and generalized convexity. *Rev. Roum. Math. Pures Appls.* 19 (2) : 213–224.

_____ and P. T. Moçanu. 1978. Second-order differential inequalities in the complex plane. *J. Math. Appl.* 65 : 289–305.

Minda, A. 1987. The strong form of Ahlfors' lemma. *Rockey Mountain J. Math.* 17 : 457–461.

Ming-Qin, Xie. 1997. A generalization of the de Branges theorem. *Proc. Amer. Math. Soc.* 125 : 3605–3611.

Moçanu, P. T. 1969. Une propriété de convexité généralisée dans la théorie de la représentation conforme. *Mathematica* (Cluj). 11 (34): 127–133. MR 42 #7881.

———, T. Bulboaca and G. Salagean. 1999. *Geometric Theory of Univalent Functions.* Cluj: Sciences Book Publishing House (in Romanian).

Montel, P. 1907. Sur les suites infines de fontions. *Annales de l'Ecole Normale* (3). 4: 233–304.

Moser, Jrgen. 1961. On Harnack's theorem for elliptic differential equations. *Communications on Pure and Applied Mathematics.* 14 (3): 577–591, doi:10.1002/cpa.3160140329.

Muir, J. R. 2008. A class of Loewner chain preserving extension operators. *J. Math. Anal. Appl.* 337: 862–879.

Nehari, Z. 1952. *Conformal Mapping.* New York: McGraw-Hill.

———. 1953. Some inequalities in the theory of functions. *Trans. Amer. Math. Soc.* 75: 256–286.

——— and E. Netanyahu. 1957. On the coefficients of meromorphic schlicht functions. *Proc. Amer. Math. Soc.* 8 : 1523.

———. 1957. On the coefficients of univalent functions. *Proc Amer. Math. Soc.* 8: 291–293.

———. 1969. Inequalities for the coefficients of univalent functions. *Arch. Rational Mech. Anal.* 34: 301–330.

———. 1970. On the coefficients of Bieberbach-Eilenberg functions. *J. Analyse Math.* 23: 297–303.

———. 1974. A proof of $|a_4| \le 4$ by Löwner's method. In *Proceedings of the Symposium on Complex Analysis, Canterbury, 1973* (J. Clunie and W. K. Hayman, eds.) London Math Soc. Lecture Note Series 12. Cambridge: Cambridge University Press. pp. 107–110.

Nevanlinna, R. 1920-1921. Über die konforme Abbildung von Sterngebebiten. *Överskt av Finska Vetenskaps-Soc. Förch.* 63 (A) Nr. 6: 1–21.

Noshiro, K. 1958. On the theory of cluster sets of analytic functions. *Amer. Math. Soc. Transl.* (2) 8 : 112.

Oros, Gh. and Georgia I. Oros. 2008. A class of univalent functions which extends the class of Moçanu functions. *Math. Reports.* 10 (60), 2 : 165–168.

Ozawa, M. 1965. On the sixth coefficient of univalent functions. *Kōdai Math. Seminar Reports.* 17: 1–9.

———. 1969. An elementary proof of the Bieberbach conjecture for the sixth coefficient. *Kōdai Math. Sem.* 21: 129–132.

Pederson, R. N. 1967. On unitary properties of Grunsky's matrix. *Tech.*

336

Report. Mellon College at Research Showcase @ CMU. www.research-showcase@andrew.cmu.edu.

_____. 1968. A proof of the Bieberbach conjecture for the sixth coefficient. *l Arch. Ratln. Mech. Anal.* 31: 331–351.

_____ and M. Schiffer. 1972. A proof of the Bieberbach conjecture for the fifth coefficient. *Arch. Ratln. Mech. Anal.* 45: 161–191.

Petermichl, S. and A. L. Volberg. 2002. Heating of the Ahlfors-Beurling operator: weakly quasiregular maps on the plane are quasiregular. *Duke Math. J.* 112 : 281–305.

Pfaltzgraff, J. A. 1971. On the Marx conjecture for a class of close-to-convex functions. *Michigan Math. J.* 18: 275–278.

_____. 1974. Subordination chains and univalence of holomorphic mappings in $\mathbb{C}^n$. *Math. Ann.* 210: 55–68.

Pommerenke, Ch. 1961/62. Über die Mittelwerte und Koeffizienten multivalenter Funktionen. *Math. Ann.* 145: 285–296.

_____. 1965. Über die subordination analytischer Funktionen. *J. reine und ungewand. Math.* 218: 159–173.

_____. 1975. *Univalent Functions*, with a chapter on quadratic differentials by Gerd Jensen. Studia Mathematica/Mathemati- sche Lehrbücher 15. Göttingen: Vandenhoeck and Ruprecht.

_____. 1985. The Bieberbach conjecture. *Mathematical Intelligencer.* 7(2): 23–25.

_____. 1992a. *Boundary Behaviour of Conformal Maps.* Berlin: Springer-Verlag.

_____. 1992b. *Boundary behaviour of conformal maps.* Grundlehren der Mathematischen Wissenschaften, Vol. 299. Berlin : Springer-Verlag.

Ponnusamy, S. 2011. *Foundations of Mathematical Analysis.* Birkhäuser.

_____ and H. Silverman. 2006. *Complex Variables and Applications.* Boston: Birkhäuser.

Reade, M. O. 1955. On close-to-convex univalent functions. *Mich. Math. J.* 3: 59–62.

Richtmyer, R. D. 1978. *Principles of Advanced Mathematical Physics,* Vol. 1. New York: Springer-Verlag.

Riemann, B. 1851. Grundlagen für eine allgemeine Theorie der Funktionen einer veränderlichen komplexe Grösse, in *Collected Works.* New York: Dover, 1953.

Robertson, M. S. 1936. A remark on the odd schlicht functions. *Bull. Amer. Math. Soc.* 366–370.

_____. 1936. On the theory of univalent functions. *Ann. of Math.* 37: 374–408.

_____. 1963. Extremal problems of analytic functions with positive real part and applications. *Trans. Amer. Math. Soc* 106: 236–253.

_____. 1969. Power series with multiply monotonic coefficients. *Michigan Math. J.* 16: 27–37.

Rogosinski., W. 1931. Über positive harmonische Sinusentwicklinugen. *Jber. Deutsch. Math. Verein.* 40: 33–35.

———. 1932. Über positive harmonische Entwicklungen und typisch-reelle Potenzreihen. *Math. Z.* 35: 93–121.

———. 1939. On subordinate functions. *Proc. Cambridge Philos. Soc.* 35: 1–20.

———. 1943. On the coefficients of subordinate functions. *Proc. London Math. Soc.* 48: 48–82.

Rohde, S. and O. Schramm. 2005. Basic properties of SLE. *Math. Ann.* 161: 883–924. arXiv:math.PR/0106036.

Roper, K. and T. J. Suffridge. 1995. Convex mappings on the unit ball in $\mathbb{C}^n$. *J. Anal. Math.* 65: 333–347.

Rosay, J. P. and W. Rudin. 1988. Holomorphic maps from $\mathbb{C}^n$ to $\mathbb{C}^n$. *Trans. Amer. Math. Soc.* 310 : 467–486.

Rudin, Walter. 1976. *Principles of Mathematical Analysis.* New York: MacGraw-Hill.

Ruscheweyh, St. and T. Sheil-Small. 1973. Hadamerd products of schlicht functions and the Pólya-Schoenberg conjecture. *Comment. Math. Helv.* 48: 119–135.

Sakaguchi, K. 1959. On a certain univalent mapping. *J. Math. Soc. Japan.* 11 : 7275.

Salvy, B. and P. Zimmermann. 1994. GFUN: A Maple package for the manipulation of generating and holomorphic functions in one variable. *ACM Trans. on Mathematical Software.* 20: 163–177.

Sansone, G. and J. Gerretsen. 1969. *Lectures on the Theory of a Complex Variable.* II. Groningen: Wolters-Noordhoff.

Schaeffer, A. C. and D. C. Spencer. 1943. The coefficients of schlicht functions. *Duke Math. J.* 10:611–635

——— and D. C. Spencer. 1950. *Coefficient Regions for Schlicht Functions.* Amer. Math. Soc. Colloquium Publication, Vol. 35. Providence, RI: American Mathematical Society.

———, M. Schiffer and D. C. Spencer. 1949. The coefficient regions of schlicht functions. *Duke Math J.* 16: 493–527.

Schiffer, M. 1938a. A method of variation within the family of simple functions. *Proc. London Math. Soc.* 44: 432–449.

———. 1938b. On the coefficients of simple functions. *Proc. London Math. Soc.* 44: 450–452.

———. 1948. Faber polynomials in the theory of univalent functions *Bull. Amer. Math. Soc.* 54: 503–517.

———. 1960. Extremum problems and variational methods in conformal mapping. *Proc. International Cong. Math.* (Edinburgh, 1958). New York: Cambridge University Press. pp. 211-231.

———. 1967. Univalent functions whose first n coefficients are real. *J. d'Analyse Math.* 18: 329–349.

338

———. 1981. Fredholm eigenvalues and Grunsky matrices. *Ann. Polon. Math.* 39 : 149–164.

——— and G. Schober. 1976. Coefficient problems and generalized Grunsky inequalities for schlicht functions with quasiconformal extensions. *Arch. Rational Mech. Anal.* 60: 205–228.

Schramm, O. 2000. Scaling limits of loop-erased random walks and uniform spanning trees. *Israel J. Math.* 118 : 221–288.

———. 2007. Conformally invariant scaling limits: an overview and a collection of problems. *International Congress of Mathematicians*, Vol. 1, Eur. Math. Soc., Zurich, pp. 513–543.

———. 2011. *Selected Works of Oded Schramm, Selected Works in Probability and Statistics*, (Itai Benjamini and Olle Häggström, eds.) New York: Springer.

Schur, A. 1921. Über die Schwarz'sche Extremaleigenschaft des Kreises unter den Kurven konstanter Krümming. *Math. Ann.* 83 : 143–148.

Schur, I. 1945. On Faber polynomials. *Amer. J. Math.* 67: 33–41.

Semes, Stephen W. 1986. The Cauchy integral, chord-arc curves, and quasiconformal mappings. In *The Bieberbach Conjecture*, Baerstein, Drasin, Duren and Marden (eds.). Math. Surveys Monograph., 21. Providence, RI: American Mathematical Society. pp. 167–183.

Sheil-Small, T. 1973. On the convolution of analytic functions. *J. reine angew. Math.* 258: 137–152.

Silverman, R. A. 1967. *Introductory Complex Analysis.* Englewood Cliffs, NJ: Prentice-Hall.

Smirnov, S. 2001. Critical percolation in the plane: conformal invariance, Hardy's formula, scaling limits. *C. R. Acad. Sci. Paris Ser. I Math.* 333: 239–244.

Špaček, L. 1933. Contributions a la théorie des fonctionns univalentes. *Časopis Pest. Mat. Fys.* 62: 12–19.

Spencer, D. C. 1947. Some problems in conformal mapping. *Bull. Amer. Math. Soc.* 53: 417–439.

Spivak, M. 1965. *Calculus on Manifolds.* Menlo Park, CA: Benjamin Cummings.

Stewart, I. 1996. The Bieberbach Conjecture, in *From Here to Infinity: A Guide to Today's Mathematics.* Oxford: Oxford University Press, pp. 164–166.

Strohhäcker, E. 1933. Beitrage zur Theorie der schlichten Funktionen. *Mat. Zeit.* 37: 356–380.

Study, E. and W. Blaschke. 1912. *Vorlesungen über ausgewählten Gegenstände der Geometrie. Vol 2: Konforme Abbildung einfach zusammenhängender Bereiche.* Leipzig: Teubner.

Szegö, G. 1941. Power series with multiply monotonic sequences of coefficients. *Duke Math. J.* 8: 559–564; also in *Collected Papers*, Vol. 2, 1982, pp. 797–802. Boston: Birkhäuser.

Todorov, P. G. 1992. A simple proof of the Bieberbach conjecture. *Bull. Cl. Sci., VI. Sér., Acad. R. Belg. 3.* 12: 335–356.

Tsuji, M. 1975. *Potential Theory in Modern Function Theory.* 2nd edition. Chelsea.

Vodă, M. 2011. Solution of a Loewner chain equation in several variables. *J. Math. Anal. Appl.* 375: 58–74.

Weinstein, L. 1991. The Bieberbach conjecture. *Internat. Math. Res. Notices.* 5: 61–64.

Wen, G.-C. 1992. *Conformal Mappings and Boundary Value Problems.* Translations of Mathematical Monographs, Vol. 106. Providence, RI: American Mathematical Society.

Wilf, H. 1994. A footnote on two proofs of the Bieberbach-de Branges Theorem. *Bull. London Math. Soc.* 26: 61–63.

———— and D. Zeilberger. 1992. An algorithmic proof theory for hypergeometric (ordinary and "q") multisum/integral identities. *Invent. Math.* 103: 575–634.

Whittaker, E. T. and G. N. Watson. 1962. *A Course of Modern Analysis.* Cambridge: Cambridge University Press.

Wong, C. 2014. Smoothness of Loewner slits. *Trans. Amer. Math. Soc.* 366: 1475–1496.

Yap, S. L. 2009. The Poincaré lemma and an elementary construction of vector potentials. *Amer. Math. Monthly.* # 3, 116.

Zeilberger, D. 1990. A fast algorithm for proving terminating hypergeometric identities. *Discrete Mat.* 80: 207–211.

Zorn, P. 1986. The Bieberbach conjecture. *Mathematics Magazine.* 59: 131–148.

Index